Springer Series in Optical Sciences

Founding Editor
H. K. V. Lotsch

Volume 256

Springer Series in Optical Sciences is led by Editor-in-Chief William T. Rhodes, Florida Atlantic University, USA, and provides an expanding selection of research monographs in all major areas of optics:

- lasers and quantum optics
- ultrafast phenomena
- optical spectroscopy techniques
- optoelectronics
- information optics
- applied laser technology
- industrial applications and
- synchrotron radiation and X-ray optics
- other topics of contemporary interest.

With this broad coverage of topics the series is useful to research scientists and engineers who need up-to-date reference books.

Thomas Kürner • Tobias Doeker •
David Humphreys • Thomas Kleine-Ostmann •
Thomas Schneider

Editors

Metrology for THz Communications

Findings from DFG FOR 2863 Meteracom

Funded by

Deutsche
Forschungsgemeinschaft
German Research Foundation

Editors

Thomas Kürner
Institut für Nachrichtentechnik
Technische Universität Braunschweig
Braunschweig, Niedersachsen, Germany

Tobias Doeker
Institut für Nachrichtentechnik
Technische Universität Braunschweig
Braunschweig, Niedersachsen, Germany

David Humphreys
Visiting Industrial Fellow
University of Bristol
Ascot, UK

Thomas Kleine-Ostmann
Hochfrequenz und Felder
Physikalisch-Technische Bundesanstalt
Braunschweig, Germany

Thomas Schneider
Technische Universität Braunschweig
Braunschweig, Niedersachsen, Germany

ISSN 0342-4111   ISSN 1556-1534   (electronic)
Springer Series in Optical Sciences
ISBN 978-3-032-01985-1   ISBN 978-3-032-01986-8   (eBook)
https://doi.org/10.1007/978-3-032-01986-8

This work was supported by Technische Universität Braunschweig.

This Springer imprint is published by the registered company Springer Nature Switzerland AG
The registered company address is: Gewerbestrasse 11, 6330 Cham, Switzerland

If disposing of this product, please recycle the paper.

Acknowledgement The work presented in this book and part of the publication was supported by the German Research Foundation (DFG) under Grant FOR 2863 (Meteracom - Metrology for THz Communications), and the publication of the book in part by the Open Access Publication Funds of Technische Universität Braunschweig. The collaborative research project DFG FOR 2863 was running between 2019 and 2026.

Contents

Contributors

Mohanad Dawood Al-Dabbagh Physikalisch-Technische Bundesanstalt (PTB), Braunschweig, Germany

Meysam Bahmanian Universität Paderborn, Schaltungstechnik (SCT)/Heinz Nixdorf Institut, Paderborn, Germany

Mounir Bensalem Technische Universität Braunschweig, Institut für Datentechnik und Kommunikationsnetze, Braunschweig, Germany

Mladen Berekovic Universität zu Lübeck, Institut für Technische Informatik, Lübeck, Germany

Enrique Castro-Camus Philipps-Universität Marburg, Physik (Fb13), Marburg, Germany
Centro de Investigaciones en Optica, Lomas del Campestre, Mexico

Ranjan Das Technische Universität Braunschweig, THz-Photonics Group, Braunschweig, Germany

Souvaraj De Physikalisch-Technische Bundesanstalt (PTB), Braunschweig, Germany

Giovanni Del Galdo Technische Universität Ilmenau, Institute of Information Technology, Ilmenau, Germany
Fraunhofer Institute for Integrated Circuits (IIS), Ilmenau, Germany

Tobias Doeker Technische Universität Braunschweig, Institut für Nachrichtentechnik, Braunschweig, Germany

André Drummond Technische Universität Braunschweig, Institut für Datentechnik und Kommunikationsnetze, Braunschweig, Germany
University of Brasilia, Federal District, Brazil

Diego Dupleich Technische Universität Ilmenau, Institute of Information Technology, Ilmenau, Germany

Alexander Ebert Institute of Information Technology, Technische Universität Ilmenau, Ilmenau, Germany

Zied Ennaceur Technische Universität Braunschweig, Institut für Datentechnik und Kommunikationsnetze, Braunschweig, Germany

Heiko Füser Physikalisch-Technische Bundesanstalt (PTB), Braunschweig, Germany

Jonas Gedschold Technische Universität Ilmenau, Institute of Information Technology, Ilmenau, Germany

Felix Gorka Philipps-Universität Marburg, Physik (Fb13), Marburg, Germany

Simon Haussmann Universität Stuttgart, Institut für Robuste Leistungshalbleitersysteme, Stuttgart, Germany

Christoph Herold Technische Universität Braunschweig, Institut für Nachrichtentechnik, Braunschweig, Germany

David A. Humphreys Physikalisch-Technische Bundesanstalt (PTB), Braunschweig, Germany

Rolf Judaschke Physikalisch-Technische Bundesanstalt (PTB), Braunschweig, Germany

Admela Jukan Technische Universität Braunschweig, Institut für Datentechnik und Kommunikationsnetze, Braunschweig, Germany

Ingmar Kallfass Universität Stuttgart, Institut für Robuste Leistungshalbleitersysteme, Stuttgart, Germany

Thomas Kleine-Ostmann Physikalisch-Technische Bundesanstalt (PTB), Braunschweig, Germany

Martin Koch Physik (Fb13), AG Halbleiterphotonik, Marburg, Germany
Philipps-Universität Marburg, Physik (Fb13), Marburg, Germany

Adam Kuchnia Physikalisch-Technische Bundesanstalt (PTB), Braunschweig, Germany

Thomas Kürner Technische Universität Braunschweig, Institut für Nachrichtentechnik, Braunschweig, Germany

Younus Mandalawi Technische Universität Braunschweig, THz-Photonics Group, Braunschweig, Germany

Nora Meyne Physikalisch-Technische Bundesanstalt (PTB), Braunschweig, Germany

Daniel M. Mittleman Brown University, School of Engineering, Providence, RI, USA

Anouar Nechi Universität zu Lübeck, Institut für Technische Informatik, Lübeck, Germany

Cao Vien Phung Technische Universität Braunschweig, Institut für Datentechnik und Kommunikationsnetze, Braunschweig, Germany

Carla Reinhardt Technische Universität Braunschweig, Institut für Nachrichten-technik, Braunschweig, Germany

J. Christoph Scheytt Universität Paderborn, Schaltungstechnik (SCT)/Heinz Nixdorf Institut, Paderborn, Germany

Thomas Schneider Technische Universität Braunschweig, THz-Photonics Group, Braunschweig, Germany

Benjamin Schoch Universität Stuttgart, Institut für Robuste Leistungshalbleiter-systeme, Stuttgart, Germany

Sebastian Semper Institute of Information Technology, Technische Universität Ilmenau, Ilmenau, Germany

Fatima Taleb Philipps-Universität Marburg, Physik (Fb13), Marburg, Germany

Reiner S. Thomä Technische Universität Ilmenau, Institute of Information Technology, Ilmenau, Germany

Maxim Weizel Universität Paderborn, Schaltungstechnik (SCT)/Heinz Nixdorf Institut, Paderborn, Germany

Dominik Wrana Universität Stuttgart, Institut für Robuste Leistungshalbleitersysteme, Stuttgart, Germany

List of Abbreviations

5G	5th generation
6G	6th generation
AP	Access point
ASIC	Application-specific integrated circuits
ACF	Autocorrelation function
AGC	Automatic gain control
AWGN	Additive white Gaussian noise
AM	Amplitude modulation
ADC	Analogue-to-digital converter
ARB	Arbitrary waveform
AWG	Arbitrary waveform generator
AI	Artificial intelligence
BBB	Beta-Barium Borate
BEOL	Back-end-of-line
BIRMN	Beijing Institute of Radio Metrology and Measurement
BPSK	Binary phase shift keying
BER	Bit error rate
CMC	Calibration and measurement capabilities
CRT	Cathode ray tube
CPU	Central processing unit
CS	Channel sounder
CW	Continuous wave
CPW	Coplanar waveguide
DNN	Deep neural network
DFG	Deutsche Forschungsgemeinschaft
DUT	Device under test
DRTO	Digital real-time oscilloscope
DSO	Digital sampling oscilloscope
DC	Direct current
DoA	Direction of arrival
DoD	Direction of departure

DRTO   Digital Real-time Oscilloscope
ENOB   Effective number of bits
EM   Electromagnetic
EO   Elector-optic
EOS   Electro-optic sampling
EVM   Error vector
EVM   Error vector magnitude
ETSI   European Telecommunications Standards Institute
FPGA   Field programmable gate array
FDTD   Finite-difference time-domain
FEM   Finite element method
FOR   Forschungsgruppe
FSPL   Free space path loss
FMCW   Frequency modulated continuous waveform
GPU   Graphics processing unit
GBSCM   Geometry-based stochastic channel model
GUM   Guide to Uncertainty in Measurements
HPBW   Half-power beamwidth
HBT   Hetero-junction bipolar transistor
HRPE   High-resolution parameter estimation
ICAS   Integrated communication and sensing
IoT   Internet of Things
IF   Intermediate frequency
IFBW   Intermediate frequency bandwidth
KCDB   International Key Comparison Data Base
ITU   International Telecommunications Union
ISO   International Organization for Standardization
ISI   Inter-symbol interference
IoT   Input/output
KRISS   Korea Research Institute of Standards and Science
LO   Local oscillator
LiDAR   Light detection and ranging
LS   Linear sweep
LOS   Line-of-sight
ML   Machine learning
MZM   Mach-Zehnder modulator
MED   Maximum excess delay
MWP   Microwave probes
MBPE   Model-based parameter estimation
MLL   Mode-locked laser
MCS   Multidimensional channel sounder
MPC   Multipath component
MIMO   Multiple-input multiple-output
NIM   National Institute of Metrology of China
NIST   National Institute of Standards and Technology

NMI	National Metrology Institutes
NPL	National Physical Laboratory
NRF	Noise reduction factor
NLOS	Non-line-of-sight
OSI	Open systems interconnection
OFDM	Orthogonal frequency-division multiplexing
OFDMA	Orthogonal frequency-division multiple access
OTA	Over-the-air
PAPR	Peak-to-average power ratio
PD	Photo-diode
PLL	Phase locked loop
PM	Phase modulation
PTB	Physikalisch-Technische Bundesanstalt
PVDF	Polyvinylidene difluoride
PSD	Power spectral density
PDP	Power delay profile
PRBS	Pseudo-random binary signals
PRN	Pseudo-random number
QAM	Quadrature amplitude modulation
QPSK	Quadrature phase shift keying
QoS	Quality of service
RF	Radio frequency
RT	Ray-tracing
RL	Reinforcement learning
Rx	Receiver
RIS	Reconfigurable intelligent surface
RC	Raised cosine
RRC	Root raised cosine
SiMoNe	Simulator for Mobile Networks
SSB	Single sideband
SINR	Signal-to-interference-and-noise ration
SINAD	Signal-to-noise-and-distortion
SFG	Sum frequency generation
SNR	Signal-to-noise ratio
SUT	Signal under test
SIMD	Single Instruction Multiple Data
SM	Streaming Multi Processo
SI	Système International d'Unités
TDMA	Time division multiple access
TUBS	Technische Universität Braunschweig
THz	Terahertz
ToF	Time of flight
THz TDS	THz time-domain spectroscopy
Tx	Transmitter
TE	Transverse electric

TM	Transverse magnetic
TRL	Through-reflect-line
VSA	Vector signal analyser
VNA	Vector network analyser
UWB	Ultra wide band
WCDMA	Wideband code division multiple access
WRC	World Radio Conference

Chapter 1
Introduction

Thomas Kürner

Abstract Terahertz (THz) communications is seen as a transmission technology to achieve wireless transmission data rates of several 10 s of Gbps up 1 Tbps. One drawback of transmission at carrier frequencies in the THz range are high path losses, which are much higher compared to that at lower carrier frequencies. This has consequences on the design of THz communication systems, which cannot be designed and characterised as a scaled and incremental version of a lower-frequency system. This is accompanied by the need for a paradigm shift for measurement procedures in predicting the performance of THz communication systems in real environments. The capability to perform measurements and the development of metrological concepts to evaluate these measurements in a well-defined way are crucial for the advance of THz communication systems. This chapter introduces the motivation for starting the project DFG FOR 2863 Meteracom (Metrology for THz Communications), describes the structure of the project, and provides the outline of this book, which is meant to summarise the main results.

1.1 Status of THz Communications

The data rates in wireless communications double about every 18 months [1]. Extrapolating on the increase described above in the long term, to the year 2030 and beyond, may lead to wireless data rates of 1 Tbit/s per data link, which even supersedes data rates of today's available fibre-optical systems. The already identified specific applications requiring such data rates are, for example, wireless back haul and front haul, kiosk downloading, wireless close proximity links, wireless data centre links, or wireless chip-to-chip communication [2]. By applying sophisticated transmission schemes, such as spatial multiplexing, such high data rates can be achieved, only, under the assumption that a bandwidth per channel

T. Kürner (✉)
Institut für Nachrichtentechnik, Technische Universität Braunschweig, Braunschweig, Germany
e-mail: t.kuerner@tu-braunschweig.de

© The Author(s) 2026
T. Kürner et al. (eds.), *Metrology for THz Communications*, Springer Series in Optical Sciences 256, https://doi.org/10.1007/978-3-032-01986-8_1

1

is available in the order of tens of GHz per link. The respective spectrum can be found at carrier frequencies beyond 100 GHz, only. In 2017, IEEE 802 has published the first standard for wireless communications operating in the frequency range above 252 GHz [2]. This standard has been revised in 2023 and covers now a frequency range of 252–450 GHz [3]. Furthermore, the regulatory activities at World Radio Conference (WRC) 2019 have led to the identification of 137 GHz of spectrum between 275 and 450 GHz for the use by fixed and mobile service [4]. This will pave the way for using carrier frequencies at several hundreds of GHz also called THz communications [5]. At the time of the writing of this book, research activities for the 6th generation (6G) of wireless communication systems have been already ongoing. THz communications has received tremendous attention in the scientific community and is seen as one of the key enabling technologies to satisfy the exponential growth of data traffic volume [6] and enabling data rates of terabits per second. Large collaborative research projects or programs have been launched, e.g. [7–10], which are dedicated to develop the required components and have demonstrated the feasibility and potential of THz communications. Channel characterisation in this frequency range has took off in the last decade with various research groups worldwide working in this area [11, 12]. Furthermore, standardisation activities have accelerated, for example, ETSI (European Telecommunications Standards Institute) has established an Industry Specification Group (ISG) THz in 2022, yielding the publication of four group reports on use cases [13] , spectrum aspects [14] , channel models [15] and RF models [16] in 2024.

Transmission at THz carrier frequencies also brings challenges, and the most notable is path loss that is – in linear scale – proportional to the square of the carrier frequency. This has consequences on the design of THz communication systems, which cannot be designed and characterised as a scaled and incremental version of a lower-frequency system. For example, high-gain antennas are indispensable to mitigate the high path loss. However, high gain comes with high directivity, only. So, in mobile scenarios, adaptive beam-forming is essential. Pure digital beam-forming however is prohibitive as the number of A/D converters is limited by reasons which result from cost and power consumption in view of the many antennas and huge bandwidth involved. What is needed are elaborated systems of high-gain antennas and phased array beam-forming; see, e.g. [17, 18]. The paradigm shift compared to lower-frequency systems comes from the inverted sequence of spatial filtering and digitisation. At lower frequencies, the receiver digitises the signal after reception from each antenna, which is omnidirectional in essence. At very high frequencies the signal at the output of the beam former of the receiving antenna array is digitised. Hence, the receiver sees the propagation channel as through a telescope. Enabling receiver and transmitter to mutually find each other under such a condition of limited viewing area has a critical impact on beam-tracking and acquisition ('device discovery'); see, e.g. [19]. Also, other system parameters (e.g. delay and Doppler spread, resp. coherence time and bandwidth) are largely influenced by the spatial characteristic of the antenna, which works as a spatial filter; see, e.g. [20]. As a consequence, channel characteristics and system design (modulation design, symbol structure and RF design) become heavily linked and can no longer be considered separately. Circuit design with compact and integrated implementation reinforces this effect due to the lack of well-defined reference interfaces. This exemplifies the required paradigm shift in measurement procedures for the performance prediction of THz communication systems in real operational environments.

The 'early-day' activities in THz communication transmission experiments in the 2000s and 2010s focused on investigating basic propagation phenomena, limited by simple measurement capabilities available at the time, and the investigation of first system concepts has been based on software simulations [21–26]. In recent years, the technological progress in semiconductor technology yielded several advanced hardware demonstrations; see, e.g. [7, 8, 10, 18, 27, 28]. These hardware demonstrations have applied both electronic and photonic approaches for the generation of radio signals at carrier frequencies above 200 GHz. Three main findings from these hardware demonstrations can be summarised as follows [29]:

- Feasibility: The principal feasibility of THz communications has been proven and has shown its potential for future wireless transmission [27, 28].
- Accuracy: Non-ideal behaviour of system components and the harsh propagation conditions require adequate and sophisticated measurement equipment, procedures and algorithms to perform measurements and to calibrate the equipment [30, 31]. This is a prerequisite to enable further optimisation of the design process yielding significant progress before the development of products could be commercially viable.
- Real-time performance: Measurements enabling the functionality of THz communications will be highly demanding due to factors such as the high carrier frequency, the high bandwidth or both. Device discovery, beam-tracking of high-gain antennas [17, 19, 19, 32, 33], real-time performance evaluation for ultra-high data rates and sampling of ultra-high bandwidths are examples where high demand on real-time performance will arise.

1.2 The Role of Metrology for THz Communications

From the findings at the end of Sect. 1.1 it is obvious that the capability to perform measurements and evaluate these measurements in a proper way are crucial for the advance of THz communication systems. In 2019, when the DFG project FOR 2863 Meteracom (Metrology for Communications) started, metrology at THz frequencies was however still in its infancy and covered mainly detector calibration to characterise ultra fast devices and to measurement uncertainty analysis of different spectrometer types available at THz frequencies. Meteracom has addressed the grand challenge of metrology in THz communication measurements systematically and in four distinct areas:

- *Area T:* Traceability to the International System of Units (SI)
- *Area A:* Characterisation of the measurement system itself

- *Area B:* Metrological characterisation of the RF components and the propagation channel
- *Area C:* Measurements required for enabling the functionality of THz communications

Area T – Traceability to the International System of Units (SI) – together with known measurement uncertainties, is essential to obtain meaningful measurement results. This can be achieved by establishing an unbroken chain of measurements to the representation of fundamental or derived units as given by the standards kept at the National Metrology Institutes (NMIs) [34]. An essential aspect, which is persistent in the whole area of THz communications, is the lack of suitable metrological methods and availability of reference measurement standards allowing an accurate and comparable metrological evaluation traceable to SI units. This is due to both the high carrier frequency and the ultra-high bandwidths, which are required to achieve the targeted data rates. New scientific challenges are resulting from new requirements to derive parameters and measure quantities, which have not been of interest or could not be derived within moderate time periods, now have to be determined with high accuracy, reliable at THz frequencies with high bandwidth usage and partly in real time or under highly dynamic channel conditions; see, e.g. [30, 35].

Area A – Characterisation of the measurement systems – includes the investigation of fundamental problems and the principally achievable measurement accuracy. The measurement uncertainty assigned is the result of a detailed modelling of the measurement process and uncertainty analysis and comprises contributions from a statistical analysis of a series of observations and other scientific knowledge about the measurement process. Even in case of a small and often negligible contribution, a measurement uncertainty budget needs to be complete for the measurement process to be understood. Together with the best estimate of the measurand, the assigned overall standard measurement uncertainty specifies the interval in which the measurand can be expected. The knowledge of measurement uncertainties and the trust in measurement results ensure comparability and interoperability. For example, an inter-comparison for material parameter measurements in the THz frequency range has been performed by Naftaly et al. [36] revealing strong discrepancies between the participants' results when measuring large absorption coefficients and the refractive index of the test samples. It shows that besides the general difficulties associated with measurements in the THz frequency range, there are still open questions about the correct way to handle data extraction. However, a comparison of measurement results from different spectrometer types shows that results are consistent, when measurement uncertainties are estimated based on metrological methods [37].

Area B – Metrological characterisation of the RF components and the propagation channel – deals with the distortions at the RF transceiver due to the ultra-high bandwidth, which can be typically neglected at lower frequencies and the specific impairments of the propagation channel and these carrier frequencies and bandwidths. On the one hand, due to the use of high-gain antennas, scenarios will exist (for example, fixed point-to-point links), where the main effect of a non-ideal transmission channel originates from the system components and not from the propagation channel. On the other hand, in highly dynamic scenarios, for example, in wireless LAN-type situations, the effect of the propagation channel will dominate. Therefore, the characterisation of the THz communication system requires a comprehensive measurement concept for all involved components including the propagation channel, which was not available at the start of Meteracom and which has at least partly to rely on emerging measurement systems as described before. History has taught us that a wireless transmission scheme always comes with long and intensive discussion on standardisation (see, for example, the early 60-GHz wireless system IEEE Std 802.15.3c-2009 and IEEE Std 802.11ad-2012). The respective decision seems to be driven by rigorous studies of the propagation phenomena (the 'channel') and the technological possibilities and limitations of key components. But experience shows that it is not unusual that economic considerations of the respective stakeholders dominate technical facts. This is made possible since the measured characterisation of the channel and the components may be too imprecise and measured results are often hard to compare because of not well-defined measurement and parameter estimation procedures and uncertainties. Decisions made on such a basis yield an absolutely unacceptable situation, since the quality of these models has a major impact on the performance of the systems. For frequencies below 6 GHz, we already have achieved an acceptable level of best practice and uncertainty assessment of multidimensional channel sounding high-resolution parameter estimation; see [38]. However, this is by far not the case for millimetre wave frequencies and even less for wireless THz systems. A metrological concept of THz sounding including the sounder architecture, multipath parameter identification and estimation of informative wireless system performance measures is not available today, and also today this is partly not available even at carrier frequencies below 100 GHz. Such a metrological concept will also contribute to a more objective assessment of different channel models and especially the measurement principles to derive these models.

Area C – Measurements required for enabling the functionality of THz communications – requires the application of sophisticated measurement procedures supporting a wide range of aspects covering setting up connections, monitoring performance and providing quantitative measures for the degree of physical layer security achieved. Challenges addressing unresolved problems are either due to the high carrier frequency, bandwidth, data rate or the requirement for real-time measurements with high accuracy in time and frequency domain. For example, the obligatory use of high-gain antennas will not allow the efficient application of concepts for device discovery as used today at carrier frequencies of 60 GHz and below, where concepts for device discovery are based on broader antenna beam widths or even omnidirectional characteristics, which are not directly applicable at THz frequencies. Ultra-high data rates require to use machine learning and coding methods to characterise channels and perform reliable real-time measurements and real-time evaluation of quality of service (QoS) parameters (e.g. bit error rate, packet error rate, delay, transmission distance, modulation, etc.). They will, for example, enable quality enhancements by the parallelization of the transmission channel [39]. Finally, the traceability for most of the measurement quantities necessary for the functionality of THz communication systems has been not available.

Although Meteracom has made use of mature measurement equipment like vector network analysers or THz time-domain (THz-TDS) spectrometers for some research questions, the main focus has been on emerging measurement equipment like the above-mentioned channel sounders and ultra-high-bandwidth sampling systems. With these two systems, we will be able to address both the high carrier frequency and the ultra-high bandwidth:

- Sampling systems are the base of analogue-to-digital conversion (ADC). Since almost all communication systems and most of the measurement equipment is digital, such an ADC functionality is essential for communications and metrology. Especially for high-bandwidth signals, common in THz communications, the typical electronic ADC are reaching their accuracy limits. This comes mainly from jitter problems of electronics, i.e. the inability of electronic systems to take the measurement value at exact times. Thus, the effective number of bit, or the resolution of ADC, decreases with the bandwidth of the signal to sample. Optical methods, however, have shown extremely low jitter values in the atto- and even zepto-second range. Thus, the investigation of optical or optically assisted sampling methods is an important new metrological task for THz communications [40–43].
- Channel sounders are important to characterise the mobile radio channel allowing a profound understanding of the electromagnetic propagation mechanisms and their interaction with antenna systems, even in dynamic environments [12, 44]. Whereas system evaluation of wired transmission systems and circuits follows very-well-defined rules of metrology, already, this is not the case for wireless systems. The reason is that the complexity of the stochastically time-variant interaction of multipath propagation, which exists especially in short-range applications (see, e.g. [45–47]) and antennas, leads to a multidimensional system characterisation which has to be broken down to useful performance measures.

Both channel sounders [44, 48–50] and sampling systems [40–43] are available at the partners participating in Meteracom enabling also experimental investigations in emerging measurement equipment without the need for high cost for buying equipment. Consequently, the channel sounder and the sampling systems have been used to provide a partial verification of the more general metrological concepts developed in this research unit.

1.3 Structure of DFG FOR 2863 Meteracom

The DFG FOR 2863 Meteracom project started on 1 August 2019 with its first 3-year phase, which was later on extended by a second 3-year phase. The project was carried out by a total of 10 principal investigators from 6 German universities:

- Technische Universität Braunschweig (Prof. Admela Jukan, Prof. Thomas Kürner, Prof. Thomas Schneider)
- Technische Universität Ilmenau (Prof. Giavanni Del Galdo, Prof. Reiner Thomä)
- Universität zu Lübeck (Prof. Mladen Berekovic)
- Philipps Universität Marburg (Prof. Martin Koch)
- Universität Paderborn (Prof. Chirstoph Scheytt)
- Universität Stuttgart (Prof. Ingmar Kallfass)

and the German National Metrology Institute PTB (Physikalisch-Technische Bundesanstalt; Prof. Thomas Kleine-Ostmann) as well as two Mercator Fellows from the National Physical Laboratory/United Kingdom (Dr. David Humphreys) and Brown University in Providence, Rhode Island (Prof. Daniel Mittleman), in the United States. The consortium was coordinated by the spokesman Prof. Thomas Kürner.

The work in the project was carried out in nine projects in each of the two phases complemented by a coordination project. The nine projects, each of them involving one to four principal investigators, have been defined along the subdivision into the four project areas derived from the grand challenges in metrology for THz communications mentioned in Sect. 1.3; see Fig. 1.1.

Project area T on traceability consisted of one project and project area A on the characterisation of the measurement systems were subdivided into three projects on accuracy, channel sounding and high-bandwidth sampling. Project area B on the characterisation of the radio channel on the hardware components consisted of three projects dedicated to the propagation channel, active transceivers and ultra-wideband sampling. The latter project accomplished its goals already in phase I and was not carried over to phase II. Project area C on the functionality of THz communication systems consisted in phase I of two projects on device discovery and parallel transmission. This project area has undergone a restructuring and continued with three projects on device discovery (continued from phase I), THz networked systems and physical impairment models. This restructuring expresses a shift towards the more system-oriented research in phase II, which is also reflected in the renaming part of this project area to 'Systems Metrology'.

1.4 Structure of the Remaining Part of the Book

The remaining part of the book follows basically the structure of the four project areas and is subdivided into five parts. An additional Part I has been included, which provides a brief summary of basic principles, techniques and measurement equipment relevant to the content of this book. Parts II, III, IV and V correspond to the four project areas defined in the previous sections. Each part consists of several chapters, whereas the chapters do not correspond 1:1 to the projects. In some of the projects various aspects have been investigated, where for each of them it was found to be worth being considered as an own chapter. On the other hand, results across multiple projects are described also in own dedicated chapters.

Phase I (2019-2022)

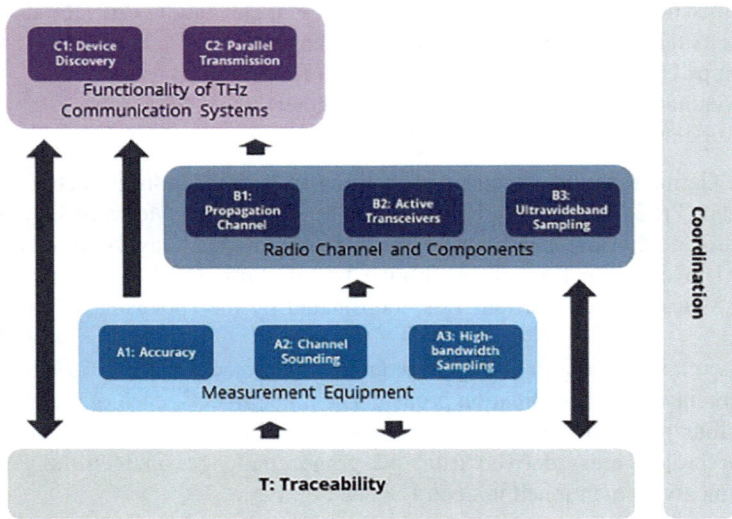

Phase II (2022-2025)

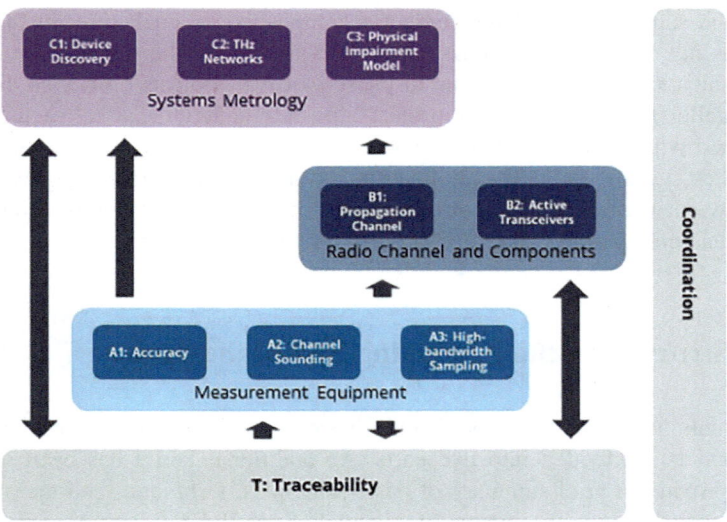

Fig. 1.1 Structure DFG FOR 2863 Metercom and breakdown in to projects for phase I (top) and phase II (bottom)

References

1. S. Cherry, Edholm's law of bandwidth. IEEE Spectr. **41**(7), 58–60 (2004)
2. V. Petrov, T. Kürner, I. Hosako, IEEE 802.15.3d: first standardization efforts for sub-terahertz band communications toward 6G. IEEE Commun. Mag. **58**(11), 28–33 (2020)
3. IEEE, IEEE standard for wireless multimedia networks. IEEE Std 802.15.3-2023 (Revision of IEEE Std 802.15.3-2016), 1–684 (2023)
4. T. Kürner, I. Hosako, *Spectrum and Standardization for THz Communications* (Springer, Singapore, 2024)
5. T. Kürner, T. Mittleman, D.M. Nagatsuma (eds.), *THz Communciations – Paving the Way Towards 1 Tbps*. Springer Series in Optical Sciences, vol. 234 (Springer, Cham, 2022)
6. I.F. Akyildiz, C. Han, Z. Hu, S. Nie, J.M. Jornet, Terahertz band communication: an old problem revisited and research directions for the next decade. IEEE Trans. Commun. **70**(6), 4250–4285 (2022)
7. V. Chinni, P. Latzel, M. Zegaoui, C. Coinon, X. Wallart, E. Peytavit, J. Lampin, K. Engenhardt, P. Szriftgiser, M. Zaknoune, G. Ducournau, Single channel 100 Gbit/s link in the 300 GHz band, in *2018 43rd International Conference on Infrared, Millimeter, and Terahertz Waves (IRMMW-THz)* (2018), pp. 1–2
8. T. Kürner, R.-P. Braun, G. Ducournau, U. Hellrung, A. Hirata, S. Hisatake, L. John, B.K. Jung, I. Kallfass, T. Kawanishi, K. Kondou, Y. Leiba, B. Napier, R. Timar, A. Renau, P. Schlegel, P. Szriftgiser, A. Tessmann, D. Wrana, THz communications and the demonstration in the ThoR–Backhaul link. IEEE Trans. Terahertz Sci. Technol. **14**(5), 554–567 (2024)
9. C. Castro, R. Elschner, T. Merkle, C. Schubert, R. Freund, Experimental demonstrations of high-capacity THz-wireless transmission systems for beyond 5G. IEEE Commun. Mag. **58**(11), 41–47 (2020)
10. J. Grzyb, P.R. Vazquez, B. Heinemann, U.R. Pfeiffer, A high-speed QPSK/16-QAM 1-m wireless link with a tunable 220–260 GHz LO carrier in SiGe HBT technology, in *2018 43rd International Conference on Infrared, Millimeter, and Terahertz Waves (IRMMW-THz)* (2018), pp. 1–2
11. C. Han, Y. Wang, Y. Li, Y. Chen, N.A. Abbasi, T. Kürner, A.F. Molisch, Terahertz wireless channels: a holistic survey on measurement, modeling, and analysis. IEEE Commun. Surv. Tutor. **24**(3), 1670–1707 (2022)
12. A. Ghosh, M. Kim, THz channel sounding and modeling techniques: an overview. IEEE Access **11**, 17823–17856 (2023)
13. ETSI, Identification of use cases for THz communication systems, ETSI ISG THz, ETSI GR THz 001 (2024) [Online]. Available: http://etsi.org/deliver/etsi_gr/THz/001_099/001/01.01.01_60/gr_THz001v010101p.pdf
14. ETSI, Identification of frequency bands of interest for THz communication systems, ETSI ISG THz, ETSI GR THz 002 (2024). [Online]. Available: http://etsi.org/deliver/etsi_gr/THz/001_099/002/01.01.01_60/gr_THz002v010101p.pdf
15. ETSI, Channel measurements and modeling in THz bands , ETSI ISG THz, ETSI GR THz 003 (2024). [Online]. Available: http://etsi.org/deliver/etsi_gr/THz/001_099/003/01.01.01_60/gr_THz003v010101p.pdf
16. ETSI, RF hardware modeling, ETSI ISG THz, ETSI GR THz 004 (2024). [Online]. Available: http://etsi.org/deliver/etsi_gr/THz/001_099/002/01.01.01_60/gr_THz004v010101p.pdf
17. S. Rey, D. Ulm, T. Kleine-Ostmann, T. Kürner, Performance evaluation of a first phased array operating at 300 GHz with horn elements, in *2017 11th European Conference on Antennas and Propagation (EUCAP)* (2017), pp. 1629–1633
18. T. Merkle, A. Tessmann, M. Kuri, S. Wagner, A. Leuther, S. Rey, M. Zink, H.-P. Stulz, M. Riessle, I. Kallfass, T. Kürner, Testbed for phased array communications from 275 to 325 GHz, in *2017 IEEE Compound Semiconductor Integrated Circuit Symposium (CSICS)* (2017), pp. 1–4

19. Q. Xia, J.M. Jornet, Expedited neighbor discovery in directional terahertz communication networks enhanced by antenna side-lobe information. *IEEE Trans. Veh. Technol.* **68**(8), 7804–7814 (2019)
20. J.M. Eckhardt, A. Schultze, R. Askar, T. Doeker, M. Peter, W. Keusgen, T. Kürner, Uniform analysis of multipath components from various scenarios with time-domain channel sounding at 300 GHz. IEEE Open J. Antennas Propag. **4**, 446–460 (2023)
21. R. Piesiewicz, T. Kleine-Ostmann, N. Krumbholz, D. Mittleman, M. Koch, T. Kürner, Terahertz characterisation of building materials. IEE Electron. Lett. **41**(18), 1002–1003 (2005)
22. R. Piesiewicz, T. Kleine-Ostmann, N. Krumbholz, D. Mittleman, M. Koch, J. Schoebel, T. Kurner, Short-range ultra-broadband terahertz communications: concepts and perspectives. IEEE Antennas Propag. Mag. **49**(6), 24–39 (2007)
23. R. Piesiewicz, M. Jacob, M. Koch, J. Schoebel, T. Kürner, Performance analysis of future multigigabit wireless communication systems at THz frequencies with highly directive antennas in realistic indoor environments. IEEE J. Sel. Top. Quantum Electron. **14**(2), 421–430 (2008)
24. C. Jansen, S. Priebe, C. Moller, M. Jacob, H. Dierke, M. Koch, T. Kürner, Diffuse scattering from rough surfaces in THz communication channels. IEEE Trans. Terahertz Sci. Technol. **1**(2), 462–472 (2011)
25. M. Jacob, S. Priebe, R. Dickhoff, T. Kleine-Ostmann, T. Schrader, T. Kurner, Diffraction in mm and sub-mm wave indoor propagation channels. IEEE Trans. Microwave Theory Tech. **60**(3), 833–844 (2012)
26. S. Priebe, M. Kannicht, M. Jacob, T. Kürner, Ultra broadband indoor channel measurements and calibrated ray tracing propagation modeling at THz frequencies. J. Commun. Netw. **15**(6), 547–558 (2013)
27. S. Koenig, D. Lopez-Diaz, J. Antes, F. Boes, R. Henneberger, A. Leuther, A. Tessmann, R. Schmogrow, D. Hillerkuss, R. Palmer, T. Zwick, C. Kroos, W. Freude, O. Ambacher, J. Leuthold, I. Kallfass, Wireless sub-THz communication system with high data rate. Nat. Photon **7**, 977–981 (2013)
28. T. Nagatsuma, G. Ducournau, C. Renaud, Advances in terahertz communications accelerated by photonics. Nat. Photon **10**, 371–379 (2016)
29. T. Kürner, THz communications – a candidate for the next generation of wireless systems? in *Opening Plenary of the International Symposium on Antennas and Propagation, ISAP 2020 [online]* (2021)
30. J.T. Quimby, D.F. Williams, K.A. Remley, D. Ribeiro, R. Sun, J. Senic, Millimeter-wave channel-sounder performance verification using vector network analyzer in a controlled RF channel. IEEE Trans. Antennas Propag. **69**(11), 7867–7875 (2021)
31. I. Kallfass, S.M. Dilek, I. Dan, Signal quality impairments by analog frontend non-idealities in a 300 GHz wireless link, in *2017 11th European Conference on Antennas and Propagation (EUCAP)* (2017), pp. 1618–1621
32. B. Peng, T. Kürner, Three-dimensional angle of arrival estimation in dynamic indoor terahertz channels using a forward–backward algorithm. IEEE Trans. Veh. Technol. **66**(5), 3798–3811 (2017)
33. B. Peng, K. Guan, T. Kürner, Cooperative dynamic angle of arrival estimation considering space–time correlations for terahertz communications. IEEE Trans. Wirel. Commun. **17**(9), 6029–6041 (2018)
34. T. Kleine-Ostmann, Introduction to special issue on terahertz metrology. J. Infrared, Millim. Terahertz Wave **35**(8), 583–584 (2014)
35. J.M. McKinnis, I. Gresham, R. Becker, Figures of merit for active antenna enabled 5G communication networks, in *2018 11th Global Symposium on Millimeter Waves (GSMM)* (2018), pp. 1–7
36. M. Naftaly, An international intercomparison of THz time-domain spectrometers, in *2016 41st International Conference on Infrared, Millimeter, and Terahertz waves (IRMMW-THz)* (2016), pp. 1–2

37. L. Oberto, M. Bisi, A. Kazemipour, A. Steiger, T. Kleine-Ostmann, T. Schrader, Measurement comparison among time-domain, FTIR and VNA-based spectrometers in the THz frequency range. Metrologia **54**(1), 77 (2017). [Online]. Available: https://doi.org/10.1088/1681-7575/aa54c2

38. M. Landmann, M. Kaske, R.S. Thoma, Impact of incomplete and inaccurate data models on high resolution parameter estimation in multidimensional channel sounding. IEEE Trans. Antennas Propag. **60**(2), 557–573 (2012)

39. X. Chen, A. Engelmann, A. Jukan, M. Medard, Linear network coding and parallel transmission increase fault tolerance and optical reach. J. Opt. Commun. Netw. **9**(4), 244–256 (2017)

40. A. Soto, M. Alem, M.A. Shoaie, A. Vedadi, C.S. Brès, L. Thévenaz, T. Schneider, Optical sinc-shaped nyquist pulses of exceptional quality. Nat. Commun. **4**, 977–981 (2013)

41. S. Preußler, N. Wenzel, T. Schneider, Flexible nyquist pulse sequence generation with variable bandwidth and repetition rate. IEEE Photon. J. **6**(4), 1–8 (2014)

42. S. Preussler, N. Wenzel, T. Schneider, Flat, rectangular frequency comb generation with tunable bandwidth and frequency spacing. Opt. Lett. **39**(6), 1637–1640 (2014). [Online]. Available: https://opg.optica.org/ol/abstract.cfm?URI=ol-39-6-1637

43. S. Preußler, G. Raoof Mehrpoor, T. Schneider, Frequency-time coherence for all-optical sampling without optical pulse source. Sci. Rep. **6**(6), 34500 (2016)

44. J.M. Eckhardt, T. Doeker, Lessons learned from a decade of THz channel sounding. IEEE Commun. Mag. **62**(2), 24–30 (2024)

45. S. Kim, A. Zajić, Characterization of 300-GHz wireless channel on a computer motherboard. IEEE Trans. Antennas Propag. **64**(12), 5411–5423 (2016)

46. D. He, K. Guan, A. Fricke, B. Ai, R. He, Z. Zhong, A. Kasamatsu, I. Hosako, T. Kürner, Stochastic channel modeling for Kiosk applications in the terahertz band. IEEE Trans. Terahertz Sci. Technol. **7**(5), 502–513 (2017)

47. A. Fricke, M. Achir, P. Le Bars, T. Kürner, A model for the reflection of terahertz signals from printed circuit board surfaces. Int. J. Microwave Wirel. Technol. **10**(2), 179–186 (2018)

48. S. Rey, J.M. Eckhardt, B. Peng, K. Guan, T. Kürner, Channel sounding techniques for applications in THz communications: a first correlation based channel sounder for ultra-wideband dynamic channel measurements at 300 GHz, in *2017 9th International Congress on Ultra Modern Telecommunications and Control Systems and Workshops (ICUMT)* (2017), pp. 449–453

49. D. Dupleich, A. Ebert, R. Müller, G. Del Galdo, R. Thomä, Verification of dual-polarized ultra-wideband channel sounder for THz applications, in *2021 IEEE 32nd Annual International Symposium on Personal, Indoor and Mobile Radio Communications (PIMRC)* (2021), pp. 1–5

50. D. Dupleich, S. Semper, M.D. Al-Dabbagh, A. Ebert, T. Kleine-Ostmann, R. Thomä, Verification of THz channel sounder and delay estimation with over-the-air multipath artifact, in *2022 16th European Conference on Antennas and Propagation (EuCAP)* (2022), pp. 1–5

Part I
Fundamentals

Chapter 2
Metrology: Definition, Use Cases, Added Value

Thomas Kleine-Ostmann and David A. Humphreys

Abstract Metrology is the art and science of measurement. Achieving precise and reliable measurements, including a quantification of their accuracy, has been the goal for the past centuries. Equipment, methods, and tools have been developed in many disciplines of science and technology to realize today's state of the art. In many cases, everyday life relies on trust in measurements. Classical examples are fair exchange of trade goods and pricing, proof of legal requirements such as compliance with safety limits, and comparability of scientific results but also technological specifications that are required for interoperability. Although much of the communication infrastructure is data-driven and defined by specification standards, metrology and traceability are essential to build and maintain the system. Also, with a growing number of applications making use of terahertz radiation, the question of the reliability of measurements in the terahertz frequency range becomes increasingly important. In this chapter we give a general introduction into metrology and its application to communication systems.

2.1 Introduction

Commercial products are usually sold with specifications listed in data sheets which are the basis for their safe use, fair trade exchange, and interoperability in technical systems. For quality control in production, measurements are the basis for quality assessment. Many technical systems rely on the precision of measurements. In science, reliable measurements are the basis for new insights. To draw the correct conclusions, measurement artefacts must be excluded and results obtained with different techniques and apparatuses need to be comparable. In communications,

T. Kleine-Ostmann (✉)
Physikalisch-Technische Bundesanstalt (PTB), Braunschweig, Germany
e-mail: thomas.kleine-ostmann@ptb.de

D. A. Humphreys
NPL, UK (Retired), University of Bristol, Ascot, UK
e-mail: david.a.humphreys@ieee.org

© The Author(s) 2026
T. Kürner et al. (eds.), *Metrology for THz Communications*, Springer Series in Optical Sciences 256, https://doi.org/10.1007/978-3-032-01986-8_2

device and channel properties need to be known to enable interoperability of devices, ensure human exposure to electromagnetic fields below the safety limit, and judge coverage and protection of communication services [1, 2].

To be sure that the underlying measurements are trustable and usable, it is not sufficient to regard a single measured value of a physical quantity as measurement result. It needs to be associated with a measurement uncertainty interval in which the true value falls with a specified probability. The determination of the measurement uncertainty following internationally accepted rules [3] is the basis for establishing traceability, an unbroken chain of successive calibrations referenced to national standards.

Following the state of the art in science, the National Metrology Institutes (NMIs) such as the *National Institute of Standards and Technology* (NIST) in the USA, the *National Physical Laboratory* (NPL) in the UK, and the *Physikalisch-Technische Bundesanstalt* (PTB) in Germany maintain standards for the representation of units with highest precision. Measurement intercomparisons are organized and evaluated to validate the standard representations within the NMIs [4].

In the past years, it has been tried to develop traceability towards higher frequencies reaching the terahertz range. Applying metrological approaches to THz systems has been referred to as *terahertz metrology* [1, 2, 6].

2.2   The International System of Units (SI)

In 1875 the "Metre Convention," formed by a treaty of 17 nations (now signed by 66 states and adopted by many more associated states), was founded to establish stable standards for mass and length based on rational principles. Defining the unit of length, the meter, as the 40 millionth part of a meridian and, subsequently, the kilogram as the mass of a single cube of 1 dm edge length of pure distilled water at 3.98 °C, where it has its greatest density, the first two international prototypes were established and copies of the representations were distributed to the member states after the first General Conference in 1889 [7]. Later on, the system of the international measurement units was continuously extended. In 1954 additional units for time, electric current, thermodynamic temperature, and luminous intensity were defined forming the so-called practical system of units. Then in 1960, the meter was redefined in terms of the wavelength of light, and the six units formed the basis for the so-called Système International d'Unités (SI). Later, in the 14th General Conference in 1971, the SI base units were completed with the mole, the unit for the amount of substance. With this, the system of units became a coherent system, which means that the derived units can be defined as the products of powers of the base units without additional numerical factor [7].

The realization of the units according to their definition and with smallest possible uncertainty is a constant challenge, since absolute measurements are limited in precision by their realization. To fulfill the ideal of a definition "for all men and all times," tracing back the SI units to fundamental constants is an optimum solution [7].

Before the redefinition of the SI in 2019, this was the case for the meter and the second, only. With the exact definition of seven defining fundamental constants with consistent fixed values (cesium hyperfine frequency $\Delta \nu_{Cs}$, speed of light in vacuum c, Planck constant h, elementary charge e, Boltzmann constant k, Avogadro constant N_A, and luminous efficacy of a defined visible radiation K_{cd}), it became possible to define the base units dependent on the fundamental constants in a quantum-based SI [5, 8].

2.3 Measurement Uncertainty

Establishing traceability to the representation of SI units can be established by using a calibrated measurement instrument with a documented measurement uncertainty that has been estimated in accordance with the *Guide to the expression of uncertainty in measurement* [3]. The calculated measurement uncertainty results from a detailed modelling of the measurement process (model function $Y = f(X_i)$) that relates N input quantities X_i (with best estimates x_i) and their uncertainties $u(x_i)$ to an output quantity Y with best estimate y and resulting measurement uncertainty $u(y)$. The uncertainties of the input quantities result from a statistical analysis of a series of observations (type A contributions) and/or other scientific knowledge about the measurement process (type B contributions, e.g., from calibration certificates).

When estimating the measurement uncertainty according to method A, a set of at least $n \geq 10$ measurements of the input quantity has to be evaluated. The arithmetic mean $x_i = \frac{1}{n} \sum_{j=1}^{n} x_{ij}$ of a set of values x_{ij} is chosen as best estimate for the input quantity, whereas the empiric standard error of the mean, $u(x_i) = \sqrt{\frac{1}{n} \frac{1}{n-1} \sum_{j=1}^{n} |x_{ij} - x_i|^2}$, is taken for the measurement uncertainty.

If method B is chosen to estimate the measurement uncertainty of input quantities, a probability density function for the input quantity has to be assumed. Its expectation value becomes the best estimate for the input quantity x_i, whereas the square root of its variance becomes its assigned measurement uncertainty. When using information from calibration certificates, a Gaussian probability density distribution is assumed. In other cases, the probability density function remains unknown. Then, for simplicity, a rectangular distribution of the input quantity is assumed based on the range of values that seems possible for the input quantity. In this case, the best estimate of the input quantity is chosen to be the center of the rectangular distribution, whereas the measurement uncertainty is determined to $u(x_i) = \frac{a}{\sqrt{3}}$ from the half-width a of the rectangular distribution.

With known uncertainties of the input quantities, the uncertainty of the output quantity can be calculated as [9]

$$u(y) = \sqrt{\sum_{i=1}^{N} c_i^2 u^2(x_i)} \qquad (2.1)$$

with sensitivity coefficients

$$c_i = \frac{\delta f}{\delta x_i} = \frac{\delta f}{\delta X_i}\bigg|_{X_i=x_i} . \qquad (2.2)$$

In combination with the best estimate of the measurand, the calculated standard measurement uncertainty of the output quantity specifies the interval in which the measurand can be expected with an uncertainty of 68% (assuming the $1 - \sigma$ interval of a Gaussian distribution). To increase the confidence to 95%, the expanded measurement uncertainty is reported which is the standard measurement uncertainty multiplied by an expansion factor ($k = 2$ in case of a sufficiently high degree of freedom). The combination of measurement uncertainties of input quantities described by different probability distributions to a resulting measurement uncertainty of an output quantity that is assumed to be Gaussian distributed is based on the central limit theorem. It assumes that the sensitivity coefficients are of comparable order. If the calculation according to Eqs. (2.1) and (2.2) fails because the partial derivatives of the model function do not exist or are zero, the analytical calculation of the resulting measurement uncertainty can be replaced by using the *"Monte Carlo"* method [3]. Here, a large number of random values for the input quantities with known probability distributions are propagated through the model function to obtain a histogram of the output quantity that is used to estimate its measurement uncertainty.

2.4 Traceability

With measurement instruments calibrated at the national standards of the NMIs, the fundamental and derived units can be disseminated to other laboratories as shown in Fig. 2.1. With only several thousands of calibrations at the NMI level for accredited calibration laboratories or other industry laboratories, measurements in production can be realized with known measurement uncertainty. Although the measurement uncertainty increases with each calibration step in the pyramid, the measurement value obtained at the product level is traceable to the SI and can be trusted. The calibration and measurement capabilities (CMC) of the NMIs are listed in the *International Key Comparison Data Base* (KCDB) [10]. The CMC entries are reviewed internationally among the NMIs and need to be validated by international key comparisons, which are listed in the KCDB as well.

2.5 Communications Metrology

All communication systems are defined by a series of detailed specification standards to ensure compatibility and interoperability. The Open Systems Interconnec-

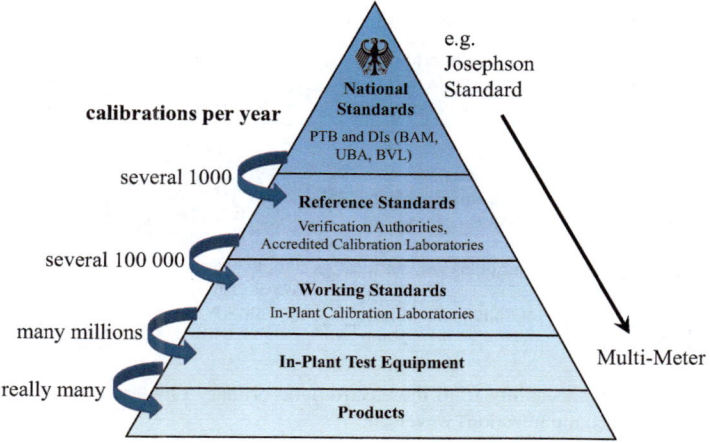

Fig. 2.1 The traceability pyramid with the national calibration standards on top

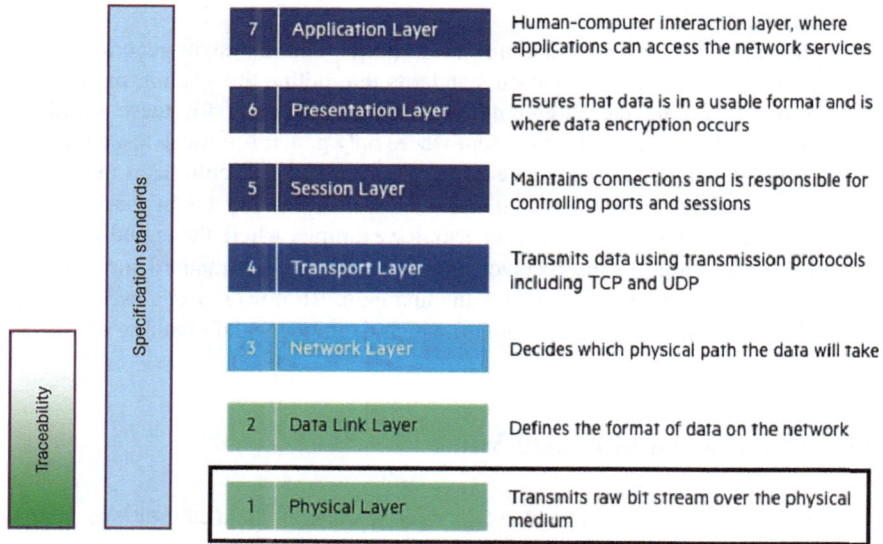

Fig. 2.2 The Open Systems Interconnection (OSI) 7-layer model

tion (OSI) 7-layer model (see Fig. 2.2) is a reference model from the International Organization for Standardization (ISO) that "provides a common basis for the coordination of standards development for the purpose of systems interconnection." [11]. This standard is also published by ITU (International Telecommunciations Union) as Recommendation X.200 [12]. The lowest layer (Layer 1 – physical layer) contains the system hardware which comprises many components and subsystems.

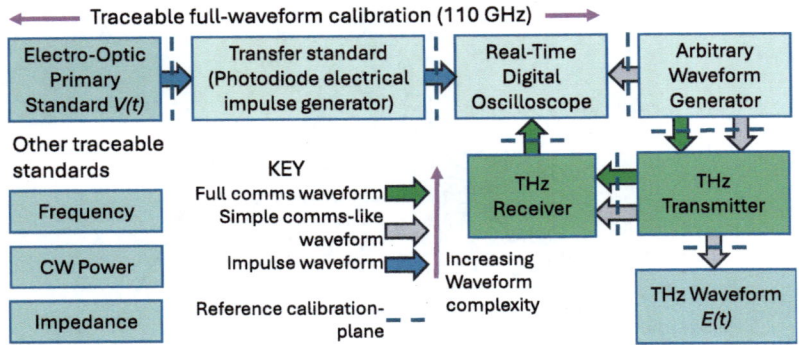

Fig. 2.3 Transfer of traceability from the electro-optic primary standard (impulse). Comments added to a complex communications waveform

In certain cases there is interaction between the components (Layer 1) and the data structures (Layer 2) [13, 14].

Each overarching standard defining the communication system generation leads to many more detailed specification standards that define the system, models, and algorithms. Traceable measurement forms an underpinning for these standards, particularly at the outset of development where only general-purpose test equipment is available. Testing dedicated test equipment presents a difficulty as it may not be compatible with the simpler stimuli used in the traceability chain (see Fig. 2.3). The manufacturer must avoid the scenario, for example, where the signal waveform generator is traceable to the receiver and vice versa [15]. Standard measurement services, such as antenna pattern and antenna gain, RF power, and power linearity, are critical to gathering accurate data on which the models and planning are based.

2.6 Use Cases and Added Value

Many routinely used electromagnetic services, such as antenna calibrations, support communications metrology indirectly but there are examples covering many generations of communication technologies, where NMIs have supported industry with difficult measurement challenges. Some of this work has been done by individual NMIs, in collaboration with industry and through Euramet European Joint Research Projects IND16 "Ultrafast" [16], IND 51 "Morse" [17] and 14IND10 "MET5G" [18], and other projects. Traceable RF peak power (2G/GSM) presented measurement difficulties [19]. Error vector magnitude (EVM) is covered in Chap. 15.

Another example is JET, a small company that was assisted by metrology to develop floating 5G base stations. Compared with the previous system, the network coverage at sea improved nearly sevenfold, giving better environmental sustainability and safety [20].

References

1. T. Kleine-Ostmann, T. Schrader, M. Bieler, U. Siegner, C. Monte, B. Gutschwager, J. Hollandt, A. Steiger, L. Werner, R. Müller, G. Ulm, I. Pupeza, M. Koch, THz metrology. Frequenz **62**(5–6), 137–148 (2008)
2. T. Kleine-Ostmann, THz metrology, in *38th International Conference on Infrared, Millimeter and Terahertz Waves (IRMMW-THz 2013)*, Mainz (2018)
3. Working Group 1 of the Joint Committee for Guides in Metrology (JCGM/WG 1), Evaluation of measurement data – guide to the expression of uncertainty in measurement, BIPM, Technical report, 2008. [Online]. Available: https://www.bipm.org/documents/20126/2071204/JCGM_100_2008_E.pdf
4. M.G. Cox, The evaluation of key comparison data. Metrologia **39**(6), 589 (2002). [Online]. Available: https://doi.org/10.1088/0026-1394/39/6/10
5. M. Naftaly, Terahertz Metrology (Artech House, Norwood, 2014)
6. E. O. Göbel, A Short Introduction to the International System of Units, in: The system of units. PTB-Mitteilungen **122**(1), 3–5 (2012)
7. R. Scharf, T. Middelmann, Paradigmenwechsel im Internationalen Einheitensystem (SI), in: Experimente für das neue Internationale Einheitensystem (SI). PTB-Mitteilungen **126**(2), 5–15 (2016)
8. E. Göbel, U. Siegner, *The New International System of Units (SI) – Quantum Metrology and Quantum Standards* (Wiley-VCH, Berlin, 2019)
9. EA Laboratory Committee, Evaluation of the uncertainty of measurement in calibration, European Accreditation, Technical report, 2022. [Online]. Available: https://european-accreditation.org/wp-content/uploads/2018/10/EA-4-02.pdf
10. The international key comparison data base KCDB, Bureau International des Poids et Mesures, Technical report, 2024. [Online]. Available: https://www.bipm.org visited on 08 Apr 2024
11. Joint Technical Committee ISO/IEC JTC 1, ISO/IEC 7498-4:1989(en) information technology – open systems interconnection – basic reference model: the basic model, ISO, Technical report, 1994. [Online]. Available: https://www.iso.org/obp/ui/#iso:std:iso-iec:7498:-4:ed-1:v1:en cited 20 Aug 2024
12. ITU-T, X.200: information technology – open systems interconnection – basic reference model: the basic model, ITU, Technical report, 1994. [Online]. Available: https://www.itu.int/rec/T-REC-X.200/en cited 19 Aug 2024
13. R. Van Tuyl, D. Ingram, D. Humphreys, D. Taylor, Optical retiming for jitter measurement calibration, in *2005 31st European Conference on Optical Communication, ECOC 2005*, vol. 4 (2005), pp. 973–974
14. ITU-T, Jitter and wander measuring equipment for digital systems which are based on the synchronous digital hierarchy (SDH), ITU, Technical report, 2005. [Online]. Available: https://www.itu.int/rec/T-REC-O.172-200504-I/en
15. D.A. Humphreys, R.T. Dickerson, Traceable measurement of error vector magnitude (EVM) in WCDMA signals, in *2007 International Waveform Diversity and Design Conference* (2007), pp. 270–274
16. M. Bieler, Euramet IND16 ultrafast, EURAMET e.V., Technical report, 2014. [Online]. Available: https://www.euramet.org/research-innovation/search-research-projects/details/project/metrology-for-ultrafast-electronics-and-high-speed-communications visited on 20 Aug 2024
17. D. Humphreys, Euramet IND51 morse, EURAMET e.V., Technical report, 2016. [Online]. Available: https://www.euramet.org/research-innovation/search-research-projects/details/project/metrology-for-optical-and-rf-communication-systems visited on 20 Aug 2024
18. T. H. Loh, Euramet 14IND10 MET5G, EURAMET e.V., Technical report, 2018. [Online]. Available: https://www.euramet.org/research-innovation/search-research-projects/details/project/metrology-for-5g-communications visited on 20 Aug 2024

19. D. Humphreys, J. Miall, Traceable RF peak power measurements for mobile communications. IEEE Trans. Instrum. Meas. **54**(2), 680–683 (2005)
20. Developing a novel over-the-air radio test solution to boost 5G connectivity at sea, NPL, Technical report, 2024. [Online]. Available: https://www.npl.co.uk/case-studies/developing-a-novel-over-the-air-radio-test-solutio visited on 20 Aug 2024

Chapter 3
RF Power Metrology

Thomas Kleine-Ostmann and Rolf Judaschke

Abstract Radio-frequency (RF) power is one of the main quantities that needs to be known accurately when designing high-frequency circuits and systems. Up to millimeter-wave frequencies, RF power traceability is established by calibrating suitable waveguide power sensors in so-called microcalorimeters, where traceability to the SI is based on DC power substitution. By using a calibrated reference sensor, the quantity RF power can be disseminated to end user devices by applying the well-known direct comparison technique. For spree-space power measurements, traceability relies on quasi-optical techniques. Comparison of both approaches in the overlapping frequency region shows good agreement within the specified measurement uncertainties. This chapter intends to give a brief overview of RF power traceability.

3.1 Introduction

Accurate power measurements are the basis for the design of many RF circuits and devices but also for the assessment of field strengths and exposure levels in personal safety, electromagnetic compatibility, and compliance with standards and regulations. Furthermore, they are the basis for the quantitative comparison of technical and scientific results, as many units of physical quantities are derived from the unit watt.

Most accurate power measurements are performed in coaxial lines and rectangular waveguides, where the electromagnetic energy is confined to distinct propagation modes such as the TEM (transversal electromagnetic) and the TE_{01} (transversal electric) or H_{01} fundamental mode in coaxial lines and metallic waveguides, respectively [1, 2]. In the former case, the measurement is done by a power sensor, acting as line termination, where in the latter case, the sensor is located at an

T. Kleine-Ostmann (✉) · R. Judaschke
Physikalisch-Technische Bundesanstalt (PTB), Braunschweig, Germany
e-mail: thomas.kleine-ostmann@ptb.de; rolf.judaschke@ptb.de

© The Author(s) 2026
T. Kürner et al. (eds.), *Metrology for THz Communications*, Springer Series
in Optical Sciences 256, https://doi.org/10.1007/978-3-032-01986-8_3

Table 3.1 Properties of different power sensor measurement principles

	Diode sensor	Thermoelectric sensor	Thermistor sensor
Sensitivity	High	Mid	Low
Speed	High	Mid	Low
Linearity	Low	High	High
Dynamic range	High	Mid	Low
Stability	Bad	High	High
Traceability	Bad	Mid	High

appropriate location in the waveguide [3]. Power sensors can be diode-based (the RF waveform is rectified), thermoelectric (the RF power absorption in a heating resistor leads to a thermal voltage according to the Seebeck effect) or thermistor based (the RF power absorption in a resistor [so-called thermistor] results in a change its resistance). Table 3.1 shows advantages and disadvantages of the three measurement principles [4].

To establish traceability, thermistors are the first choice for a primary calibration in a so- called microcalorimeter [5, 6]. The calibrated primary thermistor standard can then be used to calibrate customer sensors based on the direct comparison technique. Alternatively, thermoelectric sensors can be calibrated in a microcalorimeter [7].

3.2 Microcalorimeter Primary Power Standards in Waveguides

The primary calibration of a sensor in a microcalorimeter is based on direct current (DC) power substitution. A thermistor, acting as RF absorbing element, is part of a balanced precision measurement bridge. When switching the RF off, the required DC power P_{sub} to keep the bridge balanced can be determined.

Figure 3.1 illustrates the power distribution of an RF power measurement [4]: incident power P_{in} and reflected power $P_{refl} = \Gamma^2 P_{in}$ in the reference plane are related to the reflection factor Γ, while the sensor absorbs the power $P_{RF\,abs}$. However, power losses P_{loss} in the sensor are not detected and do not contribute to the display value $P_{display}$ of a power meter or, in case of a microcalorimeter measurement, to the substituted power $P_{DC\,sub}$.

From this, the effective efficiency η_{eff} and the calibration factor η_{cal} of a power sensor can be defined [4]:

$$\eta_{eff} = \frac{P_{display}}{P_{RF\,abs}} = \frac{P_{RF\,abs} - P_{loss}}{P_{RF\,abs}} = \frac{P_{DC\,sub}}{P_{RF\,abs}} \tag{3.1}$$

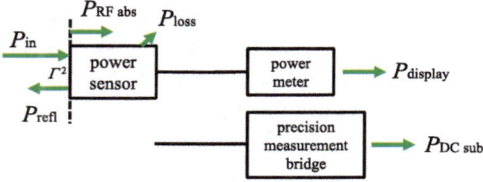

Fig. 3.1 Nomenclature for RF power measurement using a power sensor (detecting device) with a power meter as display unit or – in case of thermistor power sensors – with a precision measurement bridge for calibration based on DC substitution

Fig. 3.2 Principle of a microcalorimeter. The device under calibration together with a dummy sensor is placed into a thermally insulated vessel for measurement

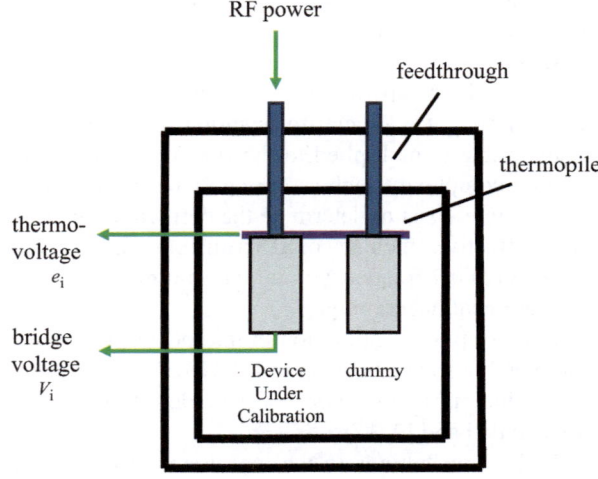

and

$$\eta_{\text{cal}} = \frac{P_{\text{display}}}{P_{\text{in}}} = \left(1 - |\Gamma|^2\right) \cdot \eta_{\text{eff}}. \tag{3.2}$$

For the determination of the effective efficiency, the thermistor power sensor is placed in the thermally insulated microcalorimeter together with a dummy sensor. Figure 3.2 shows the configuration of the microcalorimeter that allows to quantify the losses P_{loss}. It has a symmetrical setup consisting of a measurement transmission line which is terminated by the device under test (DUT) and a dummy line (terminated by the dummy sensor) acting as thermal reference. A thermopile consisting of a series connection of a larger number of thermoelements is used to measure the temperature difference between the reference planes of the DUT and the dummy sensor. The linear dependence of the thermopile voltage e from the overall dissipated calorimeter heating power is described by the heating factor k. In a first step, the calorimeter is operated with the RF power switched off, resulting in the bridge voltage V_1 and the thermal voltage e_1. The associated power dissipated in the DUT thermistor is $P_1 = U_1^2 / R = k \cdot e_1$. A second measurement with RF switched on leads to a bridge voltage V_2 and a thermal voltage e_2. Here, the dissipated power is $P_2 = U_2^2/R$. Using Eq. (3.1), the substituted power $P_{DC\,sub} = P_1 - P_2 = e_1 k(1 - U_1^2/U_2^2)$, and the relation $e_2 k = P_{loss} + P_{RF\,abs} + P_2 \approx P_{RF\,abs} + P_2$, the effective efficiency can be determined as [8]

$$\eta_{eff} = \frac{1 - \frac{V_2^2}{V_1^2}}{\frac{e_2}{e_1} - \frac{V_2^2}{V_1^2}}. \tag{3.3}$$

It should be noted that Eq. (3.3) is a simplified equation that is assuming the losses P_{loss} to be negligible. These losses take account of the fraction of the dissipated power in the feeding line in front of the reference plane that additionally increases the temperature of the thermopile assembly mount. To achieve low uncertainties, it is crucial to quantitatively account for the losses by a correction factor g that is multiplied to the fraction derived above [9]. Different methods of varying complexity such as the *offset short method*, the *short foil method*, and the *VNA method* exist to determine the correction factor [7, 10–16]. The determination of the effective efficiency of a thermistor requires a rather long measurement time of some hours per frequency point, depending on the method applied and the required measurement uncertainty.

Primary power calibration in microcalorimeters is available in different coaxial and metallic waveguides up to 170 GHz, so far. The overall expanded measurement uncertainty is in the order of 1%, depending on the frequency and the actual waveguide band [2, 17].

Main contributions to the measurement uncertainty are the uncertainty of the correction factor g, taking into account the transmission line RF losses in the feeding waveguide, nonequivalence between RF and DC heating in the substitution measurement, nonlinearities of thermopiles, and temperature instabilities in the microcalorimeter [2, 6]. Ongoing work is focusing on extending the calibration capabilities to higher frequencies, both for new coaxial connector types and for higher waveguide bands.

3.3 RF Power Calibration Based on the Direct Comparison Technique

Using the calibrated thermistor sensor, a customer power sensor (device under test - [DUT]) can be calibrated based on the external measurement comparison technique as shown in Fig. 3.3. The signal from a RF generator is divided by a power splitter. One fraction is measured with a reference power sensor, whereas the other fraction is measured consecutively with a calibrated power sensor (e.g., thermistor) and the DUT.

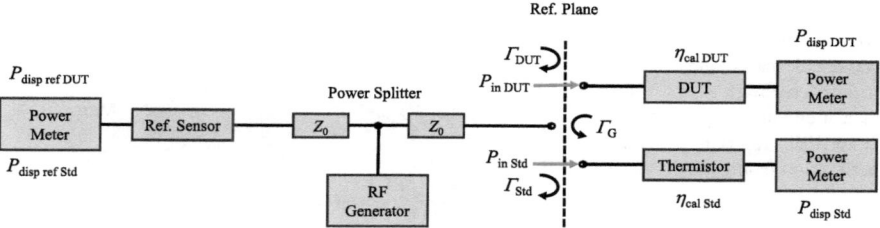

Fig. 3.3 Principle of the direct comparison technique. The generated RF power is split into a reference and a measurement branch. The former monitors the generator power stability, while the DUT and a calibrated reference power standard are connected in the reference plane, consecutively

The calibration factor of the DUT $\eta_{\text{cal DUT}}$ can then be calculated from the calibration factor of the thermistor standard $\eta_{\text{cal Std}}$ as [8]

$$\eta_{\text{cal DUT}} = \eta_{\text{cal Std}} \cdot \frac{P_{\text{disp DUT}}}{P_{\text{disp ref DUT}}} \cdot \frac{P_{\text{disp ref Std}}}{P_{\text{disp Std}}} \cdot \frac{|1 - \Gamma_{\text{G}} \Gamma_{\text{DUT}}|^2}{|1 - \Gamma_{\text{G}} \Gamma_{\text{Std}}|^2}, \qquad (3.4)$$

where $P_{\text{disp DUT}}$, $P_{\text{disp Std}}$, $P_{\text{disp ref DUT}}$, and $P_{\text{disp ref Std}}$ are the power measurements of the DUT, the standard, and the respective power meter display readings of the reference power sensor. Γ_{G}, Γ_{DUT}, and Γ_{Std} are the input reflexion factors of the generator, the DUT, and the standard, respectively.

3.4 Calibration of Free-Space Power Sensors

Especially at higher frequencies above 100 GHz, free-space space power sensors such as pyroelectric detectors or a photoacoustic detector are used to measure the total radiant power in a quasi-optical beam [2]. The detector aperture diameter has to be large enough to capture the whole beam power (2.6 times the full-width at half-maximum diameter of a Gaussian beam to detect 99% of the total power in the fundamental mode) [18].

At the Physikalisch-Technische Bundesanstalt (PTB), the German National Metrology Institute (NMI), traceability for free-space power sensors is established by using a broadband thermopile detector with a polished neutral-density glass gold plated at its back side as a reference sensor [19]. Its absorptance and reflectance is characterized by spectroscopic methods both in the THz range and at the frequency of 476.6 THz (633 nm) of a He-Ne laser. At the visible frequency, it is traced back to the system of units (SI) by comparison with a power substitution measurement in a cryogenic radiometer that serves as national radiant power standard at PTB. The calibrated reference standard allows to measure power down to frequencies of 1 THz with an expanded uncertainty of 2.4%.

Fig. 3.4 Setup for
comparison of
waveguide-based and
free-space power
measurements (Copyright
2020 IEEE, reproduced with
permission from [18])

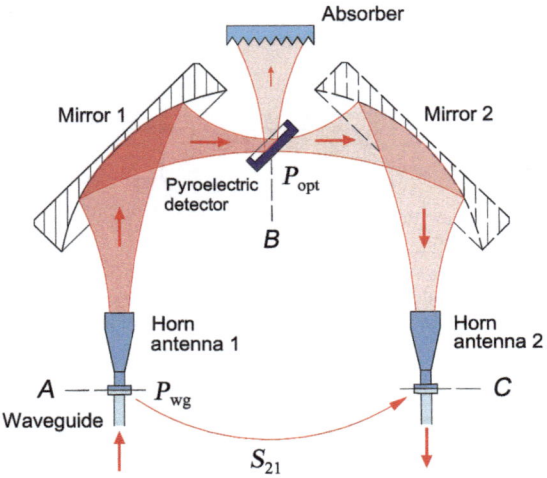

To extend the traceability to lower THz frequencies, a large-aperture pyroelectric thin-film detector is used. Its polyvinylidene difluoride (PVDF) foil, which is transparent below 2.5 THz, is coated with metal oxide layers on both sides whose conductivity has been adjusted in such a way that the absorptance of 50% is spectrally flat below 2.5 THz [20–22]. The whereabouts of the unabsorbed other half must be controlled. The transmitted part is absorbed inside the detector housing by a foam (Ecosorb®) while users must take care of the reflected part. Using a stable source such as a far-infrared gas laser at 1.4 THz, the power responsivity of the pyroelectric detector can be referenced to the optical reference standard described above [23].

3.5 Comparison of Waveguide-Based and Free-Space Power Traceability Chains

Using the calibrated pyroelectric thin-film detector with an aperture diameter of 34 mm, the power in the focus of a free-space quasi-optical setup in the WR-10 or W band (75–110 GHz) was measured to perform a comparison between waveguide-based and free-space power measurements [18]. In the setup shown in Fig. 3.4, a frequency-multiplied RF microwave synthesizer with an output power attenuated to approximately 10 dBm at the WR-10 flange was used for the comparison. Its output power P_{wg} at the flange was measured with a calibrated thermistor sensor.

For the free-space power measurement, the power is fed into a corrugated horn antenna to transform the H_{01} fundamental waveguide mode into a Gaussian beam profile. An ellipsoidal mirror produced a beam waist with approximately 10-mm diameter in the focal plane, where the power P_{opt} was measured with the thin-film detector tilted at 45° to avoid standing waves by its reflection. This is because the 90° reflection can be easily absorbed outside the beam path. For a precise comparison of waveguide and free-space power measurement, the losses between the waveguide flange and the measurement plane in the focus of the mirror have to be considered. For the accurate determination of these losses, the setup was extended symmetrically using a second mirror and a second corrugated horn antenna. Using a vector network analyzer, the losses were determined precisely by measuring the transmission scattering parameters S_{21} and S_{12} in both directions. With this, an overall expanded uncertainty of 6% was achieved for the ratio P_{opt}/P_{wg}.

Within the specified uncertainty, the ratio P_{opt}/P_{wg} exhibited a value of 1 over the whole frequency band. This is a good and valuable validation that the traceability chains, which are completely different and independent for waveguide-based and quasi-optical power measurements, are correct and valid. For the future, it is planned to repeat this experiment for other waveguide bands as well.

Acknowledgments The authors would like to thank their colleague Dr. Andreas Steiger, also with Physikalisch-Technische Bundesanstalt (PTB), for his critical review of this chapter.

References

1. T. Kleine-Ostmann, T. Schrader, M. Bieler, U. Siegner, C. Monte, B. Gutschwager, J. Hollandt, A. Steiger, L. Werner, R. Müller, G. Ulm, I. Pupeza, M. Koch, THz metrology. Frequenz **62**(5–6), 137–148 (2008)
2. E. Castro-Camus, M. Koch, T. Kleine-Ostmann, A. Steiger, On the reliability of power measurements in the terahertz band. Commun. Phys. **5**(42), 1–3 (2022)
3. A. Fantom, *Radio Frequency and Microwave Power Measurement* (Peter Peregrinus Ltd, London, 1990)
4. R. Judaschke, "HF Leistungsmessung bis 110 GHz," Physikalisch-Technische Bundesanstalt (PTB), 278. PTB Seminar, Aktuelle Fortschritte von Kalibrierverfahren im Nieder und Hochfrequenzbereich, PTB-Bericht E 103 (2014)
5. M. Sucher, H.J. Carlin, Broadband calorimeters for the measurement of low and medium level microwave power, I: analysis and design. IRE Trans. Microw. Theory Tech. **MTT-6**, 188–194 (1958)
6. G.F. Engen, A refined X-band microwave microcalorimeter. IRE Trans. J. Res. Nat. Bur. Stand. **73C**, 77–82 (1959)
7. L. Brunetti, L. Oberto, M. Sellone, E.T. Vremera, Comparison between thermoelectric and bolometric microwave power standards IEEE Trans. Instrum. Meas. **62**(6), 1710–1715 (2013)
8. R. Judaschke, Korrekturfaktorbestimmung von Hohlleiter Kalorimetern im Millimeterwellenbereich, Physikalisch-Technische Bundesanstalt (PTB), 247. PTB Seminar, Aktuelle Fortschritte von Kalibrierverfahren im Nieder und Hochfrequenzbereich, PTB-Bericht E 96 (2009)
9. J.W. Allen, F.R. Clague, N.T. Larsen, M.P. Weideman, NIST microwave power standards in waveguide, NIST technical note 1511 (1999)
10. R. Judaschke, J. Rühaak, Determination of the correction factor of waveguide microcalorimeters in the millimeter-wave range. IEEE Trans. Instrum. Meas. **58**, 1104–1108 (2009)
11. X. Cui, Y.S. Meng, W. Yuan, Y. Li, Theoretical analysis and determination of the correction factor for a waveguide microcalorimeter. Sensors **20**, 1–11 (2020)

12. D. Gu, X. Lu, B. Jamroz, D. Williams, B. Riddle, X. Cui, A self-calibrated transfer standard for microwave calorimetry, in *2018 Conference on Precision Electromagnetic Measurements (CPEM 2018)* (2018)
13. M. Celep, D. Stokes, Characterization of a thermal isolation section of a waveguide microcalorimeter. IEEE Trans. Instrum. Meas. **70**, 1–7 (2021)
14. X. Cui, T.P. Crowley, Comparison of experimental techniques for evaluating the correction factor of a rectangular waveguide microcalorimeter. IEEE Trans. Instrum. Meas. **60**, 2690–2695 (2011)
15. X. Cui, Y.S. Meng, Y. Li, Y. Zhang, Y. Shan, Design and simplified evaluation technique for waveguide microcalorimeter. IEEE Trans. Instrum. Meas. **65**, 1450–1455 (2016)
16. D. Gu, X. Lu, B.F. Jamroz, D.F. Williams, X. Cui, A.W. Sanders, NIST-traceable microwave power measurement in a waveguide calorimeter with correlated uncertainties. IEEE Trans. Instrum. Meas. **68**(6), 2280–2287 (2019)
17. The international key comparison data base KCDB, Technical report, 2024. [Online]. Available: https://www.bipm.org visted on 08 Apr 2024
18. R.H. Judaschke, M. Kehrt, K. Kuhlmann, A. Steiger, Linking the power scales of free-space and waveguide-based electromagnetic waves. IEEE Trans. Instrum. Meas. **69**, 9056–9061 (2020)
19. A. Steiger, M. Kehrt, C. Monte, R. Müller, Traceable THz power measurement from 1 THz to 5 THz. Opt. Express **21**, 14466–14473 (2013)
20. A. Steiger, W. Bohmeyer, K. Lange, R. Müller, Novel pyroelectric detectors for accurate terahertz power measurements. Technisches Messen **83**, 654–661 (2016)
21. B. Globisch, R.J.B. Dietz, T. Göbel, M. Schell, W. Bohmeyer, R. Müller, A. Steiger, Absolute terahertz power measurement of a time-domain spectroscopy system. Opt. Lett. **40**, 3544–3547 (2015)
22. A. Steiger, R. Müller, A.R. Oliva, Y. Deng, Q. Sun, M. White, J. Lehman, Terahertz laser power measurement comparison. IEEE Trans. Terahertz Sci. Technol. **6**(5), 664–669 (2016)
23. A. Steiger, B. Gutschwager, M. Kehrt, C. Monte, R. Müller, J. Hollandt, Optical methods for power measurement of terahertz radiation. Opt. Express **18**, 21804–21814 (2010)

Chapter 4
Phase Noise Metrology

**Meysam Bahmanian, J. Christoph Scheytt, Nora Meyne,
and Thomas Kleine-Ostmann**

Abstract Phase noise is one of the most important properties of oscillators
that limit the capacity of high-frequency communication systems. In heterodyne
conversion schemes, the phase noise of the local oscillator will be multiplied and
up-converted to the transmission channel. Therefore, accurate characterization of
the oscillators is highly important for the design of THz communication systems.
Especially when it comes to the characterization of high-quality oscillators with
extremely low phase noise, traceable measurement methods are not available.

In this chapter, the mathematical model and definition of the amplitude noise
(AM noise) and phase noise (PM noise) are given. Different phase noise definition
standards such as single sideband (SSB) and double sideband will also be provided.
Phase noise measurement techniques such as frequency discrimination and phase-
locked loop (PLL) technique will be discussed. The standard two-channel cross
correlation for statistical analysis of phase noise at levels below the detection limit
of the phase noise receiver will be explained with mathematical formalism.

4.1 Introduction

Fluctuations of amplitude and especially phase of a modulated signal limit its
capability to transmit high symbol rates due to the resulting time jitter in the
detector. When going to higher frequencies in terahertz communication schemes,
this becomes one of the main limiting factors for the performance of the system.
Therefore, it is of utmost importance to use high-quality oscillators in the heterodyne conversion schemes used here, as phase noise of the local oscillator (LO) is
multiplied by the frequency conversion factor in the mixing process. Fundamentals

M. Bahmanian · J. C. Scheytt
Universität Paderborn, Schaltungstechnik (SCT)/Heinz Nixdorf Institut, Paderborn, Germany
e-mail: meysam.bahmanian@uni-paderborn.de; cscheytt@hni.upb.de

N. Meyne · T. Kleine-Ostmann (✉)
Physikalisch-Technische Bundesanstalt (PTB), Braunschweig, Germany
e-mail: nora.meyne@ptb.de; thomas.kleine-ostmann@ptb.de

© The Author(s) 2026

T. Kürner et al. (eds.), *Metrology for THz Communications*, Springer Series
in Optical Sciences 256, https://doi.org/10.1007/978-3-032-01986-8_4

of phase noise metrology can be found in [1–3]. Developing high-quality electric oscillators and using optical techniques for minimum phase noise is still ongoing research.

Being able to quantify phase noise correctly, and in the best case to establish traceability to the International System of Units (SI), is crucial for system design, supplier specifications, and research in the field of THz communications. Despite the large number of phase noise analyzer solutions on the market, only very few calibration and measurement capabilities exist [4–6] and the specified measurement uncertainties are too large to qualify state-of-the-art oscillators. This is why work on phase noise measurement techniques and suitable traceability chains is part of ongoing research with the goal to establish new calibration services at the Physikalisch-Technische Bundesanstalt.

4.2   Fundamentals of Amplitude and Phase Noise

Phase noise, the random variation of the phase of a sinusoidal signal $x(t)$, can be defined as

$$x(t) = x_0 \cos(\omega_0 t + \phi_0 + \phi_n(t)). \tag{4.1}$$

In Eq. (4.1), x_0 refers to the amplitude, ω_0 is the angular frequency, ϕ_0 is the offset phase, and $\phi_n(t)$ is phase noise. This random variation is illustrated graphically in Fig. 4.1.

For random processes, it is more meaningful to evaluate the statistical properties of phase noise rather than its instantaneous value. Assuming phase noise is ergodic (that is, the temporal averages are equal to the ensemble averages), the autocorrelation function of the phase noise can be written as

$$R_{\phi_n}(\tau) = E[\phi_n(t)\phi_n(t + \tau)], \tag{4.2}$$

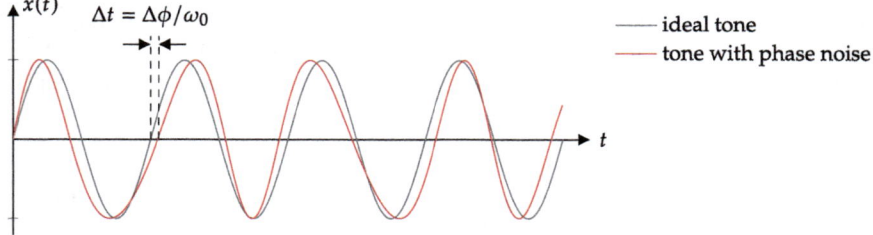

Fig. 4.1 Illustration of phase noise in a sinusoidal signal

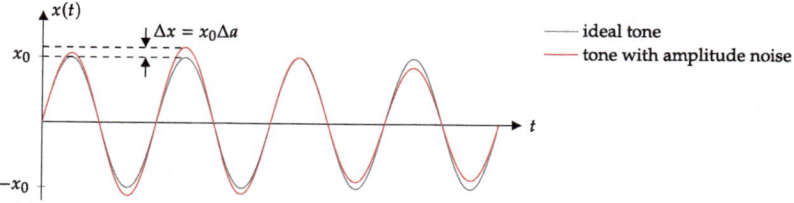

Fig. 4.2 Illustration of amplitude noise in a sinusoidal signal

where E[.] denotes the expectation value. The power spectral density of the phase noise according to the Wiener–Khintchine theorem is the Fourier transform of its autocorrelation function

$$S_{\phi_n}(f) = \int_{-\infty}^{+\infty} d\tau\, R_{\phi_n}(\tau) e^{-j2\pi f\tau}. \tag{4.3}$$

The power spectral density (PSD) of phase noise has units of rad^2/Hz. In the literature, the term phase noise is used synonymously for the PSD of the phase noise.

Amplitude noise can be defined as the random fluctuations of a sinusoidal signal amplitude, illustrated graphically in Fig. 4.2. Similar to the definition of phase noise in Eq. (4.1), amplitude noise of a sinusoidal signal can be modeled as

$$x(t) = x_0\big(1 + a_n(t)\big)\cos(\omega_0 t + \phi_0), \tag{4.4}$$

where $a_n(t)$ is the amplitude noise. Similar to what has been done for the phase noise, the autocorrelation function and PSD of the amplitude noise can be derived:

$$R_{a_n}(\tau) = E\,[a_n(t)a_n(t+\tau)], \tag{4.5}$$

and

$$S_{a_n}(f) = \int_{-\infty}^{+\infty} d\tau\, R_{a_n}(\tau) e^{-j2\pi f\tau}. \tag{4.6}$$

4.3 Conventions in Amplitude and Phase Noise Measurement

Amplitude and phase noise are described by PSDs in the frequency domain [5]. The amplitude noise or stability of the signal is described by

$$S_{a_n}(f) = \frac{\langle A_n(f)\rangle^2}{B}, \tag{4.7}$$

whereas the phase noise or stability is described by

$$S_{\phi_n}(f) = \frac{\langle \Phi_n(f) \rangle^2}{B}. \tag{4.8}$$

Here, $\langle A_n(f) \rangle^2$ and $\langle \Phi_n(f) \rangle^2$ are the mean-square normalised amplitude and phase fluctuations at an offset or Fourier frequency f from a carrier ν_0 and B is the measurement bandwidth. The unit of phase noise $S_{\phi_n}(f)$ is rad^2/Hz. Since the PSDs include fluctuations from upper and lower sidebands of the carrier ν_0, both quantities are single-sided double-sideband units of measure.

In the IEEE standard 1139-2008 [7] the quantity

$$\mathcal{L}(f) = \frac{S_{\phi_n}(f)}{2} \tag{4.9}$$

is suggested as measurand that includes the fluctuations from one sideband, only. Therefore, it is a single-sideband unit of measure. If the integrated phase noise is sufficiently small, $\mathcal{L}(f)$ may be regarded as the ratio of phase noise power per unit bandwidth (1 Hz) in a single sideband to power in the carrier. Usually, $\mathcal{L}_{dB}(f) = 10$ lg($\mathcal{L}(f)$) is expressed in dBc/Hz. However, inconsistencies of this approach are discussed in [3].

4.4 Measurement Techniques

4.4.1 Two-Channel Cross Correlation

One challenge in characterisation of an ultralow-phase noise signal is the measurement of phase noise itself. How can one measure the PSD of a signal that is smaller than the instrument noise floor by orders of magnitude? The answer is to use two pieces of hardware in parallel and extract the correlated part of the measured signals. In order to illustrate this technique, we use the simplified diagram shown in Fig. 4.3. The mathematical formalism in this section only deals with PSD measurement of a base band signal and we assume the phase noise of the carrier signal is extracted using the delay line method or the phase-locked loop method explained in next sections.

Note that we are interested in the average autocorrelation or the average Fourier transform of $x(t)$. The output signals of Channel A and Channel B can be written as

$$y_A(t) = x(t) + n_A(t) , \tag{4.10}$$

$$y_B(t) = x(t) + n_B(t) , \tag{4.11}$$

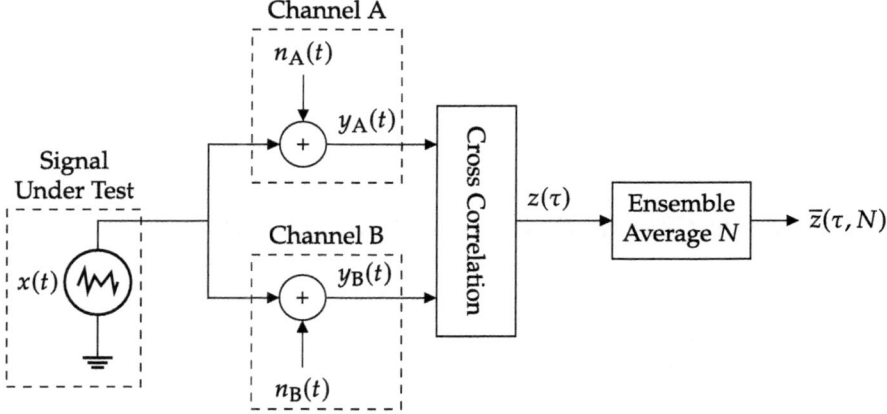

Fig. 4.3 Schematic of PSD measurement setup using cross correlation technique

where $x(t)$ is the signal under test (SUT), $n_A(t)$ and $n_B(t)$ are the equivalent input noise of Channel A and B, respectively, and $y_A(t)$ and $y_B(t)$ are the outputs of Channel A and B, respectively. Applying sample cross correlation between x_A and x_B yields $z(\tau)$

$$z(\tau) = r_{y_{A,B}}(\tau) = r_x(\tau) + r_{n_A,x}(\tau) + r_{x,n_B}(\tau) + r_{n_{A,B}}(\tau) \,, \tag{4.12}$$

with a Fourier transform of

$$Z(f) = |X(f)|^2 + N_A(f)X^*(f) + X(f)N_B^*(f) + N_A(f)N_B^*(f) \,. \tag{4.13}$$

The basic idea of the cross correlation technique is to suppress the statistically independent noise terms by averaging. The more the number of averages, the more the noise terms are suppressed. This can indeed be seen in formula (4.12). The expectation value of $r_{y_{A,B}}(\tau)$ is composed of four terms; only the first term is nonzero and all other terms are zero due to the statistical independence of $x(t)$ and $n_A(t)$, $x(t)$ and $n_B(t)$, and $n_A(t)$ and $n_B(t)$. However, expectation value means averaging over infinite sample signals which is not practically possible. Therefore, we need to find out how much suppression of the undesired noise terms is achieved as a function of number of averages. The noise reduction factor (NRF) can be defined relative to a single measurement without any averaging as

$$\text{NRF} = \frac{\sigma_z(f, 1)}{\sigma_z(f, N)} = \sqrt{N} \tag{4.14}$$

and in decibels as

$$\text{NRF}_{dB} = 10 \log_{10}(\text{NRF}) = 5 \log_{10}(N) \,. \tag{4.15}$$

Fig. 4.4 Phase detection
using double/balanced mixer

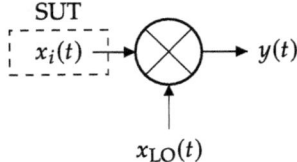

4.4.2 Single-Channel Phase Noise Measurement

In the previous section we showed how the noise floor of a baseband receiver can be
suppressed using two-channel cross correlation technique. Note that this technique
can only be used for the measurement of PSD and statistical signal analysis, not
instantaneous measurement of a signal. For phase noise analysis, the SUT has a
band-pass nature and cannot directly be applied to a baseband receiver. Therefore,
the phase noise of the SUT has to be first extracted and then applied to a two-channel
baseband receiver. The phase noise can be discriminated using a phase detector,
usually a double-balanced mixer, whose input signals are in quadrature, shown in
Fig. 4.4. For a complete two-channel phase noise analyzer, two pieces of the phase
noise measurement apparatus are necessary. The output signals of these single-
channel phase noise analyzers are then applied to a cross-correlator to suppress the
undesired noise signals by correlation and averaging.

The phase detector input, $x_i(t)$, is a single tone with an angular frequency of ω_0
and an amplitude of A_i:

$$x_i(t) = A_i \sin\left(\omega_0 t + \phi_{n,i}(t)\right), \tag{4.16}$$

where $\phi_{n,i}(t)$ is the phase noise of the SUT. Assuming the LO signal is in quadrature
with the SUT, it can be written similarly as

$$x_{\mathrm{LO}}(t) = A_{\mathrm{LO}} \cos\left(\omega_0 t + \phi_{n,LO}(t)\right). \tag{4.17}$$

Balanced mixers are switching devices and operate in saturation regime with respect
to their switching input. Hence, we assume the LO signal is close to saturation levels
of the mixer and the mixer has a constant gain independent of the level of the LO
signal. Assuming the mixing process has a gain of $2L$ (a gain of $2L$ corresponds to
a gain of L for each output mixing term), its output $y(t)$ can be written as

$$
\begin{aligned}
y(t) &= 2L A_i \sin\left(\omega_0 t + \phi_{n,i}(t)\right) \cos\left(\omega_0 t + \phi_{n,\mathrm{LO}}(t)\right) \\
&= L A_i \sin\left(\phi_{n,i}(t) - \phi_{n,\mathrm{LO}}(t)\right) + L A_i \sin\left(2\omega_0 t + \phi_{n,i}(t) + \phi_{n,\mathrm{LO}}(t)\right).
\end{aligned}
\tag{4.18}
$$

The second term on the right-hand side of (4.18) has a high frequency and can be
filtered. Assuming the variation of phase noise is small, $|\phi_{n,i}(t)|, |\phi_{n,\mathrm{LO}}(t)| \ll \pi/2$,
the phase detector output can be written as

Fig. 4.5 Phase noise measurement using delay line frequency discrimination technique

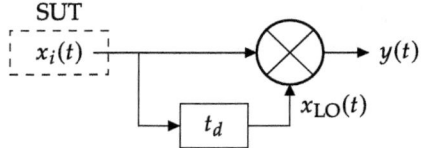

$$y(t) \approx L A_i \left[\phi_{n,i}(t) - \phi_{n,\text{LO}}(t) \right] . \tag{4.19}$$

The noise of the LO signal can then be suppressed by using a cross-correlation scheme, leading to extraction of the phase noise of SUT.

We have seen that phase detection using a balanced frequency mixer requires quadrature operation. The quadrature signal for LO is usually generated using two methods, delay line method and PLL method, which are explained in the following.

Delay Line Frequency Discriminator In the delay line method, the quadrature LO signal is a delayed version of the SUT, as illustrated in Fig. 4.5.

The LO signal, $x_{\text{LO}}(t)$, can be written as

$$x_{\text{LO}}(t) = x_i(t - t_d) = A_i \sin \left(\omega_0 t - \omega_0 t_d + \phi_{n,i}(t - t_d) \right) , \tag{4.20}$$

where t_d is the delay time and is chosen such that

$$\omega_0 t_d = \left(k - \frac{1}{2} \right) \pi , \tag{4.21}$$

for a positive integer k. With this assumption, the quadrature operation of the balanced mixer is guaranteed and the phase detector output can be written as

$$y(t) = \pm L A_i \left[\phi_{n,i}(t) - \phi_{n,i}(t - t_d) \right] . \tag{4.22}$$

The balanced mixer output is proportional to the difference between the phase noise and a delayed version of itself. This behavior is similar to a derivative operator, and since the derivative of the phase is the frequency, this technique discriminates the frequency noise, rather than the phase noise itself. The phase noise can then be calculated by integration of the frequency noise. The impulse response $h_{\text{FD}}(t)$ and its Fourier transform $H_{\text{FD}}(\omega)$ of the frequency discriminator are

$$h_{\text{FD}}(t) = \pm L A_i \left[\delta(t) - \delta(t - t_d) \right] \quad \text{and} \quad H_{\text{FD}}(\omega) = \pm 2\mathbf{j} L A_i \sin \left(\frac{\omega t_d}{2} \right) e^{-\mathbf{j} \frac{\omega t_d}{2}} .$$
$$\tag{4.23}$$

The sine terms in the transfer function of delay line frequency discriminator lead to nulls at offset frequencies of $\omega = 2k\pi/t_d$. These nulls lead to limiting the maximum offset frequency range of this technique. For the measurement of phase noise at

Fig. 4.6 Phase noise measurement using PLL technique

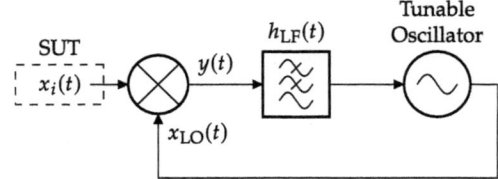

offset frequencies below the first null, $\omega \ll 2\pi/t_d$, the phase detector output can be approximated as

$$y(t) = \pm L A_i t_d \frac{\mathrm{d}}{\mathrm{d}t} \phi_{n,i}(t) \,. \tag{4.24}$$

The derivation of phase noise leads to degradation of sensitivity at low offset frequencies. In addition, (4.24) shows the gain of this technique is proportional to the delay time. This technique is especially useful for measurement of relatively noisy signals with a high drift, such as the output of an open-loop VCO. High drift of a signal prevents using more precise, however slower, methods such as phase detection using PLL technique.

Phase Detection Using PLL Technique A phase detector is an inseparable block of any phase-locked loop. If a balanced mixer is used for phase detection in a PLL with an integrator as the loop filter, the dynamics of the loop causes a zero-average signal at the phase detector output (for further details see Sect. 4.6), which is equivalent to the quadrature operation in a balanced mixer. Figure 4.6 shows the block diagram of this technique.

Using the control theory mathematical toolbox, the open-loop transfer function in the phase domain can be written as

$$H_{\mathrm{OL}}(s) = K_\phi H_{\mathrm{LF}}(s) \frac{K_V}{s} \,, \tag{4.25}$$

where K_V is the tuning sensitivity of the tunable oscillator, $H_{\mathrm{LF}}(s)$ is the transfer function of the loop filter, and K_ϕ is the phase detector gain. For a balanced mixer we showed that $K_\phi = L A_i$. The transfer function of the phase detector output to the input phase can be expressed in terms of $H_{\mathrm{OL}}(s)$ as

$$\frac{Y}{\Phi_{n,i}}(s) = \frac{K_\phi}{1 + H_{\mathrm{OL}}(s)} \,. \tag{4.26}$$

This transfer function has a high-pass behavior; this is indeed expected of any phase noise measurement system, since the LO signal of the system needs to track the signal under test. This means the LO signal has the same average phase as the SUT but with a 90-degree phase shift. The rate of tracking is also equivalent to the cutoff frequency of the transfer function given in (4.26). The noise of the tunable oscillator is also transferred to $y(t)$ with the same transfer function given in (4.26). Therefore, for phase noise measurement at offset frequencies sufficiently above the loop cutoff frequency, $|H_{OL}(\omega)| \ll 1$, the phase detector output can be approximated as

$$y(t) \approx LA_i \left[\phi_{n,i}(t) - \phi_{n,TO}(t) \right] , \qquad (4.27)$$

where $\phi_{n,TO}(t)$ is the phase noise of the tunable oscillator, and we replaced the phase detector gain with its equivalent value in balanced mixer.

4.5 Two-Channel Phase Noise Measurement

So far we have seen how to measure the PSD of a baseband signal that is below the noise the noise floor of a measurement system using two-channel cross correlation technique and also we demonstrated how to extract the phase noise of a carrier using delay line or PLL technique. We have now the necessary tools to demonstrate a two-channel phase noise measurement system. This system is composed of two single-channel phase noise analyzers and a two-channel cross correlator. Figure 4.7 shows the block diagram of this system. If both channels are perfectly isolated, their additive phase noise are uncorrelated and can be suppressed by cross correlation and averaging according to (4.15). This method is gold standard for phase noise measurement and is implemented in laboratory-grade phase noise test systems.

The frequency range of the two-channel test system can reach up to tens of gigahertz, due to technical limitation of RF components. Further enhancement of frequency range up to millimeter-wave and terahertz frequencies is possible using the two-channel down-conversion scheme illustrated in Fig. 4.8. The high-frequency signal is first down-converted to a frequency that lies in the frequency range of each phase noise measurement channel. The phase noise of the down-converted signals is then extracted by each channel and their correlated part, which is the phase noise of SUT, is extracted. The LO and the mixers also contribute to the phase noise measured by each channel. Therefore, it is necessary that the two LO be uncorrelated so their phase noise can be suppressed by cross correlation.

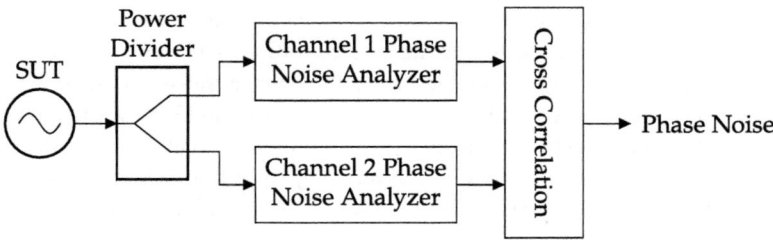

Fig. 4.7 Two-channel phase noise measurement system

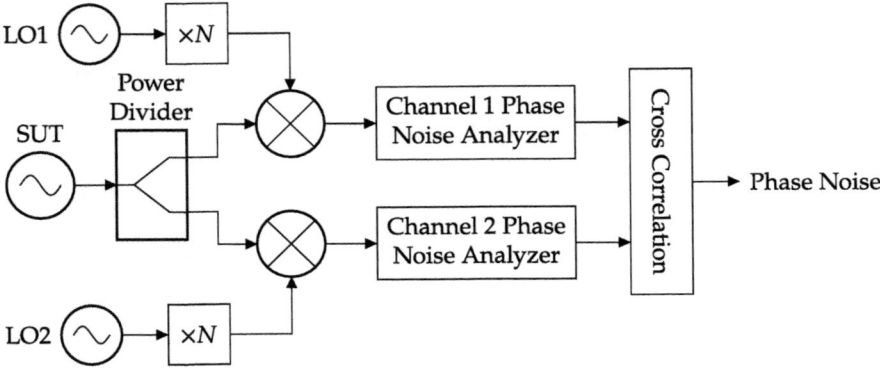

Fig. 4.8 Two-channel phase noise measurement of mm wave and terahertz signals

4.6 Traceability of Phase Noise

At the National Institute of Standards and Technology (NIST), USA, phase and amplitude noise calibration of oscillators is offered with a typical measurement uncertainty of 1 dB for carrier frequencies between 5 MHz and 110 GHz. The amplitude and phase noise standard consists of an oscillator signal that is combined with a calibrated level of bandpass-limited Gaussian noise from a noise source [5]. At Laboratoire national de métrologie et d'essais (LNE), France, calibration services for phase noise of oscillators are available with an absolute expanded measurement uncertainty of 2 dB. For carrier frequencies between 5 MHz and 18 GHz, distinct offset frequencies of 10 Hz, 100 Hz, 1 kHz, 10 kHz, and 100 kHz can be calibrated for an oscillator output power >0 dBm [4].

Traceability at industry measurement laboratories is often based on a linearity and frequency response calibration of a phase noise amplifier [6]. Therefore, the phase noise calibration is derived from the traceability of radio-frequency power and attenuation. Phase noise standards calibrated at national metrology institutes are used for verification, only.

To further approve primary standards for phase noise calibration at Physikalisch-Technische Bundesanstalt (PTB), Germany, work focusses on improving measurement conditions for calibration (electromagnetic interference and vibration-free measurement conditions, temperature stabilisation) and new, calculable noise standards (e.g., based on calculable thermal Johnson noise or artificial noise generated by arbitrary waveform generators) that can be combined with high-quality oscillators (e.g., based on a stabilized optical frequency comb). The optical high-quality oscillators can also be used to convert very-high-frequency signals down to a range where they can be measured more easily without adding too much additional noise in the mixing process.

Acknowledgments The authors would like to thank Prof. Enrico Rubiola from FEMTO-ST Institute, Dept. of Time and Frequency, Besançon, France, for his advice on the topic.

References

1. F. Walls, E. Ferre-Pikal, Measurement of frequency, phase noise and amplitude noise. Wiley Encycl. Electr. Electron. Eng. (12), Jan 1999. [Online]. Available: https://tsapps.nist.gov/publication/get_pdf.cfm?pub_id=105598
2. B. Schiek, H. Siweris, I. Rolfes, *Noise in High-Frequency Circuits and Oscillators* (Wiley, 2006). [Online]. Available: https://books.google.de/books?id=-oLQM1jxjqUC
3. E. Rubiola, F. Vernotte, The companion of Enrico's chart for phase noise and two-sample variances. IEEE Trans. Microw. Theory Tech. **71**(7), 2996–3025 (2023). [Online]. Available: https://doi.org/10.1109/TMTT.2023.3238267
4. *The International Key Comparison Data Base KCDB, CMC ID: EURAMET-EM-FR-00000EAE-1* (Laboratoire national de métrologie et d'essais (LNE), France, 2024). [Online]. Available: https://www.bipm.org
5. A. Hati, C. Nelson, N. Ashby, D. Howe, Calibration uncertainty for the NIST PM/AM noise standards, 2012-07-31 (2012)
6. A. Roth, J. Wolle, Measurement uncertainty analysis and traceability for phase noise, application note (2016)
7. IEEE standard definitions of physical quantities for fundamental frequency and time metrology – random instabilities, IEEE Std Std 1139-2008 (2009), pp. c1–35

Chapter 5
Waveform Metrology

David A. Humphreys, Heiko Füser, and Nora Meyne

Abstract Waveform metrology is the traceable characterization of dynamic electrical signals. The waveforms are measured either as voltage vs. time or as a complex (magnitude and phase) frequency spectrum, and the instrumentation calibration is traceable to SI primary standards. Waveform metrology evolved from parametric measurements of transition-duration for specification purposes. This chapter describes the development of waveform metrology and previous work on impairments and correction algorithms for digital instrumentation, such as real-time digital oscilloscopes, used in THz communication metrology. Traceability has driven the need for point-by-point uncertainties, building on the Guide to Uncertainty in Measurements (GUM) and its annexes, to link uncertainty contributions from both the time and frequency domain measurements.

5.1 Introduction and Background

The first practical commercial cathode ray tube (CRT) oscilloscopes made it possible to view high-frequency time-varying voltage waveforms. Despite the huge improvement in capability, capturing the screen image was limited to drawing or from a Polaroid™ photograph suitable for parametric descriptions, such as transition duration. Although these parametric measures were standardized [1], they are not sufficient to underpin a traceable waveform calibration [2].

In 1978, the 400-MHz Tektronix 7854 Waveform Processing Oscilloscope was introduced [3]. It could digitize waveforms from any of the 7000 series plug-in modules and transfer the results to a computer using the IEEE-488 interface [3]. Although this was not the first digital oscilloscope, it provided a practical solution

D. A. Humphreys (✉)
NPL, UK (Retired), University of Bristol, UK
e-mail: david.a.humphreys@ieee.org

H. Füser · N. Meyne
Physikalisch-Technische Bundesanstalt (PTB), Braunschweig, Germany
e-mail: heiko.fueser@ptb.de; nora.meyne@ptb.de

© The Author(s) 2026
T. Kürner et al. (eds.), *Metrology for THz Communications*, Springer Series
in Optical Sciences 256, https://doi.org/10.1007/978-3-032-01986-8_5

and electrical waveform could now be captured and post-processed to provide a corrected representation of $v(t)$.

Fully digital sampling oscilloscopes (DSOs) now have over 100-GHz bandwidth [4]. The high DSO bandwidth and the detailed models of the sampling-gate performance make them an ideal instrument to disseminate traceable waveform standards Chap. 12. Communication waveforms often have a low frame-rate, and although arbitrarily long waveforms can be measured [5, 6], the DSO is impractical for many wireless communication system measurements, but a convenient traceability route when used to calibrate high trigger-rate test-beds [7].

High-bandwidth digital real-time oscilloscopes (DRTOs) allow long-epoch, single-shot waveform acquisition for practical communication waveform measurement. Their counterpart instrument is the arbitrary waveform generator (AWG) that can create software-defined, high-bandwidth communication-like waveforms. Many AWGs have differential outputs, increasing the output drive signal but requiring careful control of cable lengths.

The fastest DRTOs have the potential to digitize mm-wave modulated RF waveforms directly but it is more common to use a mixer to lower the operating frequency and bandwidth. Determining the amplitude and group delay response of the down-conversion path is essential to verify a THz waveform. The vector signal analyzer (VSA) is an integrated instrument that combines a lower-bandwidth analogue-to-digital converter (ADC) with the RF in-phase and quadrature down-conversion. The sampling rate is higher than the Nyquist requirement (over-sampling ratio), giving more waveform detail.

The remainder of this chapter describes the correction of measurement error, from sources such as impedance match and instrument impairments. It finishes with a brief overview of point-by-point uncertainty to combine the results from time and frequency domains.

5.2 Test Instrument Behavior and Impairments

All equipment designs contain component variation and engineering/cost compromises. It is essential to understand their impact and how to correct any resulting impairments. Real-time instruments (VSA, DRTO, and AWG) have architectural similarities and will be discussed here. Sampling oscilloscopes are a key instrument for transferring traceability from the primary standard, an electro-optic sampling system (EOS) and are covered later together with their associated impairments and correction algorithms Chap. 12.

Real-time instruments can be either a waveform source or a waveform measuring instrument. Additional measurements may be required to correct for nonideal behavior, such as the RF impedance match. The following parameters and sources of impairment should be considered when making traceable waveform measurements.

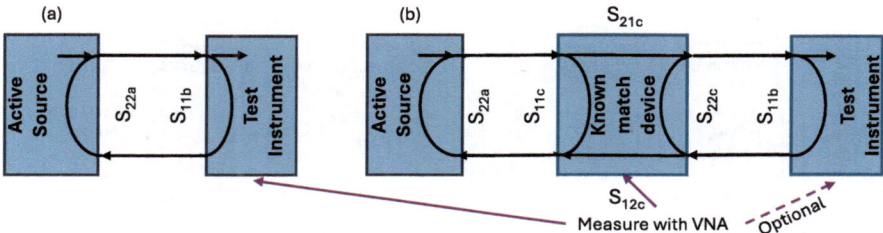

Fig. 5.1 (**a**) Correctable impedance mismatch error. (**b**) If a known insertable element is included, the active source match can be estimated by passive load pull

5.2.1 Microwave Impedance Mismatch Correction

The impedance match of the source and receiver cause significant correctable measurement errors, leading to high uncertainties [8]. At THz frequencies this is compounded by the reproducibility of the waveguide connection.

The impedance match of passive components can be measured as scattering parameters by vector network analyzers (VNAs) (see Chap. 6), but the impedance match of actively driven RF components (hot S_{22}) is often performed using active or passive load pull [9] as the device may be nonlinear. If the source is operating in the linear regime, then different insertable devices can be used to provide a degree of broadband passive load pull, providing a series of different mismatches, typically lower than $-6\,\mathrm{dB}$; see Fig. 5.1. Using scattering parameter notation, Eq. (5.1) describes the expected result.

$$V_{\mathrm{m}} = \frac{V_{\mathrm{s}} S_{21\mathrm{c}}(1 - S_{22\mathrm{c}} S_{11\mathrm{b}})}{1 - S_{22\mathrm{a}}(S_{11\mathrm{c}}(1 - S_{22\mathrm{c}} S_{11\mathrm{b}}) + S_{21\mathrm{c}} S_{11\mathrm{b}} S_{12\mathrm{c}})} \qquad (5.1)$$

The impedance match of the test instrument and insertable network, or their combination, can be separately measured using a VNA and the source match is determined as a solvable set of linear least-squares equations. If the insertable device is not present, or a combined measured [10], ($S_{11\mathrm{c}}$, $S_{22\mathrm{c}} = 0$) and ($S_{21\mathrm{c}}$, $S_{12\mathrm{c}} = 1$), this reduces to the more familiar form $V_{\mathrm{m}} = \frac{V_{\mathrm{s}}}{1 - S_{22\mathrm{a}} S_{11\mathrm{b}}}$.

5.2.2 Mixers and Up-/Down-Conversion

Although commercial DRTOs are available with $>100\,\mathrm{GHz}$ bandwidth [11], direct measurement of THz waveforms is normally achieved using mixers to up-convert and down-convert the modulated signal. This can introduce both linear and nonlinear distortion, such as in-phase/quadrature (IQ) imbalance, nonlinear group delay, and intermodulation products. Other parameters, such as impedance

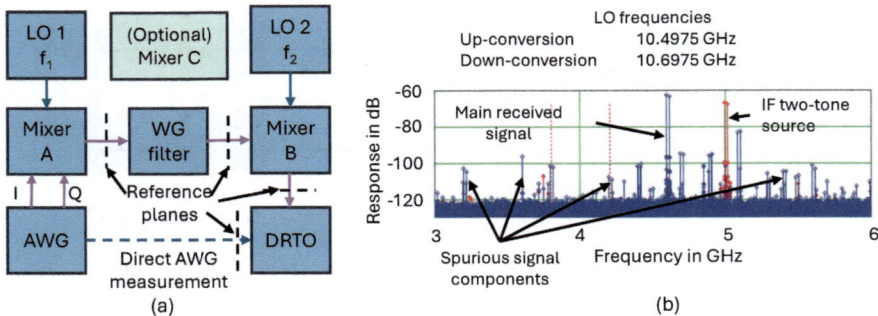

Fig. 5.2 (**a**) Mixers used for up-conversion and down-conversion can be evaluated together to estimate their linear properties such as group delay, IQ balance, and frequency response. Ideally, three or more identical mixers should be used. (**b**) Nonlinear properties may be masked if the same LO frequency is used for up- and down-conversion. This diagram was modified from Fig. 5 of [12]. The work was performed under Euramet EMPIR project reference 14IND10 MET5G – metrology for 5G communications

match and RF power levels, may alter the device characteristics and hence the resulting waveform. Typically, two identical modulators and a high-pass filter (waveguide beyond cutoff) can be used to estimate the group delay, magnitude, and IQ imbalance. The modulation bandwidth is typically a few percent of the carrier frequency (e.g., 10 GHz bandwidth at 300 GHz carrier frequency), reducing the DRTO bandwidth requirement and opening the possibility of using a higher bit-depth instrument. Intermodulation distortion products will be present from each of the mixers but may not all be visible if the same local oscillator frequency is used for both up-conversion and down-conversion [12] (Fig. 5.2). THz-modulated waveforms also exhibit phase rotations due to path changes and the stability of harmonically multiplied RF source.

5.2.3   Multiple ADC Systems and Impairments

The high sampling rates required for GHz bandwidth DRTO [11, 13] and AWG [14] instruments are achieved using multiple parallel time-interleaved ADC/sample and hold (S/H) circuits [15]. These circuits must be well matched in terms of their DC levels, gain, timing, and impulse response. Residual variation leads to impairments in the result and the pattern formed is not random and can be used to identify and correct these errors [16]. It is important to note that the ADC and S/H can contribute separately to the result. It is assumed that some internal post-processing is applied, probably using a finite impulse response filter, to achieve the desired DRTO response. The fixed and repeated sample pattern has a length equal to the number of ADCs, creating sub-Nyquist noise spurs (DC variation) and weakly nonlinear [17] modulation products (timing and gain variation) [18]. Cho

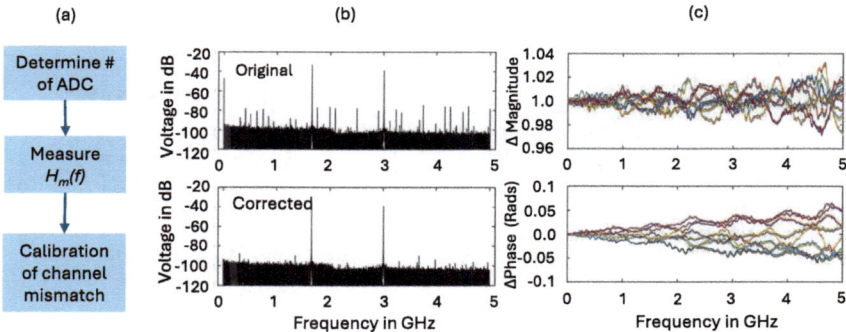

Fig. 5.3 Detection and mitigation of DRTO ADC impairments. (**a**) Calibration process. (**b**) Initial frequency components and corrected result. (**c**) Individual ADC magnitude and phase deviations (Parts (**b**) and (**c**) reproduced from [19], ©IEEE 2016)

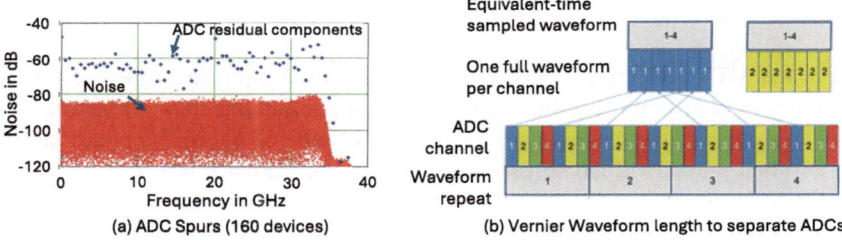

Fig. 5.4 (**a**) Shows the ADC (Reproduced from [20] ©IEEE 2015). The Vernier technique to separate the individual ADC magnitude and phase characteristics (**b**) is from the poster associated with [21]

et al. [19] developed a comprehensive correction model, sacrificing one channel to determine the ADC reference position and providing a significant reduction in the intermodulation terms; see Fig. 5.3.

A second, prime-number-based approach uses a slightly shorter epoch length, based on a different set of prime numbers (e.g., $500 \times 10^3 = 2^5 \cdot 5^6$ compared with $500 \times 10^3 - 1 = 31 \cdot 127^2$ samples and primes) spreading the interaction over many frequency points [20]. The resulting oversampling is shown in Fig. 5.5b.

The third option, shown in Fig. 5.4, is a Vernier technique where the repetition period of a test waveform is selected to cycle through each ADC, in turn forming one complete waveform set for each ADC [21]. AWGs, DRTOs, and synthesizers typically have an accessible clock reference so the instruments can be phase-locked together. The quality of the results can be verified by an Allan variance test [18].

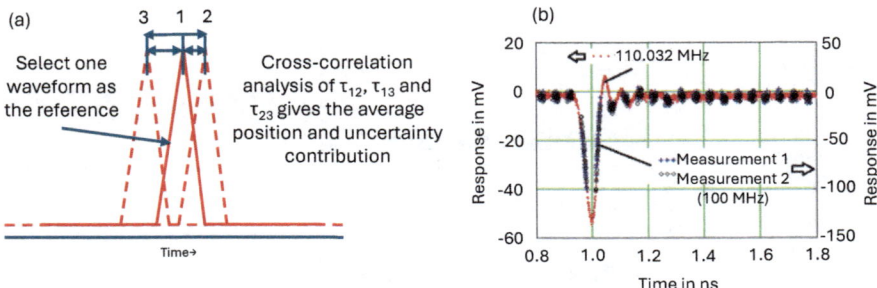

Fig. 5.5 (**a**) Cross-correlate independent measurements to optimize the time alignment. (**b**) The choice of waveform epoch and DRTO sample rate can provide equivalent-time oversampling or information about residual jitter (Reproduced from [20] ©IEEE 2015)

5.2.4 Sampling Rate and Test Data

DRTO and VSA continuously acquire the waveform and the trigger event acts as a memory marker point. For baseband signals and assuming that the instruments are phase-locked, a repetitive waveform should be stable to within the phase-locked loop residual jitter. This can be verified with an Allan variance test. It is beneficial to use different sample rates for the AWG and DRTO so that impairment spurs from the source and receiver instruments are not degenerate and can be identified and separately corrected when evaluating the equipment.

Oversampling is the ratio of the sampling rate to the Nyquist frequency. A low oversampling rate gives poor waveform definition at high frequencies. This can be an issue with VSAs which have a modest IF bandwidth allowing a higher ADC bit depth. The bit depth, effective number of bits (ENOB), and sample rate of an instrument determine the direction in which calibration should be disseminated. If the sample rates are similar, the higher ENOB instrument should calibrate the lower ENOB instrument.

5.2.5 Timebase Nonlinearity and Trigger Jitter

Unlike the DSO, which is periodically retimed, the DRTO and AWG timebases are clock disciplined and timebase distortion should be only a minor issue for phase-locked systems. A single trigger event is used to mark the start point of the record. In the absence of noise, the time uncertainty for individually captured records will be a rectangular distribution of ± 0.5 samples. For 80 GSa/s (33 GHz bandwidth) the time uncertainty $\varepsilon_t < \pm 6.25\,\text{ps}$. Electrical noise on the trigger waveform will increase the waveform-to-waveform jitter but the waveforms can be aligned in the frequency domain by power-weighted cross-correlation (see Fig. 5.5), taking account of all combinations [22]. There are advantages to selecting repetition frequencies that are not a submultiple of the sampling rate [18].

DRTO and VSA continuously acquire the waveform and the trigger event acts as a memory marker point. For baseband signals and assuming that the instruments are phase-locked, a repetitive waveform should be stable to within the phase-locked loop residual jitter. Often, it is beneficial to use different sample rates for the AWG and DRTO, related through a common lower frequency (e.g., 28 Gbaud measured at 100 GSa/s). Repeated measurements or a longer single acquisition can improve waveform detail [23] and separation of the instrument impairments. THz-modulated waveforms may also have phase rotations due to the stability of the oscillator phase and any path changes. Unlike the DSO, the time base is well disciplined and time base distortion should be a minor issue. Time alignment of the baseband waveforms can be determined by cross-correlation [22].

5.3 Point-by-Point Uncertainties

The BIPM's Guide to Measurement Uncertainty [24–26], (GUM), introduced in Chap. 2, states that "a calculation of measurement uncertainties must take correlations into account." Although the GUM is generally applied to single-parameter calibrations, such as resistance, it will also apply to waveforms as these contain a series of results that show correlation between the measured points. Waveforms comprise a series of points that can be represented in either the time domain or the frequency domain, related by the Fourier transform, which links all the time-domain and frequency-domain points. Transform properties, such as the time shift $x(t - \tau)$, give a corresponding change to all the frequency coefficients for the waveform $X(f, \tau) = X(f)\exp(-2\pi \tau)$.

Multiple waveform measurements are normally averaged to reduce noise. However, if these results are individually recorded, they provide additional statistical information such as jitter and wander, which are fast and slow timing errors in the results. The number of acquisitions is limited by practical considerations. The impact of noise and time-shift errors were discussed in Sect. 5.2.5 and Fig. 5.5. From this point in the measurement or calibration process, statistical uncertainties will be present and should be manipulated alongside the data; see Fig. 5.6.

At high frequencies, impedance mismatch is a major source of waveform errors, as described in Chap. 6. The VNA measures the S-parameters at each separate frequency and there will be associated uncertainties for each result. The impedance match correction is a frequency-dependent scaling function and therefore in the time domain the uncertainties will have a time-dependent structure [27].

Where suitable model relationships can be found, Monte Carlo [25, 28] techniques can be used to transfer the systematic uncertainty contribution between time and frequency domains. Alternatively , the GUM recommends a Taylor series approach for handling independent and correlated uncertainties, on the assumption that, because the waveform has many points, the equations have been reformulated

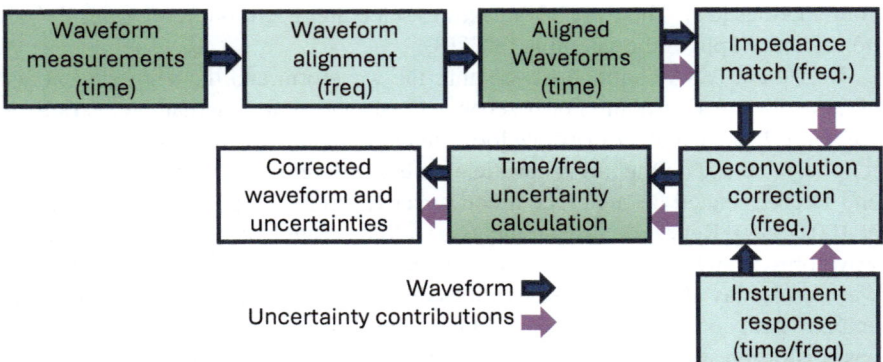

Fig. 5.6 Correcting waveform errors and calculating time/frequency uncertainties

as a Jacobean vector and a covariance matrix to transform the uncertainties between time domain and the frequency domain [29, 30]. However, the size of a covariance matrix for a trace length n grows as n^2, limiting the maximum practical length, or the number of frequency components in the waveform to a few thousand complex values.

One of the objectives of the EURAMET project IND16 "Ultrafast" was to develop an algorithm to record and manipulate and store the waveform uncertainties in a compact format. To use uncertainty information effectively, it is a necessary requirement to manipulate the waveform and uncertainties together, following the normal arithmetic rules for addition, subtraction, multiplication, and division. The resulting "compact covariance" algorithm determines the eigenvalues and eigenvectors of the covariance matrix and selects only the dominant terms to reduce the storage requirement to a few "n" times the number of frequency components. Similarly, this reduces the number of unknowns from $\frac{n^2}{2}$ to a few n. Normal scientific software can manipulate covariance matrices of a few million points without difficulty [27, 31]. In principle the eigenvalues of a large matrix can be calculated without forming the full matrix but this is slow compared with matrix manipulation.

5.3.1 Current Waveform Uncertainty Work Under EURAMET

Waveform uncertainties are still an active area of research following from EURAMET IND16 and complementing the existing standard IEC 62754 [32], which specifies methods *"for the computation of the temporal and amplitude parameters and their associated uncertainty for step-like and impulse-like waveforms."* The 3-year EURAMET research project *23NRM01 SBS Uncert*, led by PTB, started in August 2024, focusing on the sample-by-sample uncertainty computation of electrical waveforms (see https://www.ptb.de/epm2023/sbsuncert/ home). Full project details are provided in the EURAMET publishable summary [33].

5.4 Uncertainty Software

From Sect. 5.3 it is clear that software analysis tools are required for VNA and other test instruments such as the DSO and DRTO instruments. Also, to achieve the greatest benefit these software tools must interface with existing analysis routines used in business and research environments.

The NIST *Microwave Uncertainty Framework* is a Monte-Carlo-based software tool that supports several instrument types. It is important for practical handling of such uncertainties and for their dissemination to industry. A more detailed summary is provided in [34, 35].

VNA corrections for the test device and measuring instruments are essential. The uncertainty software package *VNA Tools* was developed by the METAS [36, 37]. This software is used by a number of commercial companies. Keysight and Anritsu [38] have developed software packages to handle specific instrument uncertainties.

The EURAMET IND16 *Compact Covariance* software implementations and a user guide are available in Matlab (NPL, UK) [39] and Python (VSL, Netherlands).

References

1. IEC Technical Committee TC 85, IEC 60469:2013 definitions of terms pertaining to transitions, pulses and related waveforms, The International Electrotechnical Commission (IEC), Technical repprt, April 2013. [Online]. Available: https://webstore.iec.ch/en/publication/2211
2. P.D. Hale, J. Jargon, C.M.J. Wang, B. Grossman, M. Claudius, J.L. Torres, A. Dienstfrey, D.F. Williams, A statistical study of de-embedding applied to eye diagram analysis. IEEE Trans. Instrum. Meas. **61**(2), 475–488 (2012)
3. IEEE, IEEE standard codes, formats, protocols, and common commands for use with IEEE std 488.1-1987, IEEE standard digital interface for programmable instrumentation, IEEE Std 488.2-1992 (1992), pp. 1–254
4. H. Füser, S. Eichstädt, K. Baaske, C. Elster, K. Kuhlmann, R. Judaschke, K. Pierz, M. Bieler, Optoelectronic time-domain characterization of a 100 GHz sampling oscilloscope. Meas. Sci. Technol. **23**(2), 025201 (2011). [Online]. Available: https://doi.org/10.1088/0957-0233/23/2/025201
5. D.A. Humphreys, R.T. Dickerson, Traceable measurement of error vector magnitude (EVM) in WCDMA signals, in *2007 International Waveform Diversity and Design Conference*, 2007, pp. 270–274
6. C.M.J. Wang, P.D. Hale, J.A. Jargon, D.F. Williams, K.A. Remley, Sequential estimation of timebase corrections for an arbitrarily long waveform. IEEE Trans. Instrum. Meas. **61**(10), 2689–2694 (2012)
7. K.A. Remley, D.F. Williams, P.D. Hale, C.-M. Wang, J. Jargon, Y. Park, Millimeter-wave modulated-signal and error-vector-magnitude measurement with uncertainty. IEEE Trans. Microw. Theory Tech. **63**(5), 1710–1720 (2015)

8. D.A. Humphreys, J. Howes, High accuracy optoelectronic device measurement, in *BEMC 95 7th British Electromagnetic Measurements Conference Digest*, vol. 41 (Nov. 1995), pp. 1–4. [Online]. Available: https://eprintspublications.npl.co.uk/468/

9. M.S. Hashmi, A.L. Clarke, S.P. Woodington, J. Lees, J. Benedikt, P.J. Tasker, Electronic multi-harmonic load-pull system for experimentally driven power amplifier design optimization, in *2009 IEEE MTT-S International Microwave Symposium Digest*, 2009, pp. 1549–1552

10. A.D. Gifford, D.A. Humphreys, Measurement of high-speed photodiodes using a microwave reflectometer to improve accuracy, in *IEE Colloquium on Measurements on Optical Devices*, 1992, pp. 12/1–12/7

11. *Keysight Infiniium UXR-Series Real-Time Oscilloscopes*, November 2023

12. T.H. Loh, D.A. Humphreys, D. Cheadle, K. Buisman, An evaluation of distortion and interference sources originating within a millimeter-wave MIMO testbed for 5G communications, in *2018 2nd URSI Atlantic Radio Science Meeting (AT-RASC)*, 2018, pp. 1–4

13. T. LeCroy, Serial data superpower: wavemaster 8000 HD 6-65 GHz 12 bit high-definition oscilloscope, Teledyne LeCroy, Technical report, 2024. [Online]. Available: https://cdn.teledynelecroy.com/files/pdf/wavemaster8000hd-datasheet.pdf

14. *AWG70001B and AWG70002B Arbitrary Waveform Generators Specifications and Performance Verification Includes AWGSYNC01 Synchronization Hub Specifications Technical Reference*, 2019. [Online]. Available: https://download.tek.com/manual/AWG70000B-and-AWGSYNC01-Technical-Reference-077145300.pdf

15. W. Black, D. Hodges, Time interleaved converter arrays. IEEE J. Solid-State Circuits **15**(6), 1022–1029 (1980)

16. L.P. Rao, N. Sitthimahachaikul, P.J. Hurst, Correcting the effects of mismatches in time-interleaved analog adaptive FIR equalizers. IEEE Trans. Circuits Syst. I: Regul. Pap. **59**(11), 2529–2542 (2012)

17. K.A. Remley, D.F. Williams, P.D. Hale, Chih-Ming-Wang, J.A. Jargon, Y. Park, Calibrated oscilloscope measurements for system-level characterization of weakly nonlinear sources, in *2014 International Workshop on Integrated Nonlinear Microwave and Millimetre-wave Circuits (INMMiC)*, 2014, pp. 1–4

18. D.A. Humphreys, M. Hudlicka, I. Fatadin, Calibration of wideband digital real-time oscilloscopes, in *29th Conference on Precision Electromagnetic Measurements (CPEM 2014)*, 2014, pp. 698–699

19. C. Cho, J.G. Lee, P.D. Hale, J.A. Jargon, P. Jeavons, J.B. Schlager, A. Dienstfrey, Calibration of time-interleaved errors in digital real-time oscilloscopes. IEEE Trans. Microw. Theory Tech. **64**(11), 4071–4079 (2016)

20. D.A. Humphreys, M. Hudlička, I. Fatadin, Calibration of wideband digital real-time oscilloscopes. IEEE Trans. Instrum. Meas. **64**(6), 1716–1725 (2015)

21. D.A. Humphreys, A. Raffo, G. Bosi, G. Vannini, D. Schreurs, K.N. Gebremicael, K. Morris, Maximizing the benefit of existing equipment for nonlinear and communication measurements, in *2016 87th ARFTG Microwave Measurement Conference (ARFTG)*, 2016, pp. 1–4

22. K. Coakley, P. Hale, Alignment of noisy signals. IEEE Trans. Instrum. Meas. **50**(1), 141–149 (2001)

23. M. Hudlička, C. Lundström, D.A. Humphreys, I. Fatadin, BER estimation from EVM for QPSK and 16-QAM coherent optical systems, in *2016 IEEE 6th International Conference on Photonics (ICP)*, 2016, pp. 1–3

24. Joint Committee for Guides in Metrology, JCGM GUM-1:2023 guide to the expression of uncertainty in measurement – Part 1: introduction, BIPM, Technical report, 2023. [Online]. Available: https://www.bipm.org

25. Joint Committee for Guides in Metrology, JCGM 101:2008 supplement 1 to the 'Guide to the expression of uncertainty in measurement' – propagation of distributions using a Monte Carlo method (2008), BIPM, Technical report, 2008. [Online]. Available: https://www.bipm.org

26. Joint Committee for Guides in Metrology, Evaluation of measurement data supplement 2 to the 'Guide to the Expression of Uncertainty in Measurement' extension to any number of output quantities, BIPM, Technical report, 2011. [Online]. Available: https://www.bipm.org

27. D.A. Humphreys, P.M. Harris, M. Rodríguez-Higuero, F.A. Mubarak, D. Zhao, K. Ojasalo, Principal component compression method for covariance matrices used for uncertainty propagation. IEEE Trans. Instrum. Meas. **64**(2), 356–365 (2015)
28. G. Fishman, *Monte Carlo: Concepts, Algorithms, and Applications* (Springer, New York, 2013)
29. D. Williams, A. Lewandowski, T. Clement, J. Wang, P. Hale, J. Morgan, D. Keenan, A. Dienstfrey, Covariance-based uncertainty analysis of the NIST electrooptic sampling system. IEEE Trans. Microw. Theory Tech. **54**(1), 481–491 (2006)
30. P.D. Hale, C.M.J. Wang, Calculation of pulse parameters and propagation of uncertainty. IEEE Trans. Instrum. Meas. **58**(3), 639–648 (2009)
31. D.A. Humphreys, M. Naftaly, J.F. Molloy, Effect of time-delay errors on THz spectroscopy dynamic range, in *2014 39th International Conference on Infrared, Millimeter, and Terahertz waves (IRMMW-THz)*, 2014, pp. 1–2
32. IEC Technical Committee TC 85 measuring equipment for electrical and electromagnetic quantities, "IEC 62754:2017 computation of waveform parameter uncertainties," The International Electrotechnical Commission (IEC), Technical report, May 2017. [Online]. Available: https://webstore.iec.ch/en/publication/29773
33. H. Füser, Publishable summary for 23NRM01 SBS uncert support for standardisation of sample-by-sample waveform uncertainty computation, Euramet, Technical report, 2024. [Online]. Available: https://www.euramet.org/research-innovation/search-research-projects/details/project/support-for-standardisation-of-sample-by-sample-waveform-uncertainty-computation
34. R.A. Ginley, Kicking the tires of the NIST microwave uncertainty framework, part 1, in *2016 88th ARFTG Microwave Measurement Conference (ARFTG)*, 2016, pp. 1–4
35. R.A. Ginley, Kicking the tires of the NIST microwave uncertainty framework, Part 2, in *2017 90th ARFTG Microwave Measurement Symposium (ARFTG)*, 2017, pp. 1–4
36. M. Wollensack, J. Hoffmann, J. Ruefenacht, M. Zeier, VNA tools II: S-parameter uncertainty calculation, in *79th ARFTG Microwave Measurement Conference*, 2012, pp. 1–5
37. METAS, VNA tools for reliable RF & microwave measurements, METAS, Technical report, 2024. [Online]. Available: https://www.metas.ch/metas/en/home/fabe/hochfrequenz/vna-tools.html
38. Anritsu, Exact uncertainty calculator, 2021. [Online]. Available: https://www.anritsu.com/en-us/test-measurement/support/downloads/software/dwl19572
39. P.M. Harris, D.A. Humphreys, Software for compact covariance matrix uncertainty, Nov. 2015. [Online]. Available: https://www.npl.co.uk/resources/software/covariance-matrix-uncertainty visited on 5 Sept 2015

Chapter 6
Vector Network Analysis

Mohanad Dawood Al-Dabbagh and Thomas Kleine-Ostmann

Abstract Vector network analysis is one of the fundamental measurement techniques applied in radio frequency engineering and therefore holds fundamental importance for Terahertz (THz) communications. It serves to determine the transmission and reflection properties of devices, circuits, antennas, setups or linear time-invariant networks in general. Measurands are the frequency-dependent and complex-valued scattering parameters as defined for an n-port network. The scattering parameters are used to derive component properties (e.g. filter transmission, amplification, mixer attenuation), antenna properties (gain, antenna factor) or transmission channel properties. This chapter aims to give a brief introduction to vector network analyser (VNA) measurements.

6.1 Definition of Scattering Parameters

S-Parameters, or scattering parameters, are essential in electrical engineering, especially in radio frequency (RF) and microwave engineering, to describe how RF signals behave in a linear network. These parameters characterise the transmission and reflection of RF signals within a network, such as a circuit or device like an amplifier, filter or antenna [1, 2].

For an n-port network, S-parameters are represented as an $n \times n$ matrix $\underline{S}$, where each element S_{ij} indicates the ratio of the signal emerging from port i to the signal input into port j, with all other ports matched to avoid reflections. The scattering

M. D. Al-Dabbagh (✉) · T. Kleine-Ostmann
Physikalisch-Technische Bundesanstalt (PTB), Braunschweig, Germany
e-mail: mohanad.al-dabbagh@ptb.de; thomas.kleine-ostmann@ptb.de

© The Author(s) 2026
T. Kürner et al. (eds.), *Metrology for THz Communications*, Springer Series
in Optical Sciences 256, https://doi.org/10.1007/978-3-032-01986-8_6

matrix $\underline{\underline{S}}$ maps the vector $\underline{a}$ of forward propagating waves into the vector $\underline{b}$ of backward travelling waves:

$$\underline{b} = \begin{pmatrix} b_1 \\ b_2 \\ \vdots \\ b_n \end{pmatrix} = \underline{\underline{S}} \cdot \underline{a} = \begin{bmatrix} S_{11} & S_{12} & S_{13} & \dots & S_{1n} \\ S_{21} & S_{22} & S_{23} & \dots & S_{2n} \\ \vdots & \vdots & \vdots & \ddots & \vdots \\ S_{n1} & S_{n2} & S_{n3} & \dots & S_{nn} \end{bmatrix} \cdot \begin{pmatrix} a_1 \\ a_2 \\ \vdots \\ a_n \end{pmatrix} \tag{6.1}$$

For instance, S_{11} represents the reflection coefficient at port 1, showing how much of the incoming signal at port 1 is reflected back, while S_{21} denotes the forward transmission coefficient from port 1 to port 2, indicating how much of the signal at port 1 is transmitted to port 2. Similarly, S_{12} represents the reverse transmission from port 2 to port 1, and S_{22} is the reflection coefficient at port 2. A graphical representation of a 2-port network with wave quantities is shown in Fig. 6.1.

S-Parameters are complex numbers that are dimensionless and can be represented either as amplitude and phase or as real and imaginary part. Often, the amplitude is given on a logarithmic scale as $|S_{ij}|_{dB} = 20 \cdot \log |S_{ij}|$. The magnitude of the measured S-parameter indicates the signal strength reflected from or transmitted through the device under test (DUT) network, while the phase represents the change in the incident signal's phase relative to the received signal [3]. The magnitude and phase provide essential information on signal path loss and delay caused when the measured frequency passes through the DUT network. The VNA provides the possibility of measuring in different sweep variations but typically over a wide range of frequencies, where each frequency point can provide a different response inside the measured network. This makes the S-parameters frequency dependent. These VNA measurements are performed via sending a known signal to the network and measuring the resulting signals at various ports.

In practice, S-parameters are essential for the design and analysis of RF components to ensure correct impedance matching and minimise signal loss. The use of S-parameters is found important in the simulation and modelling of new systems, as it allows researchers to test the behaviour of complex RF and microwave networks before physical prototypes are manufactured. In addition, S-parameters

Fig. 6.1 Representation of a 2-port network with wave quantities

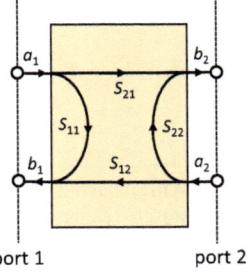

port 1   port 2

are an essential RF engineering tool to analyse and optimise signal propagation in a network, thus ensuring efficiency and effectiveness in system operation.

6.2   Architecture of Vector Network Analysers

The architecture of a VNA is designed to measure the S-parameters of a DUT across a wide range of frequencies, providing information about the transmission and reflection properties of RF and microwave networks. The VNA consists of several key components that work together to perform these measurements accurately. Figure 6.2 shows the architecture of an n-port network analyser [4].

The signal source is one of the main components of VNA, which generates a stable continuous-wave (CW) signal within a defined frequency range. This signal is directed towards DUT through a series of switches and couplers, which allow VNA to control and direct the signal as needed. The splitters and couplers are used to separate signals propagating forward and backward from the DUT; they are referred to as test sets.

Directional couplers are essential in the VNA's architecture, as they split the incident and reflected signals at each port of the DUT. The incident signal is the

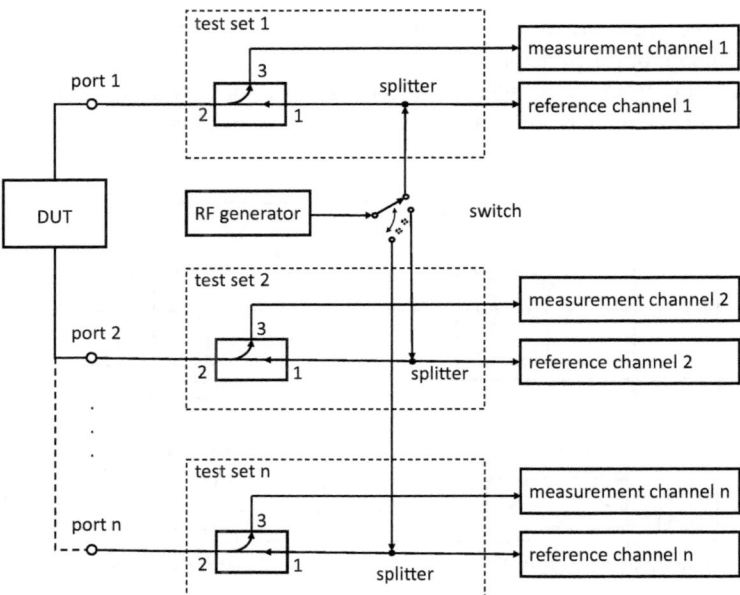

Fig. 6.2 N-port network analyser architecture. The signal from a RF generator is distributed to active ports which are switched on consecutively. Measurement and reference channel receivers serve to measure the wave quantities

one that is sent towards the DUT, while the reflected signal is the portion that is bounced back from the DUT. The VNA measures both incident and reflected signals to determine the reflection and transmission coefficients.

The high sensitivity of VNA receivers enables them to detect weak signals captured by directional couplers. When operating with frequency extensions to explore the millimetre and sub-THz range, VNA extension models perform both the up-conversion and the down-conversion of RF signals to an intermediate frequency (IF), simplifying processing.

The IF signal is then digitised by an analogue-to-digital converter (ADC), to allow the VNA's digital signal processor (DSP) to apply the data analyses. The digital amplitude and phase information are extracted through the DSP process, and the S-parameters of the DUT are calculated [5].

The calibration unit within the VNA plays an important role to ensure accurate measurements. The calibration process involves accounting for systematic errors such as cable losses, impedance mismatches and signal leakage, which can affect the accuracy of the S-parameter measurements. Calibration routines typically involve measuring known standards (e.g. open, short, match and through) to characterise the VNA's internal error terms. Generally, the upper frequency limits of the coaxial connector system used, e.g. N (18 GHz), 2.92 mm (40 GHz), 2.4 mm (50 GHz), 1.85 mm (67 GHz) or 1.0 mm (110 GHz), defines the operation frequency range of the VNA. However, the range can be extended to distinct waveguide frequency bands using external frequency extension modules, as shown in Fig. 6.3. Usually, they are operated with a 4-port VNA which has access to reference and measurement channel at two of the ports [4].

Finally, the VNA's user interface allows the operator to control the measurement process, set parameters such as frequency range and power level and view the

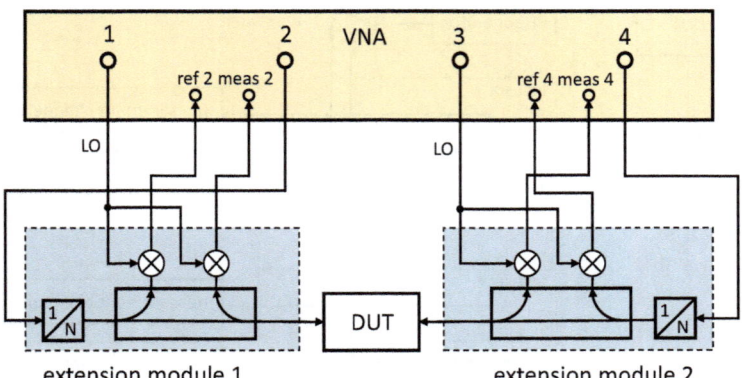

Fig. 6.3 External frequency extension modules for waveguide bands. A 4-port VNA with access to reference and measurement channels at two of the ports is necessary to operate the extension modules

results. Modern VNA systems often include advanced touch screens and software
that provide intuitive control and comprehensive data analysis capabilities.

6.3   System Error Correction

The systematic errors caused by directivity, reflection tracking, source match and
potentially the switch terms are thoroughly accounted for in calibration procedures
that characterise the quality of the test sets. These error terms can be addressed by
the procedure of systematic error correction, often denoted as the VNA calibration.

After the VNA's thermal stabilisation, we characterise the error coefficients
for each port using calibration standards with known mechanical properties. The
calibration standards are connected to the VNA and measured to calculate the error
terms as needed for numerical correction values of the measurement results.

Uncertainty budgets are essential for evaluating the performance of a calibrated
VNA system, as they quantify the contributions of systematic errors within the
measurement setup [6, 7]. This process is performed via the VNA error model,
which transforms raw measurements into meaningful corrected data. The accuracy
of error correction procedures depends on the VNA receiver's configuration,
ensuring that the linear equations defining these error are solved with minimal
uncertainty [8].

In this study, we use a four-receiver VNA to calibrate our two-port system
using established calibration methods. These methods involve characterising the
error coefficients for each port, such as directivity, reflection tracking, source match
and switch terms, which are then corrected through the calibration standards. We
also take into account other error sources, including noise floor, trace noise, drift,
nonlinearity, connection repeatability, isolation and cable movement, to ensure the
calibration is comprehensive and accurate [8].

We rely on specialised software like VNA Tools [9] to assist with the calibration
process. This software provides user-friendly interfaces for setting up the calibration, acquiring data and performing error correction, which makes the process more
efficient and accurate.

Using the four-receiver VNA, we can apply either the seven-term or the
classical 12-term error correction model. Calibration using frequency extenders and
waveguides follows the same principles as calibration with coaxial connectors, but
waveguide calibration presents some specific challenges. One of these challenges
is the need for specialised calibration standards, typically shims, to achieve optimal
results. In our case, we used two shims, one $\frac{1}{4}$ wavelength and one $\frac{3}{4}$ wavelength of
the converter's central operating range, along with a $180°$ phase shift in the reflecting
coefficient to ensure stability across the frequency band. Unlike coaxial calibration,
waveguide calibration does not have an open standard because an open waveguide
circuit lacks a well-defined reflection [8, 10].

For error correction, we use two widely recognised techniques: through, reflect
and line (TRL) and unknown through, offset short, short and match (UOSM).

These methods are known for their precision and ensure the reliability of our measurements. A comparison between these calibration methods and other error terms for sub-THz over-the-air (OTA) measurements is presented in [8] and in the Chap. 19.

TRL calibration offers simplified and efficient methodologies for waveguide systems, further enhancing measurement accuracy and repeatability. It requires mainly four different connections to encompass seven independent error terms. Two additional switch terms $\left(\Gamma_F = \frac{a_1}{b_1} \text{ and } \Gamma_R = \frac{a_2}{b_2}\right)$ are needed, totalling the number of unknowns to nine terms, which is the minimum number of independent equations to solve the TRL calibration problem effectively.

On the other hand, the UOSM calibration method, a classical 12-term error model, offers its own unique advantages. Like TRL calibration, it requires the computation of switch terms. However, unlike TRL, UOSM calibration uses the same set of calibration standards as the 12-term calibration, providing distinct degrees of freedom.

The UOSM calibration method is simple and relies on a single transmission measurement under the condition that reciprocity holds between the two transmission S-parameters. As a result, their product, S_{21}^2, assumes reciprocity [10]. The insertion loss should be well above the noise floor, and the distance between the ports should be below 90° of electrical length between the subsequent frequency steps. This is why sufficient frequency points for wider unknowns through calibration distance are necessary to obtain the correct phase values. Similar to TRL calibration, the values for Γ_F and Γ_R are measured during the unknown through calibration step.

Six additional connections are needed to complete the calibration. These connections provide two independent equations for each standard, resulting in 12 independent equations to determine the error terms. The solution has been detailed in several sources and is available in the literature [10].

We use the VNA Tools software to perform the calibration procedure. One of its advantages is the ability to test the same DUT connection across different calibrations, providing better insight into the results achieved with various calibration procedures.

6.4 Traceability of Measurements

The advancement in VNA measurement process and modelling leads to achieving measurement traceability across all fundamental RF/microwave parameters, including impedance, power, attenuation, frequency and noise.

The VNA calibration standards are traceable to the meter using precision airlines, shorts and mechanical gauges, which are among the most precise standards available. The measurement uncertainties of these mechanical systems are accurately quantified. An electrical model, based on established microwave theory, is developed to convert mechanical and physical properties into microwave S-parameters. To ensure the reliability of these models, input variable uncertainties are propagated through the S-parameter calculations using Monte Carlo simulations [6]. This approach provides a comprehensive address to the uncertainties in the S-parameters. The modelled S-parameters are then employed to calibrate a reference VNA measurement system, which in turn removes the error values related to each measured frequency point that might be present in the standards, thereby providing a direct link to fundamental units of measure such as length and other physical constants.

At sub-THz frequencies, establishing the traceability chain is increasingly challenging due to the sensitivity of measurements at such high frequencies. The uncertainties associated with these standards are typically low, yet they become more pronounced at higher frequencies [11, 12]. The complexity of propagating calibration standard uncertainties through the calibration process, error correction and ultimately the device measurement process intensifies with frequency. This would lead to measurement uncertainties in the range of 1 dB when no specialised calibrations are performed to quantify the error terms more accurately.

For calibrations beyond a basic 1-port setup, deriving sensitivity functions becomes increasingly complex. To evaluate the overall VNA system measurement uncertainties, Monte Carlo simulations are used, as even small errors can have a substantial effect on measurement accuracy.

References

1. M. Thumm, W. Wiesbeck, S. Kern, *Hochfrequenzmesstechnik – Verfahren und Messsysteme* (Vieweg + Teubner, 1998)
2. D.M. Pozar, *Microwave Engineering*, 3rd edn. (Wiley, Hoboken, 2005)
3. J. Choma, Scattering parameters: concept, theory, and applications, University of Southern California, Course Notes (2009)
4. M. Hiebel, *Fundamentals of Vector Network Analysis*, 2nd edn. (Rohde and Schwarz, Munich, 2007)
5. D. Micheli, R. Pastore, A. Vricella, A. Delfini, M. Marchetti, F. Santoni, Electromagnetic characterization of materials by vector network analyzer experimental setup, in *Spectroscopic Methods for Nanomaterials Characterization* (Elsevier, Amsterdam, 2017), pp. 195–236
6. J.C. for Guides in Metrology (JCGM), Evaluation of measurement data – guide to the expression of uncertainty in measurement.
7. E.A. of National Metrology Institutes (Euramet), Guidelines on the evaluation of vector network analysers (VNA), March 2018
8. M.D. Al-Dabbagh, D. Ulm, T. Doeker, D. Dupleich, A. Ebert, R.S. Thomä, T. Kürner, D.A. Humphreys, T. Kleine-Ostmann, Characterization of sub-THz channel sounding systems in OTA measurement scenarios using a vector network analyzer. IEEE Trans. Antennas Propag. (2025). https://doi.org/10.1109/TAP.2025.3533887
9. M. Wollensack, J. Hoffmann, METAS VNA tools – math reference V2.8.2, Federal Institute of Metrology (METAS). Visited on 30 Aug 2024. [Online]. Available: https://www.metas.ch/metas/en/home/fabe/hochfrequenz/vna-tools.html

10. J.P. Dunsmore, *Handbook of Microwave Component Measurements: With Advanced VNA Techniques* (Wiley, Hoboken, 2020)
11. T. Schrader et al., Verification of scattering parameter measurements in waveguides up to 325 GHz including highly-reflective devices. Adv. Radio Sci. **9**, 9–17 (2011)
12. D.F. Williams, 500 GHz–750 GHz rectangular-waveguide vector-network-analyzer calibrations. IEEE Trans. THz Sci. Technol. **1**, 364–377 (2011)

Chapter 7
Terahertz Time-Domain Spectroscopy

Daniel M. Mittleman and Martin Koch

Abstract This chapter briefly reviews the technique of terahertz time-domain spectroscopy (THz-TDS), a powerful method for accessing the spectral range from 100 GHz up to several THz. We first discuss the physical principles upon which the method relies. We then describe the use of THz-TDS for studies of atmospheric propagation. Finally, we discuss other types of measurements for which TDS is well suited, such as, for example, the characterisation of the relative importance of scattering and absorption when a THz beam impinges upon a rough surface.

7.1 Introduction

For most of the twentieth century, it was very laborious and cumbersome to carry out spectroscopic experiments in the terahertz frequency range [1, 2]. This changed with the advent of terahertz time-domain spectroscopy (THz-TDS). In the late 1980s groups at AT&T Bell Laboratories [3] and IBM [4] demonstrated the generation and detection of short electromagnetic pulses through the air, with frequencies ranging from 100 GHz to a few THz.

In the following 20 years, THz-TDS established itself as an elegant method for investigating the THz properties of materials, dramatically lowering the barrier for measurements in this spectral range. A wide variety of materials have been studied including semiconductors [5–7], superconductors [8], polymers [9, 10], liquids [11–13], liquid crystals [14, 15] and gases [16]. As one relevant example, the group of D. Grischkowsky [17] investigated the spectroscopic signatures of atmospheric water vapour, a topic later studied by other groups [18]. These results continue

D. M. Mittleman
School of Engineering, Brown University, Providence, RI, USA
e-mail: daniel_mittleman@brown.edu

M. Koch (✉)
Physik (Fb13), AG Halbleiterphotonik, Philipps-Universität Marburg, Marburg, Germany
e-mail: martin.koch@physik.uni-marburg.de

© The Author(s) 2026
T. Kürner et al. (eds.), *Metrology for THz Communications*, Springer Series in Optical Sciences 256, https://doi.org/10.1007/978-3-032-01986-8_7

to inform the design of wireless communication systems operating in the 100–1000 GHz range.

7.2 THz-TDS Measurement Methodology

The core component of this spectroscopic measurement technique is a femtosecond laser that emits short laser pulses in the range of 100-fs duration, with a repetition rate of typically 80 MHz. Two photoconductive antennas are gated by these laser pulses, one to generate and a second to detect the THz pulses. These antennas consist of a semiconductor substrate on which coplanar metal strip line has been deposited by photolithography. The photons of the laser pulses have a photon energy that is greater than the band gap energy of the semiconductor. The laser radiation is focused between the two strips, so that absorption of the laser radiation lifts electrons from the valence band into the conduction band, generating mobile charge carriers in the region between the metal lines. In the case of the transmitter, these lines are biased with a field strength of around 5 kV/cm. The optically excited charge carriers are accelerated in this DC field, producing a short current pulse, which is the source of an emitted THz pulse.

The pulses generated in this way propagate through a spectrometer and are then directed to a receiving antenna, which is also gated by a laser pulse. This has a similar structure, except that here no voltage is applied; instead, the optically generated charge carriers move back and forth in the field of the incoming THz pulses. This results in a very small current flow which is amplified and registered. The optical gating pulse passes through a variable delay line and can thus be delayed in relation to the incoming THz pulse. This delay can be scanned to perform a sampling measurement (see [19] for a more detailed description of THz-TDS). As is common in many ultra-fast spectroscopy systems, the measurement of a temporal process is thus reduced to the measurement of a length, namely, that of the delay line.

In the initial demonstrations of THz-TDS, researchers used a mode-locked dye laser and photo-conductive antennas based on radiation-damaged silicon on sapphire (SOS). In 1990, Grischkowsky and co-workers showed that this technique enables high-quality spectroscopy in the lower THz [5] range. In the early 1990s, mode-locked titanium sapphires became available. Because these lasers are much easier to use and offer greater stability, they immediately replaced the dye lasers. Soon, the SOS antennas were also replaced by structures based on GaAs grown at low temperature [20].

In the mid-1990s, this measurement technique was extended to an imaging configuration for the first time [21, 22]. In the years that followed, a large number of practical applications for THz imaging have been discussed and demonstrated, ranging from the quality control of industrial products and food to medical diagnostics and materials research [23–26]. Other applications, such as the nondestructive examination of cultural artefacts, have also been explored [27]. A reflection geometry is primarily used here, so that time-of-flight tomographic reconstruction methods can be employed [28, 29]. Similar to ultrasound examination, a series of reflections are produced which emanate from the various interfaces in a sample. This allows the internal structure of objects to be clarified.

In the early years of THz-TDS systems, the laser pulses were guided in free space and the optical components were permanently mounted on an optical table. This made the THz systems inflexible and the systems could not leave the protected environment of an optics laboratory. In 2000, the first fibre-coupled THz system was introduced by Picometrix [30]. This was a significant advance as it allowed the THz antennas to be freely positioned and therefore offered tremendous additional flexibility. In 2007, Wilk et al. [31] demonstrated a THz-TDS system based on a femtosecond fibre laser operating at a wavelength of 1.55 μm. These lasers are significantly less expensive than titanium-sapphire lasers, are made of highly robust telecommunications components and are designed for long-term operation. The use of this laser class was made possible by a new antenna material, based on InGaAs/InAlAs multiple quantum wells grown at low temperature, which strongly absorbs the laser radiation at this longer wavelength. The increased flexibility and ruggedness of fibre-coupled THz systems driven by fibre lasers allow THz-TDS systems to leave the laboratory and be transported to a measurement site. The THz antennas can even be attached to a robot arm, making it possible to examine uneven objects in reflection.

THz-TDS offers a number of unique advantages over other methods for accessing this spectral range. Key among these is the ultra-broad bandwidth of the THz signal. THz pulses typically manifest as single cycles of the electromagnetic field and thus contain spectral components spanning more than 1000 GHz of bandwidth (sometimes considerably more). Like a traditional blackbody source, this bandwidth is ideally suited for measurements in which one requires a broadband characterisation of a channel, a device or the property of a given material. However, unlike a blackbody, the THz signal generated in a TDS instrument is spatially coherent and can thus be focused to a diffraction-limited spot, which is advantageous for high-resolution imaging and characterisation of smaller samples and for coupling efficiently into single-mode waveguides. Also, the sampling measurement technique noted above generates a signal which is proportional to the terahertz electric field, thus preserving phase information in the measurement. As a result, spectroscopic measurements made with THz-TDS provide access to both the real and imaginary parts of the dielectric function $\epsilon(\omega)$ of a sample, without the need to use a Kramers-Kronig analysis to extract the real part from a measurement of only the imaginary part. This phase sensitivity also offers a significant advantage in signal to noise; since THz-TDS detection is insensitive to incoherent (e.g. thermal) radiation, the conventional noise source that dominates most other forms of THz measurement is absent. Indeed, a typical THz-TDS system is capable of detecting a signal whose average power level is orders of magnitude lower than that of a room-temperature thermal blackbody source in the same spectral range (despite the fact that none of the components of a THz-TDS system are typically operated below room temperature). As a result, it is possible to employ THz-TDS to study samples that are well above room temperature, such as flames [32], which would be impossible using any other technique for accessing the terahertz range.

7.3 THz-TDS Measurements Relevant for Wireless Systems

THz-TDS is useful for informing the fundamentals of THz communication systems in several different ways. Firstly, as mentioned above, the technique can be used to determine the atmospheric attenuation, which is mainly caused by the high water vapour absorption in the THz band. A series of studies have been carried out on this topic in recent years. Grischkowsky and co-workers, for example, used a THz-TDS system to transmit broadband pulses through a 137-m-long path [33, 34] and between buildings over a distance of nearly 1 km [35]. Recently, Taleb et al. presented the first study of the attenuation of THz radiation over a broad range of temperatures and humidity values [18]. The aspect of atmospheric attenuation and its characterisation with THz-TDS is discussed in more detail in Chap. 30.

Secondly, THz-TDS can be used to investigate the interaction of objects with THz beams, as could be expected to occur in a directional wireless link when an object lies close to the line-of-sight path between the transmitter and receiver. This can lead to blockage of the primary channel, as well as scattering of radiation. Scattering in unintended directions can pose a security threat, as it provides a new opportunity for eavesdropping [36]. Chapter 32 discusses recent experimental studies and simulations which explore the effects of scattering and methods for numerical modelling of these important but potentially quite complex situations.

Thirdly, the reflection and scattering properties of materials can be characterised [37, 38]. THz communication systems often rely on a line-of-sight (LOS) connection between the transmitter and receiver. However, in a case where the line-of-sight connection is temporarily blocked, for example, by people moving around in the room, non-line-of-sight (NLOS) channels which include one or more reflections from walls or furniture could be used as a backup. This concept was studied numerically twenty years ago by Kürner and co-workers [39]. A comprehensive knowledge of the reflection and scattering properties of typical indoor building materials is essential to make accurate predictions about the performance in a given situation [40, 41]. Chapter 31 presents a discussion of this type of investigation.

Fourthly, THz-TDS can be used to characterise optics and devices that can be used in communication systems for beam guidance or wavefront control. Examples include studies of lenses [42–45], components for beam steering [46, 47], filters [48, 49], mirrors [50–52] and novel antenna designs [53]. All these components could find their way into future THz communication systems.

Finally, it may be interesting to note that the first ever wireless communication link at several hundred GHz was also carried out using a THz-TDS system. In 2004 Kleine-Ostmann et al. used an electrically driven room-temperature semiconductor-based modulator to modulate the transmission through a THz-TDS system [54]. In a data transmission experiment [55], audio signals of up to 25 kHz were impressed on the 75-MHz train of broadband THz pulses. This short-range laboratory demonstration proved sufficient to transmit music at reasonable fidelity and has served as an inspiration for much subsequent research.

References

1. M.F. Kimmitt, Restrahlen to T-rays-100 years of terahertz radiation. J. Biol. Phys. **29**, 77 (2003)
2. A. Mitsuishi, Progress in far-infrared spectroscopy: approximately 1890 to 1970. J. Infrared Millim. Terahertz Waves **35**(3), 243–281 (2014)
3. P.R. Smith, D.H. Auston, M.C. Nuss, Subpicosecond photoconducting dipole antennas. IEEE J. Quantum Electron. **24**(2), 255 (1988)
4. C. Fattinger, D. Grischkowsky, Terahertz beams. Appl. Phys. Lett. **54**, 490 (1989)
5. D. Grischkowsky, S. Keiding, M. Van Exter, C. Fattinger, Far-infrared time-domain spectroscopy with terahertz beams of dielectrics and semiconductors. J. Opt. Soc. Am. B **7**(10), 2006 (1990)
6. D.M. Mittleman, J. Cunningham, M.C. Nuss, M. Geva, Noncontact semiconductor wafer characterization with the terahertz Hall effect. Appl. Phys. Lett. **71**(1), 16 (1997)
7. R. Huber, F. Tauser, A. Brodschelm, M. Bichler, G. Abstreiter, A. Leitenstorfer, How many-particle interactions develop after ultrafast excitation of an electron-hole plasma. Nature 22 **414**, 286 (2001). [Online]. Available: www.nature.com
8. T. Matsuoka, T. Fujimoto, K. Tanaka, S. Miyasaka, S. Tajima, K. Fujii, M. Suzuki, M. Tonouchi, Terahertz time-domain reflection spectroscopy for high-Tc superconducting cuprates. Phys. C: Superconductivity Appl. **469**(15–20), 982 (2009)
9. H. Hoshina, Y. Morisawa, H. Sato, A. Kamiya, I. Noda, Y. Ozaki, C. Otani, Higher order conformation of poly(3-hydroxyalkanoates) studied by terahertz time-domain spectroscopy. Appl. Phys. Lett. **96**(10), 101904 (2010). [Online]. Available: https://doi.org/10.1063/1.3358146
10. S. Sommer, T. Raidt, B.M. Fischer, F. Katzenberg, J.C. Tiller, M. Koch, THz-spectroscopy on high density polyethylene with different crystallinity. J. Infrared Millim. Terahertz Waves **37**(2), 189 (2016)
11. J.E. Pedersen, S.R. Keiding, THz time-domain spectroscopy of nonpolar liquids. IEEE J. Quantum Electron. **28**(10), 2518 (1992)
12. L. Thrane, R.H. Jacobsen, P. Jepsen, S.R. Keiding, THz reflection spectroscopy of liquid water. Chem. Phys. Lett. **240**, 330 (1995)
13. J.P. Laib, D.M. Mittleman, Temperature-dependent terahertz spectroscopy of liquid n-alkanes. J. Infrared Millim. Terahertz Waves **31**, 1015–2021 (2010)
14. T.-R. Tsai, C.-Y. Chen, C.-L. Pan, R.-P. Pan, X.-C. Zhang, Terahertz time-domain spectroscopy studies of the optical constants of the nematic liquid crystal 5CB. Appl. Opt. **42**(13), 2372 (2003)
15. N. Vieweg, M.K. Shakfa, B. Scherger, M. Mikulics, M. Koch, THz properties of nematic liquid crystals. J. Infrared Millim. Terahertz Waves **31**(11), 1312 (2010)
16. D.M. Mittleman, R.H. Jacobsen, R. Neelamani, R.G. Baraniuk, M.C. Nuss, Gas sensing using terahertz time-domain spectroscopy. Appl. Phys. B **67**, 379 (1998)
17. M. Van Exter, C. Fattinger, D.R. Grischkowsky, Terahertz time-domain spectroscopy of water vapor. Opt. Lett. **14**(20), 1128 (1989)
18. F. Taleb, M. Alfaro-Gomez, M. Dawood Al-Dabbag, J. Ornik, J. Viana, A. Jäckel, C. Mach, J. Helminiak, T. Kleine-Ostman, T. Kürner, M. Koch, D.M. Mittleman, E. Castro-Camus, Propagation of THz radiation in air over a broad range of atmospheric temperature and humidity conditions. Sci. Rep. **13**(1), 20782 (2023)
19. M. Koch, D.M. Mittleman, J. Ornik, Terahertz time-domain spectroscopy. Nat. Rev. Methods Primers **3**, 48 (2023)

20. J. Kim, S. Williamson, J. Nees, S. Wakana, J. Whitaker, Photoconductive sampling probe with 2.3-ps temporal resolution and 4-μV sensitivity. Appl. Phys. Lett. **62**(18), 2268 (1993)
21. B.B. Hu, M.C. Nuss, Imaging with terahertz waves. Opt. Lett. 20(16), 1716 (1995)
22. D.M. Mittleman, R.H. Jacobsen, M.C. Nuss, T-ray imaging. IEEE J. Sel. Top. Quantum Electron. **2**(3), 679 (1996)
23. J.A. Zeitler, Y. Shen, C. Baker, P.F. Taday, M. Pepper, T. Rades, Analysis of coating structures and interfaces in solid oral dosage forms by three dimensional terahertz pulsed imaging. J. Pharm. Sci. **96**(2), 330 (2007)
24. C. Jansen, S. Wietzke, O. Peters, M. Scheller, N. Vieweg, M. Salhi, N. Krumbholz, C. Jördens, T. Hochrein, M. Koch, Terahertz imaging: applications and perspectives. Appl. Opt. **49**, 48 (2010)
25. H. Guerboukha, S. Nallappan, M. Skorobogtivy, Toward real-time terahertz imaging. Adv. Opt. Photon. **10**, 843 (2018)
26. G.G. Hernandez-Cardoso, S.C. Rojas-Landeros, M. Alfaro-Gomez, A.I. Hernandez-Serrano, I. Salas-Gutierrez, E. Lemus-Bedolla, A.R. Castillo-Guzman, H.L. Lopez-Lemus, E. Castro-Camus, Terahertz imaging for early screening of diabetic foot syndrome: a proof of concept. Sci. Rep. **7**, 42124 (2017)
27. K. Krügener, M. Schwerdtfeger, S.F. Busch, A. Soltani, E. Castro-Camus, M. Koch, W. Viöl, Terahertz meets sculptural and architectural art: evaluation and conservation of stone objects with T-ray technology. Sci. Rep. **5**, 1 (2015)
28. D.M. Mittleman, S. Hunsche, L. Boivin, M.C. Nuss, T-ray tomography. Opt. Lett. **22**(12), 904 (1997). [Online]. Available: http://playfair.stanford.edu/wavelab/
29. S. Hunsche, D.M. Mittleman, M. Koch, M.C. Nuss, New dimensions in T-ray imaging. IEICE Trans. Electron. **E81-C**(2), 269–276 (1998)
30. J.V. Rudd, D. Zimdars, M. Wannuth, Compact, fiber-pigtailed, terahertz imaging system. Proc. SPIE Int. Soc. Opt. Eng. **3934** (2000). [Online]. Available: http://proceedings.spiedigitallibrary. org/
31. R. Wilk, M. Mikulics, K. Biermann, H. Künzel, I. Kozma, R. Holzwarth, B. Sartorius, M. Meiand, M. Koch, THz time-domain spectrometer based on LT-InGaAs photoconductive antennas exited by a 1.55 m, in *Proceedings of the Conference on Lasers and Electro-Optics* (Optical Society of America – CLEO, Baltimore, 2007)
32. R.A. Cheville, D.R. Grischkowsky, Far-infrared terahertz time-domain spectroscopy of flames. Opt. Lett. **20**(15), 1646 (1995)
33. M. Mandehgar, Y. Yang, D. Grischkowsky, Atmosphere characterization for simulation of the two optimal wireless terahertz digital communication links. Opt. Lett. **38**(17), 3437 (2013)
34. Y. Yang, M. Mandehgar, D. Grischkowsky, THz-TDS characterization of the digital communication channels of the atmosphere and the enabled applications. J. Infrared Millim. Terahertz Waves 36(2), 97 (2015)
35. G.-R. Kim, T.-I. Jeon, D. Grischkowsky, 910-m propagation of THz ps pulses through the atmosphere. Opt. Exp. **25**(21), 25422 (2017)
36. J. Ma, R. Shrestha, L. Moeller, D.M. Mittleman, Channel performance for indoor and outdoor terahertz wireless links. APL Photon. **3**(5), 51601 (2018)
37. R. Piesiewicz, T. Kleine-Ostmann, N. Krumbholz, D. Mittleman, M. Koch, T. Kürner, Terahertz characterisation of building materials. Electron. Lett. **41**(18), 1002 (2005)
38. F. Taleb, G.G. Hernandez-Cardoso, E. Castro-Camus, M. Koch, Transmission, reflection, and scattering characterization of building materials for indoor THz communications. IEEE Trans. Terahertz Sci. Technol. **13**(5), 421 (2023)
39. R. Piesiewicz, J. Jemai, M. Koch, T. Kürner, THz channel characterization for future wireless gigabit indoor communication systems. Proc. SPIE Int. Soc. Opt. Eng. **5727**, 166 (2005)
40. R. Piesiewicz, T. Kleine-Ostmann, N. Krumbholz, D. Mittleman, M. Koch, J. Schoebel, T. Kurner, Short-range ultra-broadband terahertz communications: concepts and perspectives. IEEE Antennas Propag. Mag. **49**, 24 (2007)

41. R. Piesiewicz, M. Jacob, M. Koch, J. Schoebel, T. Kürner, Performance analysis of future multigigabit wireless communication systems at THz frequencies with highly directive antennas in realistic indoor environments. IEEE J. Sel. Top. Quantum Electron. **14**(2), 421 (2008). [Online]. Available: http://ieeexplore.ieee.org
42. Y. Lo, R. Leonhardt, T. Yasui, K.I. Sawanaka, A. Ihara, E. Abraham, M. Hashimoto, T. Araki, R.M. Woodward, B.E. Cole, V.P. Wallace, R.J. Pye, D.D. Arnone, E.H. Linfield, Aspheric lenses for terahertz imaging. Opt. Exp. **16**, 15991 (2008)
43. B. Scherger, C. Jördens, M. Koch, Variable-focus terahertz lens. Opt. Exp. **19**(3), 4528 (2011)
44. E. Castro-Camus, M. Koch, A.I. Hernandez-Serrano, Additive manufacture of photonic components for the terahertz band. J. Appl. Phys. **127**(21), 210901 (2020). [Online]. Available: https://doi.org/10.1063/1.5140270
45. A. Siemion, Terahertz diffractive optics – smart control over radiation. J. Infrared Millim. Terahertz Waves **40**(5), 477 (2019)
46. Y. Monnai, K. Altmann, C. Jansen, H. Hillmer, M. Koch, H. Shinoda, Terahertz beam steering and variable focusing using programmable diffraction gratings. Opt. Exp. **21**(2), 2347–2354 (2013). [Online]. Available: https://opg.optica.org/oe/abstract.cfm?URI=oe-21-2-2347
47. B. Scherger, M. Reuter, M. Scheller, K. Altmann, N. Vieweg, R. Dabrowski, J.A. Deibel, M. Koch, Discrete terahertz beam steering with an electrically controlled liquid crystal device. J. Infrared Millim. Terahertz Waves **33**(11), 1117 (2012)
48. C. Winnewisser, F. Lewen, H. Helm, Transmission characteristics of dichroic filters measured by THz time-domain spectroscopy. Appl. Phys. A **66**, 593 (1998)
49. I.A.I. Al-Naib, C. Jansen, M. Koch, High Q-factor metasurfaces based on miniaturized asymmetric single split resonators. Appl. Phys. Lett. **94**(15), 153505 (2009)
50. I.A. Ibraheem, N. Krumbholz, D. Mittleman, M. Koch, Low-dispersive dielectric mirrors for future wireless terahertz communication systems. IEEE Microw. Wirel. Compon. Lett. **18**(1), 67 (2008)
51. W. Withayachumnankul, B.M. Fischer, D. Abbott, Quarter-wavelength multilayer interference filter for terahertz waves. Opt. Commun. **281**(9), 2374 (2008)
52. S. Park, B. Norton, G.D. Boreman, T. Hofmann, Mechanical tuning of the terahertz photonic bandgap of 3D-printed one-dimensional photonic crystals. J. Infrared Millim. Terahertz Waves **42**(2), 220 (2021)
53. A. Kludze, J. Kono, D.M. Mittleman, Y. Ghasempour, A frequency-agile retrodirective tag for large-scale sub-terahertz data backscattering. Nat. Commun. **15**(1), 8756 (2024). [Online]. Available: http://doi.org/10.1038/s41467-024-53035-5
54. T. Kleine-Ostmann, P. Dawson, K. Pierz, G. Hein, M. Koch, Room-temperature operation of an electrically driven terahertz modulator. Appl. Phys. Lett. **84**(18), 3555 (2004)
55. T. Kleine-Ostmann, K. Pierz, G. Hein, P. Dawson, M. Koch, Audio signal transmission over THz communication channel using semiconductor modulator. Electron. Lett. **40**(2), 124 (2004)

Chapter 8
Channel Sounding

Reiner S. Thomä, Jonas Gedschold, Diego Dupleich, Tobias Doeker, and Giovanni Del Galdo

Abstract A multidimensional channel sounder is necessary for the experimental characterization of time-variant multipath propagation in its joint delay, direction, and Doppler domains. In this chapter, we refer to the fundamental principles of sounding architecture design and geometric modeling of multipath wave propagation, which form the basis for model-based parameter estimation from measurements. The basic design features, requirements, and parameters of wideband spread spectrum excitation signals, mutual Tx/Rx synchronization, radio access, and data recording are explained. The specific challenges of the sounder antenna architecture for resolution and spatial filtering of propagation directions and polarization orientation in the THz frequency range are elaborated. This chapter provides an overview of the statistical analysis of spatial communication channels and introduces a parametric multipath model, which forms the basis for model-based system identification.

8.1 The Necessity of Multidimensional Channel Sounding

The transmission channel is the *heart* of any mobile communication system. It depends on the physical mechanisms of electromagnetic wave propagation. Multipath propagation goes along with dispersion in several dimensions: delay,

R. S. Thomä · J. Gedschold (✉) · D. Dupleich
Technische Universität Ilmenau, Institute of Information Technology, Ilmenau, Germany
e-mail: reiner.thomae@tu-ilmenau.de; jonas.gedschold@tu-ilmenau.de; diego.dupleich@tu-ilmenau.de

T. Doeker
Technische Universität Braunschweig, Institut für Nachrichtentechnik, Braunschweig, Germany
e-mail: t.doeker@tu-braunschweig.de

G. D. Galdo
Technische Universität Ilmenau, Institute of Information Technology, Ilmenau, Germany

Fraunhofer Institute for Integrated Circuits (IIS), Ilmenau, Germany
e-mail: giovanni.delgaldo@tu-ilmenau.de

© The Author(s) 2026
T. Kürner et al. (eds.), *Metrology for THz Communications*, Springer Series in Optical Sciences 256, https://doi.org/10.1007/978-3-032-01986-8_8

Doppler, and angle (in azimuth and elevation and on both sides of the link). The received power is fluctuating and the resulting parameters have to be considered as statistically distributed on the small and large scale. The knowledge and understanding of the time-variant channel nature allow us to predict the information-theoretic limits of communication performance on the link and system level and tell us how it can be tested and evaluated. Studying the propagation mechanisms helps us to understand how to design the radio interface and tune its parameters to achieve optimum performance.

Multipath, for instance, was originally understood as detrimental for achieving high data rates and reliable transmission since it causes fading signal power and disturbing inter-symbol interference. Later it was recognized that multipath diversity can be exploited to achieve more robust communication and even to increase channel capacity in certain situations. The prerequisite is the optimum adjustment of the radio interface to the channel statistics in delay, time, and space. The most prominent incident was the advent of multiple-input multiple-output (MIMO) channel access, which turned out to be the most powerful technology to exploit multipath.

How can we gain the knowledge and understanding of the propagation channel? In essence, there are two ways: the experimental approach by measurements and the simulation approach. While ray-tracing (RT) (Chap. 10) or related procedures are a convenient method to simulate wave propagation in large and complex environments based on the physically motivated models of ray optical propagation, measurement-based investigation of time-variant multipath propagation phenomena uses a multidimensional channel sounder (MCS). Such a system consists of a transmitter that excites the propagation channel via a dedicated antenna and a receiver with another antenna at a different position. From the recorded response the input/output (I/O) transfer characteristics are calculated and stored for offline use. The multiple dimensions are impulse (respectively frequency) response, spatial (respectively directive) structure at Tx and Rx, and "slow" time variability, which is related to Doppler shift and temporal fading. The sounder system is located and moved in a dedicated way so that it somehow represents the target mobile communication link, e.g., the radio link between access point (AP) and user equipment (UE). The stored transfer functions can be directly used to simulate the transceiver links' I/O response in both directions. Another application is the calculation of short- and long-term statistics such as spreads and fading profiles. More sophisticated processing includes estimation of the parameters of a multipath model of propagation, which we call model-based parameter estimation (MBPE). The geometry of the underlying structural model of multipath propagation corresponds directly to the ray-optical propagation model, which is used by ray tracing. The model-based approach has several advantages over the direct use of recorded data: Among others, it has superresolution capabilities that help to better understand the physical propagation mechanisms. Moreover, it offers extended possibilities in mitigation of measurement impairments and for adapting the recorded (sounder-specific) data to special radio parameters of the target application system.

The knowledge and understanding about the physics and the statistics of the propagation mechanisms are essential for standardization of radio interfaces and protocols. There are decisions and definitions that have a long lasting and far-reaching influence to the economy in many branches beyond just communication system design and manufacturing. The elements of the standards are largely defined and evaluated on the basis of channel sounding and the models derived from them. It is therefore of utmost importance that the quality of sounding devices is assessed and compared in an objective, metrological-driven manner. The greatest challenge compared to standard radio frequency (RF) lab equipment here is that propagation parameters in multiple coupled domains have to be jointly estimated in real time, e.g., directions, time delay, and Doppler, and with respect to their evolution over time. In particular, when it comes to applications in millimeter or sub-THz frequency bands, the question arises as to what makes the difference between channel sounding systems at these frequencies and the well-established MCS systems at lower frequencies. In the remainder of this chapter, we mainly give a short general introduction to MCS that serves for a better understanding of the more focused chapters, e.g., on architecture in Chap. 16 and multidimensional parameter estimation in Chap. 17 for MCS at millimeter and sub-THz frequencies.

8.2 Basic Sounder Architectures

Sounder architectures differ mainly in terms of baseband signal processing and with respect to the radio interface, which includes the up- and down-converters and the antennas. This chapter provides a general discussion on different design considerations, while specific architectures for the THz frequency range are discussed in the Chap. 16.

The baseband architecture is closely related to the waveform of the sounding signal. We need a wide band signal that has good autocorrelation properties. Such signals are well known from spread spectrum radar technology. They consist of a sequence of periodically repeated symbols with low peak-to-average power ratio (PAPR), hence constant envelope, which yields an energy-efficient use of the Tx power amplifier, i.e., close to saturation point (low-power back-off). The waveform has some inherent noise-like structure that leads to a very short peak-like autocorrelation function (ACF) when received by a matched correlation processor. This step is also known as *pulse compression*. As the correlation peak represents the total energy of one symbol of the excitation signal, we obtain a high correlation gain in signal-to-noise ratio (SNR), which is given by the signal's time-bandwidth product. Therefore, a single Tx-Rx propagation path is indicated in the delay spectrum (the magnitude squared channel impulse response (CIR)) by a sharp peak with maximum SNR. Depending on the shape of the power spectral density (PSD) there are side lobes to the ACF peaks, which may be considered disturbing for the detection of adjacent smaller propagation components.

The application of wide band signals stands in contrast to the use of vector network analyzers (VNAs) (Chap. 6) which rely on sinusoidal signals. While wide band signals allow fast, step-by-step identification of time-varying systems, a VNA would need to sequentially increase the frequency of the input signal step-by-step to cover the whole bandwidth. Each step requires a dwell time given by the settling time of the propagation channel plus the settling time of the VNA (given by the IF filter settling and LO frequency reconditioning). Since the estimation of time-variant broadband systems (for joint delay-Doppler resolution) would require a sequential application of this procedure, this would severely limit the use of VNAs for the identification of such systems.

8.2.1 Choice of the Basic Waveform Parameters

The longest delay relative to the line-of-sight (LOS) that we can observe is called the maximum excess delay (MED). This defines the minimum length of the excitation signal period T_e. At the same time, it is the shortest time interval to identify the system response. Therefore, the maximum achievable symbol repetition rate to follow the system's temporal variation on the "slow time axis" is $1/T_e$. At the same time, this sets the Nyquist bandwidth for Doppler shift estimation. The Doppler resolution is given by the number of subsequently observed symbols, respectively by the length of the coherently processed time interval on the slow time axis. The delayed multipath components, on the other hand, are resolved in the delay spectrum. The wider the excitation signal bandwidth, the better the resolution in delay. Delay is also called *Time of Flight* (ToF) or *fast time*. Therefore, for good resolution in the joint delay-Doppler plane, we need wide band signal symbols that are periodically repeated and coherently processed over some time interval. The physical limits for the identifiability of the time-variant system response are given by the so-called spreading factor (Delay spread × Doppler spread) which has to be smaller than 1. Fortunately enough, for typical mobile radio channels, it is as low as 1% … 10%. This obviously opens up some possibilities for sequential use of the hardware (e.g., time division multiple access (TDMA) or antenna multiplexing) or for coherent signal averaging.

8.2.2 Waveform Generation, Recording, and Correlation Processing

There exist several architectures to generate and receive wide band spread spectrum waveforms. Frequency Modulated Continuous Waveforms (FMCWs), for example, are very well known from automotive radar. The advantage is that relatively simple, low-power integrated analog transceiver chips are available and a very wide bandwidth can be achieved. These chips are very much aimed at the mass market and are not so suitable for general-purpose applications as they lack flexibility. Another approach is based on pseudo-random binary signals (PRBS). In this case, too, no DAC is required and we can again achieve a very wide bandwidth with low power consumption since only switching circuits are used. Waveform flexibility is limited to the choice of the feedback path in the shift register (the generator polynomial). PRBS have been extensively used for ultra wide band (UWB) sounding with GHz bandwidth below 10 GHz and with up-/down-conversion also for millimeter-wave and sub-THz frequencies [1–3]. Another potential advantage of PRBS is that they allow simple direct correlation receivers. However, this type of receiver architecture is also not so suitable for real-time sounding, as the correlation reference would have to be swept over delay to estimate all CIR samples. This would make the total measurement time similar to that of a VNA. The most promising and flexible sounder architecture consists of an Arbitrary Waveform (ARB) generator for excitation signal generation and a corresponding Nyquist sampling receiver. The ARB generator consists of a RAM containing a finite set of pre-calculated waveform samples that are periodically repeated via a DAC. Obviously, the ARB principle is most flexible in terms of the waveform to be generated. Waveform design can take place in frequency domain or in time domain. In frequency domain, we explicitly define the signal PSD which also defines the shape of the ACF. We can modify arbitrary spectral lines in phase and magnitude and even set spectral samples to zero if needed. The choice of the phases of the spectral lines is another degree of freedom that has some implicit influence on the excitation signal envelope before correlation, e.g., for minimizing the PAPR. It does not influence the shape of the ACF since this depends only on the PSD. The waveform can alternatively be defined in time domain, which has an implicit influence in frequency domain. However, in any case, the signal will be band limited by a mirror image rejection filter after the DAC. Nyquist frequency filtering is also required on the receiver side, which becomes more important if nonlinear distortions and out-of-band interference occur. The periodically repeated measurement allows identification of slowly time-varying systems (in case of moving Tx, Rx, or interacting objects). Moreover, it allows perfectly matched FFT processing as we know it from orthogonal frequency-division multiplexing (OFDM). However, here we do not need a cyclic prefix, as we always repeat the signal anyway. Coherent FFT processing of a sequence of received periods along slow time allows joint delay-Doppler estimation. The periodicity of the excitation signal has also been used to effectively reduce the sampling rate of the received signal by periodic subsampling, resulting in a longer measurement time or respectively reduced SNR [1–3].

8.2.3 RF Chain at Tx and Rx

Wide band I/Q up- and down-converter stages are required to shift the baseband signal to the target frequency band. If direct analog down conversion is used, the resulting DC component in the baseband deserves special attention because of DC offset drift or nonlinear distortion. We may need multiple Tx and Rx channels if we aim at quad polarimetric access (2 × 2, dual polarimetric at Tx and Rx). Moreover, if we like to access antenna arrays at Tx and Rx, we might ask for multiple channels on both sides. Therefore, effort is often reduced by using only single channels and switched access. This is the natural choice for the transmitter unit, as TDMA access is applied anyway to uniquely identify each individual transmit antenna contribution for full MIMO channel matrix measurement. On the receiver side, we can choose between expensive parallel and slower sequential access. The latter sets a lower limit for the observable coherence time of the channel. The transmit output power amplifier is of highest importance for sounding coverage, especially for non-LOS areas, whereas automatic gain control (AGC) can protect the receiver from being overloaded if Rx is close to Tx. Special attention is necessary for Tx and Rx LO and clock synchronization. While a VNA uses a common reference at Tx and Rx, this is normally not applicable for channel sounding since Tx and Rx are often far apart and moving. In this case, GPS-disciplined quartz or rubidium oscillators are often used. If Tx/Rx mobility is limited, fiber-optic LO reference distribution can be employed. Reference frequencies for mmW and sub-THz sounding should not be too low since large/frequency multiplication factors can cause synchronization problems. Phase noise and drift evaluation should consider the respective coherent processing interval. Therefore, any timescale expansion by sequential acquisition would increase the clock stability requirements.

8.2.4 Sounder Antenna Design

Antennas are the interface of the measurement device to the object under investigation. On the one hand, they are necessary to generate and receive waves. On the other hand, they have an unavoidable influence to the measurement result. In fact, antenna parameters are a big influence to the achievable sounding coverage because of their gain. At the same time, they act as a filter in the directional domain and are more or less polarization sensitive while polarization orientation and radiation pattern are coupled. Since the various propagation paths are defined by their multidimensional geometric parameters, the antenna radiation pattern also affects the delay and Doppler spread parameters as well as the corresponding fading. Moreover, antennas have their own impulse response (i.e., a corresponding complex frequency response) which is direction dependent. The antenna impulse response depends on the antenna design and must be taken into account, especially for very broadband operation and accurate propagation estimation because it convolves with the CIR.

Due to the specific position of the antennas between the device under test (the propagation channel) and the measurement device (the sounder), we can consider the antennas alternatively as a part of the channel or of the device. This influences the significance and generalization of the measured results. If we consider the antennas as a part of the channel, we can interpret the recorded data only within the configuration that was defined by the sounding measurement campaign. We cannot use it to represent a setup with other antenna arrangements or radiation patterns. The other option requires the knowledge of the propagation parameters "as they are" independent of the antennas used in the sounding procedure. This would allow us in a second step to incorporate any arbitrary antenna radiation patterns of some hypothetical application system to the measured data. Therefore, we look for a procedure to de-embed the influence of the antennas used throughout sounding. Since both antennas at the Tx as well as at the Rx side have influence, we can meet this requirement only if we estimate the propagation directions jointly at both sides of the link together with the other path parameters. We call this *double directional measurement* [4–6].

The standard procedures for estimating Direction of Arrival (DoA) or Direction of Departure (DoD) rely on mutual correlation of responses between spatially separated antennas of the antenna array at Rx or Tx, respectively. Eventually, this requires recording the wide band MIMO transmission matrix between all Tx and all Rx antennas, which is mostly measured in full sequential or hybrid mode (sequential at Tx and parallel at Rx side). The architecture of the respective antenna arrays determines the angular coverage and resolution performance. Full azimuth or solid angular coverage would require corresponding circular structures. Planar arrays are in general not enough for antenna de-embedding, especially at the UE side. The effective physical size of the array somehow relates to angular resolution. The distance between the antenna elements is related to the phase ambiguity and therefore often required to be less than half a wavelength. However, there are deviations. If the radiation pattern of the antennas becomes smaller, angular ambiguity reduces and the distance can become bigger. Moreover, if the bandwidth of the signal becomes wide enough, delay resolution can be used to mitigate the phase ambiguity. This can become of interest mostly for circular structures. In any case, the optimum design of the antenna geometry may be found only by considering the estimation procedure.

An alternative procedure to correlation-based estimation of the directional multipath model is *beam scanning*. This can be achieved in different ways. One is switching between highly directive antennas in a fixed array. This may be a favorable solution at millimeter or sub-THz frequencies since directive antennas tend to become small and easier to build. Another advantage is that antenna outputs are already characterized by some gain, which would help to identify weak paths. Alternatively, equivalent beamforming gain and angular filtering would be achieved from spatially sampled wave fields only after some processing. Another option is phased array beamforming. This seems obvious for the high-frequency range considered here, as analog (or hybrid) beamforming will be used anyway. This would make the corresponding chips available. However, it may be questionable if the integrated solutions are suitable for sounding. This relates to the available antenna geometry, which will be mostly restricted to planar. Moreover, the excess bandwidth, which is often desired in sounding for high path delay resolution can be a problem. Finally, yet importantly, a beamformer typically has only one output, and any one-channel radio chain with directive antennas would require sequential sweeping operation to achieve wide angular coverage. Spatial scanning by mechanical movement (linear or rotational) followed by some kind of synthetic aperture processing is another option. However, this would require spatial coherent processing for any DoA and DoD estimation.

8.3 Multidimensional Propagation and Device Data Model for Parameter Estimation

At the first glance, the structural ray-optical data model of any propagation path can be described by a superposition of multiple multidimensional Dirac deltas:

$$
\mathbf{h}_\delta(\boldsymbol{\theta}_k) = \begin{bmatrix} \gamma_{\varphi\varphi,k} & \gamma_{\vartheta\varphi,k} \\ \gamma_{\varphi\vartheta,k} & \gamma_{\vartheta\vartheta,k} \end{bmatrix} \cdot \delta(\alpha - \alpha_k) \cdot \delta(\tau - \tau_k)
$$
$$
\cdot \, \delta(\varphi_{Rx} - \varphi_{Rx,k}) \cdot \delta(\vartheta_{Rx} - \vartheta_{Rx,k}) \cdot \delta(\varphi_{Tx} - \varphi_{Tx,k}) \cdot \delta(\vartheta_{Tx} - \vartheta_{Tx,k}) \, .
$$
$$
(8.1)
$$

This *Dirac domain* corresponds to another domain, which we call the *aperture domain* that is given by a superposition of multidimensional exponential functions [7]:

$$
\mathbf{H}_e(\boldsymbol{\theta}_k) = \begin{bmatrix} \gamma_{\varphi\varphi,k} & \gamma_{\vartheta\varphi,k} \\ \gamma_{\varphi\vartheta,k} & \gamma_{\vartheta\vartheta,k} \end{bmatrix} \cdot e^{-J2\pi t\alpha_k} \cdot e^{-J2\pi f\tau_k}
$$
$$
\cdot \, e^{-J2\pi s_{Rx}\varphi_{Rx,k}} \cdot e^{-J2\pi s_{Rx}\vartheta_{Rx,k}} \cdot e^{-J2\pi l_{Tx}\varphi_{Tx,k}} \cdot e^{-J2\pi l_{Tx}\vartheta_{Tx,k}} \qquad (8.2)
$$

The path parameters in the *Dirac* domain are defined by DoD (φ_{Tx}, ϑ_{Tx}), DoA (φ_{Rx}, ϑ_{Rx}), ToF (τ), and Doppler shift (α), which have an obvious geometric meaning (compare Fig. 8.1). As the directions (DoD and DoA) are each made up of azimuth and co-elevation, there are six dimensions for each path. The corresponding aperture domain relates by a 6D Fourier transformation, whereby the spatial field distribution at the Tx and Rx antennas calculated from the directional domain involves a further geometric transformation similar to the near-/far-field transformation, which depends on the structure of the antenna array and is not given in the equation. Sometimes subsets can also be taken into account, e.g., DoD and/or DoA only in azimuth or not in the full 360° circular angle range, etc. Note that the DoD and DoA are defined in separate local (antenna-centric) polar coordinate systems that do not reveal any information about further interaction on the path between Tx and Rx. Practically, the view from the Tx and Rx is described separately. However, correct joint estimation of the parameters makes sure that the pairing of the parameters at both sides together with ToF and Doppler is correct. A similar thing applies also to polarization orientation that is defined perpendicular to the

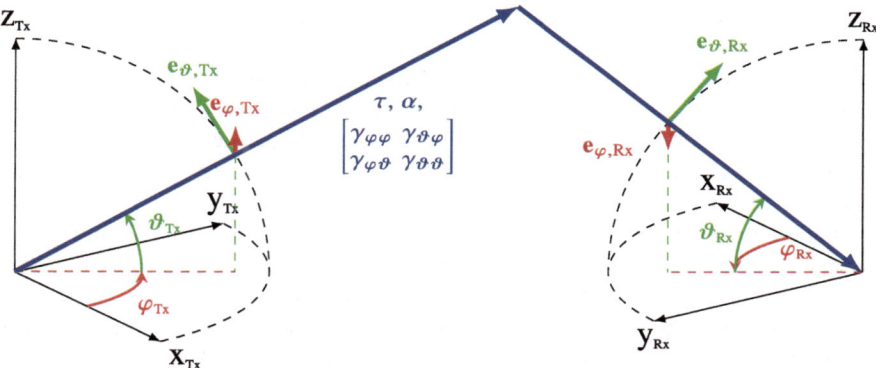

Fig. 8.1 Geometrical interpretation of the path parameters

Poynting vector of wave propagation (annotated as longitude or latitude, azimuth or co-elevation, respectively) in the corresponding local polar coordinate system. Consequently, every propagation path is characterized by a complex 2×2 path weight matrix, which relates to the two orthogonal polarization orientations in terms of the local Tx and Rx coordinates, respectively, as a function of up to 6 parameters in the corresponding joint domain. Of course, also partial Fourier transforms are possible resulting in a mixed multidimensional Dirac/aperture domain.

The estimation of this simple propagation data model can be considered as a multidimensional harmonic retrieval problem. The size of the multidimensional aperture cube limits the resolution in the Dirac domain in a predictable way if a plain Fourier transform is applied. The finite aperture size in frequency domain is given by the sounder bandwidth, in slow time it is the coherent recording time, and in the angular domain it is the size of the effective spatial observation volume related to the antenna arrays. The multidimensional Dirac delta would be smeared according to the width and weighting of the aperture distribution in the corresponding dimensions. While resolution can be significantly increased by applying the high-resolution parameter estimation (HRPE) property of the model-based estimation procedure, also the multidimensionality helps. For instance, if we have two paths with similar ToF (hence not resolved in delay) but different, e.g., opposite Doppler shifts (reflected from objects moving in opposite directions), they become very well separable in the joint delay Doppler domain. More details on modeling and estimation are given in Chap. 17.

Besides the propagation part, the model used for parameter estimation also contains the device data model. This relates to the (predominantly) linear transfer characteristics of the MCS that deviate from the ideal response. The knowledge of these characteristics gained during a calibration measurement procedure can be largely taken into account and its influence removed from the estimation results. For example, if we know that the spectral shape of the excitation signal is rectangular (or is measured to be used as a correlation reference), the sinc-shaped ACF with its side lobes need no longer be considered disadvantageous since its influence to the estimated CIR can be completely removed by the successive inference cancellation property of the estimator, assuming the SNR of the measured data is high enough. Considering and mitigating hardware impairments may be even more important for DoD/DoA estimation, as antennas are usually very imperfect devices because of manufacturing and material tolerances, mutual coupling, etc. Also, Tx vs. Rx phase drift can be mitigated to some extent.

Each model can only help as much as it is accurate. However, the multidimensional model in (8.1) (see also [7]) already contains some major simplification. Just to name a few: (i) DoA/DoD and Doppler estimation refer to narrowband signals (single sinusoids, phase only). On the other hand, time delay estimation requires wide band signals. However, if the relative bandwidth is small, we can meet both assumptions with one estimator. Otherwise, we should extend the model, e.g., introducing a linear frequency response in the array manifold [8]. Note that frequency dependence of the spatial slope is related to the size of the array and not only the difference between two antennas. Similarly, Doppler shift is frequency dependent for constant relative speed. (ii) In order to estimate the Doppler shift we need to record the time-varying channel response over some slow time to see the carrier phase variation. However, we would also observe a changing delay of the ACF envelope. Therefore, coherent integration time for one ToF resolution cell would be limited by delay-Doppler migration. This again becomes more severe for wide relative bandwidth. (iii) Wavefronts are assumed to be planar. This is of highest importance for the response of antenna arrays as it is assumed that there is no wavefront curvature that may cause phase distortion over the extent of the array. (iv) The interacting object does not introduce delay or Doppler spread. So it is assumed to be point-like or locally planar.

Especially if the narrowband assumption does not hold, the data model does not match the Khatri–Rao product in the finite discrete domain, which is a column-wise Kronecker product. In this case, the influence of different parameters (angle and delay) is no longer independent and estimation becomes more involved.

The propagation data model discussed up to now is composed of a finite number of discrete paths, which we also call *specular*. Normally, the dominant propagation paths can be assigned very well to these modeling assumptions. However, every sounding device can resolve only a certain number of specular paths depending on the amount of reliable information it can collect. Experimental propagation studies have revealed that there is a considerable amount of energy transferred between Tx and Rx by more diffuse, non-specular mechanisms [9, 10]. We call this part also dense (i.e., non-resolved) paths. The dense components are included in the propagation data model by properly shaped continuous distribution functions that are described by some (small) set of parameters [11, Ch. 2.5]. For instance, multipath interactions in a dense arrangement of interacting objects or in a confined space may be well described by exponentially decaying response functions in fast time.

References

1. R. Zetik, M. Kmec, J. Sachs, R.S. Thomä, Real-time MIMO channel sounder for emulation of distributed ultrawideband systems. Int. J. Antennas Propag. **2014**, 1–16 (2014) [Online]. Available: https://api.semanticscholar.org/CorpusID:67846887
2. A.P.G. Ariza, W. Kotterman, R. Zetik, M. Kmec, U. Trautwein, R. Muller, F. Wollenschlager, R.S. Thoma, 60 GHz-ultrawideband real-time multi-antenna channel sounding for multi giga-bit/s access, in *2010 IEEE 72nd Vehicular Technology Conference - Fall*, pp. 1–6 (2010) [Online]. Available: https://api.semanticscholar.org/CorpusID:2094380
3. R. Zetik, J. Sachs, R.S. Thoma, UWB short-range radar sensing - the architecture of a baseband, pseudo-noise UWB radar sensor. IEEE Instrum. Meas. Mag. **10**(2), 39–45 (2007)
4. R.S. Thomä, D. Hampicke, A. Richter, G. Sommerkorn, U. Trautwein, MIMO vector channel sounder measurement for smart antenna system evaluation. Eur. Trans. Telecommun. **12**(5), 427–438 (2001) [Online]. Available: https://onlinelibrary.wiley.com/doi/abs/10.1002/ett.4460120508
5. M. Steinbauer, A. Molisch, E. Bonek, The double-directional radio channel. IEEE Antennas Propag. Mag. **43**(4), 51–63 (2001)
6. V.-M. Kolmonen, P. Almers, J. Salmi, J. Koivunen, K. Haneda, A. Richter, F. Tufvesson, A.F. Molisch, P. Vainikainen, A dynamic dual-link wideband MIMO channel sounder for 5.3 GHz. IEEE Trans. Instrum. Measure. **59**(4), 873–883 (2010)
7. M. Landmann, M. Kaske, R.S. Thoma, Impact of incomplete and inaccurate data models on high resolution parameter estimation in multidimensional channel sounding. IEEE Trans. Antennas Propag. **60**(2), 557–573 (2012)
8. S. Semper, M. Döbereiner, C. Steinmetz, M. Landmann, R.S. Thomä, High-resolution parameter estimation for wideband radio channel sounding. IEEE Trans. Antennas Propag. **71**(8), 6728–6743 (2023)
9. A. Richter, J. Salmi, V. Koivunen, Distributed scattering in radio channels and its contribution to MIMO channel capacity, in *2006 First European Conference on Antennas and Propagation* (2006), pp. 1–7
10. J. Poutanen, J. Salmi, K. Haneda, V.-M. Kolmonen, F. Tufvesson, P. Vainikainen, Propagation characteristics of dense multipath components. IEEE Antennas Wirel. Propag. Lett. **9**, 791–794 (2010)
11. A. Richter, Estimation of Radio Channel Parameters, Ph.D. dissertation, Technische Universität Ilmenau (2005)

Chapter 9
Photonics-Assisted Signal Processing

Thomas Schneider

Abstract If we assume polarization multiplexing or a multiple-input multiple-output (MIMO) system with two antennas, 28% forward-error correction and a Nyquist shaping of the data signal (the maximum possible symbol rate that can be transmitted in a given bandwidth), the bandwidth of a 1 Tbit/s signal in a 16-quadrature amplitude modulation (16-QAM) format would be 80 GHz. The processing of such high bandwidths is far beyond the possibilities of todays' electronics. The requirements may be relaxed by a higher parallelization (more MIMO channels) or a higher spectral efficiency of the modulation (a 1024-QAM only needs 10 GHz). However, these solutions are accompanied by higher hardware costs and especially an increasing power consumption. Photonics may be used to down-convert the signal in the time or frequency domain. Subsequently, the down-converted signals may be processed with low-bandwidth electronics. The basic ideas behind this approach will be discussed in this chapter.

9.1 Problems of Electronic Signal Processing

The requirements for global Internet information transmission capacity are increasing at a rate of 40% per year. So, in 20 years from now, the needed capacity will be 1000 times higher than today [14]. For the next generation of wireless communications (6G and beyond) peak data rates of 1 Tbit/s are foreseen [15] for instance. Furthermore, the data rates in the worldwide communication networks and their wireless access will further increase due to new applications like the Internet of Things, autonomous driving, and many other things that might emerge in the near future. Telecommunication networks are mainly based on an optical transport network for the connection of cities, countries, and even continents and a wireless access to these networks. Therefore, new ideas are required for the optical transport

T. Schneider (✉)
Technische Universität Braunschweig, THz-Photonics Group, Braunschweig, Germany
e-mail: thomas.schneider@tu-braunschweig.de

© The Author(s) 2026
T. Kürner et al. (eds.), *Metrology for THz Communications*, Springer Series in Optical Sciences 256, https://doi.org/10.1007/978-3-032-01986-8_9

as well as for the wireless access systems. One promising solution for the wireless access might be to overcome the problems of the limited transmission bandwidth by the exploitation of new spectral regions, like the THz domain. However, a major problem for a further increase in data rates is the limited bandwidth of electronics and another important problem is the jitter and the accompanied reduced resolution and increased noise of the processed signals.

9.1.1 Jitter

Figure 9.1 shows the basic block diagram of an analog-to-digital converter (ADC).

The analogue input signal is first sampled in the time domain with a sampling rate R_s. If this sampling rate is at least twice the highest bandwidth in the signal, the sampling points perfectly represent the signal itself. However, the exact time for taking the sampling point is uncertain. This uncertainty comes from basic physical limitations and from deficiencies of the electronic devices. Figure 9.1d shows the amplitude uncertainty ΔA for a signal with a low-bandwidth (red) and for a high-bandwidth signal (black). As depicted, for the same level of time uncertainty, or jitter Δt, the black signal is sampled with a much higher amplitude uncertainty ΔA than the low-bandwidth red one. Photonics offers several very low jitter solutions. Frequency combs generated by stabilized mode locked lasers (MLLs) can show

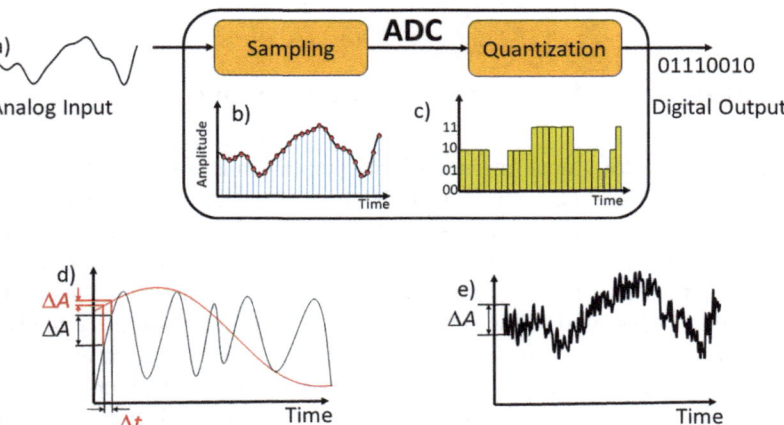

Fig. 9.1 Block diagram of an analogue-to-digital converter (ADC). In the first step the analog input signal (**a**) is sampled with an equidistant time spacing (**b**). The inverse of this time spacing is the sampling rate R_s. According to the Shannon-Nyquist theorem, R_s has at least to be twice the highest bandwidth in the signal. After this quantization in the time domain, a quantization of the amplitude follows in the second step (**c**). The predefined limited amplitude levels are assigned to a digital word, which is set to the output of the device. The reduced amplitude resolution of the sampling due to jitter is illustrated in (**d**) and the problem with noise in (**e**)

jitter values in the zeptosecond range (1 zs $= 10^{-21}$s) [1], for instance. According to the analysis by Walden [2], however, applying the Heisenberg uncertainty principle to the ADC performance gives a minimum theoretically possible jitter value of around 100 attoseconds (1as $= 10^{-18}$s). But, current state-of-the-art electronic ADCs have jitter values in the range of 100 fs (1 fs $= 10^{-15}$s), which still results in a difference of 3 orders of magnitude. The current state of the art for electronic ADC as reported in the VLSI and ISSCC conferences of 2022 to 2024 [3] is shown in Fig. 9.2.

9.1.2 Resolution

The resolution of an ADC is typically given by its effective number of bit (ENOB) value. The effective number of bit, or ENOB of an ADC, is the number of amplitude levels between the lowest and highest measurable value, which can still be distinguished from each other. This distinguishing depends on the jitter and noise level of the device and measurement, as reported in Fig. 9.1d for the jitter and Fig. 9.1e for the noise. The ENOB can be seen as the number of amplitude levels, which really make sense. This ENOB is very closely connected to the signal-

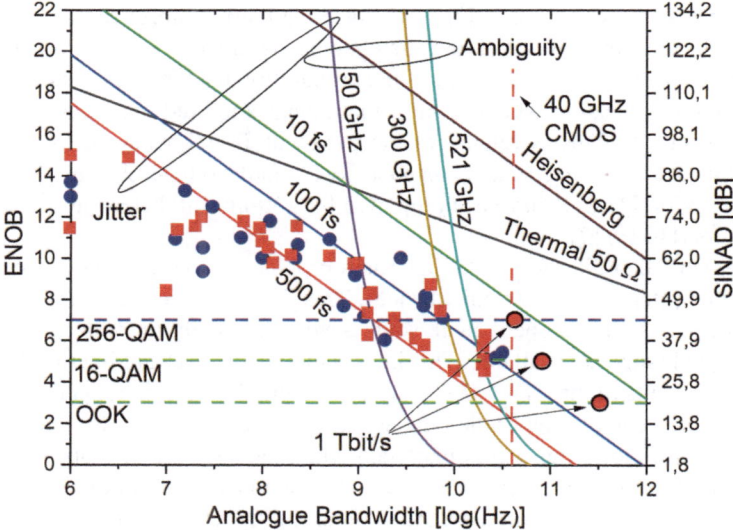

Fig. 9.2 ENOB (left) and SINAD (right) vs. analogue bandwidth of the input signal. The different limits for integrated electronics devices as jitter, thermal noise, and cutoff frequency (ambiguity) [2] are given together with recently reported ADC results from the VLSI (blue circles) and ISSCC (red squares) conferences from 2022 to 2024 [3]. The red circles show the required resolution (ENOB) and bandwidth for the reception of a 1 Tbit/s signal with different modulation formats [4]

to-noise-and-distortion (SINAD) of the measurement. As the name implies, the SINAD additionally includes possible distortions of the measurement to the signal-to-noise level. The ENOB and therefore the SINAD are as well directly connected to the maximum spectral efficiency of the modulation format which can be used for the encoding of the signal. The approximate ENOB requirements for ADCs with a 1-dB receiver sensitivity penalty and a maximum bit error rate of 10^{-3} (which corresponds to that of forward error corrected systems) is 3 bits more than the $\sqrt{M}$ amplitude levels of the constellation diagram [5, 6]. Therefore, to correctly receive a signal modulated in an M-QAM modulation format $log_2 \sqrt{M} + 3$ bit are required [5]. A 16-QAM requires 4-bit resolution (corresponding to $2^4 = 16$ amplitude levels), whereas for a 256-QAM 7 bits (128) are required.

The red circles in Fig. 9.2 give the required ENOB values for the reception of a polarization multiplexed 1 Tbit/s signal, limited to its Nyquist bandwidth, with 28% forward error correction. As can be seen, the actual electronic ADCs are far away from that goal. As shown in Fig. 9.2, the main limitation is the jitter of the electronics.

Even without any nonidealities an ADC has so called quantization noise. The amplitude of the actual sampling point can have any value between two adjacent quantization levels, but only one of them will be selected for quantization. Therefore, this noise directly comes from the quantization process and increases with decreasing resolution (or increasing bandwidth).

Another source is thermal noise. The thermal noise of a nonideal converter is given for a 50 Ω resistance in Fig. 9.1. Since the thermal noise can be assumed as white, it linearly increases with the analogue bandwidth of the input signal.

The comparator ambiguity reflects the probability that the comparator will make an ambiguous decision whether the actual signal amplitude is below or above a certain level [2]. This is related to the peak cutoff frequency of the integrated transistors, or to their speed to respond to a small voltage change. Peak cutoff frequencies of 521 GHz have been reported for 130-nm InP hetero-junction bipolar transistors (HBTs) [7].

However, according to Fig. 9.2 the main problem of toady's ADCs seems to be the normal jitter of the electronic devices. This jitter does not come from the clock of the device. As mentioned, clock sources with extremely low jitter values in the zeptosecond range are available. This jitter is mainly the aperture jitter of the sample and hold circuits in the ADC. The sample and hold device can be seen as a kind of switch, which samples and holds the signal for a certain time. However, even for a very precise clock signal it is uncertain when the switch really opens or closes. Since all electronic ADCs are based on such a sample and hold functionality, it is a limiting factor for the maximum bandwidth of the signals that can be processed.

9.2 Photonics Signal Processing

Photonics has many unique advantages, it is immune against electromagnetic interference with radio waves, optical signals can be transmitted over tenths (and with optical amplification even thousands) of kilometers in optical fibers, and the typical bandwidths of photonic signal processing devices are in the range of several tenths of GHz, much more than achievable with electronics. In the past, photonics signal processing was based on nonlinear effects in optical fibers, for instance [8]. To be effective, these optical fibers have to be several kilometers long, which has made these solutions rather impractical. However, in the last few years integrated photonics signal processing in very low-footprint CMOS technology compatible chips has been shown. This makes photonics solutions very attractive even for wireless communications and especially THz systems.

How photonics can assist in the processing of very-high-bandwidth signals is shown in Fig. 9.3. The basic idea is to down-convert the high-bandwidth optical, wireless, or whatever signal into parallel low-bandwidth ones. These low-bandwidth signals can then be further processed or directly detected with low-bandwidth electronics. If the first down-conversion does not deteriorate the signal, the reduced bandwidth leads to an increase in SINAD and ENOB for the following electronic signal processing [9]. These reduced bandwidth requirements may allow a lower power consumption of the electronics.

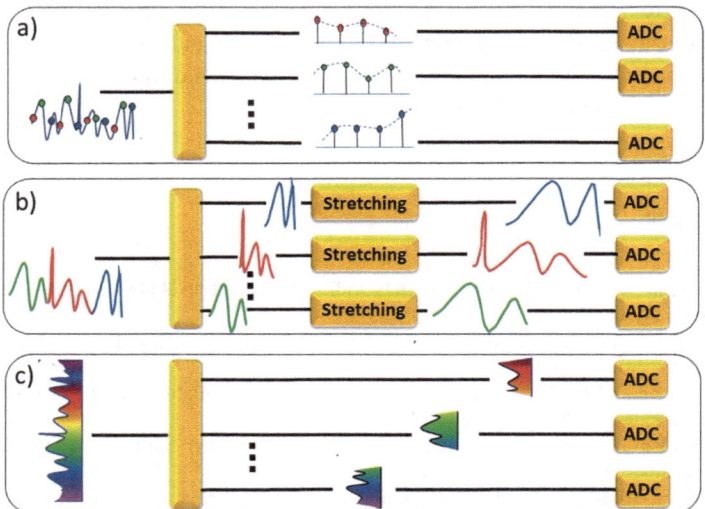

Fig. 9.3 Possibilities to measure and receive very broadband signals by a photonic down-conversion of high-bandwidth signals into parallel low-bandwidth sub-signals in the time (**a**) and (**b**) and frequency domain (**c**), after Ref. [4]

In Fig. 9.3a the basic idea is to divide all sampling points into N parallel branches. Therefore, in each branch only $1/N$th of the sampling points is available, which can be detected with electronics with $1/N$th of the sampling rate and bandwidth. For this approach photonic mode locked lasers (MLLs) can be used for the down-conversion [10]. MLL may provide very short pulses in the femtosecond range and extremely low jitter. However, low-jitter MLLs typically have quite long resonator lengths and this resonator length defines their repetition rate, which is typically in the MHz range. Thus, the sampling of real-time GHz signals would require hundreds of branches, making this solution quite impractical. Only recently integrated MLL with GHz repetition rates have been shown. However, it has yet to be shown that these MLL can provide low jitter values too.

Figure 9.3b presents another solution in which the high-bandwidth signal is divided into parallel time windows. By photonic means these time windows are then stretched in time and again processed with low-bandwidth electronics [11, 12]. Usually this time stretching is carried out with a first-order dispersion. Unfortunately, most of the dispersive elements have higher-order dispersion too, which results in a distortion of the signal.

Whereas the aforementioned methods work in the time domain, Fig. 9.3c shows a frequency domain solution [13]. By optical filters the high-bandwidth signal is divided into parallel low-bandwidth slices, which are then processed with low-bandwidth electronics. This method requires a very sophisticated post-processing of the received signal, since the filter curves of practically available optical filters are not rectangular. Additionally, the phase relationship between adjacent slices has to be correctly determined for a successful detection of the whole bandwidth signal.

In Chap. 22 we will elaborate more on that subject and we will describe a very promising method for the time-domain down-conversion of high-bandwidth signals by orthogonal sampling in detail.

References

1. X. Xie, et al., Photonic microwave signals with zeptosecond-level absolute timing noise. Nat. Photonics **11**, 44–47 (2017)
2. R.H. Walden, Analog-to-digital converter survey and analysis. IEEE J. Sel. Areas Commun. **17**(4), 539–550 (1999)
3. B. Murmann, ADC Performance Survey 1997–2022. Available: http://web.stanford.edu/~murmann/adcsurvey.html. Accessed on 15 Dec 2022
4. T. Schneider, Toward terabit receivers for optical and wireless communications. IEEE Commun. Mag. **61**, 169–174 (2023)
5. P.J. Winzer, High-spectral-efficiency optical modulation formats. J. Lightwave Technol. **30**, 3824–3835 (2012)
6. T. Pfau, et al., Hardware-efficient coherent digital receiver concept with feedforward carrier recovery for M-QAM constellations. J. Lightwave Technol. **27**(8), 989–999 (2009)
7. M. Urtega, Z. Griffith, M. Seo, J. Hacker, M.J.W. Rodwell, InP HBT technologies for THz integrated circuits. Proc. IEEE **105**, 1051–1067 (2017)

8. T. Schneider, *Nonlinear Optics in Telecommunications*. Advanced Texts in Physics (Springer-Verlag, Berlin/Heidelberg/New York, 2004). Reprint for the Peoples Republic of China, Science Press Beijing (2007)

9. Y. Mandalawi, J. Meier, K. Singh, M.I. Hosni, S. De, T. Schneider, Analysis of bandwidth reduction and resolution improvement for photonics-assisted ADC. J. Lightwave Technol. **41**(19), 6225–6234 (2023)

10. A. Khilo, et al., Photonic ADC: overcoming the bottleneck of electronic jitter. Opt. Exp. **20**(4), 4454 (2012)

11. Zhou, et al., A unified framework for photonic time-stretch systems. Laser Photonics Rev. **16**(2100524) (2022). https://doi.org/10.1002/lpor.202100524

12. A. Misra, et al., Nonlinearity- and dispersion-less integrated optical time magnifier based on a high-Q SiN microring resonator. Sci. Rep. **9**, 14277 (2019)

13. N. Fontaine, et al., Real-time full-field arbitrary optical waveform measurement. Nat. Photonics **4**, 248–254 (2010)

14. M. Nakazawa, M. Suzuki, Y. Awaji, T. Morioka, *Space-Division Multiplexing in Optical Communication Systems: Extremely Advanced Optical Transmission with 3M Technologies*, vol. 236 (Springer Nature, Berlin, 2022)

15. I.F. Akyildiz, A. Kak, S. Nie, 6G and beyond: the future of wireless communications systems. IEEE Access **8**, 133995–134030 (2020)

Chapter 10
Ray Tracing for Metrology

Christoph Herold, Diego Dupleich, Giovanni Del Galdo, and Thomas Kürner

Abstract Ray-tracing tools allow researchers to perform a deterministic analysis of the radio propagation in a given scenario. It can support channel modelling and propagation analysis by extending measurement capabilities and enhancing measurement results. Ray-tracing simulation requires a trade-off between accuracy and run-time of the simulation. The selection of the scenario, identification of features of interests, and the quality of input and environment data have a large impact on the reliability of the ray-tracing results. Compared to measurement campaigns, ray-tracing simulations offer a easy repeatability, higher flexibility, and less need for human involvement. This chapter gives an overview of fundamentals of ray tracing, the required input data, and possible applications in radio propagation predictions.

10.1 Introduction to Ray Tracing

Ray-optical methods, more commonly referred to as ray tracing, are techniques used to analyze ray propagation within a 3D environment. These techniques are commonly used in computer graphics to analyze the way light travels through an environment. Using several approximations based on the Maxwell equations, these

C. Herold · T. Kürner (✉)
Technische Universität Braunschweig, Institut für Nachrichtentechnik, Braunschweig, Germany
e-mail: c.herold@tu-braunschweig.de; t.kuerner@tu-braunschweig.de

D. Dupleich
Technische Universität Ilmenau, Institute of Information Technology, Ilmenau, Germany
e-mail: diego.dupleich@tu-ilmenau.de

G. D. Galdo
Technische Universität Ilmenau, Institute of Information Technology, Ilmenau, Germany

Fraunhofer Institute for Integrated Circuits (IIS), Ilmenau, Germany
e-mail: giovanni.delgaldo@tu-ilmenau.de

© The Author(s) 2026
T. Kürner et al. (eds.), *Metrology for THz Communications*, Springer Series in Optical Sciences 256, https://doi.org/10.1007/978-3-032-01986-8_10

approaches can also be applied to radio propagation modelling as shown by Yun and Iskander [1].

Ray-tracing simulations are a valuable tool for research in metrology: Measurement campaigns can be labor- and time-intensive. A ray-tracing tool might be able to cut down on both the amount of measurement data and the number of measurement setups needed. Although the simulation run-time can be significant, once the input data is prepared, different scenarios can be simulated with minimal human effort. The measurement capabilities can further be extended as simulations of possibly difficult or dangerous scenarios might enable the analysis of such scenarios. Relying on simulations can protect the sensitive hardware, e.g., because assembly and disassembly can be minimized and wear on components can be avoided.

Usually, ray-optical simulations separate the computation of propagation paths and the calculation of EM properties. Therefore, they can offer both, detailed information about the propagation paths, such as angles and path length, as well as the power transmitted via specific propagation effects (e.g., reflected or diffracted rays). The geometric information can help in identifying interaction points, e.g., possible sources of interference or disturbances. Information about EM (electromagnetic) properties can help to isolate single effects and their influence on the communication environment. As software offers repeatability, hypothesis and updated propagation model can be verified with less effort than a rearranged measurement campaign would entail. In summary, a well-calibrated ray-tracing tool can extend the investigation capabilities and enhance the measurement results by providing further insights for channel modelling and propagation analysis.

There is a variety of commercially tools, such as REMCOM Wireless InSite® [2] or Matlab® [3] as well as open-source software frameworks, such as Sionna [4] available. The examples given in this chapter largely stem from the Simulator for Mobile Networks (SiMoNe) [5] developed at the Institute of Communications Technology at Technische Universtät Braunschweig (TUBS) and Sionna. The necessary fundamentals for a ray-tracing tool to supply these benefits and its applications in the Meteracom project are briefly discussed in the remaining part of this chapter.

10.2   Basic Concepts of Ray-Tracing Analysis

Generally, two fundamental approaches to ray-optical analysis exist [6]: Ray launchers work according to the shout and bounce concept, launching rays in every direction with a high angular resolution. The rays and their interactions with the environment will be computed until certain abortion criteria are met. These usually include a predefined power level and a number of interactions or bounces. A ray tracer, on the other hand, analyzes the environment trying to identify interaction surfaces, blockage, and visibility relations between transmitters and receivers, e.g., by using the image method. Ray launching is more efficient for point-to-area predictions, whereas ray tracing is more efficient for point-to-point predictions. Hybrid models try to combine both approaches and find a balance between performance and accuracy.

Ray-optical methods usually work in a two-step process. The first step is to analyze the environment and find possible paths; in a second, following step, propagation effects are applied to the paths identified in the first step. This split allows for later adjustments of parameter settings and propagation models which is helpful for calibrating, modelling, and understanding effects during the analysis.

For the execution of a ray-optical analysis, two preliminary aspects are to be defined: the simulation layout and the simulation setup. For the simulation layout, basic input information are collected and prepared: The environment data and 3D models, material parameters, and transmitter and receiver positions are selected. In the simulation setup, the user needs to parameterize the simulation itself: What kind of ray-optical approach is to be used? Which propagation effects are to be considered? What resolution is needed for the prediction?

10.2.1 Propagation Effects and Types of Rays

Along the path between transmitter and receiver, the rays interact with the environment. In the most simple case, there is a line-of-sight connection and no interacting object. In this case, the direct ray's power is calculated using Friis law. Other commonly considered propagation effects for ray-optical analysis are transmission, reflections, diffraction, and scattering.

Rays that hit flat and infinitesimally large surfaces are specularly reflected off. The magnitude of the reflected field can be determined by Fresnel's equations. Reflections on a rough surface lead to scattering. Flat and rough are usually defined based on the wavelength of the respective frequency. At higher frequencies, more surfaces will be considered rough due to the smaller wavelength; see, e.g., [7].

Topological features, roofs, or edges of objects can be a source of diffraction, meaning that a ray that passes by such an edge might experience a change in direction. This effect lessens the shadowing effect of objects. Different types of rays in a 3D scenario are highlighted in Fig. 10.1.

10.2.2 Acceleration Methods

As previously mentioned, using ray-optical methods always presents a trade-off between accuracy and run-time. Algorithmic adjustments can make computations suitable for modern hardware. With its multicore processors and graphic processing units, it can highly benefit ray tracing by offering potential for a high degree of parallelization. Tasks such as line-of-sight checks, a selection of image sources, or the computation of EM properties can be efficiently done in parallel for scenarios of appropriate sizes [1]. Wherever possible, precomputation of environmental aspects

Fig. 10.1 3D view of a
meeting room scenario in
SiMoNe's ray tracer with
color-coded types of rays:
direct (red) and reflected
(blue)

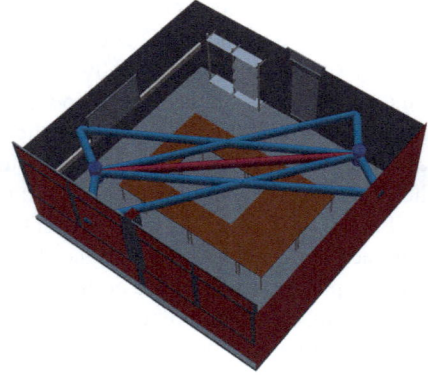

such as visibility relations can save valuable run-time later on as static transmitters, receivers, and objects do not change their position. Established relations that just have to be checked for a specific time step speed up the process. Saving the data and precomputed results in efficient data structures, such as search trees, can further cut down the run-time by allowing for faster access [8]. Apart from that, a careful selection of the necessary level of abstraction is advisable. Depending on the distance between transmitter and receiver, the effects that have significant influence can vary as [9] showed for map predictions in urban environments.

10.2.3 Limits of Ray Tracing

There are limitations to the possibilities of ray-tracing analysis. Due to the computational complexity of Maxwell equations, the electromagnetic properties during the analysis will always be an approximation by suitable models. Especially for high frequencies and short wavelength, it is practically impossible to model in a scenario to such an extent that features such as roughness and phase shifts are represented correctly.

During the digitization process, objects are often simplified, either to reduce the number of surfaces or to make it possible to consider them. Round objects are commonly approximated by polygons, for example. Small items, such as books in bookshelves, tools, and accessories, are usually not modeled; however, as shown in [10], even small objects impact the channel significantly. The comparison between measurements and ray-tracing simulations of a blockage by a forklift situation discussed in [11] shows the impact of round (real) and edge corners (model) on the simulated CIR.

10.3   Realistic Modelling of the Environment

In order to realistically predict radio propagation for a given scenario, accurate, high-quality input data and models are necessary. Depending on the scenario under investigation, input data can be manifold and vary with the needed level of detail, including but not limited to a model of the 3D environment, material properties and parameters, transmitter and receiver positions, antenna patterns, frequency range, topographic maps, weather information, and clutter data. Typically, the modelling of the 3D environment and the acquisition of material parameter are the most time-consuming steps in this process.

10.3.1   Creating 3D Environment Models

Open data initiatives and projects, such as OpenStreetMap [12] or Open Data Berlin https://www.businesslocationcenter.de/berlin3d-downloadportal/#/export   (visited 10 November 2025), can be valuable resources for building and environment data. For indoor scenarios and/or private locations, the 3D environment usually needs to be modeled separately. With the recent technological advances on tools for digitization of environments, obtaining precise maps or digital twins of different buildings and areas has become time and cost-effective. Nowadays, creating a precise replica of a room is within arm's reach since medium- to high-range cellular phones include light detection and ranging (LiDAR) modules. LiDAR operates in a fixed or in multiple directions by scanning the environment, a method known as 3D laser scanning. Laser scanners utilize, depending on the model, from ultraviolet to near-infrared light to image scenarios and objects. Professional portable tools such as the Leica^{TM} BLK 360 [13] imaging laser scanner camera utilizes an 830-nm wavelength infrared laser and offers an accuracy of 4 mm at 10-m distance and operates within a range of 0.5 m to 45 m.

The usual workflow is depicted in Fig. 10.2. The first step is to perform LiDAR scans of the scenario. This requires a static environment, and depending on the size and location of furniture and objects, multiple scanning positions must be selected to avoid being out of range and areas without environmental information due to

Material parameters

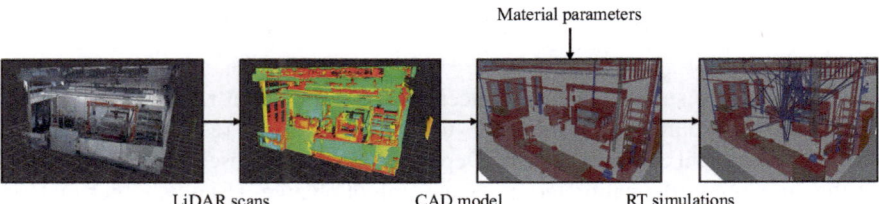

LiDAR scans   CAD model   RT simulations

Fig. 10.2 From point-cloud scans to ray-tracing simulations

blockage and shadowing by objects. It is useful to place certain tags in the place for a later identification of common areas to join different scans [14]. The results of the scans are multiple point clouds that depend on the number of scanned positions. These clouds need to be aligned and merged, which can be done with the software of the free project CloudCompare™ [15]. Once a single point cloud is obtained, multiple post-processing methods can be applied to clean the cloud, filter certain outlier or noisy points, or reduce the number of points/details.

Afterwards, with a clean point cloud, different software can be used to create the computer-aided design (CAD) model (map) of the scenario. Free or commercial software as Blender™ [16] or AUTOCAD™, respectively, can be used for this task. For that purpose, the points in the cloud are grouped considering the simplest geometrical shapes as cubes and polygons. The different objects are grouped accordingly to their constructive materials, e.g., metal, wood, glass, etc. This is later used in the ray-tracing software to assign electromagnetic properties to the objects in order to estimate the penetration, diffraction, and reflection losses, between others. Finally, this CAD models need to be exported to standard triangular language (STL) formats where the figures are represented as set of triangles.

10.3.2 Material Parameter

Radio waves behave differently when interacting with various materials. Material parameters reflect these specific properties and are usually gathered by fitting parameter values to measurements. Typical values are relative permittivity and permeability. The results are hence material and frequency specific. Multilayer and composite materials add additional challenges for modelling the behavior of materials.

The ITU Recommendation 2040 provides a variety of material parameters for the frequency range from 0.1 GHz to 100 GHz [17]. For THz frequencies, these recommendations are not valid. Within the Meteracom project, an effort has been conducted to measure and analyze simple and complex materials and provide their parameters to the research community [18]. More detailed information can be found in Chap. 31.

10.3.3 Calibration and Validation of Ray-Tracing Simulations

In order to make sure that the ray-tracer produces a faithful representation of the real-world environment, a calibration using measurement results from the real-world environment can be necessary. Depending on the purpose and the applications of the simulation, the calibrated parameter might change. Usually, either material parameter, such as the relative permittivity, or the loss created by interaction effects is adjusted based on the measurement results. For a data center environment [19], it has been shown that there is a good agreement between angular and delay information for data centers; however, deviations between predicted and measured power exist. Similarly, in an industrial environment, [11] compares measurements and ray-tracing simulations of a blockage situation showing a very good agreement in the angular and delay domain.

10.4   Applications of a Ray-Tracing Analysis

Ray-tracing simulations extend the toolbox for investigations in radio propagation. They are often employed in combination with other methods, such as measurements and link- and system-level simulations. Within the Meteracom project, ray-tracing results have been among other tasks used for the following research purposes:

Assistance for Measurement Planning   A preliminary ray-tracing simulation of a scenario can highlight interesting positions for measurements and give further insights. In [20], a ray-tracing simulation has been used to identify vulnerable positions for eavesdropping that were later verified through channel sounder measurements; see also Chap. 35.

Assistance for Measurement Interpretations   Measurements sometimes show unexpected phenomena difficult to explain at first sight. Ray tracer can assist in these cases by providing detailed information about angles, interaction effects, and delays. These details can be helpful for measurement analysis, e.g., by identifying scatters or reflection points.

Analysis of Channel Properties   Information about predicted rays are easily accessible and can be used to calculate different channel metrics, such as angular spread or delay spread, as shown in [21]. The standardized format makes reusing evaluation scripts possible.

Input for Other Simulation Tools   The ray-tracing results deliver valuable input for other simulations. Amplitude and delay can be used to create a tapped delay line for further analysis within link- and system-level simulations [22]. More information about these simulation tools can be found in Chap. 37. These simulation types have been used for the analysis of network coding concepts (see Chap. 39 and training of machine learning algorithms (see Chap. 38).

References

1. Z. Yun, M.F. Iskander, Ray tracing for radio propagation modeling: principles and applications. IEEE Access **3**, 1089–1100 (2015). Conference Name: IEEE Access [Online]. Available: https://ieeexplore.ieee.org/document/7152831/?arnumber=7152831

2. R.E.S. Software, Wireless InSiteé 3D Wireless Propagation Software. State College, Pennsylvania [Online]. Available: https://www.remcom.com/wireless-insite-propagation-software
3. T. M. Inc., Ray Tracing for Wireless Communications. Natick, MA [Online]. Available: https://de.mathworks.com/help/antenna/ug/ray-tracing-for-wireless-communications.html
4. J. Hoydis, S. Cammerer, F. Ait Aoudia, A. Vem, N. Binder, G. Marcus, A. Keller, Sionna: An Open-Source Library for Next-Generation Physical Layer Research. Preprint (2022)
5. T.U. Braunschweig, Simulator for Mobile Networks (2024) [Online]. Available: https://www.tu-braunschweig.de/ifn/forschung/simone
6. Y. Lostanlen, T. Kürner, Ray-tracing modeling, in *LTE-Advanced and Next Generation Wireless Networks: Channel Modelling and Propagation*, ed. by G. de la Roche, A. Alayón-Glazunov, B. Allen (Wiley, Hoboken, 2012), pp. 271–292
7. R. Piesiewicz, C. Jansen, D. Mittleman, T. Kleine-Ostmann, M. Koch, T. Kürner, Scattering analysis for the modeling of THz communication systems. IEEE Trans. Antennas Propag. **55**(11), 3002–3009 (2007)
8. N. Dreyer, T. Kürner, An analytical raytracer for efficient D2D path loss predictions, in *2019 13th European Conference on Antennas and Propagation (EuCAP)* (2019), pp. 1–5
9. M. Schweins, L. Thielecke, N. Grupe, T. Kürner, Optimization and evaluation of a 3D ray tracing channel predictor individually for each propagation effect. IEEE Open J. Antennas Propag. **PP**, 1–1 (2024)
10. Y.C.G. Gougeon, F. Munoz, R. D'Errico, *Ray-Tracing Calibration from Channel Sounding Measurements in a Millimeter-Wave Industrial Scenario*. 18th European Conference on Antennas and Propagation (EuCAP), Glasgow, UK (2024), pp. 1–5. https://doi.org/10.23919/EuCAP60739.2024.10501179
11. D. Dupleich, D. Sitdikov, A. Ebert, M. Boban, Measurement-based validation of ray-tracing model at sub-THz for ISAC applications of blockage in industrial scenario, in *2024 4th URSI Atlantic Radio Science Meeting (AT-RASC)* (2024), pp. 1–4
12. OpenStreetMap [Online]. Available: https://www.openstreetmap.org/
13. Leica BLK360 Imaging Laser Scanner [Online]. Available: https://leica-geosystems.com/products/laser-scanners/scanners/blk360
14. D. Dupleich, N. Han, J. Cosmas, G. Eappen, K. Ali, 6G BRAINS - D3.1: 3D Laser measurement of one factory at Bosch with 3D cloud scanner and 3D hand scanner (2021) [Online]. Available: https://doi.org/10.5281/zenodo.5786456
15. CloudCompare - home [Online]. Available: https://www.cloudcompare.org/main.html
16. Blender - a 3D modelling and rendering package, Amsterdam [Online]. Available: http://www.blender.org
17. Recommendation ITU-R P.2040-3 (08/2023) - Effects of building materials and structures on radiowave propagation above about 100 MHz. International Telecommunications Union, Geneva, CH, Standard (2023)
18. F. Taleb, G. Hernandez-Cardoso, E. Castro-Camus, M. Koch, Transmission, reflection, and scattering characterization of building materials for indoor THz communications. IEEE Trans. Terahertz Sci. Technol. **PP**, 1–10 (2023)
19. J.M. Eckhardt, T. Doeker, T. Kürner, Hybrid channel model for low Terahertz links in a data center. IEEE Open J. Commun. Soc. **5**, 4731–4745 (2024)
20. C. Herold, T. Doeker, T. Kürner, Measurements at 300 GHz in eavesdropping scenarios and first results, in *2022 Fifth International Workshop on Mobile Terahertz Systems (IWMTS)* (2022), pp. 1–4
21. C. Herold, T. Doeker, T. Kürner, Influence of channel impulse response characteristics on wireless THz-communications, in *2022 47th International Conference on Infrared, Millimeter and Terahertz Waves (IRMMW-THz)* (2022), pp. 1–2
22. J.M. Eckhardt, C. Herold, B.K. Jung, N. Dreyer, T. Kürner, Modular link level simulator for the physical layer of beyond 5G wireless communication systems. Radio Sci. **57**(2), e2021RS007395 (2022)

Chapter 11
Hardware Acceleration

Anouar Nechi and Mladen Berekovic

Abstract THz communication holds immense promise for future wireless systems due to its vast bandwidth and low latency. However, the high-frequency nature of THz signals poses significant challenges in terms of computational complexity and power consumption. Hardware acceleration emerges as a critical solution to address these challenges. The integration of specialized hardware units, such as FPGAs, ASICs, and GPUs, significantly improves the performance and energy efficiency of THz communication systems. Various hardware acceleration techniques are explored alongside case studies demonstrating their effectiveness. Future trends and challenges in the field are also discussed. Hardware acceleration paves the path for realizing practical and high-performance THz communication systems.

11.1 Hardware Acceleration for THz Communication

Pursuing advanced wireless communication systems characterized by higher data rates, lower latencies, and enhanced reliability has propelled exploration into the THz frequency band (0.1–10 THz). This underutilized spectrum, boasting a bandwidth several orders of magnitude larger than current wireless technologies, promises to deliver multi-gigabit-per-second data rates and ultralow latency communication [1, 2]. Such capabilities could revolutionize diverse applications, from real-time, high-definition video streaming and immersive virtual reality to seamless integration of the Internet of Things.

However, notable technical challenges hinder realizing THz communication's full potential. The exceptionally high frequencies and correspondingly short wavelengths intrinsic to THz signals necessitate sophisticated signal processing techniques. The computational burden associated with processing such high-bandwidth signals and the power consumption challenges of operating at THz frequencies

A. Nechi (✉) · M. Berekovic
Universität zu Lübeck, Institut für Technische Informatik, Lübeck, Germany
e-mail: anouar.nechi@uni-luebeck.de; mladen.berekovic@uni-luebeck.de

© The Author(s) 2026
T. Kürner et al. (eds.), *Metrology for THz Communications*, Springer Series
in Optical Sciences 256, https://doi.org/10.1007/978-3-032-01986-8_11

present a formidable obstacle to practical implementation. Conventional signal processing approaches, often reliant on general-purpose processors, struggle to meet the high-performance and efficiency requirements of THz systems [3].

In response to these challenges, hardware acceleration emerges as a pivotal enabler. Significant gains in performance and energy efficiency can be achieved by deploying dedicated hardware units designed explicitly for THz signal processing tasks. Field-programmable gate arrays (FPGAs), with their reconfigurable logic fabric, offer a flexible platform for prototyping and implementing THz-specific algorithms, enabling rapid adaptation to evolving standards and requirements [4, 5]. While less flexible, application-specific integrated circuits (ASICs) can provide superior performance and power efficiency once the design is finalized [6]. Additionally, graphics processing units (GPUs), originally designed for graphics rendering, have proven adept at accelerating computationally intensive signal processing tasks due to their massive parallelism [7, 8]. The strategic integration of FPGAs, ASICs, and GPUs, tailored to the specific demands of THz communication systems, can unlock the transformative potential of this burgeoning technology.

Hardware acceleration can overcome the computational and power bottlenecks currently impeding the progress of THz communication; see Chap. 40. This chapter will delve into the intricacies of hardware acceleration techniques and explore their application in THz systems. The synergy of THz communication and hardware acceleration envisions a future where ultrafast, high-capacity wireless connectivity becomes a reality, transforming how individuals interact with the digital world.

11.2 Hardware Acceleration: Platforms and Techniques

11.2.1 The Paradigm of Hardware Acceleration

In high-performance computing, hardware acceleration signifies a departure from the conventional reliance on general-purpose processors. This approach entails the utilization of dedicated hardware units that are meticulously engineered to execute specific computational tasks with exceptional efficiency. Hardware acceleration can substantially augment the overall system performance by offloading these tasks from the central processing unit (CPU), particularly in applications necessitating intensive computations. The fundamental principle underpinning hardware acceleration is task specialization. In contrast to general-purpose processors, designed to handle a wide spectrum of instructions, dedicated hardware accelerators are optimized for a particular set of operations. This specialization empowers them to attain superior performance and energy efficiency by exploiting parallelism, pipelining, and other hardware-level optimizations. However, this specialization results in decreased flexibility when compared to general-purpose processors.

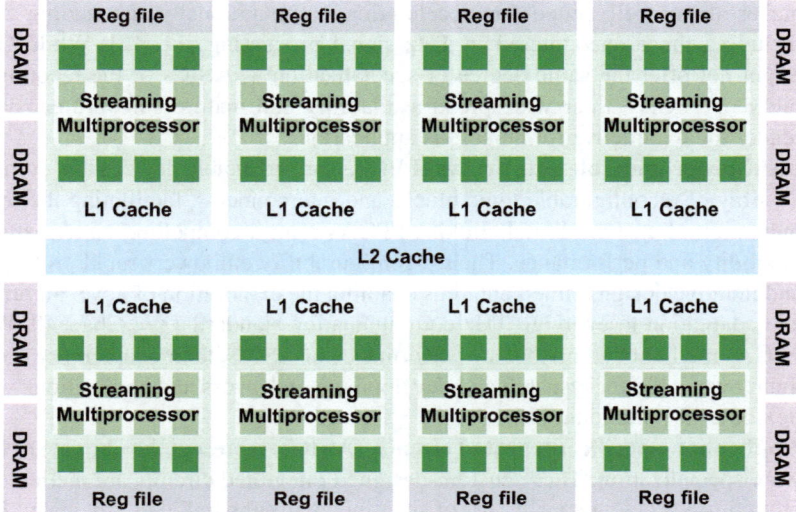

Fig. 11.1 Typical NVIDIA GPU architecture, featuring a collection of streaming multiprocessors (SMs), where each SM comprises several stream processor (SP) cores [9]

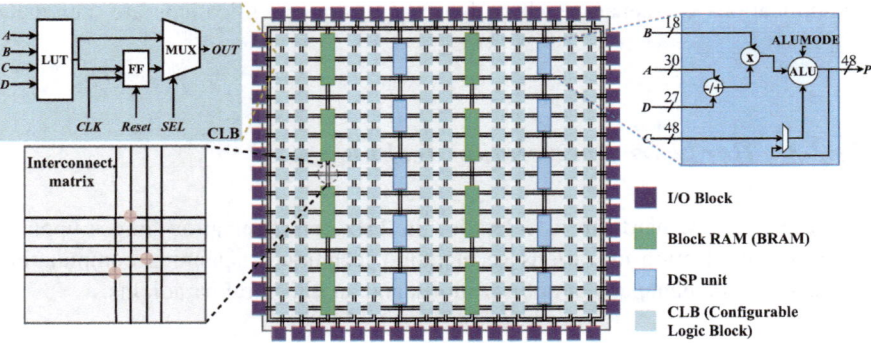

Fig. 11.2 Internal structure of modern FPGA architectures [10]

11.2.2 Hardware Acceleration Platforms

Hardware acceleration techniques are pivotal in THz communication, where the computational demands are particularly pronounced. Several promising technologies have emerged as potential candidates for expediting THz signal processing tasks (Figs. 11.1 and 11.2):

- **Graphics processing units (GPUs)** are originally conceived for graphics rendering. GPUs have metamorphosed into formidable parallel computing platforms. As shown in Chap. 40, their capacity to execute thousands of threads concurrently makes them well suited for accelerating computationally demanding tasks, including those encountered in THz signal processing [11, 12]. While GPUs might not offer the same degree of specialization as ASICs or the reconfigurability of FPGAs, their widespread availability and mature software ecosystem render them attractive for numerous applications.

- **Field-programmable gate arrays (FPGAs)** are adaptable devices that comprise an array of reconfigurable logic blocks and interconnects, facilitating the implementation of custom digital circuits. FPGAs offer a compelling combination of flexibility and performance. Their reconfigurability empowers rapid prototyping and iterative design refinement, thus enabling the exploration of novel algorithms and adaptation to evolving THz communication standards; see Chap. 40. While FPGAs might not attain the raw performance of ASICs, their adaptability renders them invaluable for research and development endeavors and applications where flexibility is paramount [13].
- **Application-specific integrated circuits (ASICs)** represent the ultimate of hardware specialization. These custom-designed integrated circuits are meticulously crafted to execute a specific set of functions with unparalleled efficiency. Upon the finalization of the design, ASICs can deliver exceptional performance and power savings, surpassing those of FPGAs. However, their inflexibility and substantial development costs render them less suitable for rapidly evolving applications or scenarios where frequent design modifications are anticipated [14–16].

11.2.3 Hardware Acceleration Techniques

The efficiency of hardware acceleration in THz communication hinges upon the judicious application of various acceleration techniques, spanning computational strategies, data management optimizations, and architectural paradigms.

11.2.3.1 Computational Optimizations

- **Pipelining:** This technique involves breaking down a complex operation into simpler stages that can be executed concurrently. By overlapping the execution of multiple instructions, pipelining enhances the system's throughput. In the context of THz communication, pipelining can be particularly beneficial for tasks such as channel equalization or forward error correction, where data streams are processed sequentially [17].
- **Parallelism:** Parallelism exploits the inherent concurrency in many signal processing algorithms [18, 19]. It can dramatically improve overall processing speed by performing multiple operations simultaneously at the instruction level (Single Instruction, Multiple Data [SIMD]) or utilizing multiple processing cores. With their massive number of cores, GPUs excel at exploiting data parallelism, making

them well suited for tasks like FFT computations and matrix operations, which are prevalent in THz communication systems.

- **Approximated Computing:** This technique deliberately reduces the precision of computations or data representation to achieve significant performance and energy efficiency gains. While some loss of accuracy might be incurred, it is often tolerable for certain THz communication applications where real-time processing is critical [20].
- **Loop Unrolling:** This technique reduces loop overhead by replicating the instructions within the loop body multiple times. By decreasing the number of branch instructions and loop iterations, loop unrolling can enhance the performance of computationally intensive loops commonly found in THz signal processing algorithms [21].

11.2.3.2 Data Management Optimizations

- **Memory Optimization:** Efficient memory management is crucial for high-performance THz signal processing. Techniques such as caching, prefetching, and data reuse can be employed to minimize memory access latency and improve data locality. Ensuring frequently accessed data resides in fast on-chip memory can significantly enhance system performance [22, 23].
- **Tiling:** This technique involves partitioning large data sets into smaller tiles that fit into on-chip memory. Processing data in tiles reduces off-chip memory accesses, enhancing performance and energy efficiency [24]. Tiling is particularly useful for THz applications dealing with large data volumes, such as beamforming or MIMO processing.

11.2.3.3 Architectural Optimizations

- **Dataflow Architectures:** Dataflow architectures prioritize the availability of data to trigger the execution of instructions [25]. By enabling fine-grained parallelism and efficient data-driven execution, they can offer significant performance advantages for THz communication tasks.
- **Systolic Arrays:** These specialized architectures consist of a regular array of processing elements, each performing a simple operation and communicating with its neighbors. Systolic arrays are highly effective for executing computationally intensive tasks with great throughput and low latency, making them well suited for THz signal processing applications such as filtering and convolution [26, 27].

The synergistic combination of these acceleration techniques, deployed strategically on appropriate hardware platforms, can lead to the realization of high-performance, energy-efficient THz communication systems capable of meeting the stringent demands of future wireless applications (Fig. 11.3).

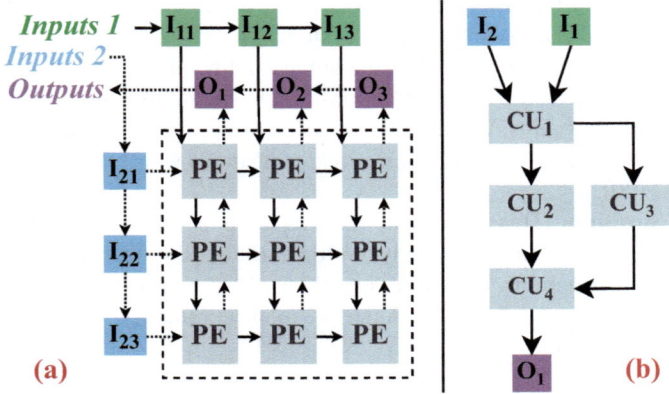

Fig. 11.3 An illustrative comparison between (**a**) systolic array and (**b**) dataflow-based accelerators [13]

References

1. T. Nagatsuma, THz communication systems, in *2017 Optical Fiber Communications Conference and Exhibition (OFC)* (IEEE, Piscataway, 2017), pp. 1–90
2. D.A. Humphreys, T. Kürner, M. Berekovic, A. Jukan, T. Schneider, I. Kallfass, J.C. Scheytt, T. Kleine-Ostmann, M. Koch, R. Thomä, An overview of the Meteracom Project, in *WWRF 43: Intelligent Applications for a 5G Connected World* (2019)
3. H.-J. Song, N. Lee, Terahertz communications: challenges in the next decade. IEEE Trans. Terahertz Sci. Technol. **12**(2), 105–117 (2021)
4. A. Batra, A. Kamaleldin, L.Y. Zhen, M. Wiemeler, D. Göhringer, T. Kaiser, FPGA-based acceleration of THz SAR imaging, in *2021 Fourth International Workshop on Mobile Terahertz Systems (IWMTS)* (IEEE, Piscataway, 2021), pp. 1–5
5. C. McDonald, T.E. Abrudan, F. Cabral, S. Kucera, H. Claussen, R. Farrell, J. Dooley, FPGA implementation of a sub-400ns 6G free-space optical wireless communications transmitter. Opt. Exp. **31**(16), 25 933–25 942 (2023)
6. O. Ferraz, S. Subramaniyan, R. Chinthala, J. Andrade, J.R. Cavallaro, S.K. Nandy, V. Silva, X. Zhang, M. Purnaprajna, G. Falcao, A survey on high-throughput non-binary LDPC decoders: ASIC, FPGA, and GPU architectures. IEEE Commun. Surv. Tutorials **24**(1), 524–556 (2021)
7. M. Gezimati, G. Singh, Terahertz data extraction and analysis based on deep learning techniques for emerging applications. IEEE Access **12**, 21174–21198 (2024). https://doi.org/10.1109/ACCESS.2024.3360930
8. F. García-Rial, L. Úbeda-Medina, J. Grajal, Real-time GPU-based image processing for a 3-D THz radar. IEEE Trans. Parallel Distrib. Syst. **28**(10), 2953–2964 (2017)
9. M. Hernández, G.D. Guerrero, J.M. Cecilia, J.M. Garcia, A. Inuggi, S.N. Sotiropoulos, in *2012 20th Euromicro International Conference on Parallel, Distributed and Network-based Processing*. Accelerating Fibre Orientation Estimation from Diffusion Weighted Magnetic Resonance Imaging Using GPUs, pp. 622–626 (2012). https://doi.org/10.1109/PDP.2012.46
10. V. Taraate, FPGA Architecture and Design Flow, in *Digital Design from the VLSI Perspective: Concepts for VLSI Beginners* (Springer Nature Singapore, Singapore, 2023), pp. 285–295. ISBN: 978-981-19-4652-3. https://doi.org/10.1007/978-981-19-4652-3_19
11. K. Świrydowicz, E. Darve, W. Jones, J. Maack, S. Regev, M.A. Saunders, S.J. Thomas, S. Peleš, Linear solvers for power grid optimization problems: a review of GPU-accelerated linear solvers. Parallel Comput. **111**, 102870 (2022)

12. J. Naghmouchi, D.P. Scarpazza, M. Berekovic, Small-ruleset regular expression matching on GPGPUs: quantitative performance analysis and optimization, in *Proceedings of the 24th ACM International Conference on Supercomputing* (2010), pp. 337–348

13. A. Nechi, L. Groth, S. Mulhem, F. Merchant, R. Buchty, M. Berekovic, FPGA-based deep learning inference accelerators: where are we standing? ACM Trans. Reconfig. Technol. Syst. **17716**(4), 1–32 (2023)

14. K. Mohammed, B. Daneshrad, A MIMO decoder accelerator for next generation wireless communications. IEEE Trans. Very Large Scale Integr. (VLSI) Syst. **18**(11), 1544–1555 (2009)

15. R. Machupalli, M. Hossain, M. Mandal, Review of ASIC accelerators for deep neural network. Microprocess. Microsyst. **89**, 104441 (2022)

16. J. Penders, B. Gyselinckx, R. Vullers, O. Rousseaux, M. Berekovic, M. De Nil, C. Van Hoof, J. Ryckaert, R.F. Yazicioglu, P. Fiorini, et al., Human++: emerging technology for body area networks, in *VLSI-SoC: Research Trends in VLSI and Systems on Chip: Fourteenth International Conference on Very Large Scale Integration of System on Chip (VLSI-SoC2006), October 16–18, 2006, Nice* (Springer, Berlin, 2008), pp. 377–397

17. K. Abdelouahab, M. Pelcat, J. Serot, C. Bourrasset, F. Berry, Tactics to directly map CNN graphs on embedded FPGAs. IEEE Embed. Syst. Lett. **9**(4), 113–116 (2017)

18. D.B. Thomas, L. Howes, W. Luk, A comparison of CPUs, GPUs, FPGAs, and massively parallel processor arrays for random number generation, in *Proceedings of the ACM/SIGDA International Symposium on Field Programmable Gate Arrays* (2009), pp. 63–72

19. A. Garcia, T. Vander Aa, Mapping of the AES cryptographic algorithm on a Coarse-Grain reconfigurable array processor, in *2008 International Conference on Application-Specific Systems, Architectures and Processors* (IEEE, Piscataway, 2008), pp. 245–250

20. L. Yue, Y. Cai, M. Zhu, P. Wang, L. Zhang, M. Sun, S. Liang, M. Lei, J. Zhang, B. Hua, et al., Improving performance of direct-detection terahertz communication system based on k-means adaptive vector quantization, in *2021 19th International Conference on Optical Communications and Networks (ICOCN)* (IEEE, Piscataway, 2021), pp. 1–3

21. W. Foudhaili, A. Nechi, C. Thermann, M. Al Johmani, R. Buchty, M. Berekovic, S. Mulhem, Reconfigurable edge hardware for intelligent IDS: systematic approach, in *International Symposium on Applied Reconfigurable Computing* (Springer, Berlin, 2024), pp. 48–62

22. Z. Wang, J. Lin, Z. Wang, Accelerating recurrent neural networks: a memory-efficient approach. IEEE Trans. Very Large Scale Integr. (VLSI) Syst. **25**(10), 2763–2775 (2017)

23. Q. Zeng, J. Liu, M. Jiang, J. Lan, Y. Gong, Z. Wang, Y. Li, C. Li, J. Ignowski, K. Huang, Realizing in-memory baseband processing for ultrafast and energy-efficient 6G. IEEE Internet Things J. **11**(3), 5169–5183 (2024). https://doi.org/10.1109/JIOT.2023.3307405

24. T. Biedert, P. Messmer, T. Fogal, C. Garth, Hardware-accelerated multi-tile streaming for realtime remote visualization, in *EGPGV@ EuroVis* (2018), pp. 33–43

25. T. Nowatzki, V. Gangadhar, N. Ardalani, K. Sankaralingam, Stream-dataflow acceleration, in *Proceedings of the 44th Annual International Symposium on Computer Architecture* (2017), pp. 416–429

26. J. Cong, J. Wang, PolySA: polyhedral-based systolic array auto-compilation, in *2018 IEEE/ACM International Conference on Computer-Aided Design (ICCAD)* (IEEE, Piscataway, 2018), pp. 1–8

27. V. Ariyarathna, A. Madanayake, J.M. Jornet, Toward real-time software-defined radios for ultrabroadband communication above 100 GHz [application notes]. IEEE Microw. Mag. **24**(8), 50–59 (2023)

Part II
Traceability

Chapter 12
Waveform Traceability Chain to Coaxial EOS

Heiko Füser and David A. Humphreys

Abstract The electro-optic effect has a potential bandwidth of a several THz. The SI-traceable primary standard for the time-domain voltage impulse response $v(t)$ is based on this property. Electro-optic sampling (EOS) systems, realized in several national measurement institutes worldwide, typical have bandwidths between 200 GHz and 500 GHz. The EOS principle and key design criteria for the PTB EOS system are described together with the sources of impairment and the necessary de-embedding to correct these errors. We describe the key results from a recent international intercomparison of EOS systems, using a 100-GHz photodiode transfer standard. The NMI participants' EOS systems are different designs, independently developed. The intercomparison results provide confidence in the agreement between the systems. Dissemination of the calibration from the EOS primary reference to industry and academia is an essential part of the traceability chain, providing confidence in the results. We describe the advantages, limitations, and impairments associated with different choices for transfer standards with coaxial connector geometry.

12.1 Introduction

Electrical waveforms underpin many fields of modern technology, either directly or through specification standards. In Chap. 5, we covered the evolution of waveform metrology, describing the problems of impairments caused by microwave impedance mismatch (Chap. 6) and arising from the instrument design. The benefits of traceability to a primary standard and the benefits that arise from a solid metrological foundation were covered in Chap. 2. The challenge lies within the

H. Füser (✉)
Physikalisch-Technische Bundesanstalt (PTB), Braunschweig, Germany
e-mail: heiko.fueser@ptb.de

D. A. Humphreys
NPL, UK (Retired), University of Bristol, Ascot, UK
e-mail: david.a.humphreys@ieee.org

© The Author(s) 2026
T. Kürner et al. (eds.), *Metrology for THz Communications*, Springer Series in Optical Sciences 256, https://doi.org/10.1007/978-3-032-01986-8_12

specification of the phase as well as the amplitude response of a system versus frequency. RF power can be measured against traceable power standards over a wide range of frequencies (Chap. 3) but this provides no phase information. An approach using "minimum phase" response with frequency assumptions was investigated by NIST [1] to estimate the instrument phase but for many years the only available solution was to use a higher-bandwidth pulse source or oscilloscope, making assumptions about this instrument's phase response.

With the development of mode-locked lasers producing picosecond and sub-picosecond (femtosecond) optical impulses, it became possible, using a photo-conductive switch [2], to generate and detect these impulses through the electro-optic effect [3]. The electrical waveform can be mapped in time by altering the relative delay between the electrical and optical sampling pulse and measuring the average electric field change. Initially, different EOS measurement systems were developed by the National Physical Laboratory (NPL) [4, 5], the National Institute of Standards and Technology (NIST) [6, 7], and later the Physikalisch-Technische Bundesanstalt (PTB) [8]. Recently, systems have been developed by the National Institute of Metrology of China (NIM), the Beijing Institute of Radio Metrology and Measurement (BIRMM), the Korea Research Institute of Standards and Science (KRISS), and others.

In this chapter we first describe the electro-optic sampling principles, focusing on the PTB EOS system, developments, and corrections to provide traceable waveform measurements in a coaxial geometry. The National Measurement Institutes focus on electrical waveforms with high-frequency components ranging up to 100 GHz and beyond as these frequency ranges are the hardest to measure. Such waveforms are used in well-established technologies such as high-speed/high-bandwidth communication systems using fiber optics, data servers, and board-to-board interconnects. In multichannel systems, there is often cross-talk between channels. Linearity, relative-delay, and calibrated vector-frequency response (system impulse response) become increasingly important to de-embed the waveforms.

A variety of standardized high-bandwidth connectors are used in industrial systems. For transfer standards the connector and cable geometry determine the maximum operating frequency. Below this frequency, only TEM_{00} can propagate along the cable, e.g., 1.85 mm (67 GHz), 1.0 mm (110 GHz), or 0.8 mm (145 GHz) coaxial standards. Good-quality connectors are essential to ensure that a low impedance mismatch is maintained.

Following from this, impedance mismatch correction is normally applied using S-parameters, measured with a VNA. However, a VNA cannot operate reliably outside the TEM_{00} frequency range. An alternative approach, based on optical time-domain sampling with the EOS, provides an alternative solution [9–11] using time-domain sampling techniques to generate and detect ultrashort voltage signals reflected from the device under test or a measurement standard, with a temporal width of a few ps, corresponding to a bandwidth of more than 500 GHz.

The second topic covered is a summary of the results from the international measurement intercomparison which also validated the performance of the PTB EOS system [12, 13].

The final topic covered in this chapter is the industrially important dissemination of the calibration to other instruments (Chap. 5). Primary standard EOS systems are complex, expensive, and relatively slow. A more convenient and robust transfer standard is needed for an industrial or research environment. We describe some of the available choices and their associated impairments.

12.2 Coaxial Electro-Optic Sampling

To realize a coaxial voltage pulse standard, coplanar waveguides (CPWs) fabricated on semiconductor substrates in combination with coplanar-to-coaxial microwave probes (MWPs) can be utilized. It is possible to design these devices such that the same samples can be used for generation, propagation, and detection of voltage pulses. At PTB [11], the substrates consist of about 500 μm-thick (100)-oriented GaAs with a thin low-temperature-grown GaAs layer. On top, a CPW structure is sputtered. The geometry of the CPW is optimized to achieve a flat characteristic impedance close to 50Ω for a wide frequency range while providing small signal attenuation. The spacing of the strip lines is matched to commercially available MWPs.

Figure 12.1 shows a schematic cross section of a CPW to illustrate the detection process via the linear electro-optic (EO) effect (also referred to as Pockels effect) occurring in the GaAs substrate [14]. The field lines of a voltage signal on the CPW reach into the material. From the backside of the structure, a 1600 nm laser beam propagates through the chip. It is reflected at the center strip line of the CPW. Due to the electro-optical effect, the polarization of the beam is changed proportional to the electrical field amplitude while propagating through the sample [15]. This polarization change can be measured via a typical electro-optic detection scheme, resulting in a detection voltage

$$V_{\text{det}} = G_{\text{det}} E_{\text{laser}}^2 \frac{\omega \Delta l}{c} n_0^3 r_{41} F_z \tag{12.1}$$

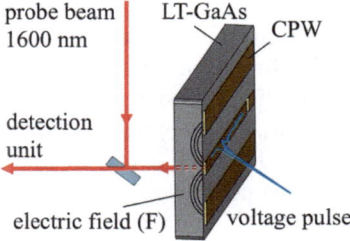

Fig. 12.1 EOS sample cross section: The substrate consists of a 500 μm-thick GaAs wafer with a thin low-temperature-grown GaAs layer. Ontop, a coplanar waveguide (CPW), is fabricated. Using (100)-oriented GaAs, the system can be used to optically probe the electric field components of a voltage signal

with the detector gain G_{det}, the electrical field of the laser E_{laser}, its center frequency $\omega = f/2\pi$, the interaction length Δl, speed of light c, refractive index n_0 and electro-optic coefficient r_{41} of GaAs, and the electrical field component of the voltage pulse along the (100)-direction F_z. For the generation of a high-bandwidth voltage pulse directly on the CPW, a photoconductive switch (PCS, also called Auston switch) is utilized [2]. A PCS consists of two metallic conductors biased by a constant voltage V_{bias} on top of a semiconductor material. A pulsed light source can temporarily increase the conductivity of the semiconductor, leading to the generation of a pulsed voltage signal. By incorporating a PCS directly in the CPW via structuring an approx. 10 μm-wide gap in the center conductor, voltage pulses can directly be launched onto the CPW. Utilizing LT-GaAs, a reduced carrier lifetime of below 1 ps is reached and, hence, voltage pulses with full-width-at-half-maximum duration in the order of 1 ps can be achieved [16]. The fast transients constitute high-bandwidth voltage pulses, significantly exceeding the bandwidth of purely electronic systems. Alternatively, commercial photodiodes (PDs) can be utilized to launch voltage pulses via MWPs onto the CPW [17, 18]. Thereby the available bandwidth is smaller, mainly restricted by the PD and MWP coaxial characteristics.

Regardless of whether the pulsed voltage signal is generated by a PCS or a PD, an accurate measurement of the signal waveform is required to utilize the voltage pulse standard as a high-bandwidth reference. This is realized by electro-optic sampling techniques, based on the precise control of the relative temporal position of the optical pulses used for signal generation and detection. For each fixed temporal relation between the laser pulses, the probe pulse is affected by the signal amplitude of a specific part of the voltage pulse travelling on the CPW. Changing this temporal relation in a controlled manner while acquiring the detection signal (refer Eq. (12.1)) allows to precisely measure the signal's waveform. There are different ways to induce such a temporal shift. The following description focuses on two different methods utilized at PTB, called synchronous and asynchronous electro-optic sampling.

In case of the synchronous sampling, the detection scheme is realized by utilizing an Er-doped fiber laser emitting at a wavelength of about 1600 nm. By frequency-doubling the emission, the frequency-doubled optical pulses with a wavelength of about 800 nm have a constant temporal correlation to the original pulses. Splitting both parts of the beam, the 800 nm emission can be used for signal generation via the PCS and the 1600 nm emission for signal detection via the electro-optic effect in the GaAs substrate. A lock-in amplifier is used to suppress noise, referenced by a pump beam modulation realized via an acousto-optic modulator. Adjusting the optical path length of one of both beam paths by, an optical delay line based on a mirror mounted on a translation stage, a well-defined temporal shift of the relative pulse positions can be introduced; refer to Fig. 12.2. With a single-pass configuration as sketched, a μm-sized motor step of the translation stage transfers into a temporal shift of several femtoseconds. In typical experimental settings, sampling intervals of about 0.5 ps are utilized, corresponding to a measurement bandwidth of 1 THz. According to the length of the translation stage or the number of consecutive passes of the

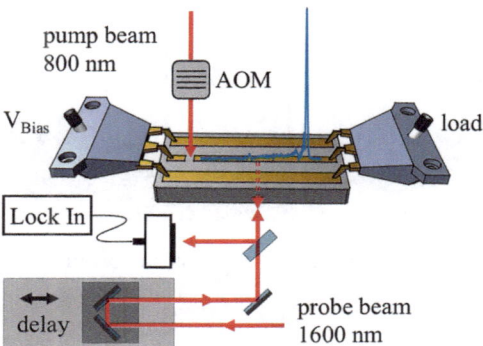

Fig. 12.2 Synchronous sampling scheme: Frequency-doubling a 1600 nm pulsed laser, the 800 nm pulses are used for the generation of voltage signals via a photoconductive switch. The 1600 nm pulses are used for signal probing utilizing the Pockels effect. By beam path length adjustments (mechanical delay stage), the relative timing of pump and probe pulses is controlled, allowing for time-resolved measurements

Fig. 12.3 Translation stage position correction. Deviation of setting and actual positioning of the translation stage within ±25 μm relate to measurement sample temporal position errors with a maximum of about ±75 fs

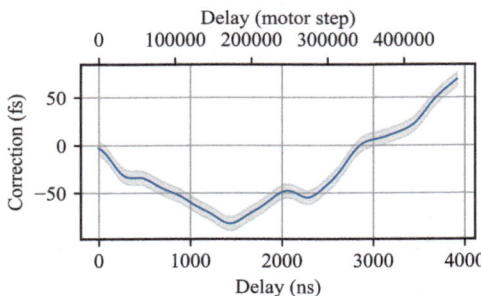

optical beam on the stage, a total delay of several nanoseconds can be achieved. This transfers to a frequency spacing on the order of several 100 MHz. Calibrating the mechanical delay line, the time axis of the synchronous sampling system is traceable to the unit of length and, thus, to the unit of time. This is realized by measuring the path length increments with a laser interferometer traceable to the SI [19]. From this measurement it is found that over a distance of 600 mm a maximal repeatable position difference of up to 25 μm occurs. Such a positional deviation translates to a temporal deviation of up to 150 fs; refer to Fig. 12.3. This can easily be corrected via sample position adjustments and interpolation techniques, resulting in a maximum uncertainty of each sample position of less than 10 fs.

Pump-probe electro-optic sampling can also be based on separate laser sources [20]. A so-called asynchronous sampling scheme, as depicted in Fig. 12.4, uses two lasers with different repetition rates f_{rep} to introduce a temporal shift between pump and probe event. A fixed difference of several 100 Hz enables fast scanning over the entire signal period without any mechanical movements. An analog-to-digital converter (ADC) clocked by the probe laser is used to record the detector signal. Subsequent data processing reconstructs the waveform and performs signal averaging to achieve a signal-to-noise ratio comparable to synchronous EOS over comparable measurement epochs. As separate sources are used, the phase-noise

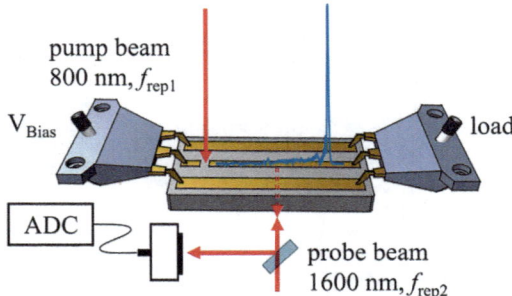

Fig. 12.4 Asynchronous sampling scheme: two pulsed lasers with different repetition rates are used for pumping (f_{rep1}) and probing (f_{rep2}). Introducing a well-controlled frequency offset to the repetition rates, pump and probe pulses perform a time-periodic sweep. Via synchronized data acquisition, a time-resolved waveform is sampled

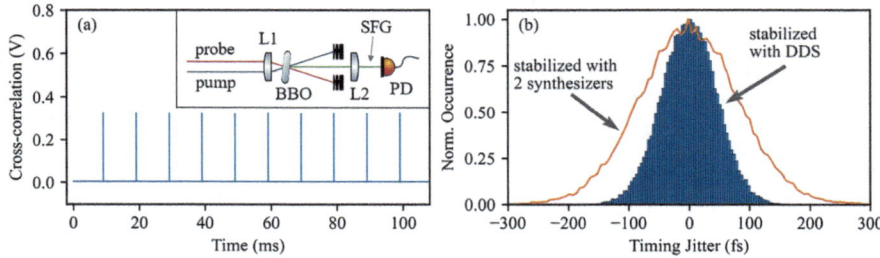

Fig. 12.5 (**a**) Cross-correlation of two lasers as a function of the measurement time at a repetition rate difference of $\Delta f = 100$ Hz. The narrow peaks indicate coincidences between pump and probe laser pulses. The cross-correlation is determined by noncollinear sum frequency generation (SFG) using a beta-barium borate (BBO) crystal as illustrated in the inset. The first lens (L1) focuses the two parallel laser beams (both center frequencies at 1600 nm on the BBO crystal). The generated SFG signal (800 nm) propagates in the center of the two original pulses and is detected by a 10-MHz PD. (**b**) Normalized jitter distribution obtained from the cross-correlation signal at $\Delta f = 100$ for the two different stabilization schemes. Based on [20]

and jitter must be considered in the analysis. Utilizing laser systems equipped with repetition rate stabilization units, external oscillators locked to frequency standards can be used to traceably define the time axis of the sampling procedure. The residual timing uncertainty is on the order of 100 fs, limited by the electronics of the stabilization units. This value can be further improved by locking one laser directly on the optical signal of the other. This stabilization principle utilizes a direct digital synthesizer (DDS) frequency shifter as shown in [20]. The advantage of this offset-locking method is that all repetition rate fluctuations of the pump laser system are passed to the probe laser, while the relative phase relationship between pump and probe laser is fixed. In this case, the residual timing jitter results mainly from the imperfection of the probe laser stabilization and electrical noise from the DDS and the environment; see Fig. 12.5.

12.3 De-Embedding

The electro-optic detection scheme as expressed by Eq. (12.1) describes a spatially and temporally static configuration. In a dynamic scenario, the measurement signal is influenced by spatiotemporal effects. This is the case for the voltage pulse traveling along the CPW while the optical pulse propagates through the electro-optical active substrate. It results in a broadening and amplitude scaling of the detected signal compared to the true voltage pulse on the CPW. The interaction properties can be expressed by complex transfer functions that describe the electric field of the waveguide, $H_F(f)$, and the optical field of the probe laser, $H_L(f)$, respectively. For the experimental geometry as sketched above, it has been shown [21] that the electric field profile inside the GaAs substrate can be well approximated by a Lorentzian function, resulting in $H_F(f) = \exp(-2\pi f t_d)$ with a time constant t_d to be determined for the specific measurement scenario. For a laser beam with a spatially and temporally Gaussian profile, the transfer function $H_L(f) = \exp(-2(\pi \sigma_t f)^2)\exp(-2(\pi \sigma_s f)^2)$ follows. Thereby σ_t and σ_s represent the temporal and spatial standard deviations, respectively. With this, the electro-optic transfer function results as

$$H_{\mathrm{EOS}}(f) = e^{-2\pi f t_d} e^{-2(\pi \sigma_t f)^2} e^{-2(\pi \sigma_s f)^2}. \tag{12.2}$$

Next to the signal alteration due to the electro-optic detection scheme, also the electronic circuit properties have to be taken into account. This includes (i) frequency-dependent reflections of the voltage pulse at each electrical transition within the setup and (ii) the frequency-dependent transmission properties of the coplanar/coaxial transition system, expressed by its scattering parameter S_{12}.

Regarding the reflection properties, the MWP-CPW junction poses a significant contribution due to the nonperfect contacting. However, also the subsequent coaxial interfaces contribute to measurable reflections as well. Hence, the voltage signal measured via Eq. (12.1) comprises both forward and backward propagating signal components. Of course only the forward propagating signal is part of the voltage pulse standard as seen by a DUT at the end of the coaxial adapter of the MWP. To correct for this and separate both propagation directions, the frequency-dependent reflection coefficient Γ at the measurement plane is used. Utilizing a time-domain approach, Γ can be adequately estimated via a series of measurements at slightly different positions on the CPW; see [11]. The forward propagating signal is then given by

$$V_{\mathrm{fwd}} = \frac{V_{\mathrm{meas}}}{(1 + \Gamma)}. \tag{12.3}$$

Since Eq. (12.3) forms an ill-posed problem [22], regularization is required to suppress noise contributions. One typical approach is given by the Tikhonov regularization [22, 23], which finds many applications in waveform metrology [24].

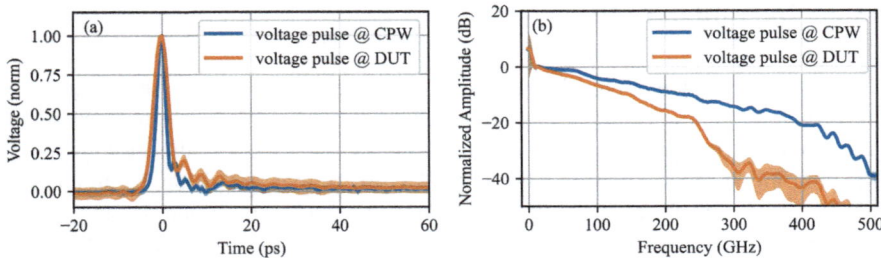

Fig. 12.6 PTB voltage pulse standard as propagating on the coplanar waveguide (CPW) and at the calibration plane as seen by the device under test (DUT) using a microwave probe with a 0.8 mm coaxial connector. (**a**) Time-domain representation and (**b**) frequency-domain amplitude spectrum. The semitransparent colors denote the 95% confidence interval

The determination of S_{12} is realized in a separate characterization measurement, where the MWP is terminated with a traceably characterized coaxial short R_s. In this case, the reflection coefficient Γ' at the CPW measurement plane is measured. Γ' can also be expressed via scattering parameters [25]:

$$\Gamma' = S_{11} + \frac{S_{12} S_{21} R_s}{1 - S_{22} R_s}. \tag{12.4}$$

Utilizing distinct time windows of Γ', S_{11} and S_{22} can be estimated. Via the reciprocity relation [26] and knowledge of R_s, Eq. (12.4) can then be solved for S_{12}.

Now combining Eqs. (12.2), (12.3), and (12.4), the final voltage pulse standard as seen by a DUT follows:

$$V_{\text{CPW}} = \frac{V_{\text{meas}}}{(1 + \Gamma) H_{\text{EOS}}}, \tag{12.5}$$

$$V_{\text{DUT}} = V_{\text{CPW}} S_{12}. \tag{12.6}$$

The difference between v_{meas} and v_{DUT} in time-domain representation as well as the frequency-domain amplitude spectrum is shown in Fig. 12.6. In time domain, a clear pulse broadening is visible, reflecting the frequency filtering as introduced by the 0.8-mm coaxial equipment mainly. Additionally, ringing artefacts occur. They can partly be attributed to the MWP low-pass filtering; partly they result from the Fourier-domain data processing with noisy input data. In frequency domain, the high spectral bandwidth of the voltage pulse standard can be seen, as well as the limiting influence of the coplanar-to-coaxial conversion.

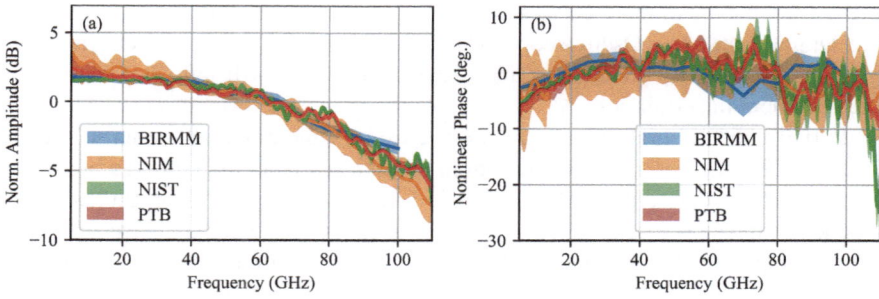

Fig. 12.7 Result of the different NMIs in waveform metrology comparison. The solid lines and semitransparent regions denote the best estimates and the coverage intervals, respectively. (**a**) Frequency-domain amplitude response of the DUT. (**b**) Frequency-domain phase response of the DUT obtained by the different groups. Based on [13]

12.4 International Comparison

Prerequisite of trustworthy and reliable primary standards realized by the National Metrology Institutes worldwide is a strong cooperation and exchange of expertise. Therefore, direct cooperations, round-robin comparisons, and joint projects within EURAMET's European Metrology Research Programme and others have been established.

The development of a traceable EOS system at PTB was accompanied by numerous national and international comparisons [13, 27, 28]. In 2020, the NMIs of China (BIRMM and NIM), USA (NIST), and Germany (Physikalisch-Technische Bundesanstalt, PTB) agreed to a comparison in ultrafast waveform metrology. A commercial ultrafast photodiode, operated at 1550 nm with a nominal bandwidth of 100 GHz (®'u2t XPDV4120R, 1.0 mm coaxial connector), was circulated as the calibration transfer standard. Despite different measurement procedures, all partners employed pump-probe EOS by exciting the PD and coupling its electrical pulse to a CPW suitable for optical detection. To calculate the electrical waveform at the PD's coaxial connector from the voltage measured by the EOS system, a change in reference plane has been performed via conventional VNAs or via the laser-based technique as described above. Figure 12.7 summarizes the result of the comparison, showing a high degree of agreement in the frequency-domain results. Thereby, the uncertainty components have been analyzed and propagated point-wise towards the final waveform, including waveform correction and reconstruction procedures.

12.5 Transferring the Calibration to End Users

As there are only a few EOS systems, worldwide dissemination of the calibration is essential to provide traceability for industry and research; see Chap. 2. The three main reference instrument classes that are regularly used to transfer the calibration are Digital Sampling Oscilloscopes (DSO) and electrical or optoelectronic pulse generators. Each has advantages and disadvantages that will influence the overall measurement uncertainties.

Initially, the calibration was transferred using standardized parametric specifications [29], such as transition-duration. The measurement epoch was chosen to be sufficient to define the 0 % and 100 % levels, typically 50 ps or 100 ps giving frequency spacings of 20 GHz and 10 GHz, respectively. Such a coarse frequency grid is impractical to transfer a full-waveform calibration, where a much finer grid spacing is required to capture the instrument response. At these finer grid-spacing, uncorrected impedance match of the source, transfer standard instruments and calibration devices are significant sources of uncertainty. For transition duration measurements the perturbations due to microwave reflections fall outside the measurement epoch and can be safely ignored. S-parameter measurements are made in the frequency domain using a vector network analyzer (VNA); see Chap. 6. The waveform measurement is made in the time domain and so the Fourier transform is heavily used in waveform and uncertainty analysis. These algorithms expect the data to fall on a uniformly spaced grid so time base distortion is important because unknown deviations from the uniform grid are equivalent to a phase modulation of the waveform, giving rise to error components in the result.

The key desirable attributes for a $v(t)$ transfer standard are stability, linearity, and immunity to impedance mismatch. In the past, the key instruments for this role were the DSOs and commercial pulse generators which were often based on step-recovery diodes or nonlinear transmission lines. Although photodiodes were identified as a potential transfer standard candidate [30], only a few research devices were available at that time. The availability of commercial photodiodes with over 100-GHz bandwidth changed the landscape. Photodiode transfer standards are a key component in the NIST EOS system [31].

It is important to note that as the VNA measures over a grid of frequencies, corrections cannot be applied outside those bounds. For high-bandwidth photodiodes, truncating the results at the correction limit of the VNA can produce a near-rectangular calibration result leading to unwanted ripples due to the sinc(f) window.

12.5.1 Digital Oscilloscopes

High-speed digital oscilloscope types with sufficient bandwidth for practical use as a transfer standard for coaxial $v(t)$ are DSOs and, more recently, Digital Real-Time Oscilloscopes (DRTOs). The impairments and strict Nyquist behavior of DRTO instruments is described in Chap. 5.

High-speed DSOs have bandwidths in excess of 70 GHz and have been used as a primary transfer instrument for many years. The DSO is suited to measuring repetitive waveforms such as electrical impulses. The behavior of the sampling gate

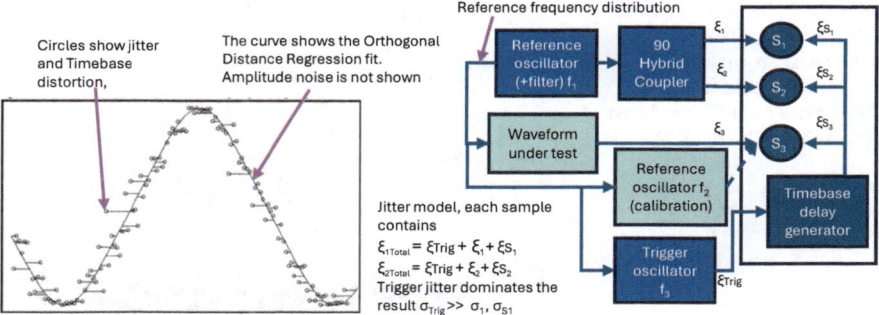

Fig. 12.8 Time base and jitter correction based on orthogonal distance regression algorithm (Reproduced from [33] ©IEEE 2006)

has been extensively studied and certain instrument designs generate an electrical pulse that mimics the sampling waveform [32], but this technique is not applicable to all DSO gate designs.

The DSO time base is part analogue and part digital. The periodic digital retiming (about every 4 ns) can give rise to a step in the waveform. If the time base distortion is considered as a static function $t_e[i]$ and random time jitter t_j is a stationary process, the $\langle t_e(i) + t_j \rangle = t_e(i)$. The implication is that the time base distortion at each point can be estimated using a simple waveform, such as a sine wave. In certain DSO designs all the channels trigger simultaneously, giving a high degree of time correlation between channels. This feature can compensate for jitter, and consequently time base distortion, using a simple in-phase and near quadrature components on two channels mapped to a circle [34]. Residual harmonic components in the reference sine wave were suppressed by hardware filtering. Hale [33] developed a more rigorous algorithm based on orthogonal distance regression algorithm (see Fig. 12.8). While both these techniques are effective and reduce the residual jitter to about 200 fs, other DSO designs have no correlation or only exhibit correlation between samplers on the same plug-in module. Good internal design layout is important as electrical interference from other signals on the DSO backplane, shown in Fig. 12.9, can affect inter-sampler correlation, giving systematic interchannel timebase errors [35]. These corrections also contribute to the measurement uncertainties as all samples contain both noise and jitter components.

12.5.2   Electrical Pulse Generators

For step generators, a key issue is determination of the various levels; particularly 0 % and 100 % is critical. NIST proposed a detailed statistical analysis approach based on the median result [36]. The step edge and comb impulse generators,

Fig. 12.9 Interference from electrical signals within some DSO designs can cause a systematic, but correctable time tbase error (Reproduced from [35] ©IEEE 2012)

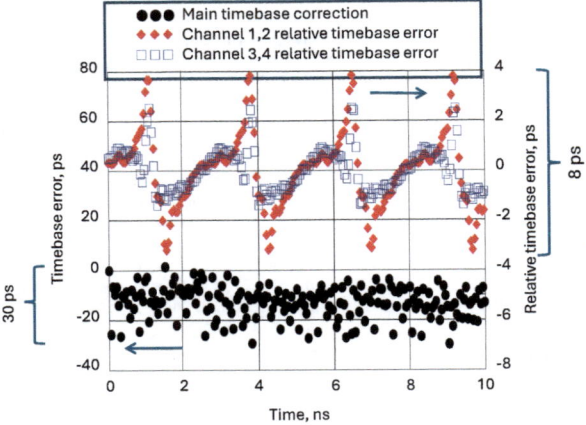

Fig. 12.10 Step-recovery diode comb generator waveform variation with drive frequency (Reproduced from [38] ©EuMA)

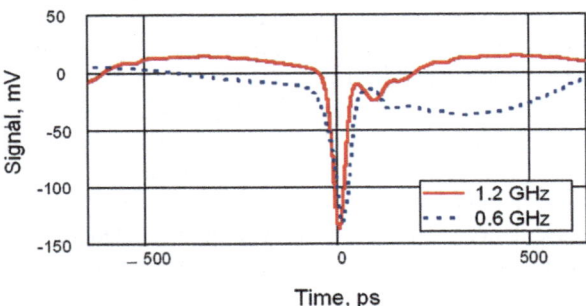

with frequency components extending to over 50 GHz, are often based on step-recovery diodes. Until the development of the nonlinear transmission line [37] these components are also used to generate the sampling gate impulse. Step-recovery diodes may have a restricted frequency range or the output waveform may change with frequency [38]; see Fig. 12.10. As these components are nonlinear, they have a higher susceptibility to feedback caused by poor impedance match of the device under test and require caution when operating as a transfer device.

Phase standards, used with nonlinear vector network analyzers (NVNA), operate over a range of comb frequencies and have bandwidths of >50 GHz. They are based on high-speed electronic logic devices and the pulse is generated from one of the edges. However, if the opposite polarity pulse is not fully suppressed, this can lead to a slight difference between the odd and even harmonic levels. Also, these devices may be locked by the manufacturer, requiring connection to a NVNA.

12.5.3   *Photodiodes*

High-speed photodiodes are typically realized as semiconductor devices with a P-I-N doping structure, where the intrinsic "I" layer has a very low doping level. For InGaAs-based devices this is typically 1 to $3 \cdot 10^{15}$ cm^{-3}. These photodiodes have the property that the device capacitance and electron drift velocity are stable over a range of bias voltages. This has the benefit of a stable impulse response, governed by the carrier transit, bond-wire inductance, and device- and stray capacitance. In addition, these photodiodes act as a current source and are less susceptible to feedback from impedance mismatch. The available photodiode bandwidth quickly increased from a few GHz, limited by device capacitance and packaging to over 100 GHz [39]. Consequently, photodiodes were proposed for use as transfer standards [40] and 20-GHz commercial photodiodes were used for an intercomparison of measurement techniques [41] and as a pulse calibration transfer standard for EOS calibration of DSO [42]. Improvements to the technology, such as uni-carrier devices, where only the electrons contribute to the device response, waveguide devices, and flip-chip structures, which have improved power handling [43] and [44]. Current commercial photodiodes are available with bandwidths of well over 100 GHz [45] from several vendors. The maximum operating frequency is limited by the co-axial connector geometry. For optical-fiber connected devices (see Fig. 12.11), standard optical fiber (SMF28) is dispersive at the high bandwidths produced by the laser source. Two options to reduce this problem are to add a tailored length of dispersion compensating optical fiber or optical filtering of the laser pulse spectrum. The high-speed photodiodes often include an internal load and the necessary bias decoupling. This both improves the output impedance match and provides some protection for the photodiode. The disadvantage of photodiode as a transfer standard is that both the calibration and recipient laboratories must have a suitable sub-picosecond optical pulse source, though turn-key solutions are commercially available.

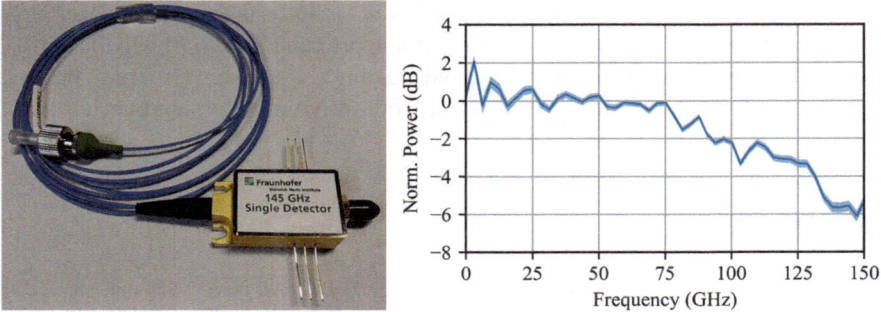

Fig. 12.11 Commercial fiber-coupled 145 GHz photodiode

Fig. 12.12 Time and
frequency uncertainty
contributions in SI traceable
EOS $v(t)$ calibration and
dissemination

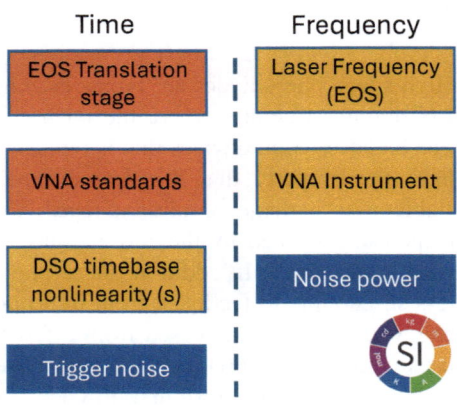

12.6   Waveform Uncertainties

This section should be read in conjunction with the "point-by-point" uncertainty overview in Chap. 5. Section 12.5.1 introduced the time base impairments associated with the Digital Sampling Oscilloscope. The objective of calibration is to provide $v(t)$, on a uniform time grid, together with the calculated uncertainties and correlation relationships at each time point.

As with the DRTO timing errors discussed in Chap. 5, the point-by-point DSO time base errors must be measured and their uncertainties calculated before impedance match de-embedding errors can be applied. The number of points in a DSO waveform is a few thousand points, and this is small when compared with a DRTO waveform. Consequently, either Monte Carlo or compact-covariance algorithms can be used.

The impedance match (S-parameter) measurements are made in the frequency domain as shown in Fig. 12.12, and consequently, the impedance mismatch correction can be calculated at the best estimate of the true measurement time. The associated residual time uncertainty provides an additional amplitude uncertainty component through a calculable Taylor series approximation. Applying the corrections in the right order is important to avoid creating a systematic error term because the timing of the DSO point and the timing of the VNA point do not match exactly.

References

1. A. Dienstfrey, P.D. Hale, D.A. Keenan, T.S. Clement, D.F. Williams, Minimum-phase calibration of sampling oscilloscopes. IEEE Trans. Microwave Theory Tech. **54**(8), 3197–3208 (2006)
2. D.H. Auston, Picosecond optoelectronic switching and gating in silicon. Appl. Phys. Lett. **26**(3), 101–103 (1975)

3. J. Valdmanis, G. Mourou, Subpicosecond electrooptic sampling: principles and applications. IEEE J. Quant. Electron. **22**(1), 69–78 (1986)
4. D. Henderson, A.G. Roddie, Calibration of fast sampling oscilloscopes. Measure. Sci. Tech. **1**(8), 673 (1990)
5. D. Henderson, A.G. Roddie, A.J.A. Smith, Electro-optic sampling for the measurement of the response of fast oscillosopes, in *Picosecond Electronics and Optoelectronics* (Optica Publishing Group, Washington, 1991), p. WE5
6. D.F. Williams, P.D. Hale, T.S. Clement, J.M. Morgan, Mismatch corrections for electro-optic sampling systems, in *56th ARFTG Conference Digest*, vol. 38 (IEEE, Piscataway, 2000), pp. 1–5
7. D.F. Williams, P.D. Hale, T.S. Clement, J.M. Morgan, Calibrating electro-optic sampling systems, in *2001 IEEE MTT-S International Microwave Sympsoium Digest (Cat. No. 01CH37157)*, vol. 3 (IEEE, Piscataway, 2001), pp. 1527–1530
8. M. Bieler, M. Spitzer, G. Hein, U. Siegner, E.O. Göbel, Ultrafast optics establishes metrological standards in high-frequency electronics. Appl. Phys. A **78**, 429–433 (2004)
9. T.-I. Jeon, D. Grischkowsky, Direct optoelectronic generation and detection of sub-ps-electrical pulses on sub-mm-coaxial transmission lines. Appl. Phys. Lett. **85**(25), 6092–6094 (2004)
10. M.Y. Frankel, R.H. Voelker, J.N. Hilfiker, Coplanar transmission lines on thin substrates for high-speed low-loss propagation. IEEE Trans. Microwave Theory Tech. **42**(3), 396–402 (1994)
11. M. Bieler, H. Füser, K. Pierz, Time-domain optoelectronic vector network analysis on coplanar waveguides. IEEE Trans. Microwave Theory Tech. **63**(11), 3775–3784 (2015)
12. M. Bieler, S. Seitz, M. Spitzer, G. Hein, K. Pierz, U. Siegner, M.A. Basu, A.J.A. Smith, M.R. Harper, Rise-time calibration of 50-GHz sampling oscilloscopes: intercomparison between PTB and NPL. IEEE Trans. Instrum. Measure. **56**(2), 266–270 (2007)
13. M. Bieler, P. Struszewski, A. Feldman, J. Jargon, P. Hale, P. Gong, W. Xie, C. Yang, Z. Feng, K. Zhao, Z. Yang, International comparison on ultrafast waveform metrology, in *2020 Conference on Precision Electromagnetic Measurements (CPEM)* (2020), pp. 1–2
14. C.-C. Shih, A. Yariv, A theoretical model of the linear electro-optic effect. J. Phys. C Solid State Phys. **15**(4), 825 (1982)
15. R.W. Boyd, Nonlinear Optics (Academic Press, New York, 2008)
16. F.W. Smith, H.Q. Le, M. Frankel, V. Diadiuk, M.A. Hollis, D.R. Dykaar, G.A. Mourou, A.R. Calawa, Picosecond GaAs-based photoconductive optoelectronic detectors, in *Picosecond Electronics and Optoelectronics* (Optica Publishing Group, Washington, 1989), p. OSDA176
17. D.F. Williams, P.D. Hale, T.S. Clement, J.M. Morgan, Calibrating electro-optic sampling systems, in *2001 IEEE MTT-S International Microwave Sympsoium Digest (Cat. No. 01CH37157)*, vol. 3 (IEEE, Piscataway, 2001), pp. 1527–1530
18. D.F. Williams, A. Lewandowski, T.S. Clement, J.C.M. Wang, P.D. Hale, J.M. Morgan, D.A. Keenan, A. Dienstfrey, Covariance-based uncertainty analysis of the NIST electrooptic sampling system. IEEE Trans. Microwave Theory Tech. **54**(1), 481–491 (2006)
19. R. Schödel, *Modern Interferometry for Length Metrology: Exploring Limits and Novel Techniques* (IOP Publishing, Bristol, 2018)
20. P. Struszewski, M. Bieler, Asynchronous optical sampling for laser-based vector network analysis on coplanar waveguides. IEEE Trans. Instrum. Measure. **68**(6), 2295–2302 (2018)
21. P. Struszewski, *Laser-based Measurements of Complex High-Frequency Signals* (Fachverlag NW in der Carl Schünemann Verlag GmbH, Bremen, 2021)
22. A.N. Tikhonov, *Solutions of Ill-Posed Problems* (VH Winston and Sons, Washington, 1977)
23. W.L. Gans, The measurement and deconvolution of time jitter in equivalent-time waveform samplers. IEEE Trans. Instrum. Measure. **32**(1), 126–133 (1983)
24. P.D. Hale, A. Dienstfrey, Waveform metrology and a quantitative study of regularized deconvolution, in *2010 IEEE Instrumentation & Measurement Technology Conference Proceedings* (IEEE, Piscataway, 2010), pp. 386–391
25. P. Struszewski, K. Pierz, M. Bieler, Time-domain characterization of high-speed photodetectors. J. Infrared Millimeter Terahertz Waves **38**, 1416–1431 (2017)

26. R.B. Marks, D.F. Williams, A general waveguide circuit theory. J. Res. Natl. Inst. Stand. Tech. **97**(5), 533 (1992)
27. P.D. Hale, D.F. Williams, A. Dienstfrey, J. Wang, J. Jargon, D.A. Humphreys, M. Harper, H. Füser, M. Bieler, Traceability of high-speed electrical waveforms at NIST, NPL, and PTB, in *2012 Conference on Precision electromagnetic Measurements* (IEEE, Piscataway, 2012), pp. 522–523
28. M. Bieler, U. Arz, Comparison between time-and frequency-domain high-frequency device characterizations, in *2016 Conference on Precision Electromagnetic Measurements (CPEM 2016)* (IEEE, Piscataway, 2016), pp. 1–2
29. IEC, 60469:2013 Definitions of terms pertaining to transitions, pulses and related waveforms. Technical report, The International Electrotechnical Commission (IEC) (2013)
30. D.A. Humphreys, A.J. Moseley, GaInAs photodiodes as transfer standards for picosecond measurements. IEE Proc. J. (Optoelectr.) **135**(2), pp. 146–152 (1988)
31. T.S. Clement, P.D. Hale, D.F. Williams, C.M. Wang, A. Dienstfrey, D.A. Keenan, Calibration of sampling oscilloscopes with high-speed photodiodes. IEEE Trans. Microwave Theory Tech. **54**(8), 3173–3181 (2006)
32. D.R. Larson, N.G. Paulter, The effects of offset voltage on the amplitude and bandwidth of kick-out pulses used in the nose-to-nose sampler impulse response characterization method. IEEE Trans. Instrum. Measure. **50**(4), 872–876 (2001)
33. P.D. Hale, C.M. Wang, D.F. Williams, K.A. Remley, J.D. Wepman, Compensation of random and systematic timing errors in sampling oscilloscopes. IEEE Trans. Instrum. Measure. **55**(6), 2146–2154 (2006)
34. D.A. Humphreys, F. Bernard, Compensation of sampling oscilloscope trigger jitter by an In-phase and quadrature referencing technique. Technical report, ARMMS (2005)
35. D.A. Humphreys, M. Akmal, Channel timebase errors for digital sampling oscilloscopes, in *2012 Conference on Precision Electromagnetic Measurements* (2012), pp. 520–521
36. N.G. Paulter, D.R. Larson, The 'median' method for the reduction of noise and Trigger Jitter on waveform data. J. Res. Natl. Inst. Stand. Tech. **110**, 511–527 (2005)
37. M.J.W. Rodwell, S.T. Allen, R.Y. Yu, M.G. Case, U. Bhattacharya, M. Reddy, E. Carman, M. Kamegawa, Y. Konishi, J. Pusl, R. Pullela, Active and nonlinear wave propagation devices in ultrafast electronics and optoelectronics. Proc. IEEE **82**(7), 1037–1059 (1994)
38. D.A. Humphreys, M. Harper, J. Miall, D. Schreurs, Characterization and behavior of comb phase-standards, in *2011 41st European Microwave Conference* (2011), pp. 926–929
39. S.Y. Wang, D.M. Bloom, 100 GHz bandwidth planar GaAs Schottky photodiode. Electron. Lett. **19**, 554–555 (1983)
40. D.A. Humphreys, A.J. Moseley, GaInAs photodiodes as transfer standards for picosecond measurements. Optoelectr. IEE Proc. J. **135**(2), 146–152 (1988)
41. D.A. Humphreys, T. Lynch, D. Wake, D. Parker, C.A. Park, S. Kawanishi, M. McClendon, P. Hernday, J. Schlafer, A.H. Gnauck et al., Summary of results from an international high speed photodiode bandwidth Intercomparison, in *High-Speed/High-Frequency Optical Fibre Measurement Conference, 17–18 September 1991, York* (1991)
42. D.F. Williams, P.D. Hale, T.S. Clement, J.M. Morgan, Calibrating electro-optic sampling systems, in *2001 IEEE MTT-S International Microwave Sympsoium Digest (Cat. No.01CH37157)*, vol. 3 (2001), pp. 1527–1530
43. C. Wei, X. Xie, Z. Wang, Y. Chen, Z. Zeng, X. Zou, W. Pan, L. Yan, 150 GHz High-Power Photodiode by Flip-Chip Bonding. J. Lightwave Tech. **41**(23), 7238–7244 (2023)
44. J.S. Morgan, F. Tabatabaei, T. Fatema, C.W. Tang, K. Sun, K.M. Lau, A. Beling, Bias-insensitive GaAsSb/InP CC-MUTC photodiodes for mmWave generation up to 325 GHz. J. Lightwave Tech. **41**(23), 7092–7097 (2023)
45. P. Runge, F. Ganzer, J. Gläsel, S. Wünsch, S. Mutschall, M. Schell, Broadband 145 GHz photodetector module targeting 200GBaud applications, in *2020 Optical Fiber Communications Conference and Exhibition (OFC)* (2020), pp. 1–3

Chapter 13
Characterising VNA Measurements of Time-Varying Sub-Terahertz Transmissions

Mohanad Dawood Al-Dabbagh, David A. Humphreys, and Thomas Kleine-Ostmann

Abstract Vector Network Analyser (VNA) is primarily designed for analysing time-invariant networks, where measurements are conducted one frequency point at a time, with the duration for each point determined by parameters such as the intermediate frequency bandwidth (IFBW) and dwell time. The total measurement time is further influenced by the number of frequency points included in the total sweep. This conventional measurement technique becomes problematic when dealing with time-varying signals, as any changes in the signal must be tracked throughout the entire sweep period. This chapter investigates strategies to adapt VNAs for measuring time-varying transmissions, focusing on their feasibility and limitations. Specifically, it evaluates sweep times using segmented waveforms. We apply experimental validation to construct a measurement scenario investigating dynamic conditions. These measurements and validations will enhance the understanding of time-varying signal measurements and facilitates applications to more complex channel sounder (CS) systems, allowing for improved estimation of associated uncertainties. The work we present in this chapter details the experimental setup, mathematical modelling to simulate a time-varying scenario using a VNA, and the analysis of time intervals recorded for each measured segment under varying velocities. Additionally, it examines the role of optimised reflector positioning for achieving accuracy at each frequency step.

M. D. Al-Dabbagh (✉) · T. Kleine-Ostmann
Physikalisch-Technische Bundesanstalt (PTB), Braunschweig, Germany
e-mail: mohanad.al-dabbagh@ptb.de; thomas.kleine-ostmann@ptb.de

D. A. Humphreys
NPL, UK (Retired), University of Bristol, UK
e-mail: david.a.humphreys@ieee.org

© The Author(s) 2026
T. Kürner et al. (eds.), *Metrology for THz Communications*, Springer Series in Optical Sciences 256, https://doi.org/10.1007/978-3-032-01986-8_13

13.1 Time-Varying Transmission

Time invariance in a system implies that its response remains relatively constant over time. For example, if an input signal is delayed or attenuated, the system's output will simply be a delayed and attenuated version of the original, with considerations to noise and measurement drift deviations over time. This concept is straightforward when applied to static systems. However, in time-varying systems, the response characteristics change over time, introducing variability in the output. This variability can significantly impact signal measurement and analysis, especially in dynamic environments where conditions are continuously changing.

An intuitive and physically meaningful way to characterise time-varying channels is through delays and Doppler shifts. Delays result from multipath propagation and time dispersion, while Doppler shifts occur due to mobility, as well as carrier frequency offsets and oscillator drift [1]. In real-world wireless communications, mobility is a common feature, with signals affected by non-stationary time-delay variation due to movement of transmitters, receivers, or reflections from moving objects. This fast-fading channel presents major challenges for accurate signal measurement and analysis, particularly in dynamic environments where channel conditions are constantly evolving [2, 3].

The VNA measures the frequencies sequentially and the receiver averages the vector voltage over a period corresponding to the IFBW. However, CS of moving targets will return a Doppler-shifted signal where the vector-voltage phase will vary over the averaging period. Higher IFBW measurements allow a higher rate of phase change (velocity). Besides, the VNA hardware takes time to store measurements and for the synthesisers to change frequencies. This limits the highest practical IFBW as the storage time and acquisition time become comparable; see Fig. 13.1.

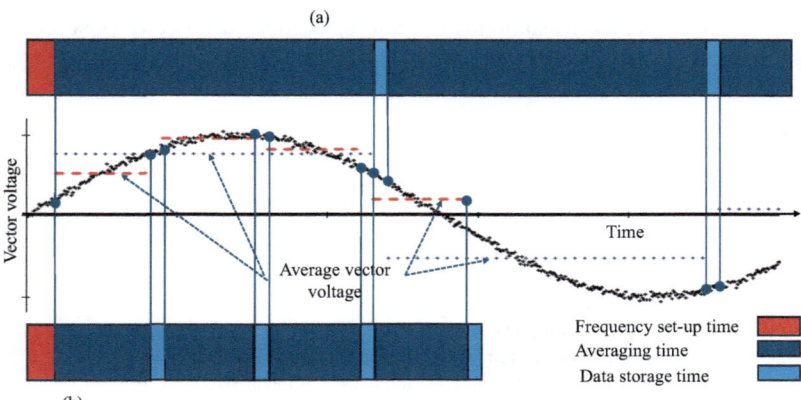

Fig. 13.1 Reflection from a moving object will cause a rotation of the vector voltage phase. This phase rotation can be captured using a higher VNA IFBW. (**a**) Low IF bandwidth. (**b**) High IF bandwidth

As newly deployed millimetre-wave communication systems and future mobile networks investigate higher frequencies, they will need to address these challenges. Additional considerations arise in dynamic scenarios where moving objects further influence signal behaviour through the Doppler effect and variations in path length. Managing these factors is essential to ensure accurate data collection and meaningful measurements. Effective handling of such dynamic elements enables reliable analysis and interpretation of signal behaviour, which is critical for operating at high frequencies and under variable conditions [4, 5].

13.2 Traceability Challenges of Time-Varying Signals

Achieving traceability of time-varying transmissions presents a challenge at terahertz (THz) frequencies because of a lack of suitable standards and traceable measuring equipment. Traceability ensures that measurement results are reliable and comparable by linking them to national or international standards through a chain of certified calibrations. The definition of metrology and its general application to communication systems for achieving traceability are presented in the Chap. 2. Measurement uncertainty plays a key role in this process, expressed statistically to indicate the degree of confidence that the true value lies within a defined range. Substantial progress has been made in extending traceability for VNAs up to 1.1 THz, enabled by precise waveguide measurements and electromagnetic simulations that minimise uncertainties [6]. In dynamic environments where signal propagation involves mobility, measurement uncertainties face an additional challenge: the reproducibility of measurements.

VNAs are fundamental instruments for measuring the reflection and transmission characteristics of device under test (DUT) across a wide range of frequencies, including radio frequency (RF), microwave, millimetre-wave, and THz bands. Measurements at higher frequencies introduce significant challenges. Above 100 GHz, small waveguide dimensions increase alignment uncertainties, requiring sophisticated error models and control software developed by national metrology institutes (NMIs). The VNA calibration corrects for error coefficients such as directivity, reflection tracking, and source match alongside other terms like noise floor, trace noise, drift, non-linearity, and cable movements. These calibration strategies are designed to ensure accurate corrections and meaningful data in all measurement scenarios [7]. The general principles of the VNA systems are presented in the Chap. 6.

Operating at millimetre-wave and THz frequencies also requires special attention to wave attenuation, which complicates over-the-air (OTA) measurements and increases the need for precise calibration techniques. Verification devices, such as flush short circuits, short waveguide sections, and cross-connected waveguides, provide reference standards for maintaining calibration accuracy. These techniques help ensure that measurements remain traceable despite the challenges associated with high-frequency transmission. An investigation of OTA measurement artefacts is presented in the Chap. 19 and in [8].

When measuring time-varying transmissions, ensuring repeatability adds another layer of complexity. Repeatable measurements require precise control over factors such as the velocity, position, and direction of moving objects, as well as the performance of the VNA, frequency points, and step durations. Slight variations in these parameters can lead to different measurement outcomes, affecting the measured magnitude and phase. Therefore, simplified and controlled measurement scenarios are essential for accurately characterising dynamic systems.

The importance of traceability for time-varying measurements extends beyond the individual VNA to other CS systems, especially for future wireless communication technologies. CSs operating at sub-THz frequencies, which aim to characterise real-world propagation channels, rely on accurate and traceable measurements to evaluate system performance under dynamic conditions. Establishing traceable time-varying measurements helps align the results across different systems, improving confidence in system designs and measurement methodologies. This alignment ensures that future wireless networks can operate effectively, even in environments with moving transmitters, receivers, or reflecting objects, and supports the development of next-generation communication standards.

In this chapter, we analyse time-varying signals by examining the reflections from a moving object at various velocities. Unlike traditional CSs, the VNA measures frequencies sequentially, requiring a distinct approach to handle reflections from moving objects. To estimate both the velocity and position of the object, we use a waveform consisting of mixed segments of continuous wave (CW) and linear sweep (LS) signals. Measurements were conducted across frequencies from 220 GHz to 300 GHz (within the WR03 waveguide band) using various IFBWs. This method highlights the VNA's ability to precisely characterise time-varying signals at sub-THz frequencies [9].

13.3   Waveform Selection

The selection of waveforms plays a critical role in the accurate analysis of time-varying signals, particularly in complex measurement scenarios. Different waveforms, such as CWs and LSs with varying frequency steps, offer unique advantages and are used to provide a comprehensive understanding of signal behaviour and measurement accuracy [9].

CW waveforms are invaluable in applications where precise velocity estimation is required. Due to their constant frequency, CW waveforms allow for the continuous monitoring of an object's movement, making them particularly effective in detecting changes in velocity, such as acceleration or deceleration, within a measurement setup. The simplicity of CW signals enables high temporal resolution, which is essential for tracking rapid changes in velocity. This makes CW waveforms ideal for scenarios where understanding the dynamics of moving objects is crucial, such as in radar systems or Doppler-based measurements [10].

On the other hand, LS waveforms are designed to provide detailed information about an object's position over a defined range. By sweeping the frequency linearly over time, LS waveforms can map the distance to a target with high accuracy [11]. This ability to estimate starting and ending positions is crucial in applications where precise localisation is needed, such as in range-finding or imaging systems. The LS's capacity to cover a broad frequency spectrum also makes it useful for capturing reflections from multiple targets, allowing for a more detailed analysis of the environment.

However, LS waveforms often require prior knowledge of the object's velocity to maximise their effectiveness. This is because the relationship between the frequency sweep and the object's motion can lead to complex interpretations of position and velocity data. In practice, position and velocity can be estimated through various combinations of waveform segments, which may include both CW and LS components. By strategically combining these segments, it is possible to enhance the accuracy of the measurement results and reduce the uncertainty associated with dynamic scenarios [9].

Segmented waveforms, composed of a sequence of discrete frequency steps, offer a balance between temporal resolution and frequency coverage. These waveforms are particularly useful in scenarios where both velocity and position information are needed simultaneously. By carefully designing the frequency steps and segment durations, segmented waveforms can capture detailed information about the object's motion while maintaining high accuracy across the frequency spectrum. This approach is advantageous in applications where a comprehensive understanding of both velocity and position is essential, such as in advanced radar systems or in-depth signal analysis [9].

Table 13.1 outlines the chosen waveform, which consists of four segments—three CW segments with 100 points each and one LS segment covering a measurement bandwidth of 30 GHz. Figure 13.2 presents a graphical configuration of the segmented waveform. The waveform was tested using IFBWs of 1 kHz and 10 kHz. The combination of CW and LS segments, along with varying IFBWs, provides distinct advantages by offering a more comprehensive understanding of signal behaviour and enhancing measurement accuracy.

By analysing the performance of these waveforms in various dynamic scenarios, we can identify the optimal configurations for specific applications. This enables the precise measurement of time-varying signals, providing valuable insights into the behaviour of objects in motion and the interactions between signals and moving targets.

Table 13.1 Frequency sweep configuration (adopted from [9])

Start (GHz)	Stop (GHz)	Step (MHz)	Points
220	220	–	100
240	240	–	100
260	260	–	100
270	300	50	601
		Total points	901

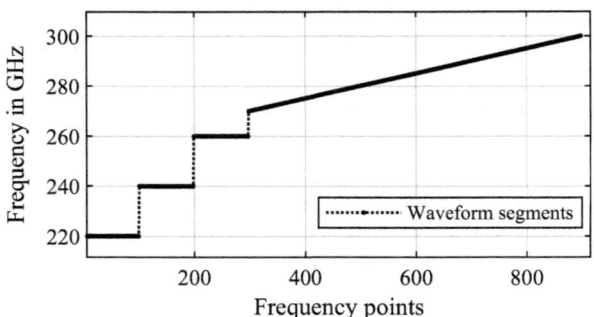

Fig. 13.2 Combined CW and LS segments of the measured waveform (adopted from [9])

13.4 Experimental Design

13.4.1 Building the Measurement Scenario

The measurement setup involved configuring the VNA and associated equipment to accurately capture time-varying signals. Key elements included selecting appropriate antennas, calibration standards, and maintaining environmental control to ensure reliable and high-precision results. We presented this measurement scenario in [9].

Calibration was performed using the through-reflect-line (TRL) method to correct systematic errors and accurately reflect the characteristics of the DUT. Calibration accuracy is critical, especially at high frequencies. Regular verification was essential when operating near the equipment's upper frequency limits, where non-linearities and imperfections become more pronounced.

Environmental factors such as temperature, humidity, and electromagnetic interference were controlled to maintain measurement consistency. Temperature variations affect VNA cables and connectors, altering signal phase and amplitude, while humidity affects the dielectric properties of air. Measurements were performed in a temperature-controlled laboratory to minimise these effects.

The setup employed a Rohde & Schwarz ZVA50™ VNA with VDI-WR3.4 VNAX™ frequency extension modules, covering 220 GHz to 325 GHz, along with 20-dBi standard gain horn antennas for consistent radiation patterns and gain. The antennas were aligned towards a flat metallic reflector mounted on a Newport™ motorised linear stage, the selected velocities ranged between 10 mm/s and 200 mm/s. This setup enabled the investigation of the influence of the time-varying signal on the stability of the S-parameters under varying velocities. Considering the antenna phase centre at different frequency points improves the accuracy of phase measurements, which are crucial for characterising the system. The fixed initial and final positions of the reflector can serve as reference points for this purpose [8, 12, 13].

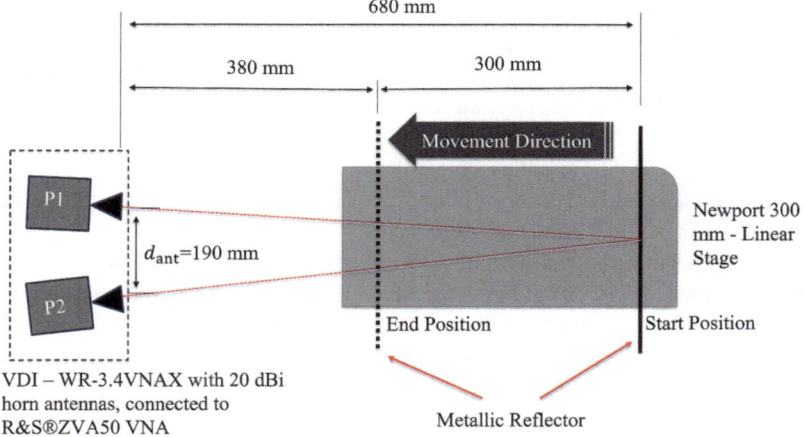

Fig. 13.3 Illustration of the measurement setup (adopted from [9])

Fig. 13.4 Picture of the measurement setup

Continuous measurement sweeps were recorded using the METAS VNA Tools software, with the reflector moving along a 300-mm travel range. The motion was directed towards the antennas, with initial and final positions at 680 mm and 380 mm from the antenna apertures, respectively. Additionally, sweeps were taken before and after the movement of the reflector to capture the dynamic impact on the signal. A schematic of the setup is shown in Fig. 13.3, and a photograph of the measurement setup is shown in Fig. 13.4.

13.4.2   Simulating the Measurement Scenario

To accurately measure moving objects using a VNA, it is necessary to account for the continuous change in position. This requires determining the duration of each frequency step and evaluating its effect on measurement precision. Adjusting the path length for each frequency step ensures that the frequency response remains accurate, even in dynamic conditions [9].

For CW segments, the step duration δt_{CW} is adjusted between 0.1 ms and 15 ms in increments of 0.1 ms. The capture times for each frequency step are calculated as [9]:

$$t_{\text{Capture\_CW}} = [0, \delta t_{CW}, 2\delta t_{CW}, \ldots, (p-1) \cdot \delta t_{CW}], \tag{13.1}$$

where p represents the number of CW points at a given frequency. The initial capture distance, d_0, is varied between 380 mm and 680 mm in 0.1-mm increments to optimise the measurement process. The actual distance of the moving object during capture is given by Al-Dabbagh et al. [9]:

$$d_{\text{capture}} = d_0 - v \cdot t_{\text{Capture\_CW}}, \tag{13.2}$$

where v is the velocity of the moving reflector, determined by the linear stage, with speeds ranging from 0 mm/s to 200 mm/s. The effective path length for each measurement is calculated as [9]:

$$Path = 2 \cdot \sqrt{\left(\frac{d_{\text{ant}}}{2}\right)^2 + d_{\text{capture}}^2}, \tag{13.3}$$

where d_{ant} is the fixed separation between the horn antennas, set at 190 mm. The theoretical phase and its unwrapped counterpart are computed using [9]:

$$\varphi = -2\pi \left(\frac{Path \cdot f}{c}\right) + \varphi_0, \tag{13.4}$$

where φ represents the computed phase, φ_0 is the initial phase, c is the speed of light, and f is the frequency. The unwrapped phase is then calculated as [9]:

$$\varphi_{\text{Uphase}} = \text{unwrap}(\varphi) \cdot \frac{180°}{\pi}. \tag{13.5}$$

To analyse phase variations, the difference between the start and end of a CW segment is evaluated as [9]:

$$\delta\varphi_{\text{Uphase\_CW}} = \varphi_{\text{Uphase}}(i) - \varphi_{\text{Uphase}}(j), \tag{13.6}$$

where i and j correspond to the start and end points of the CW segment, respectively. The deviation between measured and theoretical phase shifts is determined using [9]:

$$\varphi_{\text{CW}} = \delta\varphi_{\text{Uphase\_CW,meas}} - \delta\varphi_{\text{Uphase\_CW,theor}}. \tag{13.7}$$

For LS segments, a similar process is followed, focusing on optimising the step duration δt_{LS} over a range of 0.1 ms to 15 ms in 0.1-ms increments. A grid search is conducted to minimise the phase error, defined as [9]:

$$\text{err} = \sum_f \left(\varphi_{\text{Uphase\_LS,meas}}(f) - \varphi_{\text{Uphase\_LS,theor}}(f) \right)^2, \tag{13.8}$$

where $\varphi_{\text{Uphase\_LS,meas}}$ and $\varphi_{\text{Uphase\_LS,theor}}$ correspond to the measured and theoretical unwrapped phase at each frequency f for the LS segment.

13.5 Measurement Results and Discussion

Following the TRL calibration of the segmented waveform, the acquired S_{21} measurements are examined and compared with simulated data to assess the precision and consistency of the experimental setup.

13.5.1 CW Segment Analysis

The optimisation process began with an analysis of three CW segments at 220 GHz, 240 GHz, and 260 GHz, tested across different velocities. During this evaluation, φ_{CW} was computed using the optimised values of δt_{CW} and d_0. Since CW segments do not inherently provide position information, the fixed separation distance d_{ant} between the horn antennas caused a non-linear variation in path length as the reflector moved. To compensate for these variations and improve alignment with the actual motion, d_0 was optimised [9].

Figure 13.5 presents the optimised CW segments at different δt_{CW} values for IFBWs of 1 kHz and 10 kHz. The measurement results in Fig. 13.5a correspond to segmented CW signals recorded at velocities of 10 mm/s, 20 mm/s, and 50 mm/s. The best fit for φ_{CW} across all tested velocities yielded an optimised δt_{CW} of 2.07 ms at 1 kHz IFBW.

Figure 13.5b shows the 10-kHz IFBW measurements, where faster sweeps allow for velocity assessments up to 200 mm/s. The figure illustrates the φ_{CW} calculations for three CW segments at various δt_{CW} values. Across all measured velocities and frequencies, the best fit resulted in an optimised δt_{CW} of 0.4 ms [9].

13.5.2 LS Segment Analysis

The evaluation of the LS segment provides insight into the reflector's position at each frequency point, which can also be interpreted through the power delay profile

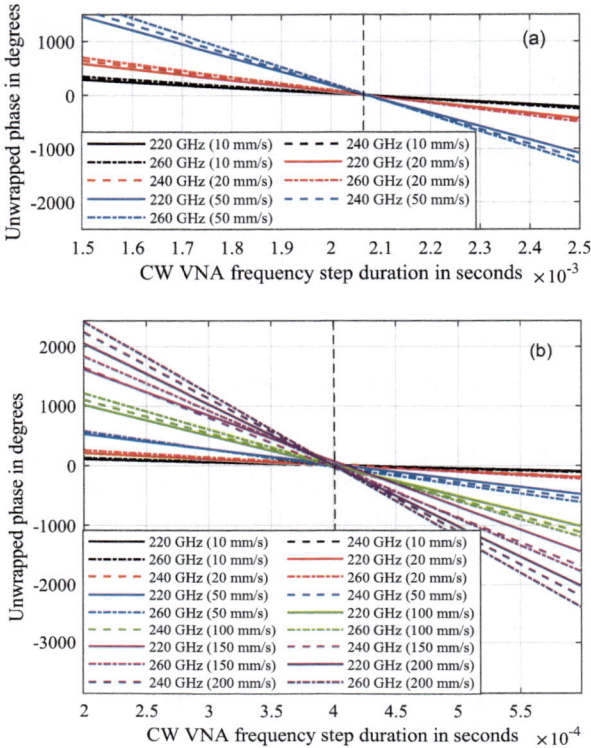

Fig. 13.5 Unwrapped phase agreement between the measured and simulated time-varying positions, represented by φ_{CW} vs. different CW step durations δt_{CW} at various velocities: (**a**) IFBW $= 1$ kHz and (**b**) IFBW $= 10$ kHz (adopted from [9])

(PDP). Figure 13.6 presents an example of the PDP for a time-varying signal, highlighting the agreement between measured and simulated data. This agreement is observed when the reflector moves at a velocity of 10 mm/s, with the VNA IFBW set to 1 kHz in Fig. 13.6a and 10 kHz in Fig. 13.6b.

The optimisation was applied to determine δt_{LS} and d_0 at various measurement velocities using both IFBWs. Figure 13.7 presents the results for measurements taken at 1 kHz IFBW, with tested velocities of 10 mm/s, 20 mm/s, and 50 mm/s. The optimised δt_{LS} values ranged between 2.8 ms and 3 ms. The parameter d_0 exhibited a progressive decrease across captured segments, starting from 680 mm and reducing to 380 mm, showing the expected change in the reflector's position.

A similar optimisation was conducted at 10 kHz IFBW, with the findings shown in Fig. 13.8. The results indicate δt_{LS} values ranging from 1.1 ms to 1.5 ms. For velocities between 50 mm/s and 200 mm/s, δt_{LS} stabilised around 1.1 ms to 1.2 ms, demonstrating greater consistency at higher speeds due to reduced multiple fitting occurrences. The trend in d_0 followed the expected decline across captured

Fig. 13.6 Measurement and simulation agreement of the LS segment for the frequency band between 270 GHz and 300 GHz, PDP of the moving reflector at different positions with a velocity of 10 mm/s (**a**) IFBW = 1 kHz and (**b**) IFBW = 10 kHz

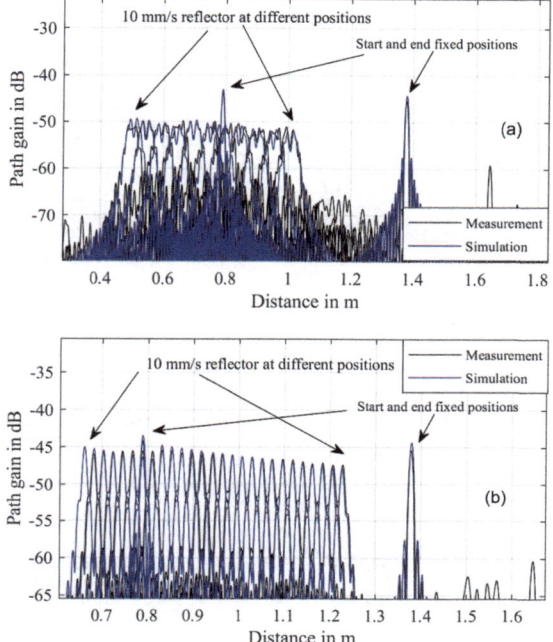

segments, confirming a steady reduction in distance until reaching the final fixed location [9].

13.6 Conclusion

The analysis presented in this chapter represent a unique investigation using the VNA, where time-varying signals present unique challenges to consider, as VNAs are traditionally designed for static frequency analysis. To extend their capabilities, it is necessary to adapt the measurement approach by introducing waveforms suitable for dynamic conditions. By carefully selecting segmented waveforms and optimising acquisition parameters, a VNA can be used to capture limited Doppler shifts, providing a reference for comparing and calibrating other CS systems.

Segmented waveforms, which consist of discrete frequency steps, offer a practical compromise between time resolution and frequency coverage. These waveforms are particularly valuable when both velocity and position information need to be captured simultaneously. Segmented waveforms enable high-accuracy motion analysis by precisely tuning the frequency steps and segment durations while maintaining broad spectral coverage.

In this chapter, a compact measurement setup was developed to investigate time-varying signal propagation using a VNA. The proposed testbed provides a controlled

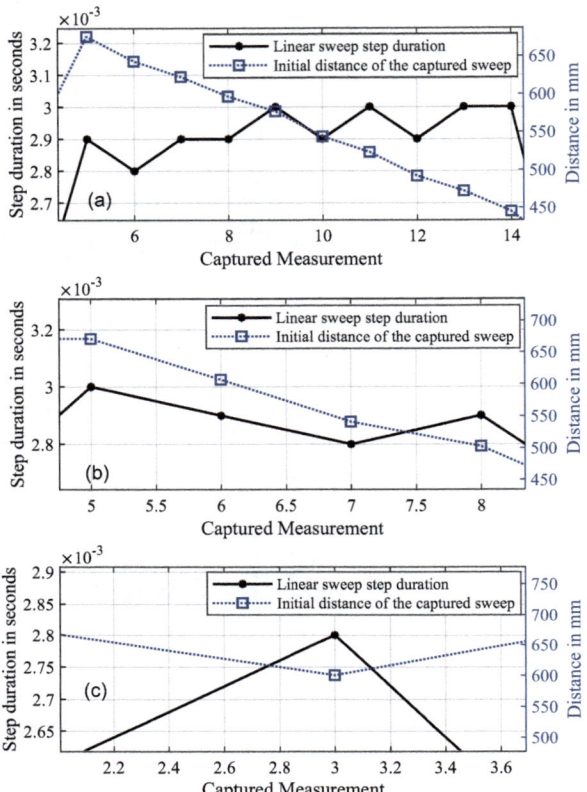

Fig. 13.7 Optimised δt_{LS} and d_0 at 1 kHz IFBW, at different velocities: (**a**) 10 mm/s, (**b**) 20 mm/s, and (**c**) 50 mm/s (adopted from [9])

environment to characterise CS systems operating at sub-THz frequencies. This investigation will be extended to find the optimal waveform selection to find the best parameter optimisation capabilities.

The results highlight the importance of directive antennas in ensuring sufficient gain to overcome noise, mainly when higher IFBWs are employed for faster sweeps.

Ensuring measurement accuracy in dynamic conditions also requires a stable setup. Securing the frequency extension modules and minimising cable movement are essential to prevent errors introduced by mechanical shifts. Maintaining a consistent and controlled environment also mitigates external factors that could affect signal integrity.

The investigation was restricted to single-direction movement to simplify the analysis, with clearly defined start and end positions throughout the measurement sequence. The results revealed variations in time intervals across different IFBWs, along with distinct behaviours in CW and LS segments. Optimised acquisition

Fig. 13.8 Optimised δt_{LS} and d_0 at 10 kHz IFBW, at different velocities: (**a**) 10 mm/s, (**b**) 20 mm/s, (**c**) 50 mm/s, (**d**) 100 mm/s, and (**e**) 200 mm/s (adopted from [9])

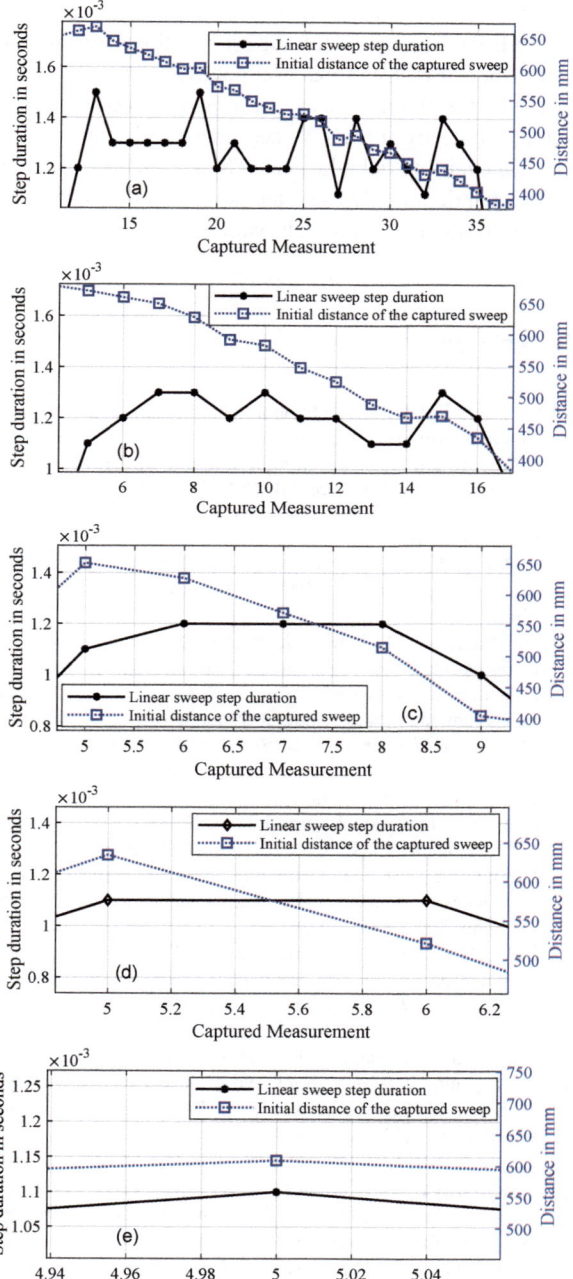

parameters were identified for velocities of 50 mm/s and above, demonstrating the feasibility of using a VNA for precise time-varying signal analysis.

Further investigations are necessary to effectively address the uncertainties associated with each measured position and velocity. This includes integrating the velocity data from the CW segments with the positional information derived from the LS, enabling calculations across the entire segmented waveform. Additionally, an independent analysis of the stage's velocity, performed using separate measurement equipment, would help validate the velocity values obtained by the VNA.

References

1. G. Matz, F. Hlawatsch, Fundamentals of time-varying communication channels, in *Wireless Communications over Rapidly Time-Varying Channels* (Elsevier, Amsterdam, 2011), pp. 1–63
2. J. Wu, P. Fan, A survey on high mobility wireless communications: challenges, opportunities and solutions. IEEE Access **4**, 450–476 (2016)
3. F. Hlawatsch, G. Matz, *Wireless Communications over Rapidly Time-Varying Channels* (Academic Press, Cambridge, 2011)
4. J. Wang, W. Zhang, Y. Chen, Z. Liu, J. Sun, C.-X. Wang, Time-varying channel estimation scheme for uplink MU-MIMO in 6G systems. IEEE Trans. Veh. Tech. **71**(11), 11 820–11 831 (2022)
5. G.P. Sharma et al., Toward deterministic communications in 6G networks: state of the art, open challenges and the way forward. IEEE Access **11**, 106 898–106 923 (2023)
6. N. Ridler, R. Clarke, M. Salter, A. Wilson, Traceability to national standards for S-parameter measurements in waveguide at frequencies from 140 GHz to 220 GHz, in *2010 76th ARFTG Microwave Measurement Conference* (IEEE, Piscataway, 2010), pp. 1–7
7. M. Zeier, D. Allal, R. Judaschke, *Euramet Calibration Guide No. 12: Guidelines on the Evaluation of Vector Network Analysers (VNA)*, vol. 3 (European Association of National Metrology Institutes, Braunschweig, 2018)
8. M.D. Al-Dabbagh, D. Ulm, T. Doeker, D. Dupleich, A. Ebert, R.S. Thomä, T. Kürner, D.A. Humphreys, T. Kleine-Ostmann, Characterization of sub-THz channel sounding systems in OTA measurement scenarios using a vector network analyzer. IEEE Trans. Antennas Propag. **73**(6) pp. 3943–3958 (2025)
9. M.D. Al-Dabbagh, D.A. Humphreys, T. Kleine-Ostmann, Investigating time-varying signal propagation at sub-THz frequencies using a VNA, in *2025 19th European Conference on Antennas and Propagation (EuCAP)* (2025), pp. 1–5
10. M. Zhao et al., Measurement of the rotational Doppler frequency shift of a spinning object using a radio frequency orbital angular momentum beam. Opt. Lett. **41**(11), 2549–2552 (2016)
11. S. Hemour, N. Barbot, Backscattering modulation 101: VNA measurements, in *2023 IEEE 13th International Conference on RFID Technology and Applications (RFID-TA)* (IEEE, Piscataway, 2023), pp. 169–172
12. D. Jung et al., Terahertz antenna-in-package design and measurement for 6G communications system. IEEE Trans. Antennas and Propag. **72**(2), 1085–1096 (2024)
13. M.D. Al-Dabbagh, D. Ulm, T. Kleine-Ostmann, D. Humphreys, Horn antenna phase center position influence on sub-THz measurements uncertainties, in *2024 18th European Conference on Antennas and Propagation (EuCAP)* (2024), pp. 1–5

Chapter 14
THz Waveform Traceability

David A. Humphreys, Heiko Füser, Adam Kuchnia, and Dominik Wrana

Abstract Communication systems are defined in detail by specification standards but traceability, 'an unbroken chain of measurements from the device under test to the relevant primary SI standards', is essential to specify, buy, and test components and systems. Above 110 GHz, two or more near identical mixers can be used together to estimate conversion loss and group delay. Here we investigate a free-space approach, based on a photoconductive sampling detector and a frequency comb generated by a femtosecond mode-locked laser. The technique has been demonstrated experimentally at 100 GHz using a pre-existing device and optical pulses from the PTB primary standard electro-optic sampling system. The existing antenna and higher-frequency designs (100 GHz/300 GHz) have been modelled for verification and to identify potential design issues. In operation, the high-frequency signal components are down-converted to a 38 MHz frequency space. We have developed selection rules to allow direct down-conversion and identification of all the signal components of a pseudo random-number QAM waveform. This will allow traceable characterisation of free-space complex modulated THz waveforms.

D. A. Humphreys (✉)
NPL, UK (Retired), University of Bristol, UK
e-mail: david.a.humphreys@ieee.org

H. Füser · A. Kuchnia
Physikalisch-Technische Bundesanstalt (PTB), Braunschweig, Germany
e-mail: heiko.fueser@ptb.de; adam.kuchnia@ptb.de

D. Wrana
Universität Stuttgart, Institut für Robuste Leistungshalbleitersysteme, Stuttgart, Germany
e-mail: dominik.wrana@ilh.uni-stuttgart.de

© The Author(s) 2026
T. Kürner et al. (eds.), *Metrology for THz Communications*, Springer Series in Optical Sciences 256, https://doi.org/10.1007/978-3-032-01986-8_14

14.1 Introduction

We understand the importance of traceability from Chap. 2 and that different traceability paths can be applied to test equipment and components for THz communication. Although RF power is a scalar parameter, it is important for calibration of nonlinear devices and amplifiers Chap. 3.

Section 14.2 extends the analysis of RF mixer cross-calibration method outlined in Chap. 5. At frequencies over 100 GHz, the dimensional tolerances on waveguide components are very tight and small misalignments introduce connection repeatability uncertainties; see Chap. 13. Most VNA systems operate with continuous-wave (CW) signals but some recently developed instruments include modulators to create complex waveforms; see Chap. 24. The ability to calibrate the conversion loss and phase delay together with a supporting uncertainty analysis provides a traceability route for guided-wave vector closed-loop measurements of passive and active components.

Traceable measurement solutions for independent modulated sources in a free-space environment are an open-loop problem where the source and detector do not share the same RF oscillator source; see Chap. 5. Possible options, based of electro-optic sampling and fs photoconductive devices are discussed in Sect. 14.3, demonstrating the operating principle in Sect. 14.4. We outline preliminary work using a picosecond photoconductive switch as a mixing element. The existing prototype device is modelled in Sect. 14.5 and this model is extended to a 100 GHz/300 GHz design in Sect. 14.6.

Communication signals are narrow-band, typically 1% to 3% of the carrier frequency, but at 300 GHz this represents a significant modulation bandwidth, for example, 3% is a 9 GHz complex modulation. Restrictions on the waveform design for traceable measurements of a complex modulated waveform using a photoconductive sampler are discussed in Sect. 14.7.

Fig. 14.1 Test configuration for mixer calibration using a DSO, after [1]. The third synthesiser has been added; otherwise, the default trigger rate is 10 MHz

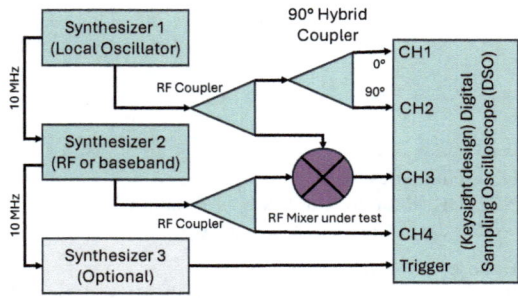

14.2   Mixer-Based Strategies

Vector network analysers use mixers to up-convert and down-convert RF signals. Mixers can be calibrated at lower frequencies where a calibrated digital sampling oscilloscope (DSO) [1] or a phase-standard and a non-linear vector network analyser (NVNA), such as a Keysight PNA-X, is available [2]. The DSO approach requires timebase linearity correction and separate measurements of the system and DUT S-parameters; see Fig. 14.1. The measurement epoch must be long enough to capture at least one period of the lowest frequency signal. All the frequency components (baseband, LO, and RF) must be harmonics of the trigger frequency such that $f_1 = n_1 \cdot f_3$ and $f_2 = n_2 \cdot f_3$ where n_1 and n_2 are natural numbers. The original paper used a 10 MHz trigger rate (100 ns measurement epoch). Here, a third synthesiser has been added to allow greater flexibility in the choice of operating frequencies and measurement epoch. As discussed in Chap. 5, the timebase correction algorithm requires synchronous operation across all the DSO channels. Although stitching methods can be used to extend the time epoch, the number of measurement points needs to be sufficient to represent adequately the highest frequency measured. For example, a calibrated 70 GHz bandwidth DSO (1.85-mm coaxial connectors) with a RF of 65 GHz and 5120 points in the epoch used with a 20-ns epoch (50 MHz) would give an adequate representation of the RF signal (3.938 points per cycle).

The VNA approach can measure the mixer performance and the necessary S-parameters. This is described in depth in a comprehensive application note [3].

At higher frequencies where a phase standard is not available, a pair of nominally identical mixers can be measured to determine conversion loss and group delay using up-conversion to THz followed by filtering to remove unwanted harmonic components and down-conversion to baseband. This assumes that the mixers are reciprocal.

Mixers are sensitive to RF power levels and impedance match; see Chap. 5. The passive technique to determine and correct for active device impedance match can be included with the harmonic filter component, provided that the mixer performance remains linear over the range of impedance match changes. This two-port network must be characterised with a VNA for all states.

The key identities that link the up-conversion (sum) and down-conversion (difference) frequency components are based on the sine and cosine identities which derive from $\exp(j\theta)\exp(j\phi) = \exp(j(\theta + \phi))$ and $\exp(j\theta) = \cos(\theta) + j\sin(\theta)$. These govern the relationship between the intermediate frequency (IF), RF, and local oscillator (LO) frequencies [4].

An individual pair of reciprocal and nominally identical mixers, designated A and B, provide two sets of result for the frequency-dependent up-conversion and down-conversion as the components are assumed to be reciprocal, $X_{AB}(f_i) = A(f_i) \cdot B(f_i)$ and $X_{BA}(f_i) = B(f_i) \cdot A(f_i)$, where i is the frequency index and X is the measured complex response at f_i. This can be rewritten as a linear algebra problem by taking the logarithm of the products; see Eq. 14.1, where $\phi_A(f_i) = \arg(A(f_i))$

and $\phi_B(f_i) = \arg(B(f_i))$.

$$\log(X_{AB}(f_i)) = \log(|A(f_i)|) + \log(|B(f_i)|) + j \cdot \phi_A(f_i) + j \cdot \phi_B(f_i) \qquad (14.1)$$

'Phase wrapping' leads to ambiguous results where the sum of the phase delays exceeds 2π radians. As measurements are typically performed over a range of frequencies, calculating the group delay $\tau(f_i)$ is more robust, Eq. 14.2.

$$\tau_A(f) = -\frac{1}{2\pi} \frac{\mathrm{d}\phi(f)}{\mathrm{d}f}(\phi_A(f)) \qquad (14.2)$$

and in discrete form (Eq. 14.3),

$$\tau_A(f_i) + \tau_B(f_i) = -\frac{1}{2\pi(f_i - f_{i-1})}\arg\left(\frac{X_{AB}(f_i)}{X_{AB}(f_{i-1})}\right) \qquad (14.3)$$

As reciprocity, $\tau_A(f_i) = \tau_B(f_i)$ and $|A(f_i)| = |B(f_i)|$ are assumed, the average complex conversion loss can be calculated. With three or more devices, a better estimate of the complex conversion loss and confidence interval can be made. However, because of the cost of these components, it is unlikely that individuals will have sufficient devices to do this, but a manufacturer will have access to a larger number of similar components permitting a more detailed analysis.The results are measured at baseband, so traceability is through lower-frequency references, such as phase standards.

14.3 Free-Space Direct Down-Conversion

Direct down-conversion of the modulated THz radiation to baseband provides a free-space, physics-based approach to traceability. This solution imposes certain restrictions on the THz RF carrier and modulation frequency stability, the modulated waveform and carrier frequency, for example, the length of the pseudo-random data sequences.

14.3.1 Heterodyne and Sampling

The optical source is a mode-locked laser with a 76 MHz repetition rate, producing 140 fs optical pulses. In the frequency domain this can be treated as a comb of frequencies, starting at dc with a 76 MHz frequency spacing. The strict Nyquist sampling restricts the baseband to 38 MHz. A CW THz signal will be down-converted to this frequency band by the closest comb harmonic to the CW frequency; see Fig. 14.2a. This is a vector process and the phase of each THz frequency component is retained but it is important to know if the baseband signal f_b component in

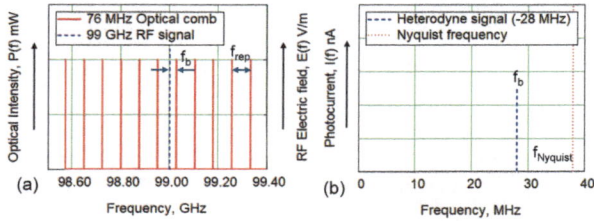

Fig. 14.2 Heterodyne mixing (**a**) 76 MHz optical comb (red) and THz CW signal (blue). (**b**) shows the real baseband difference frequency component at 28 MHz

Fig. 14.2b originates from the upper harmonic (Fig. 14.2a), which is a negative frequency with conjugate phase, or from the lower harmonic, corresponding to a positive frequency.

From the time-domain sampling viewpoint, the optical pulse and detection system has sufficiently short sampling aperture to sample within a single cycle of the modulated RF signal. The recovered and low-pass filtered result will contain both the sampled, modulated waveform, and noise. For a CW signal, the measurement epoch, a fixed number of samples, should also be an integer number of THz RF cycles.

The rules governing epoch duration for a modulated waveform are set out in Sect. 14.7. As synthesisers are subject to drift, additional information about the oscillator frequency and phase stability can be recovered from successive measurement blocks (see Sect. 14.7.1) [5].

14.3.2 Receiver Antenna

The ideal antenna element should be broadband, free from additional resonances and have a near uniform radiation pattern to give a high-quality wavefront. A simple dipole antenna meets these criteria with the exception of a limited bandwidth. Bi-conic antennas also have a dipole-like radiation pattern and a broader bandwidth but are not physically practical from a fabrication perspective. The compromise is to use a "bow-tie" antenna which is a planar antenna deposited directly onto the semiconductor substrate. This antenna geometry was introduced in [6] and theoretically described in [7]. Studies at 2.4 GHz have shown good radiation patterns and detailed modelling of surface currents for triangular and circular arc bow-tie designs [8]. In this application, the GaAs substrate thickness ($500\,\mu$m) corresponds to 1.8λ at 300 GHz.

The permittivity of the GaAs is 12.9 and for a semi-infinite material thickness the effective permittivity, assuming a semi-infinite substrate is given by Eq. 14.4.

$$\epsilon_{\text{eff}} = \frac{1 + \epsilon_{\text{GaAs}}}{2} \tag{14.4}$$

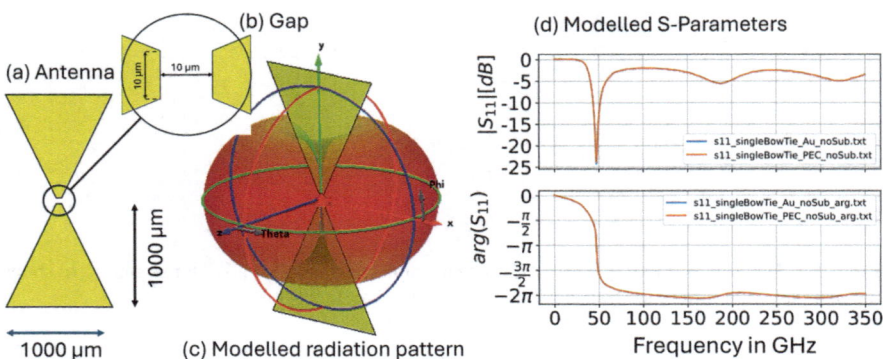

Fig. 14.3 Bow-tie antenna (**a**) and gap (**b**) dimensions. The modelled results are shown for the radiation pattern (**c**) and S-parameters (**d**) of the antenna in a uniform (air) dielectric. The main antenna resonance occurs at 47.25 GHz, and higher-frequency harmonic resonances are indicated by dips in the S_{11} response. Note that there is little sensitivity along the 'y'-axis

The bow-tie antenna and gap dimensions are shown in Fig. 14.3a, b. This design was modelled in a uniform (air) dielectric (see Sect. 14.5) and the results for the far-field radiation pattern and S-parameters are shown in Fig. 14.3c, d, respectively. The radiation pattern is uniform meeting the the radiation uniformity and wavefront criteria. In operation, the received THz radiation produces a voltage across the gap between the triangular antenna elements. This electric field can be sensed either via the electro-optic effect (EOS) in the GaAs or by photoconductive sampling using a low-temperature-grown GaAs epitaxial layer. The effect of the GaAs substrate is explored in Sect. 14.5.

14.3.3 Sampler Options to Recover the Waveform

There are two options to down-convert the THz radiation to baseband. These are to directly probe the electric fields using the electro-optic effect in the GaAs substrate or to use the photoconductor as a mixer to provide down-conversion to frequencies below 38 MHz.

The EOS approach is passive, samples the THz signal directly and requires no connection to the antenna elements. Although the electro-optic process is fast, the overall speed is limited by the overlap time between the electrical and optical waveforms. Also, a substrate light-entry EOS design has multiple optical reflections that must be corrected [9]. This is covered in detail in Chap. 12. As a passive device, verification of the complex frequency response is more difficult.

In principle, both EOS and LT-GaAs photoconductive mixing have sufficient bandwidth to measure a THz repetitive waveform; see Chap. 12. The main discriminating factor is that EOS is purely passive but the photoconductor approach can operate as a transmitter or as a receiver, depending on the antenna bias conditions. Two or more near identical antennas can be used to 'self-calibrate' and determine the antenna frequency and phase response. The analysis approach is the same as used in Sect. 14.2. For these reasons, the high-speed LT-GaAs photoconductor has been chosen for use as a traceable RF detector.

14.3.4 THz Photoconductive Device

To understand a photoconductive detector it is worthwhile to start with the carrier dynamics, governed by the rate equations covering non-thermal equilibrium behaviour; see Eqs. 14.5 and 14.6. The photon generates an electron-hole pair, expressed by the generation term $G(t)$ which is the same for both electrons (n) and holes (p). The carrier lifetimes τ_p, τ_n determine the recombination rate.

$$\frac{d}{dt}n = G(t) - \frac{n}{\tau_n}, \tag{14.5}$$

and

$$\frac{d}{dt}p = G(t) - \frac{p}{\tau_p}, \tag{14.6}$$

On the sub-picosecond timescales diffusion contributions can be ignored [10] reducing the familiar Shockley current-density relations to Eq. 14.7.

$$J(x) = q(n(x)\mu_n + p(x)\mu_p)E(x) \tag{14.7}$$

where q is the electronic charge $E(x)$ is the electric field generated across the $(10\,\mu\text{m})$ photoconductive gap between the antenna elements, and $n(x)$ and $p(x)$ are the carrier densities, generated as electron-hole pairs, by the mode-locked laser pulse. The electron mobility (μ_n) is far greater than the hole mobility (μ_p), and although the holes still account for half the optically generated carrier population, their contribution is about 4.7 %.

Photoconductors have ohmic contacts, with very low barriers at the metal to semiconductor interface, allowing bidirectional current flow. Consequently from Eq. 14.7, the photo-carriers generated by the short optical pulse sample the instantaneous voltage across the antenna gap, causing a current flow that is proportional to the electric field across the gap.

GaAs is normally grown at a temperature of about 600 °C and has a carrier lifetime of about 32 ns [11], but gallium arsenide grown by molecular beam epitaxy (MBE) on a GaAs substrate at low temperatures of about 200 °C (LT-GaAs) is crystalline but contains a high density $(n_t > 10^{18}/\text{cm}^3)$ of point defects, which act as recombination and trapping centres greatly shortening the carrier lifetime; see

Fig. 14.4
Low-temperature-grown
GaAs has many
recombination centres, giving
a carrier lifetime of <1 ps,
compared with high-quality
GaAs epitaxial films which
have a carrier lifetime of
≈32 ns

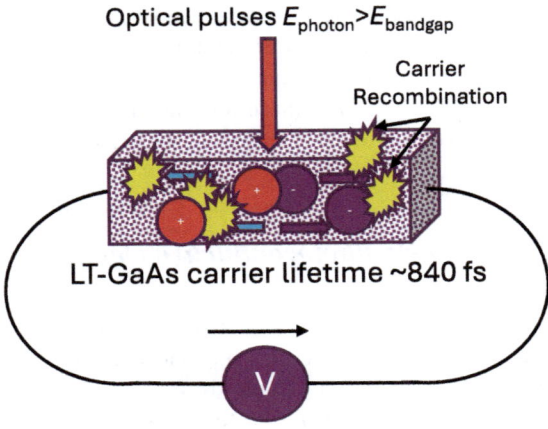

Fig. 14.4. For example, the LT-GaAs material grown at PTB has a carrier lifetime of 840 fs, determined by pump-probe reflectivity change [9]. In THz time-domain spectroscopy (TDS), photoconductive switches generate sufficient high-frequency components to measure material properties over more than 2 THz Chap. 7. The antenna and photoconductor can be used both as a transmitter and as a receiver. In transmitter mode (constant bias voltage), a photoconductor will generate a current that is proportional to the product of the applied voltage and the optical illumination, radiating the RF components in the passband of the antenna.

14.4   Existing Prototype Device and Concept Verification

In previous work [12] bow-tie antennas, designed for operation at around 100 GHz, were deposited on a 500 μm-thick GaAs/LT-GaAs substrate. The dimensions of the antenna elements are shown in Fig. 14.3a. The high electric field region between the electrodes is shown in Fig. 14.3b. At the gap, the substrate thickness/gap width has a high ratio (50:1), but at the widest dimensions of the antenna, this falls to a ratio of 1:2 which may have implications for modelling as the semi-infinite substrate approximation will not apply. Six devices were fabricated and these are connected by bond wires, as shown in Fig. 14.5.

14.4.1   Initial Measurements to Prove the Operating Principle

The down-conversion was evaluated using the pre-existing photoconductive antenna [12]. At this point, it was unknown whether the antenna had sensitivity near 100 GHz. The femtosecond laser source produces 140 fs optical pulses at a nominal

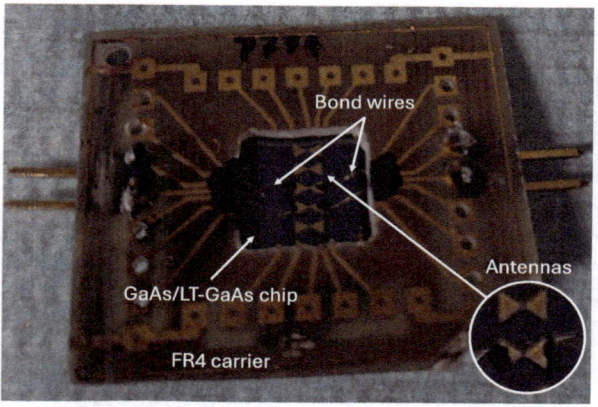

Fig. 14.5 Close-up photo of prototype (P774) photoconductive bow-tie antennas. The design comprised of six antennas on a GaAs/LT-GaAs substrate and FR4 carrier. Two antennas are connected with bond wires

repetition rate of 76 MHz. As the laser is absorbed by both the GaAs substrate and the LT-GaAs epitaxial layer, the optical illumination must be from the top surface. For convenience, the THz radiation illuminates the antenna through the substrate. The THz source, shown in Fig. 14.6, used a synthesised RF source and a × 6 harmonic multiplier. To maintain phase stability this is phase-locked to the laser frequency reference. In this case, the THz electric field across the gap results in a photocurrent of a few nanoampere, which can be measured with commonly available measurement electronics [13].

Two frequencies of 6 × 16.5000 GHz (99 GHz) and 6 × 16.5005 GHz (99.003 GHz) were evaluated for the test. As the exact RF and laser frequencies are known, the closest laser comb harmonic number (1303) can be found by rounding the result from Eq. 14.8. The expected beat frequency and sign can be calculated using Eq. 14.9, and this process can be repeated for all the frequency components in a complex waveform.

$$\frac{6 \cdot 16500\,\text{MHz}}{76\,\text{MHz}} = 1302.63 \tag{14.8}$$

$$6 \cdot 16500\,\text{MHz} - 1303 \cdot 76\,\text{MHz} = -28\,\text{MHz} \tag{14.9}$$

It was noted that other frequency components, due to interference and unintended mixing components, were present in the results, shown in Fig. 14.7.

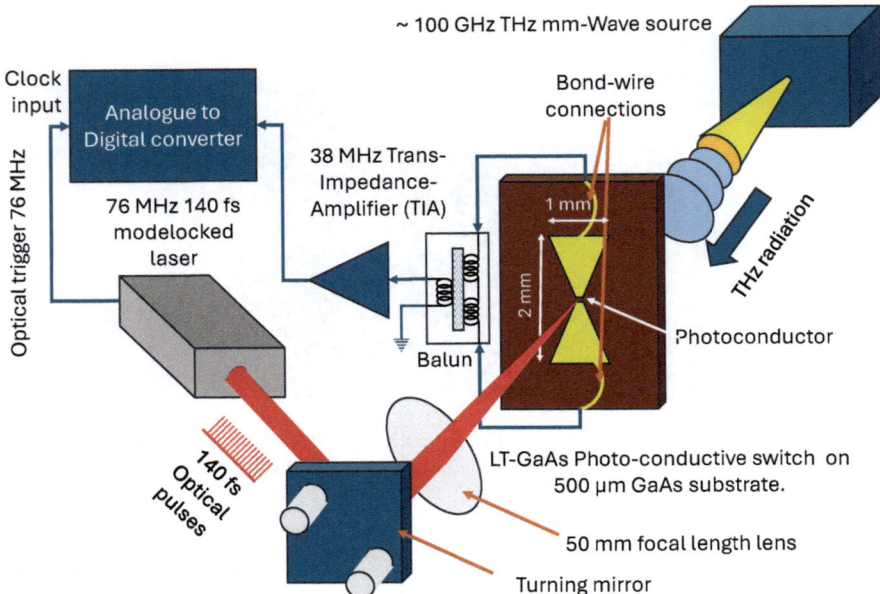

Fig. 14.6 Proof-of-principle heterodyne mixing experiment configuration using the pre-existing photoconductive antenna [12]. The balun transformer was added to improve the system signal-to-noise ratio. The laser repetition frequency and RF synthesiser are phase-locked (not shown). Several turning mirrors were used to ease the optical alignment to the $10\,\mu m \times 10\,\mu m$ photo-conductor (only one mirror shown).The reference distribution to phase-lock the laser repetition frequency and RF synthesiser is not shown

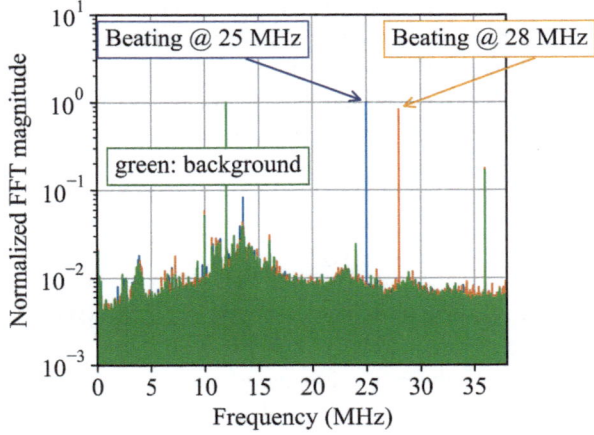

Fig. 14.7 Experimental verification results for the photoconductive mixer [12] at 99 GHz and 99.003 GHz. The green (background) infers a typical signal-to-noise ratio of 40 dB

14.5   Antenna Modelling

Modelling the existing antenna behaviour allows self-verification against measurements and identification of potential issues for a 300 GHz antenna design. The properties of a bow-tie antenna modelled in a uniform dielectric (air) shows a uniform radiation pattern; see Fig. 14.3c.

The antenna model works on the principle that the antenna elements are lossless (perfect electrical conductor [PEC]) or have a low loss (evaporated gold antenna elements), so any power that is not radiated (all modes) will be reflected back to the source. In terms of the travelling waves into a_i and from b_i, a lossless network has the property that $\Sigma_i |a_i|^2 = \Sigma_i |b_i|^2$. The models assume a discrete 50 Ω impedance port connecting the two electrodes of the antenna. The port is excited with a pulse and the software calculates the field distributions and overall electric properties of the structure for a list of given frequencies.

The normalised reflected wave is expressed by $S_{11} = \frac{b_1}{a_1}$. The antenna will radiate in a number of modes and the model includes the substrate material and other elements, which couple power back into the antenna and will alter the S_{11} result. The sum of the power in all the radiated modes can be considered as the second port, and therefore, the normalised total radiated power will be $1 - |S_{11}|^2$. The peak RF radiation occurs when the antenna is at its resonant frequency. The maximum power transfer occurs when the source and antenna impedances form a conjugate matched pair.

14.5.1   Modelling of Existing Test Device

A simple free-floating bow-tie antenna, based on the dimensions of the existing prototype (Fig. 14.3a, b), was modelled as a test device with both PEC and gold as the antenna material. $|S_{11}|$ has a minimum with a value of about -24.2 dB at a frequency of 47.25 GHz; see Fig. 14.3c.

The photograph of the prototype device (Fig. 14.5) shows six antenna elements on the GaAs substrate. As these have been arranged in parallel, rather than in series, there may be some coupling between antenna elements. Although bond wires are shown on two antennas in the photograph, these were not included in the model, which comprises the six bow-tie antennas on LT-GaAs grown on a 500μm-thick GaAs substrate with only a single active antenna which would be illuminated through the GaAs substrate (see Fig. 14.7). Assuming reciprocity, the radiation pattern from the substrate will dictate the antenna performance. The results (Fig. 14.8) show a complex radiation pattern at 100 GHz. Although the gold conductor will have some associated loss, the difference between PEC and gold antennas was negligible. Similarly, the difference between a single antenna element and the full six antennas was also negligible. Analysis of the same antenna design on a semi-infinite GaAs substrate gave a more uniform response, similar to Fig. 14.3c

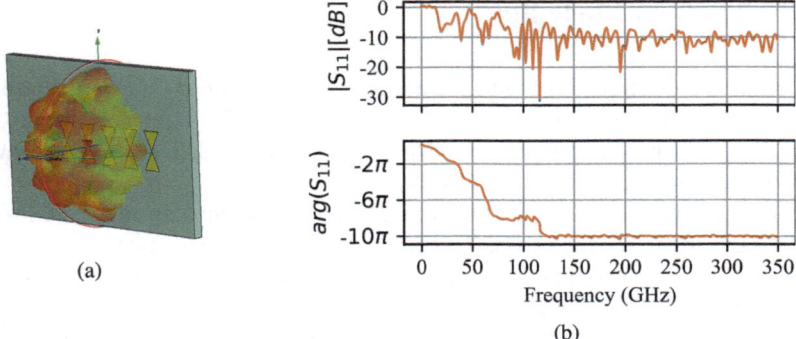

Fig. 14.8 Radiation pattern and S-parameter simulation results for P774 prototype bow-tie antennas on the LT-GaAs/GaAs substrate. (**a**) Modelled radiation pattern at ($f = 100\,\text{GHz}$) from one of six bow-tie antennas on LT-GaAs/GaAs substrate. (**b**) Simulated antenna S_{11}. magnitude and phase based on a characteristic line impedance of 50 Ω. No significant phase changes occur above 125 GHz

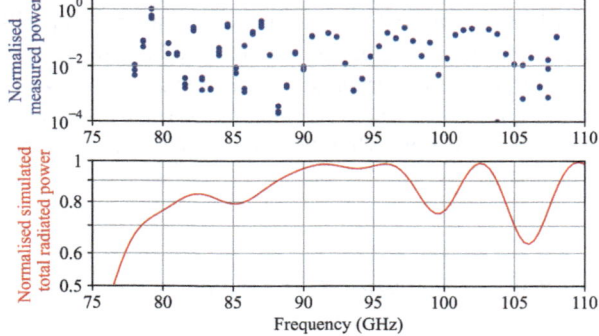

Fig. 14.9 Comparison of measurement and simulation results. The upper plot shows measurements of CW power vs. frequency. The lower plot shows a simulation of the total radiated power vs. frequency. In this plot, the variations are much lower, suggesting that the radiation pattern is frequency dependent

but with the first resonance at 19 GHz due to the high permittivity of the substrate ($\epsilon_{GaAs} = 12.9$). The experimental and modelling results suggest that 100 GHz is the 5th harmonic of the first resonance.

14.5.2 Modelling Outcomes

The following conclusions can be drawn from the measured and modelled results, shown in Fig. 14.8. The complexity of the antenna pattern and the S_{11} appears to be linked to the substrate thickness; Figs. 14.9 and 14.10. Firstly, the substrate must be

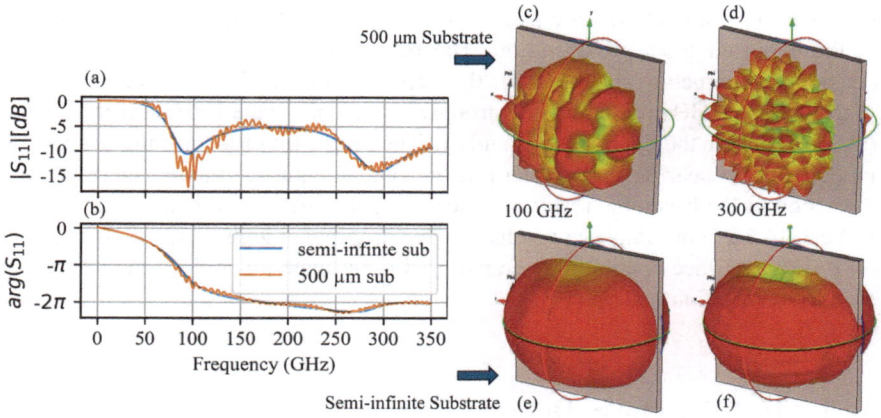

Fig. 14.10 Simulated response, (**a**) and (**b**) and antenna pattern for a 100 GHz/300 GHz antenna on LT-GaAs on 500 μ mthick GaAs substrate (**c**) and (**d**) and LT-GaAs on semi-infinite GaAs substrate (**e**) and (**f**)

thick enough to behave as semi-infinite material. The 3–4 mm-thick GaAs optical windows are commercially available. The radiation pattern for odd harmonics has been modelled for a semi-infinite structure. This shows an increasingly uneven radiation pattern with harmonic number. The original device used bond-wire connections to the antenna but as the antenna becomes smaller this will be impractical for smaller high-frequency devices so thin connection lines must be printed on the substrate.

14.6   Modelling and Design of 300 GHz LT-GaAs PCAntenna

The measurement and simulation results for the existing prototype show that although the fundamental operating frequency is the best and gives the most uniform response on a semi-infinite GaAs substrate, it may be feasible to fabricate a larger (lower-frequency) antenna that has sensitivity at 300 GHz. The existing lower-frequency antenna showed some sensitivity at 100 GHz (5th harmonic), and consequently, an antenna with a fundamental frequency of 100 GHz will be significantly smaller. A secondary effect is that coupling the low-frequency mixing components from the antenna must be connected by narrow <10 μm wide gold evaporated lines, rather than wire bonds. Also, some additional optics will be needed to focus the THz radiation onto the antenna element as the radiation from the horn coupled test equipment will be divergent.

The bow-tie antenna design has a length of 450 μm, a width of 150 μm and a gap of 10 μm. The simulation results are shown for both 500 μm-thick GaAs substrate and for a semi-infinite GaAs substrate are shown in Fig. 14.10. The modelled antenna patterns are shown from the substrate side, which corresponds to the expected mode of operation. The results for the standard substrate show considerable structure, in line with the previous result. This structure is finer at (harmonic) 300 GHz (d), but is still strongly in evidence at the 100 GHz fundamental (c) and visible in the S_{11} results (a) and (b). The results for the semi-infinite substrate model show a near-uniform pattern at the fundamental (c) though there is some distortion of the harmonic result (d). Increasing the substrate thickness, as outlined in Sect. 14.5.2, may improve results but as GaAs has a high dielectric constant at 300 GHz; a surface coating may be required to mitigate this reflection and provide a uniform radiation pattern [14].

14.7 Test Waveform Design

The frequency mapping of THz frequencies to the 38 MHz window by mixing the THz frequency component with the closest optical sampling comb frequency was shown in Sects. 14.3 and 14.4. The implication is that multiple frequency components can be mapped to the same frequency window with the caveat that some carrier and modulation frequencies will lead to degenerate results. It also means that the same frequency pairs can be generated using widely separated or closely separated CW THz signals to verify the system linearity.

Pseudo-random number (PRN) sequences are used to exercise all the constellation points. Each constellation point is represented by multiple bits, and the symbol sequence is chosen by including a delay between a common pseudo-random bit sequence (PRBS) for each bit in the symbol (see Chap. 15). If multiple symbol sequences are used, then each set will contain different delays to each bit sequence to minimise the cross-correlation terms.

A quadrature amplitude modulated symbol has m states (mQAM) (see Chap. 15), so 16QAM has 16 states. The length of a pseudo-random number sequence, PRNn, is $2^n - 1$ symbols. Equation 14.10 gives the criterion to populate all the (mQAM) states:

$$2^n - 1 > m \tag{14.10}$$

A longer sequence length (Eq. 14.11) is required to exercise all the transitions between the states.

$$2^n - 1 > m^2 \tag{14.11}$$

In cases where the PRN sequence must be kept short, it is still necessary to meet the population criterion (Eq. 14.10).

The PRN sequence is normally filtered by a two-part root raised-cosine (RRC) filter [15, 16] that limits the bandwidth spread on transmission and the received noise bandwidth. When sampling the radiated THz signals, only the transmission

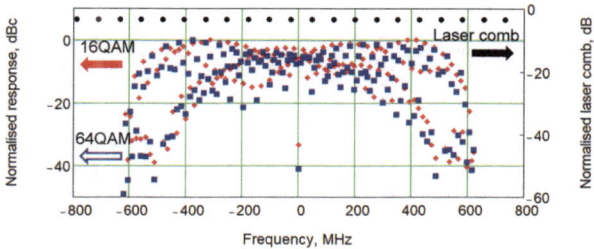

Fig. 14.11 Pseudo-random number sequence (PN7) mapped as 16QAM and 64QAM at 1 GSymb/s centred at 300 GHz. The 76 MHz laser comb frequencies are shown. The dynamic range is greater than 40 dB

RRC filter will have been applied, altering the signal components. The result may also contain additional signal components due to distortion and circuit impairments. The RRC filter truncates all the out-of-band frequency components, so in principle the data set is limited to $(2^n - 1) \cdot (1 + \alpha)$ components where α is the roll-off parameter in the *Joost* RRC filter. In practice, noise and distortions will increase the number of frequency components at a low level.

The different density constellations, 16QAM (4 bits) and 64QAM (6 bits), contain the same frequency components and the number is set by the PRN sequence length; see Fig. 14.11. The frequency components are spaced on a uniform frequency grid about the RF centre frequency. The complex modulation means that the cosine frequency components are symmetric about the carrier frequency and the sine frequency components are anti-symmetric about the carrier frequency. Consequently, all the frequency components must be down-converted as real signals.

14.7.1 Mapping the Signal Components to Baseband

When using a photoconductive sampling receiver, the full waveform is not directly accessible and must be reconstructed from the individually identified frequencies. The laser used in this system has a nominal operating frequency $f_{Laser} = 76$ MHz. The optical length of the laser cavity dictates the repetition rate ($f_{laser} = f_{Laser} + \delta f$). There is a small degree of length control and the laser is locked to an external synthesiser. The exact pulse repetition rate is recovered using a photodiode which drives the ADC clock. The other oscillators governing the THz carrier and the symbol rate must include this frequency correction.

The THz carrier and modulation frequencies must be tailored to avoid discontinuities at the end of the Fourier transform window. So, for a nominal carrier frequency of $f_{CF} = 300$ GHz and a symbol rate f_{Symb} of 1 GSymb/s, this will require frequency correction so that the actual values used, f_{cf} and f_{symb}, are adjusted to

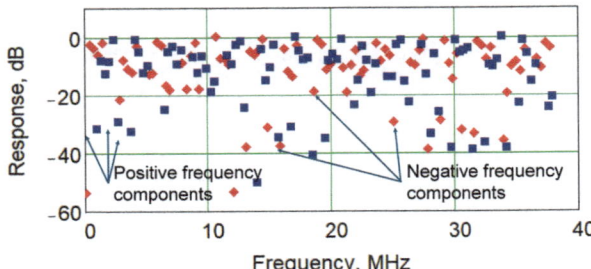

Fig. 14.12 Simulation of 16QAM 1 Gsymb/s at 300 GHz, from Fig. 14.11, mapped to 38 MHz frequency space. Each frequency component will be above (positive) or below (negative) the nearest optical comb frequency

$f_{CF} \cdot \frac{f_{laser}}{f_{Laser}}$ and $f_{Symb} \cdot \frac{f_{laser}}{f_{Laser}}$, respectively, to maintain synchronisation with the measurement epoch.

To avoid the possibility of degeneracy where the carrier frequency or modulation frequency components are mapped to the same frequency component in the 38 MHz window, the frequencies must be chosen so that neither the carrier nor the symbol rate is harmonic of the laser repetition rate. There are several factors, such as the length of the data acquisition and the synthesiser stability, that will affect the choices available. The highest common factor between f_{Data} and f_{Laser} is 4 MHz. The repeat period of data is set by the symbol rate and the length of the PRN sequence, so that a PRN7 sequence, which is a Mersenne prime, at 1 GSymb/s, has a repeat period T_{Data} of 127 ns but the number of frequency components is higher because of the filter parameter α. Analysis of the separation between the down-converted frequencies sets a number of periods. For example, for $f_{CF} = 300$ GHz and $f_{Symb} = 1$ GSymb/s, the minimum integer number of blocks is 4, giving an acquisition length of 31.75 µs. Acquisition of multiple blocks, e.g. 32 (1.016 ms), would provide information about the phase stability of the oscillators and allow some limited phase correction. A simulation of the original and mapped RF spectra are shown in Figs. 14.11 and 14.12 respectively. Although some frequencies are quite closely spaced, there is no degeneracy.

References

1. D. Williams, H. Khenissi, F. Ndagijimana, K. Remley, J. Dunsmore, P. Hale, J. Wang, T. Clement, Sampling-oscilloscope measurement of a microwave mixer with single-digit phase accuracy. IEEE Trans. Microwave Theory Tech. **54**(3), 1210–1217 (2006)
2. J. Dunsmore, A new calibration method for mixer delay measurements that requires no calibration mixer, in *2011 41st European Microwave Conference* (2011), pp. 480–483
3. D. Ballo, PNA-X Application Note 1408-23: active component test: Mixers and frequency converters, Keysight, Tech. Rep. (2023)

4. J.P. Dunsmore, *Mixer and Frequency Converter Measurements*, Ch. 7 (John Wiley & Sons, Ltd, 2020), pp. 435–531 [Online]. Available: https://onlinelibrary.wiley.com/doi/abs/10.1002/9781119477167.ch7

5. D.A. Humphreys, M. Hudlička, I. Fatadin, Calibration of wideband digital real-time oscilloscopes. IEEE Trans. Instrum. Measure. **64**(6), 1716–1725 (2015)

6. O. Lodge, Electric Telegraphy. Patent No. 609,154. United States Patent Office (1898)

7. R. Compton et al., Bow-tie antennas on a dielectric half-space: theory and experiment. IEEE Trans. Antennas Ans. Propag. **Ap–35**(6), pp. 622–631 (1987)

8. R. Gonçalves Licursi de Mello, A.C. Lepage, X. Begaud, The bow-tie antenna: performance limitations and improvements. IET Microw. Antennas Propag. **16**(5), 283–294 (2022) [Online]. Available: https://ietresearch.onlinelibrary.wiley.com/doi/abs/10.1049/mia2.12242

9. M. Bieler, H. Füser, K. Pierz, Time-domain optoelectronic vector network analysis on coplanar waveguides. IEEE Trans. Microwave Theory Tech. **63**(11), 3775–3784 (2015)

10. A. Othonos, Probing ultrafast carrier and phonon dynamics in semiconductors. J. Appl. Phys. **83**, 1789–1830 (1998)

11. R.K. Ahrenkiel et al., Minority-carrier lifetime in gaas thin films. Appl. Phys. Lett. **53**, pp. 598–599 (1988)

12. H. Füser, Terahertz frequency combs and terahertz steering, Ph.D. dissertation, Technische Universität Braunschweig (2014)

13. Y.-S. Lee, *Principles of Terahertz Science and Technology* (Springer, Berlin, 2009)

14. J. Kröll, J. Darmo, K. Unterrainer, Metallic wave-impedance matching layers for broadband terahertz optical systems. Opt. Exp. **15**(11), 6552–6560 (2007) [Online]. Available: https://opg.optica.org/oe/abstract.cfm?URI=oe-15-11-6552

15. M. Joost, Theory of Root-Raised Cosine Filter (2010), https://michael-joost.de/rrcfilter.pdf. Accessed 26 Sep 2024

16. SC of the IEEE Microwave Theory and Techniques Society, IEEE Recommended Practice for Estimating the Uncertainty in Error Vector Magnitude of Measured Digitally Modulated Signals for Wireless Communications, in *IEEE Std 1765–2022* (2022)

Chapter 15
Constellation Diagrams, EVM, and Error Correction

David A. Humphreys, Nora Meyne, and Dominik Wrana

Abstract Digital communication systems transmit multiple bits, encoded as magnitude and phase at a specific time point or at a specific sub-carrier frequency, depending on the type of modulation used. Filtering is used to limit the bandwidth and to minimise receiver noise. This chapter outlines the formation of the constellation diagram and the use of 'error vectors', deviation from the ideal constellation points, containing contributions from both the transmitter and the receiver. Error vectors can be used to characterise impairments or reported as a single-parameter 'error vector magnitude' (EVM), which can be used directly as a measurement tool to assess parameters and systems. The IEEE 1765 standard uses known data, but in general the data is not known in advance, and the nearest constellation point is selected, leading to a saturation of the EVM value. We describe statistical methods to recover the true constellation point of a repeated data sequence.

15.1 Introduction

The constellation diagram is an essential tool for assessing components and system performance in high-speed communications systems. Section 15.2 describes the constellation diagram formation and time-domain representations are used for illustration. The concepts are applicable to frequency-domain schemes such as OFDMA, used in 4G and 5G systems. Added noise is often considered to have

D. A. Humphreys (✉)
NPL, UK (Retired), University of Bristol, Ascot, UK
e-mail: david.a.humphreys@ieee.org

N. Meyne
Physikalisch-Technische Bundesanstalt (PTB), Braunschweig, Germany
e-mail: nora.meyne@ptb.de

D. Wrana
Institut für Robuste Leistungshalbleitersysteme, Universität Stuttgart, Stuttgart, Germany
e-mail: dominik.wrana@ilh.uni-stuttgart.de

© The Author(s) 2026
T. Kürner et al. (eds.), *Metrology for THz Communications*, Springer Series
in Optical Sciences 256, https://doi.org/10.1007/978-3-032-01986-8_15

Gaussian statistics (AWGN), but at THz frequencies, phase noise due to movement and oscillator tracking becomes a greater problem [1].

Error vector magnitude (EVM) is used in many telecommunication system specification standards. Section 15.3 also overviews the IEEE 1765 Recommended Practice for EVM Uncertainty estimation [2]. When the data is unknown, EVM results are underestimated because the nearest constellation point may be incorrect. In normal data transmission, the true symbol value is recovered using forward error-correction (FEC) algorithms. We discuss a statistical approach to recover the true symbol sequence from repetitive test data in Sect. 15.4. A data-centric approach to recovering error vector data is outlined in Sect. 15.5. This may allow error vector estimation within the receiver.

15.2 Constellation Diagram

There are many formats for digital modulation of an RF or optical carrier [1]. For this part of the discussion we will consider time-domain modulation schemes. The simplest and most robust digital scheme is Binary Phase-Shift Keying (BPSK), in which the modulation phase is reversed ($\pm \pi$), giving 1 bit per symbol. This scheme can be expanded to Quadrature Phase-Shift Keying (QPSK) using the Fourier orthogonality relationship between sine and cosine, $\int_0^{2\pi} \sin(\theta) \cdot \cos(\theta) \mathrm{d}\theta = 0$, to provide two independent data streams (2 bits per symbol). QPSK constellations are normally offset by $\frac{\pi}{4}$ and mapped to the corners of the constellation diagram, sharing the power equally between sine and cosine components.

The basis for the constellation diagram is a plot of magnitude and phase for $a(t_s) \cdot \exp(j\theta(t_s))$ where $j = \sqrt{-1}$, and t_s is the symbol time. In the normal mathematical convention for a plot of complex numbers, the horizontal and vertical axes show the normalised *real* and *imaginary* components, respectively, mapped to $\pm 1, \pm j$. At each symbol time point, the waveform should pass through one of the mapped points defined for the modulation scheme. Overlaying many of these symbol points as a *constellation diagram* is a convenient diagnostic tool to view the deviation of real systems from the ideal modulation states. The complex modulation waveform can also be viewed as real and imaginary components in the same way as an eye diagram; see Fig. 15.1.

Filtering is essential to limit the bandwidth as the discontinuities created in the magnitude and phase at symbol boundaries would greatly increase the transmission bandwidth. A key property that the centre of the symbol should be unchanged was initially achieved in early systems using a Gaussian filtering, but this causes inter-symbol interference (ISI). The concept of an efficient two-part filter that limits transmission bandwidth spreading caused by discontinuities and receiver noise bandwidth was proposed in 2008 [3]. This is commonly implemented as the root raised-cosine (RRC) [4]. The combination of two RRC filters forms a raised-cosine

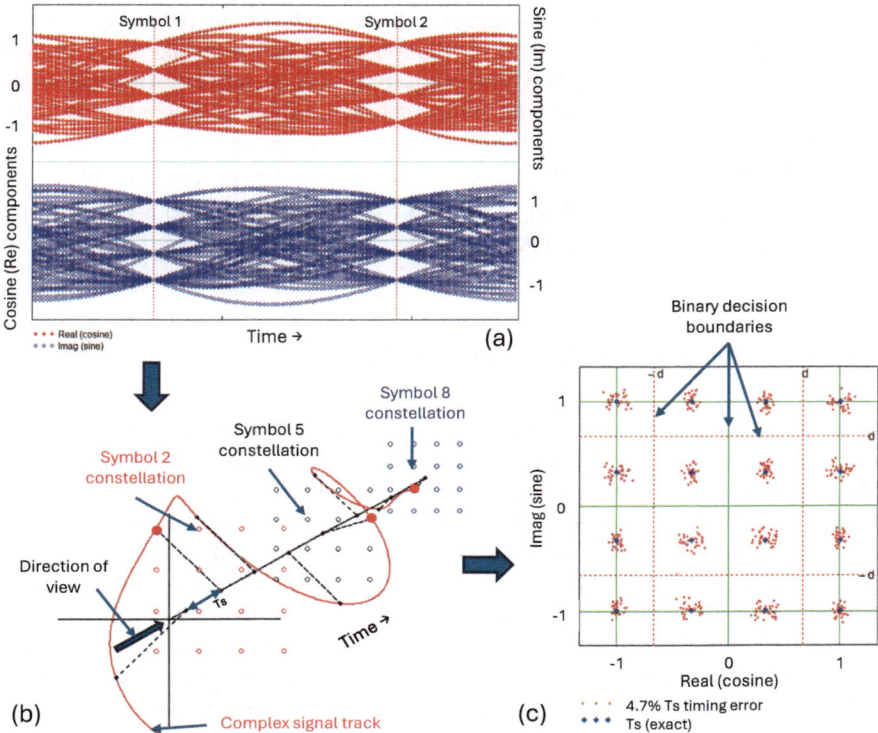

Fig. 15.1 Evolution of the constellation diagram from complex modulation waveform. (**a**) Real and imaginary vs. time showing symbol points. (**b**) Complex track evolution vs. time, showing symbol points. (**c**) Resulting constellation diagram for imperfect and perfect timing alignment. Ts is the time between two symbols

(RC) filter, which has similar properties to the sinc function and causes no ISI as the zero-crossing points fall on the adjacent symbol positions.

Noise and signal impairments can be seen as deviations from the ideal constellation positions. A binary boundary marker is found between the states. The separation of these modulation states is directly related to the noise immunity of the scheme. For BPSK (Binary Phase Shift Keying) mapped to ± 1, the boundary will be at zero on the real axis. QPSK (Quadratue Phase Shift Keying) has two boundary lines, one for the bit states on the real axis and one for the bit states on the imaginary axis. The signal power is equally divided between the sine and cosine components in QPSK, but for BPSK, only the cosine noise component makes a contribution to bit errors, so the QPSK/BPSK capacity is double and the probability of error is also double but the SNR is the same. The probability of an error P_e (unit-less) and the bit-error ratio (BER) is therefore the same for both. Note: For the symbols, the error rate is the product of the *BER* and the *symbol rate* and has units s^{-1} and each symbol error may contain one or more incorrectly recovered bits.

Grey coding [5] is a minimum-change code in which two neighbouring symbols differ by only a single bit. In the constellation diagram this means that a symbol error on the real or imaginary axis will change only a single bit of the data stream. In the 16-QAM (16 quadrature amplitude modulation) example shown in Fig. 15.1, the Euclidean distance to exceed the decision boundary and create a single-bit error is $\frac{1}{3}$. If AWGN gives an error probability of 1 % for each of the two possible transitions (2 % total), then for a Euclidean distance of $\frac{\sqrt{2}}{3}$ corresponding to a two-bit error (1-state) the probability of error is 0.05 %.

15.2.1 Constellation Recovery and Alignment

Recovering and aligning the constellation can contribute to the errors in the result, due to symmetry/orientation, phase-locked loop tracking error or movement. Direct connection of the generator, device under test (DUT) and receiver (back-to-back testing) will generally give a stable result. Using the same RF source, or phase-locked RF sources, minimises phase-tracking and frequency-tracking errors, yielding very-well-defined constellation points. This is ideal for evaluating components and instruments in design or production.

The majority of cellular communications uses carrier frequencies below 6 GHz, wavelengths >5 cm, and the operation above these frequencies has been studied by ITU [6] and [7]. Carrier phase noise, Doppler frequency offset and the associated tracking bandwidth are proportional to the carrier frequency, $f_{\text{Doppler}} = \frac{V_r f_c}{c}$, where c is the speed of light and V_r is the relative velocity of the source and transmitter. Incorrect tracking will appear as symbol phase noise and increased bit errors.

The complex components of a low-noise waveform are very tightly bound at the symbol point (see Fig. 15.1a). A small timing error in the sampling point, for example, an interpolation error or the closest test instrument time sample, will increase the apparent symbol noise (see Fig. 15.1c).

15.2.2 Constellation Assignment and Orientation

It is important to note that constellation has $\frac{\pi}{4}$ rotational symmetry and so the data must be aligned or coded to overcome this problem. For example, header information is normally encoded over several symbols. The data can be aligned by selecting header or training code sequences from a set that does not support $\frac{\pi}{4}$ rotational symmetry. There are many image representations of code assignment, and the two commonest place '0000' is the top left-hand corner, as one would read text on a page (conjugate $(1, -j)$ representation) and where '0000' is the bottom left-hand corner, corresponding to a mathematical Cartesian plot $(1, +j)$.

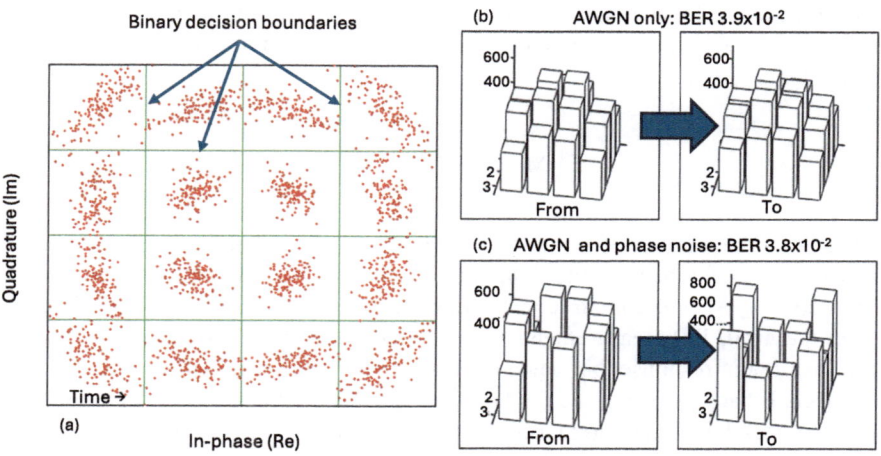

Fig. 15.2 (**a**) shows a constellation diagram simulation with added AWGN and phase noise. (**b**) shows the constellation errors for AWGN with a symbol error ratio of 3.8%. (**c**) shows the constellation errors for AWGN and phase noise with a symbol error ratio of 3.9%. Note the difference in the distribution of changes when phase noise is added [8]

Noise is a consideration when designing a training sequence. Figure 15.2 shows a simulated result for 100 runs of a PN11 pseudo-random number (PRN) sequence with AWGN and a combination of AWGN and residual phase noise errors. As expected, AWGN affects all symbols proportional to the number of adjacent constellation points. So for 16-QAM, the central four points have the highest probability of error (P_e) and the corners have half the number of adjacent states $\frac{1}{2}P_e$. In millimetre-wave and THz systems, residual phase noise is a greater problem as oscillator phase noise and phase-tracking errors are expected to be more significant. The central four constellation points have the lowest P_e. RF power compression and other design imperfections will also alter the expected distribution of errors.

15.3   EVM, Metrology and Standards

Error vector magnitude (EVM) is a quality metric with many applications. In the field it is a valuable tool to optimise link transmission. In the laboratory or production it is used to assess distortion introduced into a measured digitally modulated communication signal at the component level or after it has passed through a physical system, such as an electronic component or a transmission channel. The error vector (EV) is the difference between the measured symbol point and the expected or reference symbol point. Optimisation of symbol position, normalisation and corrective filters must be applied before the EV is determined. This gives valuable information about the component and system behaviour but too much detail for regular use. EVM is a simple parameter calculated as the root magnitude square of the error vectors, divided by the root magnitude square of the symbol values (Eq. 15.1):

$$\text{EVM} = \sqrt{\frac{\sum_{i=1}^{n} |y_i - x_i|^2}{\sum_{i=1}^{n} |x_i|^2}} \tag{15.1}$$

where n is the number of symbols, y_i is the time-aligned and normalised symbol and x_i is the reference symbol at the point i, drawn from a palette of symbols x_{ref}. Measured EVM will contain contributions from both the source and the receiver. Over-the-air (OTA) measurements typically have a higher EVM that may mask the test instrument response, but when the source and receiver instruments are directly connected (back-to-back), the EVM measurements are much lower. Consequently, back-to-back EVM measurements of a receiver made with different manufacturers' source instruments show considerable variation, as shown in Fig. 15.3. The objective of this work [9] was to demonstrate this variation and show that good agreement could be obtained for the estimate of the receiver EVM contribution using different unmodified sources. A digital real-time oscilloscope (DRTO) was used as the low frame-rate (100 Hz) and long acquisition epoch rendered the DSO impractical. A DSO is a practical traceable solution up to 110 GHz for high symbol-rate data with relatively short PRN sequences, such as PN-9 (511). Longer waveforms can be measured by aligning many shorter epoch measurements.

At millimetre-wave frequencies, impedance match and the DSO, timebase correction and vector frequency response corrections are essential for predistortion

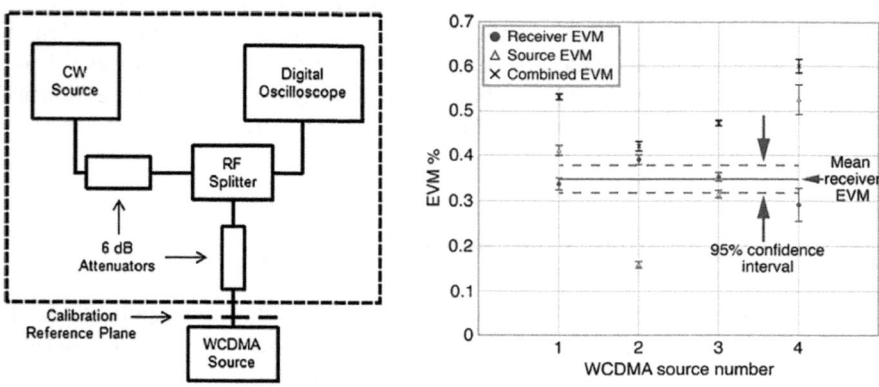

Fig. 15.3 Reference receiver (**a**) used for WCDMA (wide-band code division multiple access) receiver evaluation. The CW source was included to improve the DRTO timing accuracy. (**b**) Uncorrected receiver measurement results show large EVM variations due to impairments in the different measurement sources and manufacturer-supplied test waveforms. Predistortion of the waveforms was not an option but the agreement between the unknown receiver results suggest that correlated contributions were not significant (Figure reproduced from [9] ©IEEE 2013)

corrections and uncertainty budget calculations. The uncorrected EVM (8%), close to the EVM saturation limit for 64-QAM, was reduced to about 1.5% using pre-distortion of the AWG waveform using a reference receiver with known (traceable) characteristics. The objective of this work was to create a reference source with minimum EVM, defined by the receiver uncertainty. This approach allows traceable characterisation of receivers [10]. EVM is used extensively in communication standards to ensure commonality of methods. There are slight differences in the definition and parameters for TDMA (time division multiple access) [11] and OFDMA (5G) [12] standards. The IEEE prepared a generic EVM recommended practice standard, developed from a metrological perspective. This uses a defined data-set (data-directed) approach to avoid incorrect symbol assignment [2].

Error vector magnitude (EVM) represents the normalised rms value of the residual error vector at each constellation point. EVM has a reciprocal relationship with signal-to-noise ratio (SNR) [1, 13] which are directly related through Eq. 15.2:

$$\text{EVM}_{\text{RMS}} = \sqrt{\frac{1}{\text{SNR}} + 2 - (1 - g_t)\sqrt{\frac{1 + \cos(\phi_t)}{1 + g_t^2}}} \qquad (15.2)$$

where g_t is the gain imbalance and ϕ_t is the quadrature error.

15.3.1 Overview of IEEE Standard 1765–2022

The goal of the IEEE 1765 approach is to verify the validity of the EVM obtained from a given lab's measurement and/or post-processing procedures. Methods to estimate the EVM arising from distortion introduced by a DUT or channel, described in many textbooks and standards, do not include uncertainty estimation or describe methods to check the validity of the results. This is not surprising for two reasons: firstly, the objective of these standards is to maintain a working telecommunication system and so EVM is a tool often used as a pass/fail metric. Equipment-related errors are likely to be small compared with EVM levels introduced by the RF channel.

The IEEE standard was written from a metrological perspective as a recommended practice to provide a standardised approach to estimate the EVM from measurements and to facilitate the evaluation of uncertainties related to the receiver hardware by comparison to a reference measurement of a standardised 'IEEE 1765 Reference Waveform'. This approach provides insight into the impact of distortion, introduced by the receiver hardware, on the estimate of EVM. Baseline EVM and waveform pre-processing algorithms have been defined. These allow the isolation of receiver hardware effects on the reported EVM by eliminating algorithmic corrections of the signal that can minimize the output EVM, a technique used in Chap. 24 to evaluate concatenated components. The IEEE Baseline EVM Algorithm includes only the minimum set of steps needed to calculate EVM.

IEEE 1765 provides a dissemination route for EVM from the National Measurement Institutes (NMI) to calibration laboratories by comparing nominal (computed) values of EVM and their associated uncertainties allow the user lab to validate its measurements and uncertainty analyses with respect to those made by the reference laboratory. Such measurement comparisons for validation of a measurement-derived quantity such as EVM are well accepted as best-practice approaches (see note 6.5.3 in [14]), *'When metrological traceability to the SI units is not technically possible, the laboratory shall demonstrate metrological traceability to an appropriate reference, e.g.: a) certified values of certified reference materials provided by a competent producer; b) results of reference measurement procedures, specified methods or consensus standards that are clearly described and accepted as providing measurement results fit for their intended use and ensured by suitable comparison'.*

15.3.1.1 Data-Aided and Non-Data-Aided Symbol Recovery

In many applications the ideal symbol value is not known, the 'non-data-aided' scenario, and a 'recovered symbol', is used in Eq. 15.1 for the x_i value. The recovered symbol location for x_i is the geometrically closest measured symbol and gives the minimum value for $|y_i - (x_{ref})_j|$, where j is the reference symbol index. However, for high EVM values this may not be the correct symbol choice, leading to saturation of the EVM results; see Fig. 15.4. Non-data-aided EVM algorithms are commonly used in test instrumentation and to optimise operational scenarios. To achieve a valid comparison result between measurement system in different laboratories requires the tests to be performed with the same reference data set.

Data-aided methods either must start with a known dataset or be able to recover the true dataset in the presence of errors; see Sect. 15.4.

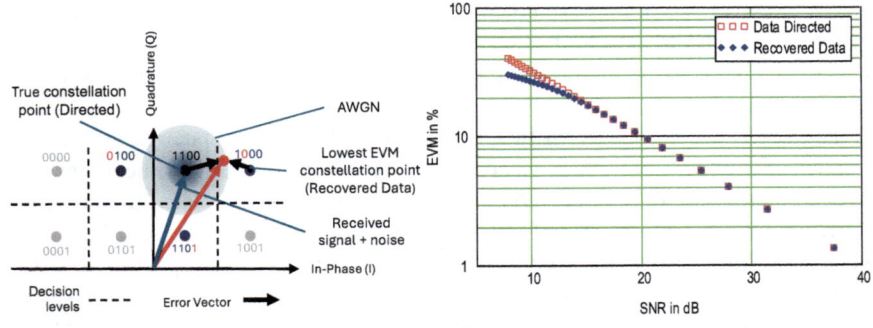

(a) Upper half of 16-QAM constellation diagram   (b) EVM roll-off due to incorrect symbol identification (1-bit error)

Fig. 15.4 (**a**) Minimum error vector may not correspond to the true constellation point. This leads to EVM saturation due to bit errors (**b**). Pre-existing source material similar to Fig. 1 of [2]

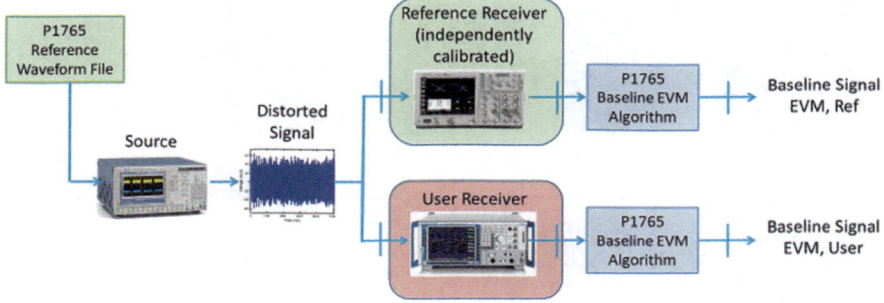

Fig. 15.5 IEEE 1765 approach to identify contributions of receiver hardware and EVM algorithm to uncertainty in a user's estimate of EVM in a distorted signal (Reproduced from [2], ©IEEE 2022)

15.3.1.2 Baseline EVM Algorithm

For a viable metrological standard the data structure must be defined. The purpose of the *Baseline EVM Algorithm* in the IEEE 1765 standard is to estimate the EVM of the physical signal incident on the receiver's reference plane, without correcting for any signal impairments. Consequently, the Baseline EVM Algorithm is deliberately designed not to correct for, and thereby obscure, impairments in the measured signal, such as in-phase/quadrature (I/Q) offset and frequency drift. Many commercial EVM algorithms are designed to correct for numerous impairments in both the signal and receiver hardware simultaneously with the goal of minimising EVM and best determining the transmitted bit stream. The generalised approach is shown in Fig. 15.5. Predistortion is used to compensate for imperfections introduced by the source hardware.

The Baseline EVM Algorithm is data-directed, so the true symbol values are known in advance. Reference data sets and algorithms are provided for both single-carrier time-domain waveforms and OFDMA. The flow of the baseband EVM algorithm for the single-carrier case is shown in Fig. 15.6.

The measurement and analysis comprises two stages: pre-processing, which covers measurement, instrument-related corrections and resampling to ensure that the measured and reference waveforms are sampled at the same rate and the two waveform vectors are the same length.

For RF measurements, the oversampling rate should be sufficient to provide at least four samples per cycle of the RF carrier. Reference instruments, such as a DSO, have significant time base distortion and, although not stated in the standard, this correction is normally applied in the time domain prior to conversion to the frequency domain. Other corrections, such as impedance match and time drift, are best applied in the frequency domain.

Instruments such as a vector signal analyser (VSA) down-convert the RF signal using RF mixers before sampling the waveform to give a complex base band result. Interpolation may be required as the oversampling ratio of a VSA is often lower than

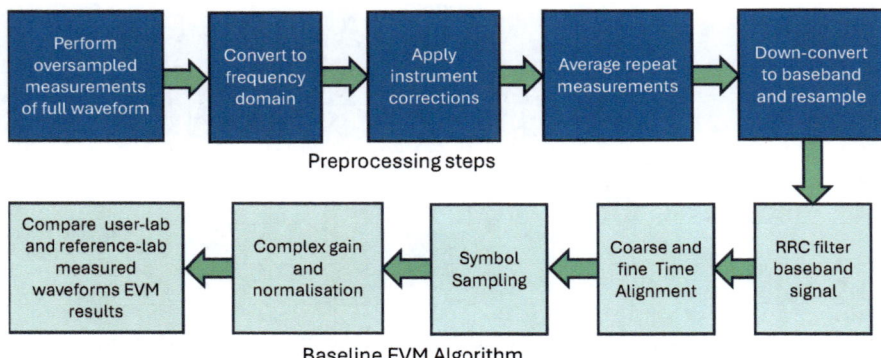

Fig. 15.6 Pre-processing and baseline EVM analysis following the IEEE 1765 standard

the target specified by the standard. At THz frequencies the receiver may comprise a DRTO and broadband RF down-converter, which will also need to be characterised.

Time alignment is performed in two stages. A coarse time-alignment algorithm, based on a cross correlation, that provides a starting point for minimising the error function (Eq. 15.3) for frequency-domain fine time alignment

$$E(G, \tau) = \sum_{i=1}^{K} |X_i - GZ_i \, e^{(j \frac{2\pi(i-1)\tau}{T_s}))}|^2 \qquad (15.3)$$

where G is the complex gain, τ is the delay to be optimised, Z_i and X_i are the frequency-domain components of the measured and reference data set, respectively, T_s is the symbol period and K is the number of waveform-sampled points in the finite duration signals. This is reformulated and solved by an iterative approach.

$$Im \left[\left(\sum_{i=1}^{K} X_i^* Z_i \, e^{(j \frac{2\pi(i-1)\tau}{T_s})} \right) \left(\sum_{i=1}^{K} X_i Z_i^* \, e^{(-j \frac{2\pi(i-1)\tau}{T_s})} \right) \right] = 0 \qquad (15.4)$$

and gain G_0 at τ_0 is given by

$$G_0 = \frac{\sum_{i=1}^{K} X_i Z_i^* \, e^{(-j \frac{2\pi(i-1)\tau}{T_s})}}{\sum_{i=1}^{K} Z_i Z_i^*} \qquad (15.5)$$

15.3.1.3 EVM Systematic Errors

Distortion, IQ imbalance and other systematic errors mean that, in the absence of noise, there will be a residual error vector component. Unlike the error vector, EVM is a scalar and so, assuming linearity, the concatenation or comparison of

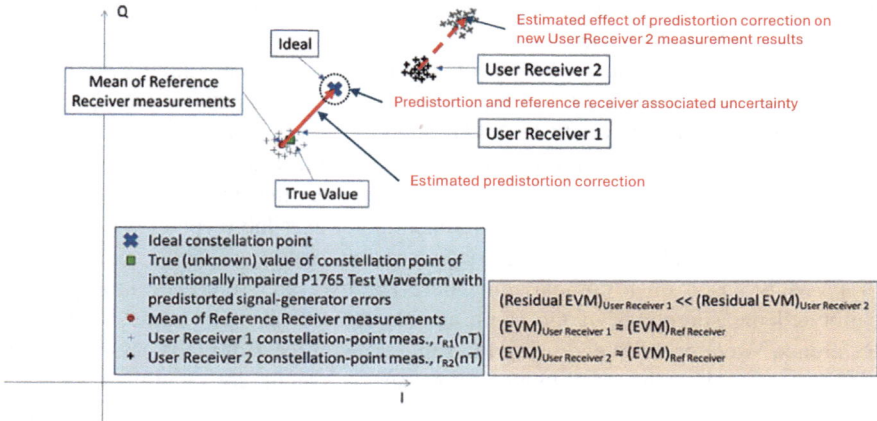

Fig. 15.7 EVM with intentional or source error impairment and with corrective predistortion to the ideal constellation point (Modified from Fig. 10 of [2], which is ©IEEE 2022)

two instruments with similar EVM values is not predictable as the relative phases of the residual error vectors are unknown. Figure 15.7 shows that the true value of the constellation point can differ from the ideal constellation point due to deliberately introduced impairments and/or signal-generator error. User receivers 1 and 2 have comparable EVM. Applying a corrective predistortion, optimised using the reference receiver measurements, would bring the source result as close to the ideal symbol point as possible. Note that the final *User Receiver 1* EVM result contains the uncertainty contributions of both the reference and user receivers. Consequently, the minimum reported EVM cannot be lower than the uncertainty limit set by the reference receiver.

15.3.1.4 Uncertainties and Comparison

Uncertainties contain both correlated (Method 1) and uncorrelated components (Method 2). The NIST Uncertainty Framework [15] propagates correlated uncertainties by Monte Carlo methods and contains detailed uncertainty models. This was used to calculate the uncertainty estimates of EVM and associated uncertainties for a 1 GSymbol/s 64-QAM signal with a centre frequency of 44 GHz [10] were used as an example and compared with a simple sensitivity analysis. Errors introduced by uncorrected impedance mismatch are correlated with the incident data. Although a correction can be applied, the result will have associated correlated uncertainties.

In laboratory measurements the signals may not be dominated by noise, though there will be a noise component in the results. The reference receiver noise can be reduced by averaging multiple sets of error-vector measurements, giving a statistical uncertainty component $u_{\text{noise}} = \frac{\sigma_{\text{noise}}}{\sqrt{n}}$, the averaged standard error of the mean noise component for the reference receiver in the source EVM estimate. The source EVM decreases as the number of waveforms is increased. Using this uncertainty contribution to back-off the EVM value maintained a convergent EVM result, independent of the number of waveforms acquired, so that fewer reference measurements could be taken at the expense of the reference receiver uncertainty and hence the limiting EVM value. The uncertainty components of both reference-lab receiver and signal source are propagated through to the final value of uncertainty in EVM measured by a reference-lab or user-lab receiver [9].

The approach taken in the IEEE standard for assessing a lab's measurement of EVM is based on a comparison where the user and reference labs perform calibrated measurements of the same signal, derived from the same IEEE 1765 Reference Waveform file and using the same transmitter. Both labs compute EVM with the same algorithm to eliminate differences in calculations and then apply the associated measurements uncertainties The EVM and uncertainties are compared using Eq. 15.6:

$$E_n = \frac{\text{EVM}_{\text{lab}} - \text{EVM}_{\text{ref}}}{\sqrt{U_{\text{lab}}^2 + U_{\text{ref}}^2}} \tag{15.6}$$

where U_{lab} and U_{ref} are the user lab and reference expanded uncertainties, respectively, assuming the same coverage factor and degrees of freedom. The ISO/IEC-defined E_n score [14] is recommended as the criteria for assessing agreement between two quantities.

15.3.1.5   IEEE 1765 Open Source Project

Reference datasets and software are available from the IEEE1765 Open Source Project [16]. This open-source project contains material referenced by IEEE 1765. It contains software (Matlab (2012b and 2020a) and GNU Octave (6.3.0)), reference waveforms and the associated calculated results and for single-carrier and OFDM EVM calculation. In addition, there is information about how to contribute and licensing.

Data is available for two specific communication modulation signal classes, single-carrier (square) 64-QAM and OFDM (square) 64-QAM, covering EVM up to 19.6% and 7.1%, respectively. In addition, there is a separate class of artificially distorted square 64-QAM signal constellations, with analytically calculable EVM values that can be used to assess EVM algorithms and EVM measurement equipment. Data files are provided for complex baseband (with real-valued in-phase and quadrature data) and RF waveforms (with real-valued samples as a function of time). The data is a 511 symbol sequence from a pseudo-random number (PRN) algorithm (PN9) and symbols are Grey coded, numbered 0 to 63 and conjugate mapped $(1, -j)$—'0000' is the top-left symbol—to the constellation.

All material in the repository is subject to change. The material in the IEEE 1765 repository is presented 'as is' and with all faults. Use of the material is at the sole risk of the user. IEEE specifically disclaims all warranties and representations with respect to all material.

15.4 Recovering True Error Vector

EV and EVM are useful diagnostic tools, particularly if data-directed. Waveform errors, such as IQ imbalance and amplifier saturation, cause specific and quantifiable errors that can help with design and testing. For a n-QAM square constellation, all locations will be exercised if the PRN sequence length exceeds n and all transitions will be exercised if the PRN sequence length exceeds n^2. Exercising all transitions of a high-density constellations with a PRN sequence may be impractical.

From Sect. 15.3.1 we have seen that there are benefits to having access to the true constellation points. In IEEE 1765, there was prior knowledge of the data set but this luxury is not always available. The simplest strategy is based on the statistics of the data. For a symbol error ratio of $1 : 10^{-2}$, 99% of the data is correct. Although the overall incidence of a symbol error is low, Fig. 15.2 shows that, depending on the type of noise and distortion, some constellation points have a higher incidence of errors. In the case of significant noise or distortion, where the constellation point falls close to a bit decision boundary, the number of samples may not provide sufficient confidence for the correct symbol selection. Other assessment metrics such as the mean value, or a polar coordinate approach, can be helpful. A $P_e < 0.5$ is required for all constellation points. Additional verification may be possible from the type of data. Sufficient copies of the data sequence are required for a good statistical estimate. At each constellation point, the mode of the distribution should be unique and define the true constellation point. The error vectors or EVM can then be calculated in accordance with the IEEE 1765 Baseline algorithm.

The example shown in Fig. 15.8 used 48 repeats of a 1040 symbol 16-QAM data sequence. The measurements were made at 83 GHz, using phase-locked synthesiser and phase-locked loop oscillator in the receiver. The symbol point recovery is based on the IEEE 1765 method with x20 oversampling for the coarse assessment. Unlike the IEEE case, the symbols are not known in advance but 16-QAM constellation points fall onto three circles. Coarse and fine minimisation of $\sum_{i=1}^{n} ||y_i'(k)| - rQAM_{jmin}|^2$ where $y_i'(k)$ is the symbol y_i sampled at the kth oversampling point and $rQAM_{jmin}$ is the constellation, e.g. 16-QAM, 64-QAM, etc., gives the symbol timing. Phase and frequency offsets can be recovered. The quality of the phase-tracking algorithm affects the final EVM recovered [1]. There is no restriction on the repetitive data format, other than practical limitations of length. As the data is repetitive, the mean EV values can be extracted to evaluate systematic effects and the variance/covariance of each data set provides information about residual noise and the performance of the data-recovery algorithm used.

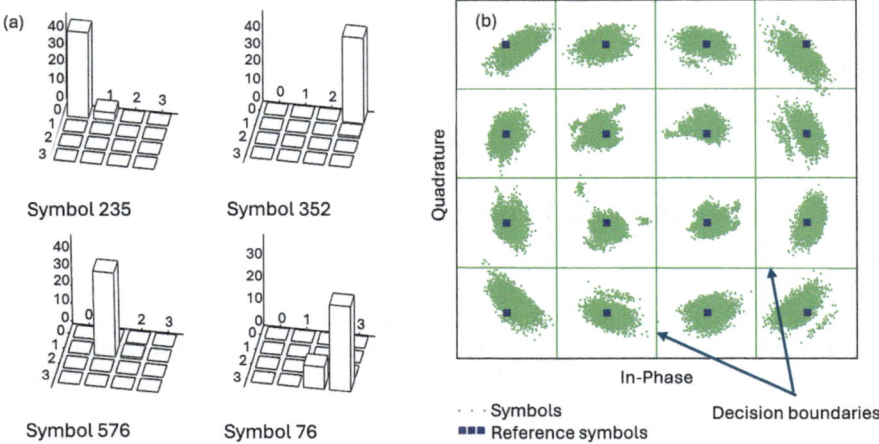

Fig. 15.8 Statistical evaluation of constellation EVM 10.45%. (**a**) shows four examples of statistically identified symbol errors and (**b**) shows the constellation diagram

15.5   Data-Only Approach

Test equipment typically has a higher resolution than a receiver but both can typically recover several different QAM constellation densities. There are test scenarios where the equipment is not co-located, and although EVM is useful, error vector data can be passed back to the operator in real time and may provide valuable insights. An estimate for EV can be achieved by altering the scaling and recovering the data as a higher-order QAM sequence. The mapping scale factor is given by Eq. 15.7:

$$QAM_{scale} = \frac{\sqrt{n1QAM} - 1}{\sqrt{n2QAM} - 1} \sqrt{\frac{n2QAM}{n1QAM}} \tag{15.7}$$

where $n1QAM$ and $n2QAM$ are the initial and mapped QAM states. Higher QAM orders are recommended for greater detail and reduced additional EVM contribution caused by rounding (see Fig. 15.9). In (c), the plot shows results for a 4096-QAM (additional 4 bits per ADC) mapping. The rounding error contribution for 4096-QAM (1.7 %) increases the estimated EVM from 10.85 % to 11.97 %. Receiver or test instrument effective number of bits (ENOB) will limit this approach for high-bandwidth THz communications.

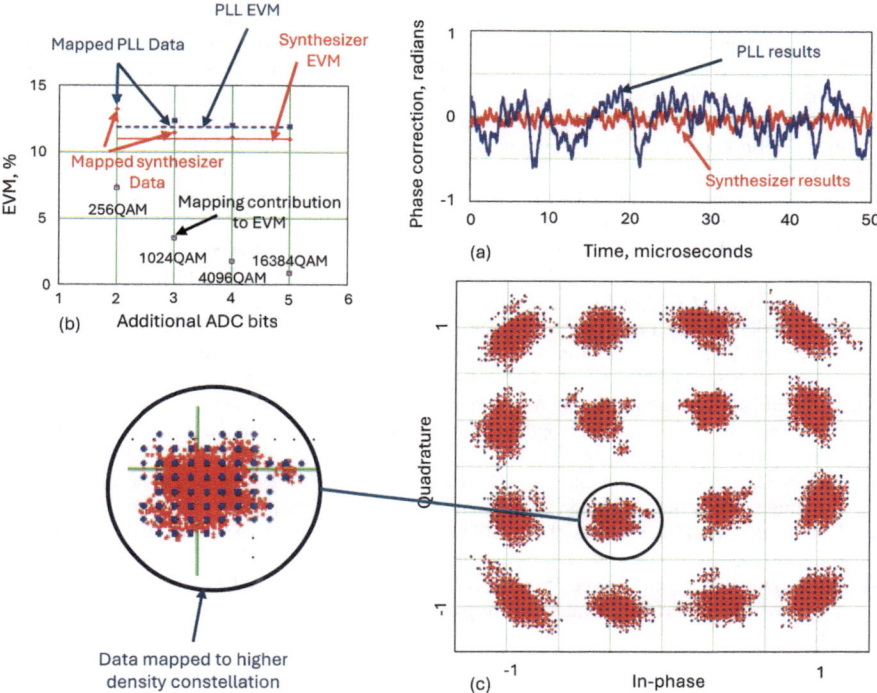

Fig. 15.9 Recovering 16-QAM EV estimates using a higher-order QAM constellation (**a**) Phase corrections for constellation recovery show greater instability for the PLL (phase-locked loop); this results in a higher EVM. (**b**) The point spacing of the higher-order QAM increases the resulting EVM, a 4096-QAM constellation contributes about 1.7% to the EVM budget and providing symbol-specific EV data. (**c**) 16-QAM measurement results rescaled and mapped to 4096-QAM

References

1. D.A. Humphreys, Modulation formats, in *THz Communications: Paving the Way Towards Wireless Tbps*, ed. by T. Kürner, D. M. Mittleman, T. Nagatsuma (Springer, Berlin, 2022), Ch. 29
2. SC of the IEEE Microwave Theory and Techniques Society, IEEE Recommended Practice for Estimating the Uncertainty in Error Vector Magnitude of Measured Digitally Modulated Signals for Wireless Communications, in *IEEE Std 1765–2022* (2022)
3. B. Farhang-Boroujeny, A square-root nyquist (M) filter design for digital communication systems. IEEE Trans. Signal Proces. **56**(5), 2127–2132 (2008)
4. M. Joost, Theory of Root-Raised Cosine Filter (2010), https://michael-joost.de/rrcfilter.pdf. Accessed 26 Sep 2024
5. F. Gray, Pulse-Code Communicator, U.S. patent 2,632,058, Serial No. 785697, Bell Telephone Laboratories
6. Report ITU-R M.2541-0, Technical feasibility of IMT in bands above 100 GHz (2024), https://www.itu.int/pub/R-REP-M.2541-2024. Accessed 8 Sep 2024
7. Report ITU-R M.2376-0, Technical feasibility of IMT in bands above 6 GHz (2015), https://www.itu.int/pub/R-REP-M.2376-2015. Accessed 8 Sep 2024

8. D.W. David Humphreys, H. Füser, C.V. Phung, Progress towards traceability for THz communications waveforms and the use of "data-enabled analysis" in testing, in *Workshop: DFG Meteracom: Introduction to DFG FOR 2863 Meteracom* (2024) [Online]. Available: https://meteracom.de/workshop-12-march-2024/

9. D.A. Humphreys, J. Miall, Traceable measurement of source and receiver EVM using a real-time oscilloscope. IEEE Trans. Instrum. Meas. **62**(6), 1413–1416 (2013)

10. K.A. Remley, D.F. Williams, P.D. Hale, C.-M. Wang, J. Jargon, Y. Park, Millimeter-wave modulated-signal and error-vector-magnitude measurement with uncertainty. IEEE Trans. Microwave Theory Tech. **63**(5), 1710–1720 (2015)

11. LTE; Evolved Universal Terrestrial Radio Access (E-UTRA); Base Station (BS) radio transmission and reception (3GPP TS 36.104 version 8.3.0 Release 8), in *6.5.2 Error Vector Magnitude, ETSI TS 136 104 V8.3.0* (2008–2011), p. 17

12. LTE; Evolved Universal Terrestrial Radio Access (E-UTRA); User Equipment (UE) conformance specification; Radio transmission and reception; Part 1: Conformance testing, in *E4.1 EVM. 3GPP TS 36.521-1 version 18.5.0 Release 18* (2024), p. 6129

13. M. Hudlička, C. Lundström, D.A. Humphreys, I. Fatadin, BER estimation from EVM for QPSK and 16-QAM coherent optical systems, in *2016 IEEE 6th International Conference on Photonics (ICP)* (2016), pp. 1–3

14. ISO 17025:2017, General requirements for the competence of testing and calibration laboratories, https://www.iso.org/standard/66912.html. Accessed 22 Sep 2024

15. R.A. Ginley, Kicking the tires of the NIST microwave uncertainty framework, Part 2, in *2017 90th ARFTG Microwave Measurement Symposium (ARFTG)* (2017), pp. 1–4

16. IEEE SA 1765 Open Source Project, IEEE 1765 Recommended Practice for EVM Measurement and Uncertainty Evaluation, https://opensource.ieee.org/1765/1765 Accessed 27 Sep 2024

Part III
Metrology for Measurement Systems

Chapter 16
Architecture for Channel Sounding

Jonas Gedschold, Tobias Doeker, Mohanad Dawood Al-Dabbagh, Alexander Ebert, Giovanni Del Galdo, and Reiner Thomä

Abstract This chapter deals with architectures for wideband channel sounders suitable for characterising THz wave propagation. A maximum-length binary-sequence (MLBS) sounder, a vector network analyser, and an arbitrary waveform generation-based sounder are presented as implementation examples of THz channel sounders. Sounding is usually concerned with several measurement dimensions. We are discussing impacts on the architecture if directional channel information is to be captured. This includes improvements in dynamic range when using mechanically rotated antennas. Finally, a front-end design for capturing polarimetric channel information is presented.

16.1 Introduction to THz Channel Sounders

Measuring the wave propagation in a transmission channel is an important tool for understanding the characteristics of possible communication links and developing radio interfaces. As such, measurement-based channel characterisation, i.e. channel sounding in Chap. 8, complements simulation-based approaches such as ray tracing

J. Gedschold (✉) · A. Ebert · R. Thomä
Institute of Information Technology, Technische Universität Ilmenau, Ilmenau, Germany
e-mail: jonas.gedschold@tu-ilmenau.de; alexander.ebert@tu-ilmenau.de;
reiner.thomae@tu-ilmenau.de

T. Doeker
Institut für Nachrichtentechnik, Technische Universität Braunschweig, Braunschweig, Germany
e-mail: t.doeker@tu-braunschweig.de

M. D. Al-Dabbagh
Physikalisch-Technische Bundesanstalt (PTB), Braunschweig, Germany
e-mail: mohanad.al-dabbagh@ptb.de

G. Del Galdo
Institute of Information Technology, Technische Universität Ilmenau, Ilmenau, Germany

Fraunhofer Institute for Integrated Circuits (IIS), Ilmenau, Germany
e-mail: giovanni.delgaldo@tu-ilmenau.de

© The Author(s) 2026   181
T. Kürner et al. (eds.), *Metrology for THz Communications*, Springer Series
in Optical Sciences 256, https://doi.org/10.1007/978-3-032-01986-8_16

in Chap. 10. After sufficient processing, the measurement results of channel sounding build a necessary foundation for channel model development tailored to the frequency ranges to be explored.

However, especially at (sub)-THz frequencies, the architecture design of channel sounders faces some challenges. The multidimensional nature of wave propagation requires a simultaneous and critical (Nyquist) sampling in space and time to resolve propagation delays, direction-dependent propagation, and the time variance of the channel. However, spatial sampling via antennas is more complex for THz since the effective antenna apertures for a fixed directivity decrease with increasing frequency [1]. This leads either to the design of high-gain antennas with larger effective apertures or to a reduced power transmission, which negatively impacts the signal-to-noise ratio (SNR). Hence, parallel spatial sampling would require many high-gain antennas to cover the area of interest or even an omnidirectional view of the channel. Only a coherent omnidirectional view on both link ends would allow a de-embedding of the sounder hardware and the derivation of so-called double-directional channel models [2, 3]. However, such a setup is limited by the integration sizes of antennas and signal chains and, of course, limited budgets. Consequently, alternative approaches, such as building synthetic apertures by mechanically rotating high-gain radio interfaces, are required (Sect. 16.3.1). Such sequential approaches demand intense acquisition time for a single snapshot and severely limit the applicability of such an architecture for time-variant channels. This is again most severe for THz channels since the Doppler shifts, resulting from movements of objects or the radio interfaces themselves, are proportional to the carrier frequency. This requires sufficient Doppler bandwidth, i.e. measurement rate in slow time competing with the sequential directional sampling. Hence, current THz sounder architectures usually focus on a subset of the propagation dimensions such as delay-Doppler or delay-angle which does not allow a de-embedding of the hardware and, hence, no hardware-independent channel description.

Especially due to these difficulties, a metrological assessment of implemented sounder architectures is inevitable. Otherwise, hardware limitations, simplified processing algorithms, or a varying calibration quality may significantly affect the comparability between channel sounders [4] and resulting uncertainties may propagate to the derived channel models. Useful concepts for sounder verification are benchmarking measurements with test channels provided by over-the-air (OTA) artefacts (see Chap. 19) or the use of a vector network analyser (VNA) as a traceable reference device (see Chap. 6 and Sect. 16.2.2).

In this chapter, we focus on specific implementations of THz channel sounders used throughout this book. Comprehensive overviews of available channel sounder architectures are, for example, given in [5–7]. Common architectures are time-domain-based sounders with hardware-generated sounding signals [8–11], as well as more flexible concepts built on arbitrary waveform generators (AWGs) and digitisers [12–14]. The remainder of this chapter is organised as follows. In Sect. 16.2 we discuss three different sounder architectures that are used throughout the book: a time-domain-based sounder (Sect. 16.2.1), a VNA-based architecture (Sect. 16.2.2), and the use of an AWG and a signal analyser as baseband units for a sounder (Sect. 16.2.3). In Sect. 16.3 we discuss further aspects of multidimensional sounding: directional sounding with rotated high-gain antennas (Sect. 16.3.1), as well as the design of polarimetric sounding interfaces (Sect. 16.3.2).

16.2 Implementations of THz Sounder Architectures

16.2.1 MLBS-Based Sounder

One possible implementation of a channel sounder is a correlation-based channel sounder. This type of sounder exploits the relationship between the cross-correlation of an input and an output signal with respect to the auto-correlation of the input signal. It can be shown that [15]

$$\psi_{xy}(\tau) = \int\limits_{-\infty}^{\infty} x(\tau) \cdot y(t + \tau) \, dt = \int\limits_{-\infty}^{\infty} s(\tau - t) \cdot \psi_{xx}(t) \, dt \,, \qquad (16.1)$$

where ψ_{xy} and ψ_{xx} represent the cross-correlation and auto-correlation functions, respectively. For channel sounding, the input signal, denoted by x, is the transmitted signal, and the output signal y is the measured signal at the receiver (RX). The term s in (16.1) denotes the impulse response (IR), which in the context of channel sounding, before any further processing, is the convolution of the IR of the transmitter (TX), the RX, and the channel. If the auto-correlation of the input signal ψ_{xx} equals the Dirac function, the cross-correlation then equals s in (16.1). Therefore, not surprisingly, correlation-based channel sounders typically employ a transmit signal whose auto-correlation nearly equals the Dirac function, and the received signal is cross-correlated with the transmitted signal. Consequently, the correlation-based channel sounder directly determines the measured IR in the time domain.

Regarding channel sounding, pseudo-noise sequences, especially maximum length sequences, are widely used [10, 16, 17]. These sequences can be efficiently implemented in both software and hardware using maximal linear-feedback shift registers. With a register of m, the randomly generated sequence has a length of $2^m - 1$ and is repeated periodically. Hence, the sequence length and the duration of each chip determine the maximum measurable delay and distance, expressed as

$$T_{\text{rep}} = (2^m - 1) \cdot T_{\text{chip}} \,, \qquad (16.2)$$

where T_{chip} represents the chip duration. The clock frequency f_{CLK} for generating the sequence determines the delay resolution $T_{\text{chip}} = \frac{1}{f_{\text{CLK}}}$ and the bandwidth of the signal. To ensure synchronised transmission and sampling of the received signal, the clock frequency should also be distributed to the RX. With synchronisation between

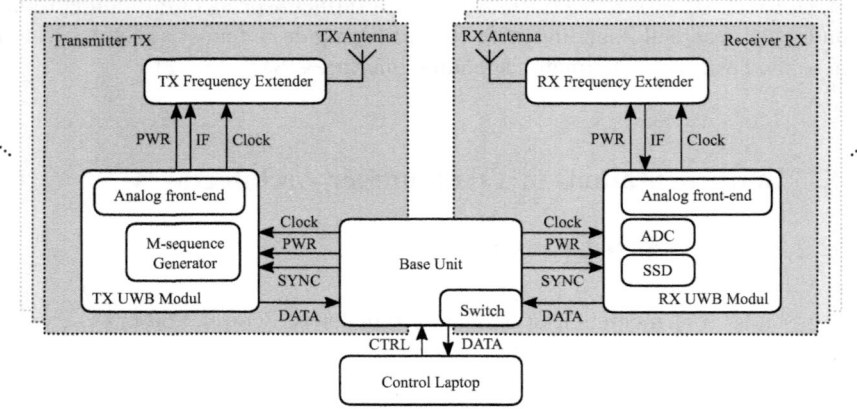

Fig. 16.1 Schematic illustration of a correlation-based MIMO channel sounder (modified after [8])

the TX and RX, the system can be easily extended, to multiple transceivers. In fact, using a correlation-based channel sounder, multipe input-multiple output (MIMO) measurements are feasible, provided the TX and RX are synchronised and the same transmit sequence is employed. To avoid interference, time-division multiplexing can be used to accurately determine the IR of multiple TXs. Alternatively, near-orthogonal codes may be employed. Figure 16.1 illustrates a schematic of a correlation-based channel sounder in a MIMO setup.

For channel sounding in the THz frequency range, the baseband signal must be upconverted. Typically, this is achieved through multiple stages of multipliers based on the clock frequency. The same principle applies to downconversion at the RX end. The received signal is then sampled by analog-to-digital converters (ADCs) in in-phase and quadrature. However, ADCs with high enough sampling rates and bit depths may be either extremely expensive or simply commercially not available, yet. To overcome this problem, subsampling may be conducted. If the channel is assumed to be static for a short period – a quasi-static channel assumption – the periodically reproduced received signal will be sampled multiple times. In each period, only a part of the received signal is sampled according to the maximum sampling rate of the ADC, with the sample points shifting for each period. Using a subsampling factor of, for example 4, the signal is combined using the differently sampled periods as shown in Fig. 16.2.

To calibrate a correlation-based channel sounder, either a calibration using waveguides (back-to-back (B2B) calibration) or a calibration performing line-of-sight (LOS) measurements (OTA calibration) is possible [18]. In both cases, the measured reference IR is used to eliminate the influence or IR of the measurement system on the desired channel impulse response (CIR). Since IRs are concatenated by convolution in the time domain, a deconvolution approach is commonly used for calibration [18]. To increase the SNR of the measured signal, multiple measured IRs

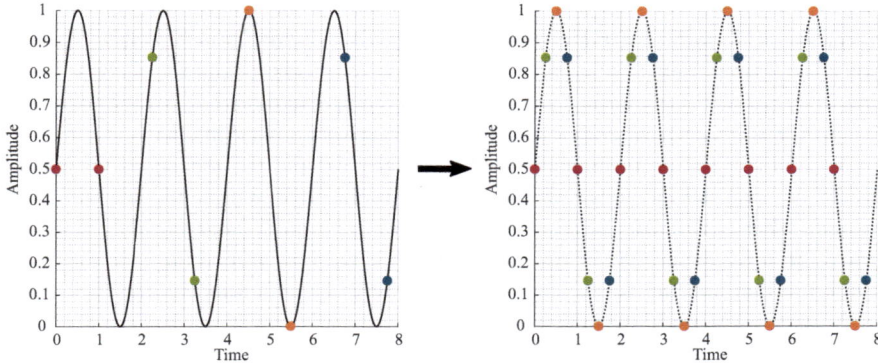

Fig. 16.2 Example illustration of the subsampling principle for a subsampling factor of 4 and a period length equivalent to 8 samples. All required samples can be acquired over four periods (left). The waveform is reconstructed by combination of the subsampling values (right)

can be averaged, as long as the channel can be assumed to remain static during the averaging period.

16.2.2   Using Vector Network Analysis for Channel Sounding

The VNA is a fundamental measurement device used to characterise and emulate measurement networks, including static wireless channels [19]. This makes it essential for calibrating and validating the scattering parameters of other channel sounding systems. VNAs offer flexibility in terms of frequency range selection, frequency step size, and range resolution. They also provide accuracy by detecting small variations in the measurement environment, thanks to their low noise floor, which can be controlled by setting a suitable narrow intermediate frequency bandwidth (IFBW). This bandwidth can be as low as a few hertz, allowing the VNA to elevate the investigated signal above the noise and enhance confidence in the measured values.

Compared to other channel sounding systems, VNAs have well-established calibration procedures that resolve systematic errors using linear equations via mechanically characterised calibration standards. This ensures that measurements are repeatable with known uncertainty contributions for each error term, establishing traceability in the measured values. The VNA's architecture, measuring the frequencies in sequential steps, makes it unsuitable to characterise rapidly changing channels. To overcome this challenge an investigation into time-varying transmission using a VNA is presented in Chap. 13.

Our work focuses on the THz frequency range. Using VNAs, we perform high-frequency measurements with the aid of frequency extender modules. These modules convert and amplify the signals of the original VNA frequency range to cover the extended frequency bands. Two waveguide bands were investigated in our work: WR05 (140 GHz–220 GHz) and WR03 (220 GHz–325 GHz). System error corrections were applied using different calibration procedures, accounting for various error terms, including noise floor, trace noise, and drift. The architecture of a 4-port VNA with frequency extension modules is discussed in Chap. 6.

To evaluate the performance of VNA calibration and the measured values, we create an uncertainty budget for each systematic error within the measurement system. These error terms are integrated into error models that translate raw data into corrected data. The error terms and models vary depending on the VNA configuration, and the set of accounted error term calculations changes according to the applied calibration procedure.

The error coefficients considered during the VNA calibration process include directivity, reflection tracking, source match, and switch terms. These terms are corrected using mechanically characterised calibration standards, represented by shims of varying lengths, short standards, and match standards. Additionally, our work incorporates extended error terms caused by environmental and hardware influences on the measured values, such as noise floor, trace noise, drift, non-linearity, and cable movement, to ensure a comprehensive and reliable calibration process [20].

The VNA calibration procedure may appear time-consuming, and calculating uncertainties for each error term requires significant effort. However, specialised software, such as VNA Tools, simplifies this process by facilitating the creation of uncertainty budgets through a calibration workflow. The software also allows the investigation of the same calibration standards with different calibration procedures and supports the determination of over-determined error terms. This is especially necessary at THz frequencies, where connection repeatability and non-linearity investigations pose challenges [21].

Using VNAs for radio channel investigations requires careful metrological considerations, particularly when the measurement range exceeds a few metres. Signal attenuation, cable movements, and reflections from unaccounted paths can impact the studied frequencies. Due to their sensitivity to small changes, VNAs require accurate positioning and alignment to ensure reliable measurements. This makes VNAs more suitable for controlled measurement conditions, while outdoor measurements are generally impractical. For extended distances, one solution is to use optical-based interfaces [22], which replace long VNA cables. While effective, this approach increases complexity and cost and may introduce signal-to-noise ratio challenges due to active components in the signal path. Additionally, RF-to-optical converters are often limited to specific frequency ranges, requiring multiple assemblies to cover the entire range of interest. Another solution involves signal amplification to compensate for cable losses. Both optics-based and amplification-based solutions have unique advantages and challenges, as discussed in [23].

16.2.3 Arbitrary Waveform Generation-Based Sounder

AWGs provide the flexibility to generate *arbitrary* waveforms in software. As time passes, AWGs with increasing larger bandwidths become available and cheaper, so that they represent a viable alternative to, e.g. MLBS-based sounders introduced in the previous section. AWGs contain digital-to-analog converters (DACs) with high sampling rates and corresponding analog anti-aliasing filters to generate the baseband signals. Additionally, they may include IQ-mixers to mix the baseband signal to an intermediate frequency. The IF signal has to be fed afterward to an appropriate frequency frontend. AWGs may either *play* stored waveform files (which may require a resampling routine to match the clock rate of the DACs) or provide software applications to generate certain types of waveforms directly on the device. The receiver at IF or baseband has to provide similar flexibility. For example, signal acquisition can be performed with signal analysers providing ADCs with a sufficient sampling rate.

Of course, sounding waveforms will not be completely arbitrary but follow basic design considerations to maintain sufficient SNR such as high time-bandwidth products, a low peak-to-average power ratio (PAPR), and good correlation properties to allow impulse compression see Chap. 8. As such, AWGs are mostly in use for waveforms that are unfeasible to be generated in hardware, such as Frank-Zadoff-Chu sequences [8, 24] or phase optimised multicarrier signals [12, 25].

Furthermore, multicarrier (or multisine/multitone) signals in combination with an AWG are of special interest for channel sounding applications. These signals allow direct control over power and phase spectra represented by the carrier weights

$$c_\mu = \alpha_\mu \cdot \exp j\phi_\mu \qquad (16.3)$$

using the discrete Fourier transform (DFT):

$$x[n] = \sum_\mu c_\mu \cdot \exp\left(j\frac{2\pi\mu n}{N}\right), \qquad (16.4)$$

with frequency index $\mu = -N/2, \ldots, N/2-1$ and time index $n = 0 \ldots N$. Similar to the concept of Orthogonal Frequency-Division Multiplex (OFDM) in communications, the carrier weights can be adjusted to allow a precoding of the waveform. Instead of a strictly sequential 2-step sounding process where measurement and (post-)processing are usually decoupled, AWG architectures allow to adapt the waveform to the measurement scenario or to conduct a sequence of measurements with different waveform attributes [26].

The optimisation of power and phase spectra does have different design goals. The phase spectrum defines how the different sinusoids superpose in the time domain and, hence, indirectly controls the amplitude distribution of the waveform and its PAPR. For example, Schroeder [25] developed a phase spectrum design approach which provides a low PAPR of the resulting waveform:

$$\phi_\mu = \phi_0 - 2\pi \sum_{j=0}^{\mu-1} (\mu - j) \frac{\alpha_\mu^2}{\sum_\mu \alpha_\mu^2} . \tag{16.5}$$

This design reduces for a frequency-flat power spectrum to the well-known quadratic phase spectrum (also known from a Newman multitone signal):

$$\phi_\mu = \phi_0 - \frac{\pi \mu^2}{N} . \tag{16.6}$$

The power spectrum defines the autocorrelation function (ACF) of the waveform. Additionally, the direct control over the power spectrum results in a strict band limitation of multicarrier signals, as no power is transmitted outside the selected carriers. An equal power distribution over the carriers allows the creation of a frequency-flat power spectrum exploiting the full bandwidth. This may seem inappropriate for sounding since the resulting ACF is sinc-shaped leading to a superposition of complex weighted sinc functions in the channel impulse response. This results in constructive or destructive interferences due to the low sidelobe damping if path delays are close to each other relative to the reciprocal of the bandwidth. However, if model-based parameter estimation is applied as post-processing step, the data model can correctly account for the ACF of the waveform which allows to coherently model and estimate the superpositions of the ACFs in the channel impulse response (see Chap. 17 for more details).

A further aspect when considering model-based parameter estimation is that the power spectral density of the waveform is directly related to a statistical performance limit of parameter estimators, namely, the Cramér-Rao bound [27]. This bound is a lower limit for the mean-squared error any unbiased estimator can achieve. This knowledge allows for designing a multicarrier waveform that improves the Cramér-Rao bound. The optimised power distribution allows using available (and especially for THz frequencies limited) transmit power optimally. This is exemplified in Fig. 16.3a for improved delay estimation performance (again refer to Chap. 17 or [28] for details). However, this optimisation is not universal but is built on previous knowledge about the propagation channel. Hence, such an optimisation requires an online adaption of the waveform during a measurement campaign which is possible given the flexibility of an AWG as baseband unit.

Furthermore, multicarrier signals provide a means of coping with nonlinear distortions if amplifiers are operated outside their linear operating range to maximise output power. This can easily impact the quality of the sounding results. Multicarrier signals provide a straightforward way of analysing and detecting in-band as well as out-of-band distortions. If the transmit power for a carrier is switched off, the receiver can sense the noise and distortion power at the position of this carrier [29, Ch. 3.5] as shown in Fig. 16.3b. This corresponds to noise-power ratio tests where a

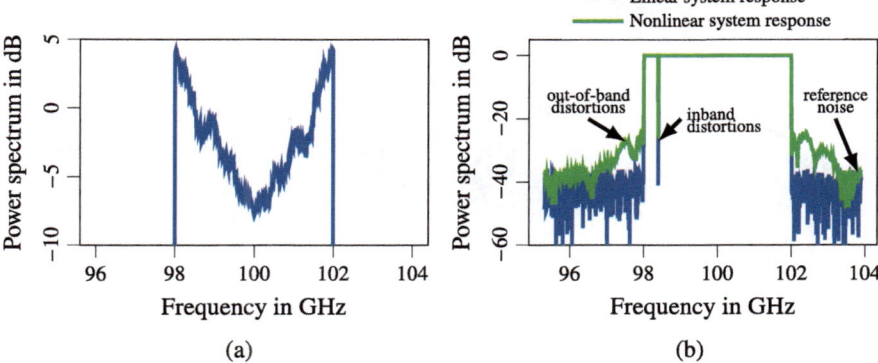

Fig. 16.3 Two different applications of multicarrier precoding: (**a**) numerically optimised power spectrum to improve the MSE of high-resolution path delay estimation in a multipath scenario. (**b**) Implementation of the noise-power ratio test by deactivating individual subcarriers (originally from [26])

notch filter is applied to the transmit signal to introduce such gaps in the spectrum. In the case of a linear system response, the sensed power level corresponds to the noise level. If the system responds non-linearly, harmonic and intermodulation distortions of the other carriers may sum up at the switched-off carrier position leading to an increase in the sensed power level relative to the (linear) noise floor. The assumption is that switching off a single carrier has negligible influence on the amplitude distribution of the resulting waveform such that the observed nonlinear behaviour is a good approximation for a fully occupied multicarrier signal. The strict band limit of a multicarrier signal furthermore allows detecting out-of-band distortions which lead to a so-called spectral regrowth at the edges of the occupied bandwidth. With enough distance to the occupied band, out-of-band distortions are usually negligible such that these areas allow sensing the (linear) noise level as a reference.

16.3 Multidimensional Sounding

16.3.1 Directional Sounding via Mechanically Rotated Antennas

One possibility to evaluate the directional structure of the propagation channel is to rotate single high-gain antennas on the transmitter and receiver sides. This can be achieved by placing the frequency frontends and antennas on positioners as sketched in Fig. 16.4. The rotation step size can be chosen such that the combined pattern from all scan positions has a nearly flat power characteristic considering

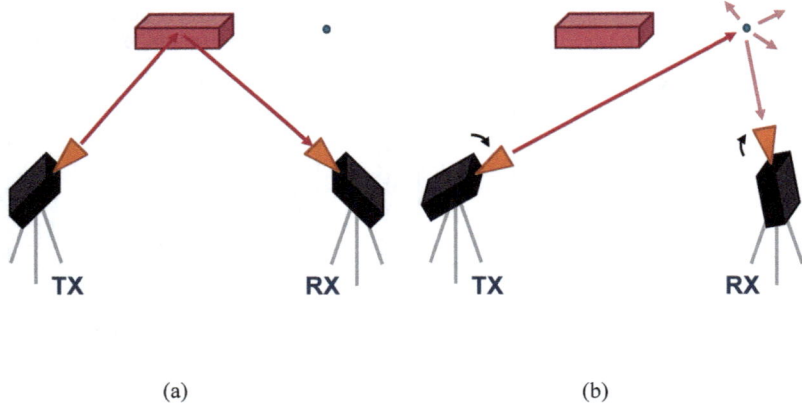

(a)   (b)

Fig. 16.4 Directional sounding via mechanically rotated antennas: (**a**) antennas facing a strong specular reflector. (**b**) Antennas facing a weak scatterer. With proper gain control the receiver can compensate for the different receive powers improving the overall dynamic range of the system

the scanning angle. Usually, a phase coherent acquisition is difficult with such a setup since the mechanical rotation will influence the positioning and bending of RF cables, and the long acquisition time for a full scan may result in severe phase drifts between the transmitter and receiver. Hence, angular spectra are usually derived from an incoherent superposition of the received powers.

Directional sounding with rotated high-gain antennas provides a benefit that is most prominent for spatially sparse channels as often encountered at THz frequency bands. If strong propagation paths (e.g. from the line-of-sight) and weak ones are spatially well-separated relative to the antenna beamwidth, the receiver can compensate for the difference in the received power levels. This issue is sketched in Fig. 16.4a, b. The compensation can be realised by automatic gain control which introduces additional, damping stages if the received signal exceeds specific power levels. Hence, it plays a significant role in omitting non-linear distortions in the presence of strong propagation paths while improving the overall dynamic range of the system. However, combining different scan positions requires a prior compensation of the different gain stages. This will inevitably lead to varying noise floor levels for the different scans. For example, a scan position with a line-of-sight path encounters a high damping during the measurement, and its compensation significantly boosts the noise floor. Hence, after combination, this noise floor may mask weak propagation paths impinging from other directions, eventually ruining the previous improvement in dynamic range. In the case of an incoherent power combination, one way of coping with this issue is to apply denoising strategies on the single scans before combining [30]. For example, a detector could consider the statistics of the received power to distinguish between pure noise samples from samples containing propagation paths. This way, the propagation channel's double-directional structure (based on received power) can be analysed with a suitable dynamic range.

Similar problems arise in the case of coherent processing using model-based high-resolution parameter estimation to estimate parameters of a double-directional channel model (see Chap. 17). This method benefits from joint processing of all observations (scan positions or antenna positions in an array) resulting in an increased SNR for parameter estimation. However, the estimator must either account for varying noise floors at the different antenna ports if the gain stages are compensated beforehand or include the varying gain stages into the model, mapping the parameters of the propagation channel to the measurement. These problems, however, are still not solved yet.

16.3.2 Polarimetric Sounding

As introduced in Chap. 8, wave polarisation may be characterised by a 2-dimensional orientation vector perpendicular to the Poynting vector. This concept originates from the Jones formalism. The propagation channel may change the polarisation status and typically the coordinate system used at the RX will not match the one at the TX. Therefore, the polarisation status at the RX may be obtained as a vector transformation using a 2×2 scattering matrix. Hence, an unambiguous characterisation of the polarisation requires access to two orthogonal polarisations at the transmitter and receiver, respectively. Additionally, channel measurements for each combination of them are required to determine all elements of the scattering matrix. One common approach is to define vertical and horizontal polarisations representing the two base vectors of a Cartesian coordinate system perpendicular to the Poynting vector (compare Fig 8.1, Chap. 8). If polarisation is not considered during sounding and only single polarised antennas are in use, a significant proportion of the channel may remain *unobservable*.

Orthogonal access to polarisation is maintained by the antenna design. Hence, antennas are required with two feed points each exciting a different operation mode with a significant sensitivity for either the horizontal or vertical polarisation. At THz frequencies, feeds and antennas have to be designed based on waveguides. Orthomode transducers allow the feeding of a quadratic waveguide with the two orthogonal polarisations simultaneously. The further frontend design up to the antennas (including the orthomode transducers) is visualised in Fig. 16.5. Two identical RF chains at the transmitter and receiver are connected to the orthomode transducers. At the transmitter on the top, the RF chain consists of a mixer and a bandpass filter. At the receiver on the bottom, additional amplifiers at RF and IF frequencies are used. To distinguish the received power that originated from the horizontal or vertical polarisation at the transmitter, an orthogonal access strategy is required. In this case, we are using time-division multiple access (TDMA) via switching the baseband signal to either signal chain. At the receiver, both channels can be recorded in parallel. The mixers require an local oscillator (LO) frequency of

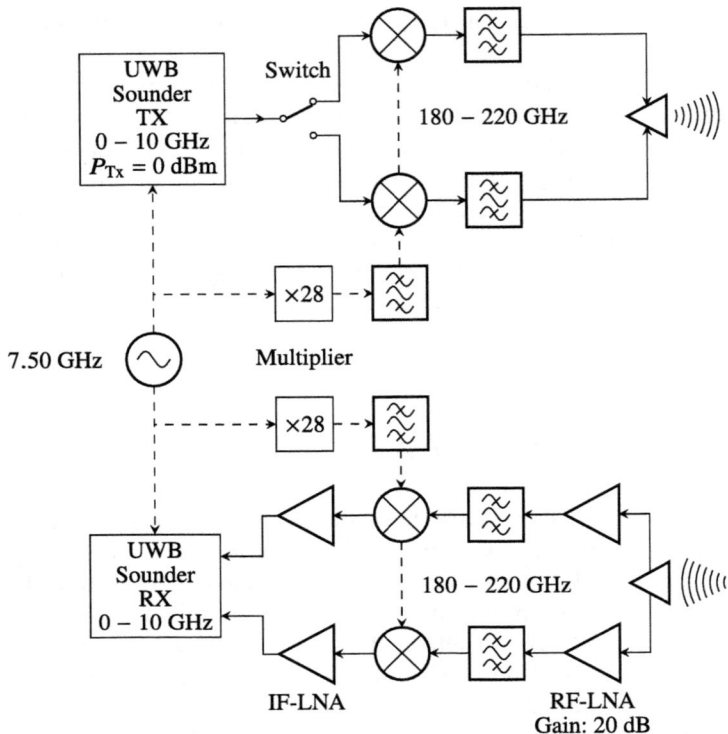

Fig. 16.5 Block diagram of dual-polarised transmit and receive frontends for a 180 GHz to 220 GHz frequency range

210 GHz which is achieved by an LO source of 7.5 GHz and multipliers placed close to the mixers. To assure phase stability for a coherent measurement, the LO source is shared between all mixers as well as the baseband units. As baseband units, ultra-wideband (UWB) M-sequence units similar to Sect. 16.2.1 are in use. A challenge for this design is the LO distribution over long distances between TX and RX, as well as a possible mechanical rotation of the frontends to achieve a directive channel representation (compare Sect. 16.3.1) making the LO prone to interference. Hence, it is beneficial to use RF-over-fibre connections which are less prone to interference. The presented polarimetric frontend design has been used and verified, for example, in [31].

References

1. H. Friis, A note on a simple transmission formula. Proc. IRE. **34**(5), 254–256 (1946)
2. M. Steinbauer, A. Molisch, E. Bonek, The double-directional radio channel. IEEE Antennas Propag. Mag. **43**(4), 51–63 (2001)
3. R. Thoma, M. Landmann, G. Sommerkorn, A. Richter, Multidimensional high-resolution channel sounding in mobile radio, in *Proceedings of the 21st IEEE Instrumentation and Measurement Technology Conference (IEEE Cat. No.04CH37510)*, vol. 1 (2004), pp. 257–262
4. C. Gentile, A.F. Molisch, J. Chuang, D.G. Michelson, A. Bodi, A. Bhardwaj, O. Ozdemir, W.A.G. Khawaja, I. Guvenc, Z. Cheng, F. Rottenberg, T. Choi, R. Müller, N. Han, D. Dupleich, Methodology for benchmarking radio-frequency channel sounders through a system model. IEEE Trans. Wirel. Commun. **19**(10), 6504–6519 (2020)
5. A. Ghosh, M. Kim, THz channel sounding and modeling techniques: an overview. IEEE Access **11**, 17 823–17 856 (2023)
6. W. Jiang, Q. Zhou, J. He, M.A. Habibi, S. Melnyk, M. El-Absi, B. Han, M.D. Renzo, H.D. Schotten, F.-L. Luo, T.S. El-Bawab, M. Juntti, M. Debbah, V.C.M. Leung, Terahertz communications and sensing for 6G and beyond: a comprehensive review. IEEE Commun. Surv. Tutor. **26**(4), 2326–2381 (2024)
7. I.F. Akyildiz, C. Han, Z. Hu, S. Nie, J.M. Jornet, Terahertz band communication: an old problem revisited and research directions for the next decade. IEEE Trans. Commun. **70**(6), 4250–4285 (2022)
8. J.M. Eckhardt, A. Schultze, R. Askar, T. Doeker, M. Peter, W. Keusgen, T. Kürner, Uniform analysis of multipath components from various scenarios with time-domain channel sounding at 300 GHz. IEEE Open J. Antennas Propag. **4**, 446–460 (2023)
9. S. Rey, J.M. Eckhardt, B. Peng, K. Guan, T. Kürner, Channel sounding techniques for applications in THz communications: a first correlation based channel sounder for ultra-wideband dynamic channel measurements at 300 GHz, in *2017 9th International Congress on Ultra Modern Telecommunications and Control Systems and Workshops (ICUMT)* (2017), pp. 449–453
10. R. Müller, R. Herrmann, D.A. Dupleich, C. Schneider, R.S. Thomä, Ultrawideband multichannel sounding for mm-wave, in *The 8th European Conference on Antennas and Propagation (EuCAP 2014)* (2014), pp. 817–821
11. N. Bräunlich, C.W. Wagner, J. Sachs, G. Del Galdo, Configurable pseudo noise radar imaging system enabling synchronous MIMO channel extension. Sensors **23**(5), (2023) [Online]. Available: https://www.mdpi.com/1424-8220/23/5/2454
12. R. Takahashi, K. Shibata, M. Kim, Development and verification of double-directional channel sounder at 300 GHz, in *2022 International Symposium on Antennas and Propagation (ISAP)* (2022), pp. 523–524
13. T. Eichler, R. Ziegler, White paper: fundamentals of THz technology for 6G. Rohde & Schwarz, Technical report (2023) [Online]. Available: https://www.rohde-schwarz.com/
14. White paper: a new sub terahertz testbed for 6G research. Keysight, Technical report (2023) [Online]. Available: https://www.keysight.com/us/en/assets/7120-1082/white-papers/A-New-Sub-Terahertz-Testbed-for-6G-Research.pdf
15. J.G. Proakis, D.G. Manolakis, *Digital Signal Processing: Principles, Algorithms, and Applications*, 4th edn. (Pearson Education, New Jersey, 2007)
16. A.F. Molisch, *Wireless Communications*, 2nd edn. (John Wiley & Sons, West Sussex, 2011)
17. R. Zetik, M. Kmec, J. Sachs, R.S. Thomä, Real-time MIMO channel sounder for emulation of distributed ultrawideband systems. Int. J. Antennas Propag. **2014**(1), 317683 (2014) [Online]. Available: https://onlinelibrary.wiley.com/doi/abs/10.1155/2014/317683
18. T. Doeker, J.M. Eckhardt, C.E. Reinhardt, T. Kürner, Time-domain channel sounder calibration at low terahertz band. IEEE Open J. Antennas Propag. **5**(6), 1598–1611 (2024)

19. M.D. Al-Dabbagh, D. Ulm, T. Kleine-Ostmann, D. Humphreys, Horn antenna phase center position influence on sub-THz measurements uncertainties, in *2024 18th European Conference on Antennas and Propagation (EuCAP)* (IEEE, Piscataway, 2024), pp. 1–5
20. M. Zeier, D. Allal, R. Judaschke, Guidelines on the evaluation of vector network analysers (vna). EURAMET Calibration Guide **3**(12), 507–521 (2018)
21. M. Wollensack, J. Hoffmann, J. Ruefenacht, M. Zeier, VNA Tools II: S-parameter uncertainty calculation, in *79th ARFTG Microwave Measurement Conference* (IEEE, Piscataway, 2012), pp. 1–5
22. M. Bengtson, Y. Lyu, W. Fan, Long-range VNA-based channel sounder: design and measurement validation at mmwave and sub-THz frequency bands. China Commun. **19**(11), 47–59 (2022)
23. L. Carslake, J. Skinner, T.H. Loh, Design and preliminary indoor assessment of a long-range sub-THz VNA-based channel sounder between 500 GHz and 750 GHz, in *2024 18th European Conference on Antennas and Propagation (EuCAP)* (IEEE, Piscataway, 2024), pp. 1–5
24. D. Chu, Polyphase codes with good periodic correlation properties (corresp.). IEEE Trans. Inf. Theory **18**(4), 531–532 (1972)
25. M. Schroeder, Synthesis of low-peak-factor signals and binary sequences with low autocorrelation (corresp.). IEEE Trans. Inf. Theory **16**(1), 85–89 (1970)
26. J. Gedschold, D. Dupleich, S. Semper, M. Döbereiner, A. Ebert, G.D. Galdo, R.S. Thomä, Metrology of multicarrier-based delay-doppler channel sounding for sub-THz frequencies. IEEE Open J. Antennas Propag. **6**(4), 1175–1187 (2025). https://doi.org/10.1109/OJAP.2025.3566473
27. E. Van den Eijnde, J. Schoukens, On the design of optimal excitation signals. IFAC Proc. Vol. **24**(3), 1139–1144 (1991). *9th IFAC/IFORS Symposium on Identification and System Parameter Estimation*, Budapest
28. J. Gedschold, S. Semper, M. Döbereiner, R.S. Thomä, Excitation signal design for THz channel sounding and propagation parameter estimation, in *18th European Conference on Antennas and Propagation (EuCAP)* (2024), pp. 1–5
29. R. Pintelon, J. Schoukens, *System Identification: A Frequency Domain Approach* (Wiley, London, 2012)
30. D. Dupleich, R. Müller, R. Thomä, Practical aspects on the noise floor estimation and cut-off margin in channel sounding applications, in *2021 15th European Conference on Antennas and Propagation (EuCAP)* (2021), pp. 1–5
31. D. Dupleich, A. Ebert, R. Müller, G. Del Galdo, R. Thomä, Verification of dual-polarized ultra-wideband channel sounder for THz applications, in *2021 IEEE 32nd Annual International Symposium on Personal, Indoor and Mobile Radio Communications (PIMRC)* (2021), pp. 1–5

Chapter 17
Multidimensional Parameter Estimation for Channel Sounding

Jonas Gedschold, Sebastian Semper, Giovanni Del Galdo, and Reiner Thomä

Abstract This chapter deals with parametric multidimensional channel models, with the estimation of the corresponding parameters (such as delays, path weights and angles) from channel sounding data and with the metrological aspects associated to this processing chain. Deterministic and stochastic components of the channel model are discussed, focusing on decomposing the channel into a superposition of discrete multipath components and non-resolvable dense multipath components (DMC). Terahertz frequency bands readily enable the deployment of wide bandwidths; hence, models designed for narrow bandwidths need to be revisited and adapted. From a metrological point of view, the limits of parameter estimation need to be considered. It is shown how these limits relate to the choice of the sounding waveform and how they can aid in determining a reasonable number of propagation paths (model order selection).

17.1 A Metrological Perspective on Model-Based High-Resolution Parameter Estimation in Channel Sounding

One of the main goals of metrology is the characterization of uncertainties of measurement devices with respect to systematic and stochastic errors. This includes to setup traceable verification chains from measurements to a physical *ground truth*. In the case of channel sounders (Chap. 16), such verification chains include

J. Gedschold (✉) · S. Semper · R. Thomä
Institute of Information Technology, Technische Universität Ilmenau, Ilmenau, Germany
e-mail: jonas.gedschold@tu-ilmenau.de; sebastian.semper@tu-ilmenau.de; reiner.thomae@tu-ilmenau.de

G. Del Galdo
Institute of Information Technology, Technische Universität Ilmenau, Ilmenau, Germany

Fraunhofer Institute for Integrated Circuits (IIS), Ilmenau, Germany
e-mail: giovanni.delgaldo@tu-ilmenau.de

© The Author(s) 2026
T. Kürner et al. (eds.), *Metrology for THz Communications*, Springer Series in Optical Sciences 256, https://doi.org/10.1007/978-3-032-01986-8_17

measurement artifacts with known properties creating *reference channels* (Chap. 19) or the use of a traceable reference device such as a vector network analyser (VNA) (Chap. 6). However, channel measurements are seldomly used without subsequent postprocessing. Especially the derivation of channel models builds upon extracted parameters such as decomposing the channel into a superposition of plane waves with attributes like propagation delays, propagation angles, or Doppler shifts. Hence, it is required to include the postprocessing steps into the verification chain to, e.g., ensure a certain quality of the channel models, or to enable comparability of processed sounding results. The metrological assessment of model-based parameter estimation has three main perspectives, as explained in more detail in the following three sections.

17.1.1 Hardware Modeling

Postprocessing algorithms can partly account for hardware characteristics. This can, for example, relate to using calibration measurements and appropriate equalization filters [1] to compensate for the device transfer function of the sounder. A more sophisticated approach is using model-based parameter estimation. In this case, one can set up a model that considers, on the one hand, the propagation parameters of interest but, on the other hand, also a more or less complex model of the measurement device allowing a mapping of the propagation parameters to the actual measurement. In this chapter, we consider model-based parameter estimation based on the well-known double-directional channel model [2]. Such a model understands the propagation channel as a superposition of plane waves traveling on multiple paths from the transmitter to the receiver. Such paths are parameterized by propagation delays, Doppler shifts, and directions of departure and arrival in the local coordinate systems of the transmitter and receiver, respectively (compare Chap. 8). This model already imposes a certain structure on the channel which is built on prior knowledge about the wave propagation but also includes certain simplifications and approximations to maintain computational feasibility. The device model accounts, e.g. for the device transfer function and band limitation of the sounder, and allows mapping the highly resolved propagation parameters to the measurement of the channel impulse response (CIR). To a full extent, also the directional antenna response can be considered if the sounder and algorithm architecture allow for a double-directional resolution of departure and arrival angles at the transmitter and receiver. In this way, the correct influence of the antenna on a specific propagation path can be considered. The separation of a device model from a propagation model ultimately leads to hardware de-embedding, hence, a hardware-independent description of the propagation channel via the estimated propagation parameters. Additionally, model-based high-resolution parameter estimation schemes can potentially go beyond the nominal resolution limits of the hardware (e.g., regarding bandwidth). This property is sometimes referred to as *super-resolution*.

Metrologically, model-based parameter estimation shifts the problem of quantifying the accuracy of the measurement hardware to the device model and the calibration measurements contributing to these models. However, sounder architectures to allow double-directional resolution are still under development for THz frequencies and have to overcome some challenges (Chap. 16). Hence, hardware de-embedding is only partly possible. This introduces significant uncertainties when comparing sounding results from different channel sounders since they will be influenced by the characteristics of the actual measurement hardware and the partial compensation by the involved algorithms. Incomplete observations or data models may introduce a significant estimation error, e.g., when ignoring elevation during angle estimation [3]. Additionally, the simplifying assumptions for the propagation models such as far-field approximations or narrowband models for Doppler and angle estimation may impact the quality of the results. From a metrological perspective, oversimplified models can also lead to a model mismatch and ultimately to the estimation of so-called pseudo-true parameters. These parameters may be optimal considering the cost function or optimization objective but lack a physical meaning [4].

See Sect. 17.2 of this chapter for more details about device modeling and hardware de-embedding.

17.1.2 Statistical Properties of the Estimator

The task of estimation algorithms is to derive the unknown model parameters from noisy observations of the CIR. As such, the estimator itself introduces both systematic and stochastic errors, often characterized by the bias and variance of the estimates. Evaluating these properties experimentally is difficult since it requires the knowledge of the *true* model parameters to compare estimates versus ground truth. Alternatively, one can rely on theoretical performance analyses of a certain estimator type. For example, we are considering maximum likelihood parameter estimation in this chapter. Such estimators are known to be unbiased and efficient. Efficiency means that the mean-squared error of the estimates meets the Cramér-Rao bound (CRB), a theoretical performance limit for every unbiased estimator [5, Ch. 6.4]. However, there is no closed-form solution for maximum likelihood estimation of the propagation parameters. Hence, actual implementations are only approximations, making an experimental performance verification inevitable. Also, model mismatch and incomplete data models as discussed in the previous section will impact the estimation bias.

See Sect. 17.3 of this chapter for more details about parameter estimation.

17.1.3   Interplay Between Hardware and Algorithms

As already mentioned, if the objective of a channel measurement is the derivation of parameters of a channel model, these parameters should be the target of metrological assessment. Hence, hardware and algorithm must be considered a joint pipeline, because the algorithm's capability to account for the specific hardware is a key factor for the accuracy of the estimates. Additionally, a channel sounder always requires a propagation channel between the transmitter and receiver. This channel may be just a wire or waveguide in the case of a back-to-back measurement or include over-the-air propagation when using the antennas. However, already a basic verification of the propagation delay between transmitter and receiver requires some kind of estimator, e.g., by performing a peak search in the magnitude-squared CIR with or without interpolation. Hence, the derived accuracy of the propagation delay is indivisibly connected to the chosen algorithm.

Therefore, test or reference channels can be used to evaluate the interplay between hardware and software. Such channels can be created, e.g., by propagation artifacts (Chap. 19). A ground truth can now be achieved by employing simple propagation models and the knowledge of the geometry of the artifact to predict the propagation paths in the channel.

The algorithm design also plays a major role when it comes to resolving multiple (probably closely spaced) propagation paths in the channel. The hardware responses to these paths may, e.g., due to band limitations of the hardware, overlap in the CIR. Hence, mutual interference between paths will occur, which can only be resolved by correct modeling of this overlap [6].

17.2   Hardware Modeling and De-Embedding

In this section, we discuss the double-directional channel model (Sect. 17.2.2) which allows for a de-embedding of the sounder hardware including the antennas. Hence, a device-independent description of the propagation channel is possible if the parameters of such a model are available/estimated. The core of this model is a mapping from the parameters of the propagation model, i.e., the multidimensional Dirac delta function in delay (Sect. 17.2.2.1), Doppler (Sect. 17.2.2.2), and angles of arrival and departure (Sect. 17.2.2.3), to the measured CIR by considering the hardware response to the individual propagation paths. Additionally, we discuss the necessity of stochastic model components to characterize measurement noise but also to cope with the resolution limits of the sounder (Sect. 17.2.3).

17.2.1 Notation

Although the problem of modeling a channel and estimating channel parameters has a multidimensional structure, the algorithmic formulation is often reduced to two dimensions by vectorizing higher-dimensional structures to enable the use of matrix and vector products. The remainder of this chapter makes use of a simplified Einstein notation which decouples the mathematical formulation from its implementation via vector/matrix products [7]. In this way, the formulation of the multidimensional modeling and estimation problem is more convenient and concise.

Let us assume a tensor product between $\mathbf{a} \in \mathbb{C}^{K \times N \times I}$ and $\mathbf{b} \in \mathbb{C}^{N \times K \times J}$ to yield the result $\mathbf{c} \in \mathbb{C}^{I \times J}$. The elements of $\mathbf{c}$ are defined as

$$c_{i,j} = \sum_k \sum_n \mathbf{a}_{k,n,i} \cdot \mathbf{b}_{n,k,j} . \tag{17.1}$$

Instead of (17.1) we refer to the whole tensor $\mathbf{c}$ with

$$\mathbf{c}_{i,j} = \mathbf{a}_{k,n,i} \cdot \mathbf{b}_{n,k,j} . \tag{17.2}$$

Hence, $\mathbf{a}$ and $\mathbf{b}$ are multiplied elementwise over those axes denoted with the same indices while the summation happens over those axes, whose index is present on the right-hand side but not on the left-hand side of the equation.

Alternatively, one had to vectorize $\mathbf{a}$ and $\mathbf{b}$ over the dimension K and N to form $\mathbf{A} \in \mathbb{C}^{K \cdot N \times I}$ and $\mathbf{B} \in \mathbb{C}^{K \cdot N \times J}$ and represent (17.2) as a dot product

$$\mathbf{C} = \mathbf{A}^T \cdot \mathbf{B} . \tag{17.3}$$

17.2.2 The Double-Directional Channel Model

The double-directional channel model [2] interprets the channel as a superposition of plane waves with parameters such as delay, Doppler shift, and angles of arrival and departure formulated as a multidimensional Dirac delta function (Chap. 8). As such, it assumes a certain structure of the channel that is derived from *a priori* knowledge about the measurement. However, to maintain a manageable complexity of the estimation problem, the model is built on approximations and simplifying assumptions which have to be kept in mind when applying the model to a channel sounding problem.

A foundation for hardware de-embedding is the accurate mapping from the parametric channel description to the observation of the channel sounder, i.e., the measurement of the electric field at the receiver antenna ports. This mapping has to account, e.g., for the band limitation of the respective measurement aperture in time and space. To formulate the mapping function, we first gather all parameters of a propagation path k into vector $\boldsymbol{\theta}_k$

$$\boldsymbol{\theta}_k = \left[\tau_k, \alpha_k, \varphi_k^{\mathrm{Tx}}, \vartheta_k^{\mathrm{Tx}}, \varphi_k^{\mathrm{Rx}}, \vartheta_k^{\mathrm{Rx}}, \gamma_{\varphi\varphi,k}, \gamma_{\varphi\vartheta,k}, \gamma_{\vartheta\varphi,k}, \gamma_{\vartheta\vartheta,k} \right], \qquad (17.4)$$

which is composed of a propagation delay τ, a Doppler shift α, azimuth and elevation angles of departure and arrival φ, ϑ, and the polarimetric path weights γ for the respective polarization orientation in the local coordinate systems of transmitter and receiver perpendicular to the pointing vector of the wave propagation (Fig. 8.1, Chap. 8). Now, we can write the mapping from the parameters to the observation $\mathbf{s}(\boldsymbol{\theta}_k) \in \mathbb{C}^{N_{\mathrm{Tx}} \times N_{\mathrm{Rx}} \times N_{\mathrm{f}} \times N_{\mathrm{t}}}$ at $N_{\mathrm{T_x}}$ and $N_{\mathrm{R_x}}$ antenna ports with N_{f} frequency samples and N_{t} slow time samples for a *single* propagation path via so-called atomic functions $\mathbf{a}$ [7] for the individual parameters:

$$\mathbf{s}_{t_x,r_x,f,t}(\boldsymbol{\theta}_k) = \gamma_{p_1 p_2,k} \cdot \mathbf{a}_f^{\mathrm{f}}(\tau_k) \cdot \mathbf{a}_t^{\mathrm{t}}(\alpha_k) \cdot \mathbf{a}_{p_1,t_x}^{\mathrm{tx}}(\vartheta_k^{\mathrm{tx}}, \varphi_k^{\mathrm{tx}}) \cdot \mathbf{a}_{p_2,r_x}^{\mathrm{rx}}(\vartheta_k^{\mathrm{rx}}, \varphi_k^{\mathrm{rx}}). \qquad (17.5)$$

The subscript indices are the tensor indices according to the Einstein formulation (seen in Sect. 17.2.1) and visualize the Kronecker or outer-product structure of the data model. The antenna ports at the transmitter and receiver are indexed by t_x and r_x, while the frequency and time axis are indexed by f and t, corresponding to the Fourier transform of the fast time axis and the slow time axis, respectively. The polarimetric path weights are connected to the antenna responses at the transmitter and receiver via p_1 and p_2. The superscript indices uniquely identify the individual atomic functions formulated in the following sections. In order to obtain a complete multipath model, we superimpose K distinct propagation paths, via

$$\mathbf{s}_{t_x,r_x,f,t}(\boldsymbol{\theta}) = \sum_{k=1}^{K} \mathbf{s}_{t_x,r_x,f,t}(\boldsymbol{\theta}_k), \qquad (17.6)$$

where $K \in \mathbb{N}$ is the so-called model order, i.e., the number of observed specular propagation paths.

17.2.2.1 Frequency Response Function

The measurement in the fast time domain results from a convolution of the transmitted waveform and the hardware transfer functions of the transmitter and receiver with the Dirac deltas of the delay parameters. At this point, the hardware transfer function is considered to be direction-independent and, hence, captures the device response up to the antenna ports. The antenna response mapping is discussed in Sect. 17.2.2.3. When formulating the model in the frequency domain, $\mathbf{a}^{\mathrm{f}}$ maps the delay parameters to the frequency response via

$$\mathbf{a}_f^{\mathrm{f}}(\tau_k) = \mathbf{g}_f \cdot \exp\left(-\jmath 2\pi \mathbf{f}_f \tau_k\right) \in \mathbb{C}^{N_{\mathrm{f}}}, \qquad (17.7)$$

where $\mathbf{g} \in \mathbb{C}^{N_f}$ accounts for the transfer function of the signal chain and the transmitted waveform. We consider the impact of the waveform as well as the signal chain jointly as *device transfer function* in the following. The exponential function accounts for the aperture domain of the Dirac delta function for frequencies $\mathbf{f} \in \mathbb{R}^{N_f}$ for the number of samples N_f. If the device (excluding the antennas) is considered to be linear time-invariant (LTI), it can be sufficiently characterized by a frequency transfer function that can be identified by a (wired) back-to-back measurement. In this case, the transfer function of the cable or waveguide used to connect the transmitter to the receiver has to be known or measured beforehand by a reference device (e.g., a VNA).

Usually, there are two ways of dealing with the device transfer function:

1. The device transfer function can be equalized/calibrated at the receiver via filtering [1]. Consequently, the device model can assume a frequency flat device transfer function.
2. The measured device transfer function from the calibration measurement can be directly included in the model (17.7) to map the delay parameter to the measurement of the CIR [8].

An equalization at the receiver comes to its limits if notches or regions of high attenuation exist in the measured transfer function and, therefore, the signal-to-noise ratio (SNR) is low. In this case, filter-based equalization at the receiver also boosts additive noise components (from amplification or quantization) which has side effects on the power spectral density (PSD) of the measurement noise. Inverse filtering, for example, may result in a strong noise amplification. In this case, a good alternative is a Wiener filter approach [9], i.e., a Wiener deconvolution considers regions of low SNR via a damping of the filtered spectrum. Still, there may be a signification model error when assuming a frequency flat device transfer function after calibration. Hence, it is beneficial to include the device transfer function into the model [8] which also provides more flexibility in designing the waveform (see Sect. 17.3.3).

17.2.2.2 Time Response Function

Similar as for the delay parameters, the Doppler parameters can be mapped to slow time via

$$\mathbf{a}_t^t(\alpha_k) = \exp\left(\jmath 2\pi \mathbf{t}_t \alpha_k\right) \in \mathbb{C}^{N_t} \tag{17.8}$$

for time steps $\mathbf{t} \in \mathbb{R}^{N_t}$ for the number of samples N_t. This Doppler shift-based model relies on the so-called narrowband assumption. It assumes that the Doppler effect on a wideband signal can be represented solely by the Doppler shift on the carrier frequency, hence, as for a single sinusoid. This is a sufficient approximation for small relative bandwidths but can also be extended to a *wideband* model if required [10] as

$$\mathbf{a}^t_{t,f}(\alpha_k) = \exp\left(j2\pi\left[(\mathbf{f} - f_c)/f_c + 1\right]\mathbf{t}\alpha_k\right) \in \mathbb{C}^{N_t \times N_f}. \tag{17.9}$$

It must be noted that this model does not account for any change in the signal's envelope due to the Doppler effect but just relaxes the assumption that the Doppler effect on the whole bandwidth can be approximated by a frequency shift of the carrier f_c. However, this already introduces a significant drawback in computational complexity since the Kronecker structure of the model is violated due to the frequency dependency of the Doppler atom. Nevertheless, this model nicely shows that if the range of $\mathbf{f}$, (i.e., the bandwidth) is small compared to the carrier frequency, (17.9) can be sufficiently approximated by (17.8). Hence, the narrowband assumption for Doppler estimation is justified for a small relative bandwidth which is typically the case for THz applications due to the high carrier frequencies.

17.2.2.3 Antenna Response Function

A major benefit of model-based parameter estimation is that the model can account for the angle-dependent antenna responses due to the decomposition of the channel into propagation paths with individual angles of departure and arrival. Hence, when formulating the mapping from the angles to the measurements at the antenna ports, the antenna response corresponding to the individual angle can be considered. The mapping functions $\mathbf{a}^{tx} \in \mathbb{C}^{2 \times N_{Tx}}$ and $\mathbf{a}^{rx} \in \mathbb{C}^{2 \times N_{Rx}}$ account for the polarimetric radiation patterns of the antenna elements at the transmitter and receiver, respectively. The polarization is defined with respect to the unit vectors $\mathbf{e}_\varphi$ and $\mathbf{e}_\vartheta$ of the local spherical coordinate system of the antenna arrays (refer to Fig. 8.1, Chap. 8). Hence, the polarimetric path weight $\boldsymbol{\gamma}$ in (17.5) accounts for the mapping of the wave polarization to the polarimetric-dependent responses of the individual antennas.

The mapping can be expressed efficiently by the so-called effective aperture distribution function (EADF) [11]. This function is based on calibration measurements usually performed in an anechoic chamber on a certain scanning grid in elevation and azimuth. Afterwards, the measurement data for a single polarization is represented as $\mathbf{B}_p \in \mathbb{C}^{N_\vartheta \times N_\varphi}$ where N accounts for the number of grid points in elevation and azimuth. To be applicable to model-based high-resolution parameter estimation, the EADF provides two features:

1. Beampattern interpolation to provide a *grid-free* pattern for the high-resolved angles of the propagation paths
2. Beampattern compression to only store as much data of the calibration measurement as necessary

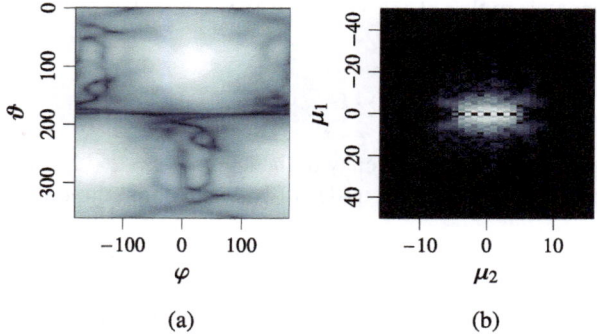

Fig. 17.1 EADF-based antenna modeling. (**a**) Beampattern. (**b**) Aperture domain

The basic idea is to use a 2D fast Fourier transform (FFT) to transform the measurements into the aperture domain:

$$\mathbf{B}_p \xrightarrow{\text{2D FFT}} \mathbf{G}_p . \qquad (17.10)$$

Figure 17.1 visualizes the magnitude of $\mathbf{B}_p$ in (a), as well as the aperture domain in (b). It can be anticipated from Fig. 17.1b that $\mathbf{G}_p$ usually has compact support; hence, only a subset of the coefficients has to be stored allowing significant data compression. Furthermore, the aperture representation allows to efficiently calculate an interpolated beampattern, e.g., for the transmitter as

$$\mathbf{a}_{p_1,t_x}^{\text{tx}}(\vartheta^{\text{tx}}, \varphi^{\text{tx}}) = \mathbf{d}_{g_1}^{\text{ele}}(\vartheta) \cdot \mathbf{G}_{g_1,g_2,p_1,t_x} \cdot \mathbf{d}_{g_2}^{\text{azi}}(\varphi) . \qquad (17.11)$$

The vectors $\mathbf{d}$ are essentially discrete Fourier transform vectors (more precisely discrete-time Fourier transform vectors) calculating the interpolated beampattern for arbitrary angles ϑ and φ.

The model (17.11) is again built on the narrowband assumption, representing the antenna beampattern as frequency flat over the considered bandwidth. However, antenna elements will introduce a severe direction-dependent frequency response distortion due to the possibly high absolute bandwidths in Terahertz (THz) sounding. While the narrowband assumption is still valid for the Doppler model in the THz range (Sect. 17.2.2.2), narrowband antenna modeling may introduce severe model mismatch impeding the performance of parameter estimation [4]. The calibration measurement can easily be extended to capture each angle's frequency response. However, the measured frequency steps during calibration may not match the required frequency grid for parameter estimation. For this reason, a frequency interpolating function $\boldsymbol{\xi} \in \mathbb{R}^{Q \times N_f}$ can be added to (17.11) which essentially maps from Q frequency samples of the EADF to the required N_f frequency samples of the estimation model [7]:

$$\mathbf{a}^{\text{tx}}_{f,p_1,t_x}(\vartheta^{\text{tx}}, \varphi^{\text{tx}}) = \mathbf{d}^{\text{ele}}_{g_1}(\vartheta) \cdot \mathbf{G}_{q,g_1,g_2,p_1,t_x} \cdot \mathbf{d}^{\text{azi}}_{g_2}(\varphi) \cdot \boldsymbol{\xi}_{q,f} . \tag{17.12}$$

Again, adding a frequency dependency to the antenna model violates the Kronecker structure of the model leading to a trade-off between computational complexity and model accuracy.

While the correct modeling of the antenna beampattern allows a perfect device de-embedding in theory, it must be considered that this method is based on calibration measurements including the associated uncertainties limiting the accuracy of the device model [3]. Furthermore, correctly estimating the angles of departure and arrival requires sufficient angular resolution (e.g., through an antenna array) which is not yet achievable with current THz sounder architectures (Chap. 16). Hence, up to now, device models can only partly account for the device response, e.g., just for the direction-independent part up to the antenna ports.

17.2.3   Stochastic Channel Components

Besides the highly resolvable propagation paths as discussed in Sect. 17.2.2, there may be a significant proportion of the received power that cannot be reliably resolved as specular components by the sounder hardware and the estimation process. It has been proven to be beneficial to describe these components as a colored noise process $\mathbf{n}_{\text{DMC}}$ called dense multipath components (DMC) [12, Ch. 2.5]. Hence, the observation model at the receiver is formulated as

$$\mathbf{y} = \mathbf{s}\left(\boldsymbol{\theta}\right) + \mathbf{n}_{\text{DMC}}(\boldsymbol{\psi}) + \mathbf{n} \in \mathbb{C}^{N_{\text{Tx}} \times N_{\text{Rx}} \times N_f \times N_t} , \tag{17.13}$$

which also accounts for measurement noise $\mathbf{n}$ related to the sounding system usually modeled as i.i.d. circular symmetric Gaussian distributed. Physically, the DMC can be interpreted as a superposition of many paths resulting from diffuse scattering processes in the channel arriving within the Rayleigh resolution of the sounding hardware. Due to the central limit theorem, each delay bin (excluding specular components) is modeled as a complex Gaussian distribution with zero mean. Since the power delay profile (PDP) usually has an exponential decay over fast time, the DMC is commonly modeled as a parametric exponentially decaying function, comparable to Room Impulse Responses in acoustics. Hence, a joint estimation of dense and specular components is required, which considerably enhances the estimation quality for the specular parameters [3]. It must be noted that this propagation-based interpretation of the parametric colored noise process as an attribute of the channel is only true if model mismatch and estimation errors are sufficiently small [3]. Otherwise, the estimator could use the additional degree of freedom of the noise model to compensate for these errors.

17.3 High-Resolution Parameter Estimation

Once an appropriate model has been formulated, its parameters must be estimated from the channel sounding measurements. Here, we consider maximum likelihood estimation as a prominent example. Thanks to the information provided by the channel model, we achieve higher resolution than the one that would have been obtainable from the raw channel measurements, e.g., following a nonparametric DFT-based approach (hence the term "high resolution").

17.3.1 Maximum Likelihood Estimation

In this chapter, following the Bayesian approach to the estimation problem, we limit ourselves to discussing maximum likelihood estimation, which, assuming no model mismatch, leads to optimal estimates by maximizing the *a posteriori* probability, i.e., considering all available *a priori* information and the observations. Maximum likelihood estimators allow jointly estimating specular paths and noise parameters and provide some desired statistical properties, i.e., unbiasedness and efficiency. The RIMAX [12] and pyMAX [7] estimators are examples for propagation parameter estimation.

The likelihood of a set of parameters given an observation is formulated in the log-likelihood function considering a complex Gaussian noise process with structured covariance $\mathbf{R}$:

$$\lambda(\boldsymbol{\theta}|\mathbf{y}) = -\ln(|\mathbf{R}|) - \mathbf{r}_{\mathbf{q}_1}^*(\boldsymbol{\theta}) \cdot \mathbf{R}_{\mathbf{q}_1,\mathbf{q}_2}^{-1} \cdot \mathbf{r}_{\mathbf{q}_2}(\boldsymbol{\theta}), \qquad (17.14)$$

where we use $\mathbf{q}_{1,2}$ as a shortcut for the indices (tx, rx, f, t) and subscripts $1, 2$ to denote the occurrence of the corresponding indices in $\mathbf{R}$. Equation (17.14) is the logarithm of a multivariate Gaussian distribution. The mean of the distribution is represented by the specular path components; hence, the residual $\mathbf{r}$ given a set of parameters $\boldsymbol{\theta}$ is

$$\mathbf{r}(\boldsymbol{\theta}) = \mathbf{y} - \mathbf{s}(\boldsymbol{\theta}). \qquad (17.15)$$

The parameter estimator can now maximize λ given an observation $\mathbf{y}$:

$$\max_{\boldsymbol{\theta}} \lambda(\boldsymbol{\theta}|\mathbf{y}) \qquad (17.16)$$

which results in a parameter estimate $\hat{\boldsymbol{\theta}}$.

If only measurement noise has to be considered and no DMC, the noise covariance $\mathbf{R}$ is a scaled identity matrix and (17.16) reduces to a nonlinear least squares problem

$$\min_{\boldsymbol{\theta}} \|\mathbf{r}(\boldsymbol{\theta})\|_2^2. \qquad (17.17)$$

Unfortunately, there is no closed-form solution for (17.16); hence, actual implementations are only approximations of the maximum likelihood estimator, e.g., by implementing an iterative gradient-based maximization of (17.16).

17.3.2 Bias and Variance of Parameter Estimates

The simplest way to assess the performance of an estimator is by quantifying its bias and its variance. The bias refers to a systematic error in estimation; hence, if, on expectation, the estimated parameters differ from the true parameters:

$$\mathbf{E}\left(\hat{\boldsymbol{\theta}}\right) - \boldsymbol{\theta} = \boldsymbol{\epsilon} .\tag{17.18}$$

The variance refers to the mean-squared error (MSE) of the parameter estimates considering a statistical ensemble of independent realizations of the observation, e.g., in case of an unbiased estimator it is

$$\mathbf{E}\left(\hat{\boldsymbol{\theta}} - \boldsymbol{\theta}\right)^2 = \mathrm{MSE}\left(\hat{\boldsymbol{\theta}}\right) .\tag{17.19}$$

An estimator is termed efficient if its variance asymptotically reaches the CRB, a theoretical performance limit for the MSE of (unbiased) parameter estimators. This is the case for maximum likelihood estimators; however, as discussed in Sect. 17.3.1, there is no closed-form solution for the discussed propagation parameter estimation problem. Actual implementations are only approximations of a maximum likelihood estimator; hence, those properties have to be treated with caution and need to be verified metrologically for the implemented estimators. If the estimator can be considered unbiased and efficient, the CRB allows, e.g., quantifying how given antennas arranged in a specific geometry perform in the context of high-resolution parameter estimation. These insights become a precious tool for the design of antenna arrays, as they guide the choice of the radiation patterns, number of elements, and array geometry. Similarly, the CRB allows assessing the impact of choosing a certain waveform for the sounding step, as explained in the following section.

17.3.3 Influence of the Waveform's PSD on Delay Estimation Performance

When evaluating how a certain waveform design impacts the parameter estimation performance, one first must decide on a quantity rating for this performance. The CRB can be a suitable quantity when considering an unbiased and efficient estimator. The CRB is calculated directly from the signal model just considering the model and its parameters. As such, it assumes a *perfect* model matching the structure of the observations. See Sect. 17.3.4 for a discussion on a lower bound in case of model misspecification.

The CRB corresponds to the main diagonal elements of the inverse of the Fisher information matrix (FIM) $\mathbf{F} \in \mathbb{R}^{N_\theta \times N_\theta}$ [5, Ch. 6.4]:

$$\text{MSE}\,(\theta_i) \geq \text{CRB}(\theta_i) = \left(\mathbf{F}^{-1}\,(\theta)\right)_{ii}. \tag{17.20}$$

The Fisher information has a useful property, namely, it is additive with respect to the observation samples $N_{\mathrm{T_x}}$, $N_{\mathrm{R_x}}$, N_f, and N_t [13]; hence, each sample contributes linearly to the information about the propagation parameters. Therefore, evaluating the Fisher information matrix for each frequency bin provides a measure of the information distribution over frequency (i.e., the spectrum of the Fisher information). Evaluating one element on the main diagonal of $\mathbf{F}$ for a delay parameter τ_k considering an i.i.d. Gaussian noise process with noise power σ^2 results in [14].

$$\mathbf{F}_{\tau_k,\tau_k,f} = 2 \cdot \underbrace{\frac{\left|\gamma_{p_1,p_2,k}\right|^2 \cdot N_\mathrm{t}}{\sigma^2}}_{\text{SNR}} \cdot \underbrace{\left|(2\pi \mathbf{f}_f)\right|^2}_{\substack{\text{frequency}\\\text{distribution}}} \cdot \underbrace{\left|\mathbf{g}_f\right|^2}_{\substack{\text{device}\\\text{transfer}\\\text{function}}}$$

$$\cdot \underbrace{\left|\mathbf{a}^{\mathrm{tx}}_{f,p_1,t_x}(\vartheta^{\mathrm{tx}}_k,\varphi^{\mathrm{tx}}_k)\right|^2 \cdot \left|\mathbf{a}^{\mathrm{rx}}_{f,p_2,r_x}(\vartheta^{\mathrm{rx}}_k,\varphi^{\mathrm{rx}}_k)\right|^2}_{\substack{\text{antenna}\\\text{responses}}}. \tag{17.21}$$

It depends on the SNR of the propagation path and the frequency responses of the device and antenna. Intuitively, if the device transfer function has a dip at a certain frequency or the antennas do not *see* a specific path, the information provided by the observation will be low. Most importantly, the factor $(2\pi \mathbf{f}_f)^2$ favors frequency bins at the edges of the bandwidth B, considering a definition of $\mathbf{f}$ from $-B/2$ to $B/2$. Since the device transfer function also includes the PSD of the waveform, the power spectrum of the waveform directly influences the information content about the delay parameter. This insight can also be used to specifically design waveforms that provide a low CRB for delay parameter estimation [14]. Since a multipath scenario includes more than one propagation path, a joint waveform can be found by maximizing the determinant of $\mathbf{F}$ [15]. An exemplary power spectrum is shown in Fig. 17.2. It also exhibits a concentration of transmit power on the edges of the bandwidth. Of course, the flexibility to generate such waveforms requires a sounder architecture with software-based waveform generation, e.g., with an arbitrary waveform generator (AWG) as baseband unit (Chap. 16).

Fig. 17.2 Example for a
Fisher information-optimized
power spectrum to improve
delay estimation performance
in a multipath scenario
(modified after [14])

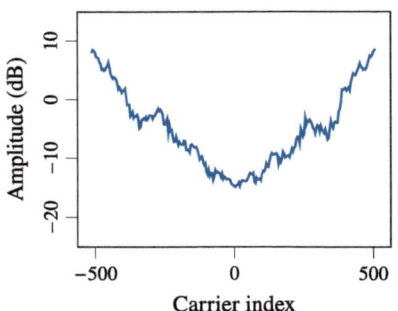

17.3.4 Model Order Selection Under Misspecification

The correct estimation of the number of propagation paths is one of the central problems in the type of parameter estimation task we consider in (17.17). More precisely, we are concerned with the correct estimation of K in (17.6). From a pure optimization point of view, one can simply increase K as long as we have enough observations in $\mathbf{s}$ in order to estimate each of the $\boldsymbol{\theta}_k$. Since the fit of $\mathbf{s}(\boldsymbol{\theta})$ to $\mathbf{y}$ improves for increasing K, we would obtain a better solution to (17.17) in terms of a lower objective function value for the estimated parameters. But not only does this pose tremendous numerical challenges, we will also see that this is not a useful approach from a channel estimation perspective.

From a system identification standpoint, we are interested in obtaining a *realistic* description of the measured radio channel. To this end, the principle of maximum likelihood only serves as a *proxy* and is only a means to another end. As the model order K is also a parameter of the observed radio channel, we cannot rely on maximum likelihood alone, but need to invoke some suitable kind of regularization that can also be traced back to statistical properties of the measured data.

Additionally, we have to consider the influence of possible errors we introduce by employing our specific choice of signal model $\mathbf{s}$. The validity for the question of the *true* model order hinges on the degree to how much the assumptions that serve to formulate $\mathbf{s}$ are violated. For less accurate $\mathbf{s}$, i.e., a *misspecified* model, the question as to what the correct estimate for K should be is getting more involved.

One tool that is grounded in statistical properties of the observations is to use performance bounds for the MSE during the estimation process. It is possible to obtain estimates for the lower bound on the MSE for each individual propagation path that is currently being estimated. If the lower bound on the MSE exceeds a certain predefined threshold for some paths, we remove those from the estimation, as they have been deemed to be unreliable given the model and the SNR. Not only are they too unreliable, but at the same time they also influence the numerical stability of the optimization problem, which in turn leads to a *global* deterioration of the estimated propagation paths.

A straightforward and well-studied tool is the CRB (see (17.20)), which in case of maximum likelihood estimation is a *tight* lower bound on the estimator's

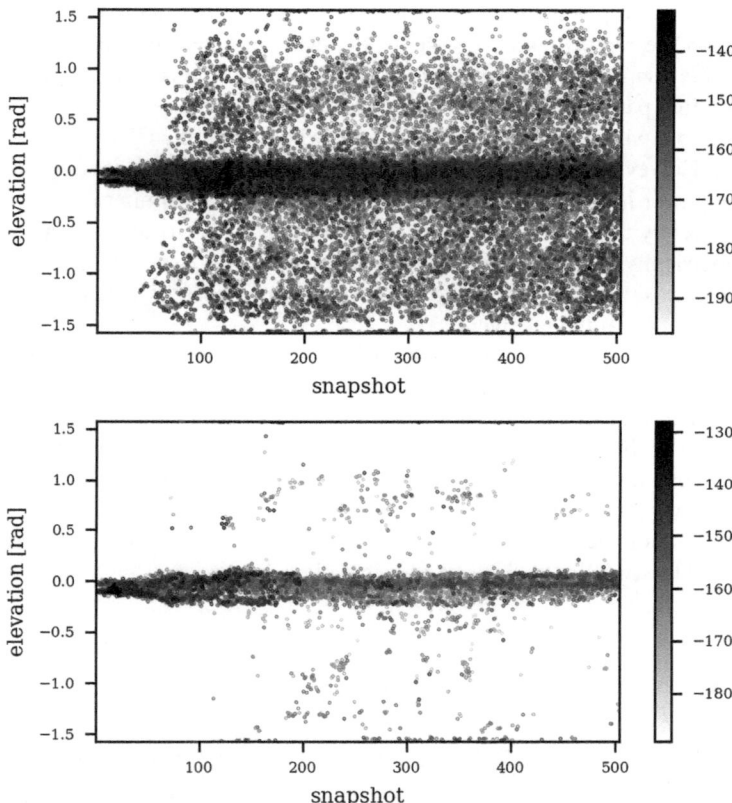

Fig. 17.3 Model order selection via CRB versus model order selection via MCRB for estimated elevation angles. Employing the MCRB significantly reduces the number of unphysical paths from the estimation

MSE. However, as, for instance, shown in [4], the CRB is not correctly predicting the estimators performance, if model misspecification is becoming increasingly dominant. Luckily in such cases the maximum likelihood estimator still produces an optimal solution in terms of a suitable distance measure on probability spaces [16], but the CRB has to be replaced with the so-called Misspecified Cramér-Rao Bound (MCRB). It follows a similar calculation for the Fisher information but also incorporates the model mismatch introduced by **s** into the prediction of the estimator's variance.

In Fig. 17.3 we show estimation results from a maximum likelihood estimator applied to real multiple-input multiple-output (MIMO) measurement data [17] acquired in an urban scenario [18]. Both nodes make use of a 8×2 stacked polarimetric circular patch array (two differently polarized ports per elements) operating at a frequency of 2.53 GHz and a bandwidth of 20 MHz and an IR length of 12.8 μs. Hence, we collected a 32×32 polarimetric MIMO matrix. The TX was located on a rooftop at 30 m above ground level and the RX approximately at 2.3 m on a car on the road. Figure 17.3 shows the estimated elevation parameters at the TX over the snapshots, where the color value encodes the respective power of the corresponding path in dB.

On the top, we see the algorithm's output while using the CRB for model order selection. However, since the TX is located above ground on a rooftop, almost no scattered power is expected at elevation angles $\gg 0$. With similar reasoning, paths arriving directly from the ground (i.e., the *south pole* of the local coordinate system) are geometrically questionable. Finally, the antenna array is constructed from patch antennas, whose beamwidth in elevation technically does not yield enough SNR close to the *poles* at the receiver.

However, the estimated elevation angles almost uniformly cover the whole angular range. Consequently, the estimator clearly produces *un-physical* paths. As an alternative, we can apply model order selection based on the MCRB. The results can be seen in Fig. 17.3 on the bottom. As a result, the paths that deviate much from the equatorial circle are efficiently dropped from the estimation. Although some spurious paths remain, the distribution of those around the equator shows a significantly more deterministic behavior. Consequently, these estimation results are much more trustworthy in terms of physical plausibility.

17.4   Conclusion

In this chapter, we have discussed that model-based high-resolution parameter estimation has the potential to de-embed the hardware from the measurements and describe the propagation channel by a set of highly resolved propagation parameters independent of the measurement device. However, this device de-embedding is subject to several sources of errors: uncertainties of the calibration measurements used to describe the hardware, model errors due to wrong assumptions about the propagation mechanisms or hardware, oversimplifications of the model in favor of less computational complexity (e.g., narrowband vs. wideband modeling), or incompleteness of the observation or model (e.g., an unavailable measurement dimension such as elevation). Furthermore, the impact of these errors may not be obvious during the parameter estimation process, since the estimator may still be able to significantly minimize its cost function, but consequently, the estimated parameters lose their physical meaning. Since the estimated model parameters are subsequently used to draw conclusions about the channel (e.g., to derive channel models), a metrological assessment of the full processing pipeline including hardware, models, and algorithms is important. A way of doing so is to evaluate the statistical estimation performance either theoretically/numerically or experimentally versus a physical ground truth as derived, e.g., from over-the-air artifacts as discussed in Chap. 19. Useful statistical quantities for this purpose are the bias and the variance, i.e., MSE of the estimates. Furthermore, we have discussed that analytical lower bounds for the MSE can aid in selecting useful waveforms for channel sounding or even detecting and discarding unphysical paths from the estimates in the case of model misspecification.

References

1. T. Doeker, J.M. Eckhardt, C.E. Reinhardt, T. Kürner, Time-domain channel sounder calibration at low terahertz band. IEEE Open J. Antennas Propag. **5**(6), 1598–1611 (2024)
2. M. Steinbauer, A. Molisch, E. Bonek, The double-directional radio channel. IEEE Antennas Propag. Mag. **43**(4), 51–63 (2001)
3. M. Landmann, M. Kaske, R.S. Thoma, Impact of incomplete and inaccurate data models on high resolution parameter estimation in multidimensional channel sounding. IEEE Trans. Antennas Propag. **60**(2), 557–573 (2012)
4. S. Semper, E. Pérez, M. Landmann, R. Thomä, Misspecification under the narrowband assumption: a cramér-rao bound perspective, in *2023 31st European Signal Processing Conference (EUSIPCO)* (2023), pp. 1524–1528
5. L. Scharf, C. Demeure, *Statistical Signal Processing: Detection, Estimation, and Time Series Analysis*, ser. Addison-Wesley Series in Electrical and Computer Engineering (Addison-Wesley Publishing Company, Boston, 1991)
6. J. Gedschold, D. Dupleich, S. Semper, M. Döbereiner, A. Ebert, G.D. Galdo, R.S. Thomä, Metrology of multicarrier-based delay-doppler channel sounding for sub-thz frequencies. Under Review at the IEEE Open J. Antennas Propag. (2024) [Online]. Available: https://doi.org/10.36227/techrxiv.173337536.61771937/v1
7. S. Semper, M. Döbereiner, C. Steinmetz, M. Landmann, R.S. Thomä, High-resolution parameter estimation for wideband radio channel sounding. IEEE Trans. Antennas Propag. **71**(8), 6728–6743 (2023)
8. D. Dupleich, S. Semper, M.D. Al-Dabbagh, A. Ebert, T. Kleine-Ostmann, R. Thomä, Verification of THz channel sounder and delay estimation with over-the-air multipath artifact, in *2022 16th European Conference on Antennas and Propagation (EuCAP)* (2022), pp. 1–5
9. R. Kessel, Wiener filter estimation of transfer functions. J. Exp. Anal. Behav. **81**(3), 289–296 (2004). [Online]. Available: https://onlinelibrary.wiley.com/doi/abs/10.1901/jeab.2004.81-289
10. J. Gedschold, S. Semper, R.S. Thomä, M.Döbereiner, G. Del Galdo, Dynamic delay-dispersive UWB-radar targets: modeling and estimation. IEEE Trans. Antennas Propag. **71**(8), 6814–6829 (2023)
11. M. Landmann, G. Del Galdo, Efficient antenna description for mimo channel modelling and estimation, in *7th European Conference on Wireless Technology* (2004), pp. 217–220
12. A. Richter Dr. Ing., Estimation of radio channel parameters. Ph.D. Dissertation, Technische Universität Ilmenau, 2005
13. G. Jávorszky, S. Boyd, I. Kollár, L. Vandenberghe, S. Wu, Optimal excitation signal design for frequency domain system identification using semidefinite programming, in *8th IMEKO TC4 Symposium on Recent Advances in Electrical Measurements* (1996), pp. 192–197
14. J. Gedschold, S. Semper, M. Döbereiner, R.S. Thomä, Excitation signal design for THz channel sounding and propagation parameter estimation, in *18th European Conference on Antennas and Propagation (EuCAP)* (2024), pp. 1–5
15. E. Van den Eijnde, J. Schoukens, On the design of optimal excitation signals. IFAC Proc. Vol. **24**(3), 1139–1144 (1991). *9th IFAC/IFORS Symposium on Identification and System Parameter Estimation*, Budapest
16. S. Fortunati, F. Gini, M.S. Greco, C.D. Richmond, Performance bounds for parameter estimation under misspecified models: fundamental findings and applications. IEEE Signal Process. Mag. **34**(6), 142–157 (2017)

17. G. Sommerkorn, M. Käske, C. Schneider, S. Häfner, R. Thomä, Full 3D MIMO channel sounding and characterization in an urban macro cell, in *2014 XXXIth URSI General Assembly and Scientific Symposium (URSI GASS)* (2014), pp. 1–4
18. S. Semper, M. Döbereiner, M. Landmann, J. Gedschold, R. Thomä, Estimating angular diffuse components and model misspecification in MIMO channel sounding. IEEE Trans. Antennas Propag. (2024). https://doi.org/10.1109/TAP.2025.3613529

Chapter 18
Variable Reference Structure for THz Channel Sounding

Tobias Doeker, Carla Reinhardt, Daniel M. Mittleman, and Thomas Kürner

Abstract To facilitate the comparison of different channel sounding measurements across the globe, there is a compelling need for common reference models and calibration techniques. One potential solution for achieving consistent measurement results is the implementation of a standardised reference scenario for calibration and post-processing purposes. This chapter introduces a common reference structure that is compact enough to be easily portable, enabling its use in various locations worldwide. The chapter presents and compares measurement results obtained using both a correlation-based channel sounder and a time domain spectroscopy (TDS) system.

18.1 Reference Structure Design

In order to serve as a reference structure for achieving comparable results across different channel sounding measurement systems, the structure should have the following characteristics. The structure should be:

1. Easily portable
2. Adaptable for use with different types of channel sounding systems
3. Configurable to generate different and reproducible measurement results

The first point limits the dimension of the reference structure. However, an easily portable waveguide artefact, for example, would not be feasible because it would not be connectable to a TDS system. As a TDS system uses lenses in connection with optically exited emitters and a channel sounder uses antennas with waveguide

T. Doeker (✉) · C. Reinhardt · T. Kürner
Technische Universität Braunschweig, Institut für Nachrichtentechnik, Braunschweig, Germany
e-mail: t.doeker@tu-braunschweig.de; carla.reinhardt@tu-braunschweig.de;
t.kuerner@tu-braunschweig.de

D. M. Mittleman
School of Engineering, Brown University, Providence, RI, USA
e-mail: daniel_mittleman@brown.edu

© The Author(s) 2026
T. Kürner et al. (eds.), *Metrology for THz Communications*, Springer Series
in Optical Sciences 256, https://doi.org/10.1007/978-3-032-01986-8_18

Fig. 18.1 Schematic top-down view of the realisation of the reference structure: front part drawn in black and rear part drawn in dark blue. Outer dimensions are given (based on [1])

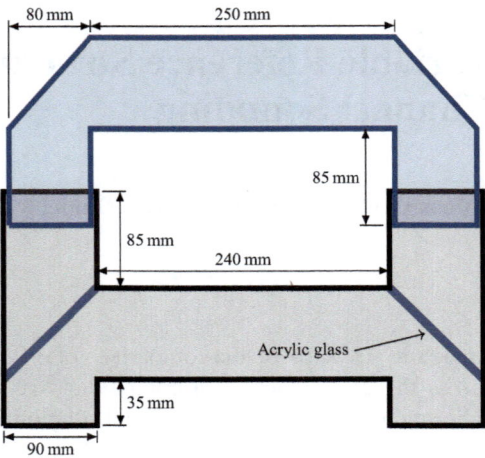

connections, the second point eliminates the possibility of a specific connection to the reference structure. One possible realisation is a hollow metallic structure, comparable to a small ventilation shaft. Due to the specular and strong reflection properties of metal, the signal path is clearly defined, allowing the expected channel impulse response (CIR) to be precisely designed. Furthermore, this design allows for easy redirection of the signal and the creation of additional paths/multipath components (MPCs). A schematic view of such a structure is shown in Fig. 18.1.

The structure has maximum dimensions of 42 cm by 38.5 cm and a height of 9 cm. As indicated by the different colours in Fig. 18.1, the structure consists of two parts, where the rear part fits into the front part. The structure is made of aluminium, and due to the walls angled at 45° in the rear part of the structure, the signal is redirected by about 180° relative to the openings in the front part as the signal is redirected by 90° due to the reflection at the walls angled at 45° at each side of the rear part. The rear part is not permanently mounted to the front part but fits in exactly, ensuring no gaps between the two parts. This design allows the rear part to be slid in and pulled out of the front part of the structure, enabling different path lengths to be achieved. Additionally, acrylic glass is inserted at a 45° angle in the front part of the structure as indicated in Fig. 18.1. This configuration causes part of the signal to reflect off the acrylic glass and be redirected by 180° (at each side of the structure the signal is redirected by 90° due to the reflection at the glass), while another part is transmitted into the rear part of the structure where it is also redirected by 180°. Theoretically, the structure therefore has a CIR with two MPCs. The transmitter (TX) and the receiver (RX) can be placed in front of the openings on the left and right sides of the structure. To ensure reproducible results, the TX and RX are placed within a holder structure in front of the openings, as shown in Fig. 18.2.

Due to the design of the structure, the path length of the MPC reflected off the acrylic glass is 49 cm, resulting in a first MPC delay of approximately 1.63 ns. For

Fig. 18.2 Photo of the antenna holders (left) and mounted at the opening of the structure (right)
[1]

the second MPC, the path length can be varied between 82 cm and 99 cm, yielding
delays ranging from 2.74 ns to 3.30 ns.

18.2   Channel Sounder Measurement Results

Channel sounder measurements are conducted using a correlation-based channel
sounder available at Technische Universität Braunschweig (TUBS) [2, 3]. After
post-processing, the measured and calibrated CIR has a bandwidth of 4 GHz with
a centre frequency of 304.2 GHz [4]. The TX and the RX are equipped with horn
antennas with a half power beam width (HPBW) of 8.5° and maximum antenna
gain of 26.3 dBi. They are placed centred in front of the opening of the structure.
Figure 18.3 shows the measurement results for horizontal and vertical polarisation,
with the rear part of the structure fully slid into the front part, i.e. the minimal path
length of the second MPC is realised.

As expected due to the material characteristics of acrylic glass, the transmission
and reflection losses change for the different polarisations. In terms of delay, the
measured MPCs fit very well with the theoretically expected MPCs, as the deviation

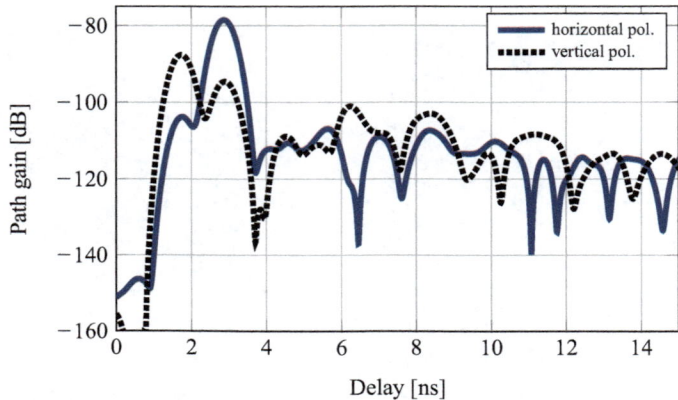

Fig. 18.3 Comparison of the CIRs for vertical and horizontal polarisation with 2-mm-thick acrylic glass [1]

between measured and expected delay is less than 150 ps [1]. For the first MPC, i.e. the reflected signal at the acrylic glass, the amplitude also matches very well between measurement and theory: For vertical polarisation −87 dB is expected, and −88 dB is measured, and for horizontal polarisation −105 dB is expected, and −104 dB is measured [1]. For the second MPC, however, the results of theory and measurement deviate by about 10 dB for vertical polarisation and about 3 dB for horizontal polarisation [1]. Measurements without the acrylic glass also reveal a deviation between the measured and expected path gain, indicating that the deviation is caused by effects arising in the rear part of the structure. This needs to be further investigated.

Figure 18.4 shows measurement results for horizontal polarisation but for different insertion depths of the rear part, i.e. for different path lengths of the second MPC. As expected, the delay becomes longer and the path gain decreases if the length of the second MPC is decreased. The difference in the delay between expected and measured delay is again less than 150 ps, but due to the above-mentioned issue, the measured path gain deviates slightly from the expected path gain [1].

18.3   TDS Measurement Results

The TDS measurements are conducted using the *TeraFlash pro* from *Toptica Photonics*. The measurable delay ranges from 200 up to 4200 ps with a resolution of 0.05 ps, which leads to a bandwidth of 20 GHz. To calculate the desired CIR, the measured signal is deconvolved using a line-of-sight (LOS) reference measurement (comparable to an over-the-air (OTA) calibration) referred to as 'pre-processing' in this context. During the pre-processing, the data is windowed to select specific

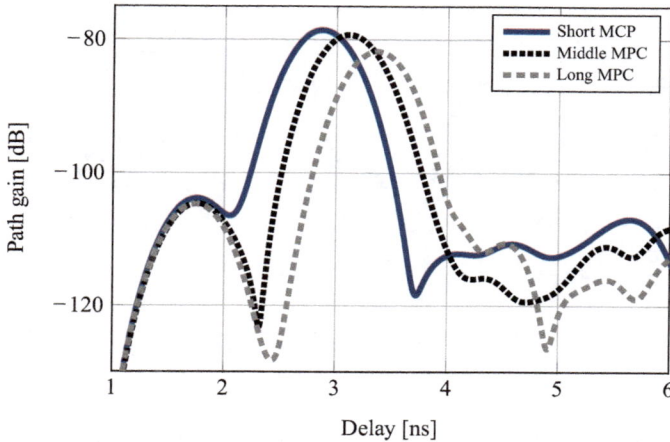

Fig. 18.4 Comparison of the CIRs for different length of the second MPC for horizontal polarisation with 2-mm-thick acrylic glass [1]

frequency ranges of interest. The first frequency band ranges from 150 up to 450 GHz, achieving a delay resolution of approximately 3 ps. This frequency range is a compromise between delay resolution and information of interest in the CIR. The second frequency range matches the frequency range of the channel sounder, from 300 up to 308 GHz, resulting in a delay resolution of 125 ps. It should be noted that the calculated CIRs are interpolated afterward so that the final delay resolution is 0.05 ps in both cases, comparable to the basic delay resolution of the measurement system.

For the second frequency range (300–308 GHz), Fig. 18.5 shows the comparison measurements for horizontal and vertical polarisation. Similar to the measurement shown in Fig. 18.3, a thin acrylic glass with a 2-mm thickness is chosen, and the rear part of the structure is fully slid into the front part of the structure. In terms of delay, the measured values fit very well to the expected values, with the deviation between theory and measurement being less than 50 ps. As only relative path gains can be provided in this measurement setup, the path gain cannot be compared directly to the expected values. However, it can be observed that the path gain of the first MPC changes by about 11 dB between horizontal and vertical polarisation, which is less than the expected difference of approximately 18 dB. On the other hand, the change in polarisation results in a difference of 7 dB in path gain for the second MPC, whereas theory predicts only 2.5 dB. Once more, an influence of the rear part of the structure on the measured path gain can be observed, similar to what was seen in the channel sounder measurements.

A noteworthy advantage of TDS is its ability to perform wideband measurements, resulting in high-resolution outcomes in the time domain. This capability enables a more detailed investigation of the properties of the reference structure. Figure 18.6 shows a portion of the measurements described above, within the

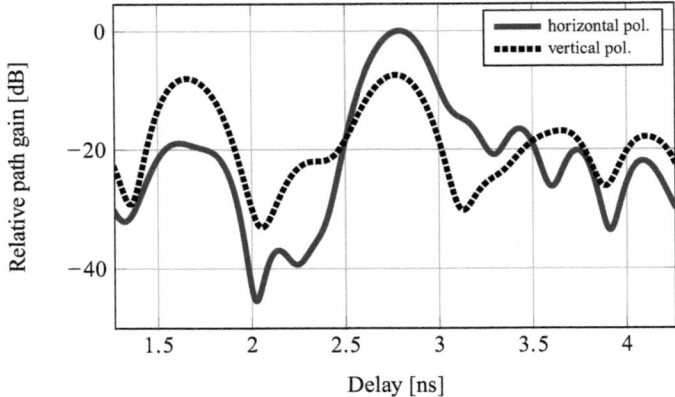

Fig. 18.5 Comparison of the CIRs for vertical and horizontal polarisation with 2-mm-thick acrylic glass in the frequency range 300–308 GHz

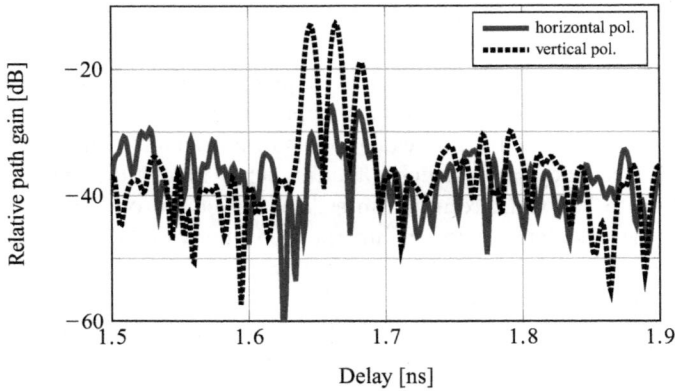

Fig. 18.6 Comparison of the CIRs for vertical and horizontal polarisation with 2-mm-thick acrylic glass in the frequency range 150–450 GHz

frequency range of 150 to 450 GHz. It can be observed, especially for vertical polarisation, that the first MPC, which appears as a single component using 8 GHz of bandwidth, is actually composed of three different MPCs. The delay differences between the peaks are 18.7 and 18.2 ps, which correspond to additional path lengths of 5.6 and 5.5 mm, respectively. These components are probably caused by the acrylic glass. With a 2-mm thickness and an angle of incidence (AOI) of 45°, as specified in the measurement setup, the path length is 5.66 mm longer if the signal is reflected at the backside of the acrylic glass compared to a reflection at the front side of the glass. It should be noted that the slightly different transmission angle within the acrylic glass due to Snell's law is negligible in this context.

The three different MPCs can, therefore, be explained as follows:

1. The first MPC arises from reflections at the front side of the glass on both the TX and the RX side of the structure.
2. The second MPC results from a reflection on the backside of the glass on one side of the structure, combined with a reflection on the front side of the glass on the other side of the structure.
3. The third MPC originates from reflections on the backside of the glass on both sides of the structure.

Since the first MPC is merely a reflection at the glass, the path gain does not suffer from transmission losses, resulting in the highest path gain in this case. However, the second MPC has a comparable path gain even though it is attenuated by transmission losses. This is because the path occurs in two ways: reflection at the front side of the TX glass and the backside of the RX glass and reflection at the backside of the TX glass and the front side of the RX glass. These reflections result in constructive interference, enhancing the path gain. As expected, the third MPC suffers from transmission losses on both acrylic glasses, resulting in a lower path gain compared to the other MPCs.

For different insertion depths of the rear part of the structure and for horizontal polarisation, Fig. 18.7 shows the measurement results comparable to Fig. 18.4 of the channel sounder measurements. Once again, the expected behaviour of the structure can be illustrated by the measurement results. The delay of the second MPC increases with a lower insertion depth of the rear part of the structure. The deviation between the measured delay and the expected delay is, once again, less than 50 ps. Furthermore, the path gain decreases slightly with increasing path length.

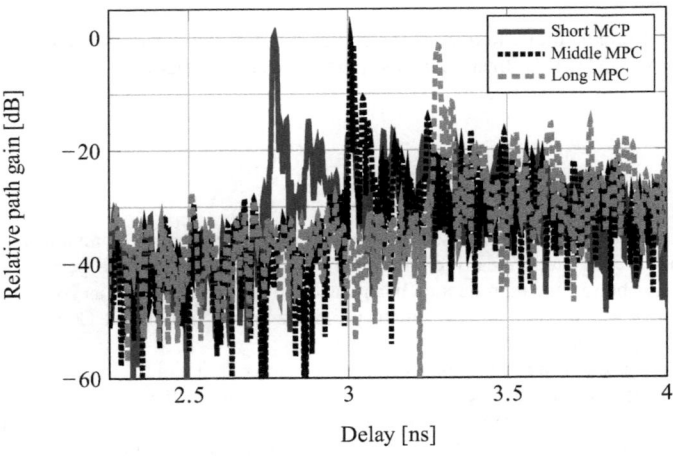

Fig. 18.7 Comparison of the CIRs for different length of the second MPC for horizontal polarisation with 2 -mm-thick acrylic glass in the frequency range 150–450 GHz

18.4 Summary

To compare different channel sounding techniques, a variable reference structure is introduced. Various requirements for such a structure are identified and incorporated into a structural model. This model features two paths leading to a CIR with two distinct MPCs, where the delay of the second MPC can be varied due to a movable part of the structure. Channel sounder measurements and measurements with the TDS system reveal comparability of the results for different polarisations as well as varying lengths of the second path. The implementation and use of the structure as a comparative tool for different systems have thus been demonstrated. Moreover, through high-resolution results in the time domain obtained with the TDS system, it has been shown that the measured CIR behaves as expected from a theoretical point of view, particularly concerning the measured delays of the MPCs.

References

1. T. Doeker, C.E. Reinhardt, C. Herold, U. Hellwung, D.M. Mittleman, T. Kürner, Variable reference structure for channel sounding in the low terahertz range, in *Proceedings of 19th European Conference on Antennas and Propagation (EuCAP)* (Stockholm, 2025)
2. J.M. Eckhardt, A. Schultze, R. Askar, T. Doeker, M. Peter, W. Keusgen, T. Kürner, Uniform analysis of multipath components from various scenarios with time-domain channel sounding at 300 GHz. IEEE Open. J. Antennas Propag. **4**, 446–460 (2023)
3. S. Rey, J.M. Eckhardt, B. Peng, K. Guan, T. Kürner, Channel sounding techniques for applications in THz communications: a first correlation based channel sounder for ultra-wideband dynamic channel measurements at 300 GHz, in *Proceedings of 9th International Congress on Ultra Modern Telecommunications and Control Systems (ICUMT)* (Munich, 2017), pp. 449–453
4. T. Doeker, J. Eckhardt, C. Reinhardt, T. Kürner, Time-domain channel sounder calibration at low terahertz band. IEEE Open J. Antennas Propag. **01**, 1–1 (2024)

Chapter 19
Artefacts for Channel Sounding Performance Evaluation

**Mohanad Dawood Al-Dabbagh, Diego Dupleich, Tobias Doeker,
Reiner Thomä, Thomas Kürner, Thomas Kleine-Ostmann,
and David A. Humphreys**

Abstract This chapter investigates measurement artefacts used to verify the performance of different channel sounding systems. Some artefacts were selected based on waveguide fabrication and cascades, while others were based on over-the-air (OTA) measurements under a controlled environment. We will discuss the principle behind each of the selected artefacts and their setup configurations, including considerations for equipment calibration and environmental factors. The measurements examine single and multipath reflections. We discuss the influence of antennas on the measurement scenarios and the uncertainties related to different error terms we addressed during our measurement investigations.

19.1 Introduction

The wide spectrum availability at sub-terahertz (THz) frequencies promises high data rates for the future generations of mobile communication systems [1–5]. Developing fully operational sub-THz communication systems can help to shape realistic channel models at such high-frequency band. To achieve this goal, it would require high accuracy measurements, precise component characterisations, and advanced signal processing techniques. Testing the sub-THz using an ultra-

M. D. Al-Dabbagh (✉) · T. Kleine-Ostmann
Physikalisch-Technische Bundesanstalt (PTB), Braunschweig, Germany
e-mail: mohanad.al-dabbagh@ptb.de; thomas.kleine-ostmann@ptb.de

D. Dupleich · R. Thomä
Technische Universität Ilmenau, Institute of Information Technology, Ilmenau, Germany
e-mail: diego.dupleich@tu-ilmenau.de; reiner.thoma@tu-ilmenau.de

T. Doeker · T. Kürner
Technische Universität Braunschweig, Institut für Nachrichtentechnik, Braunschweig, Germany
e-mail: t.doeker@tu-braunschweig.de; t.kuerner@tu-braunschweig.de

D. A. Humphreys
NPL, UK (Retired), University of Bristol, UK
e-mail: david.a.humphreys@ieee.org

© The Author(s) 2026
T. Kürner et al. (eds.), *Metrology for THz Communications*, Springer Series
in Optical Sciences 256, https://doi.org/10.1007/978-3-032-01986-8_19

wideband (UWB) real-time channel sounder (CS) system would provide a great advantage for researchers willing to investigate and advance the possibilities of emulating different measurement scenarios and exploring potential future applications that put the next generation of mobile communications a reality [6].

In this chapter, we tested two correlational-based pseudo-noise (PN) UWB CS operating in the 183.75–191.25 GHz and 300–308 GHz frequency ranges under different controlled measurement scenarios. These CS systems were investigated using a vector network analyser (VNA) at WR05 and WR03 frequency bands to evaluate the feasibility and repeatability of the tests. We analysed the measured values, among the different measuring devices, and evaluated their performance in comparison with theory. Some of the measurements were conducted using waveguides, and other measurements were performed OTA with line-of-sight (LoS) and non-line-of-sight (NLoS) measurement scenarios. These artefacts allow for an in-depth analysis of errors related to linearity, phase deviations, drifts, in-phase and quadrature (IQ) imbalance, and other factors affecting system performance. The architecture of the CS measurement equipment, including its design and functionality, is explained in detail in the Chap. 16. The Chap. 6 provides an overview of the basic principles and operational concepts of the VNA. This chapter focuses on static measurement conditions. To extend the investigation to dynamic measurement scenarios, please refer to the Chap. 13.

19.2   Connected Waveguide Artefact

To accurately characterise different measurement devices, reference measurement artefacts would be the obvious choice to investigate the reference plane, linearity, and phase stability. This is applied when studying measurement devices operating at sub-THz such as a CS or a VNA.

To characterise the CS systems, a well-characterised single waveguide or a cascade of several waveguides consisting of attenuators and lines with different attenuations and lengths is used. However, these waveguides face the challenge of achieving measurement repeatability, as they suffer from significant misalignment the smaller the waveguide size, i.e. the higher the frequency band. This necessitates investigating the repeatability of the link, which can lead to large measurement uncertainties [7].

A cascade of waveguide components made up of attenuators and lines is typically used for a CS reference measurement to establish a reference phase and magnitude. This setup helps specify the calibrated delay and path loss for other measurements. In Fig. 19.1a, a cascade of two attenuators and two lines was tested to measure the transmission and reflection S-parameters using a VNA and two WR03 band frequency extension modules.

The CS propagates an instantaneous bandwidth spanning several GHz, resulting in a frequency-dependent transfer function. The frequency response is then equalised via deconvolution using a well-known artefact. This process, known as

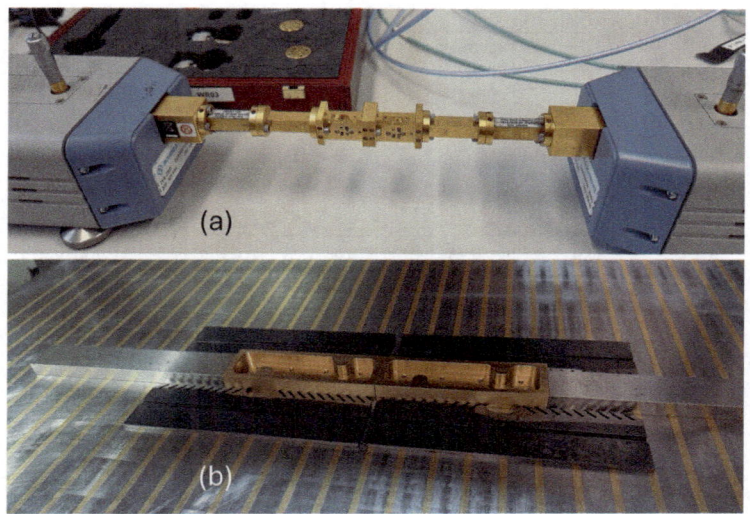

Fig. 19.1 Component characterisation: (**a**) Attenuators and lines cascaded in the WR03 band. (**b**) Waveguide artefact fabrication in the WR05 band

back-to-back calibration, is widely used in channel sounding systems. However, it can be challenging at sub-THz frequencies, where waveguide components require careful consideration to achieve repeatability.

A waveguide line in the WR05 band has been simulated and fabricated by the Scientific Instrumentation Department at PTB, as shown in Fig. 19.1b. The line was tested using a VNA. Variations in the geometry of connectors can induce mechanical stress and unwanted deformations during the mating process, leading to changes in electrical performance when reconnected in different orientations. This effect is specific to each connector pair combination.

Characterising flange repeatability involves testing the same device under test (DUT), such as matched loads or shorts, multiple times. To ensure reliable results, at least four reconnections at different azimuthal positions for all components connected to the test port must be performed. Additionally, flange repeatability can be estimated by considering the tolerances of the alignment mechanism. Applying the correct torque when tightening screws is crucial, though torque specifications may vary by manufacturer. Several factors influence torque, including the base material of the flange, its thickness, the quality of the threaded holes and screws, and surface properties such as material, flatness, and roughness.

Ensuring precise and consistent waveguide connections requires accounting for flange quality, appropriate torque application, and any additional weight or alignment tools used during setup, as these factors can impact the effective torque at the interface [8, 9].

We characterised the measurement repeatability of the waveguide artefacts using our VNA measurements and tested them in terms of reconnection, flipping, and

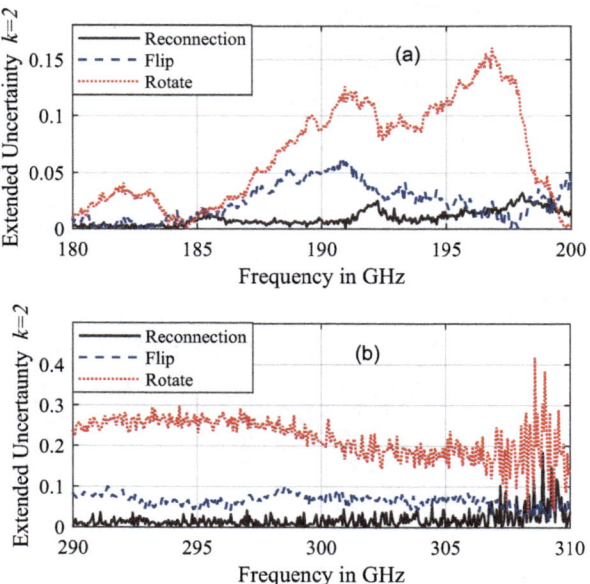

Fig. 19.2 Magnitude expanded measurement uncertainty of waveguide components in dB, applied using waveguides reconnection, flipping, and waveguide artefact rotation by 180°: (**a**) WR05 and (**b**) WR03

rotation. Figure 19.2 shows an example of the expanded measurement uncertainty of the tested waveguides at 180–200 GHz and 290–310 GHz.

The VNA characterisation of the selected waveguide artefacts plays an essential role in matching the magnitude and phase variations and shifting the reference plane between the VNA and the CS. This process facilitates matching the VNA and CS measurements. The CS frequency response can be calculated as [10]:

$$H_{\text{deconv}}(f) = A e^{-\text{j}2\pi f d/c} \frac{H_{\text{mean}}(f)}{H_{\text{B2B}}(f)}, \qquad (19.1)$$

where f denotes the frequency following down-conversion, A signifies the quantified magnitude of the characterised artefact, d represents either the effective electrical length of the examined waveguide or the precise time delay derived from a LoS propagation path, and c corresponds to the speed of light. $H_{\text{mean}}(f)$ is the frequency response of the mean complex received impulse response of K recorded sequences, and $H_{\text{B2B}}(f)$ is the frequency response of the back-to-back measurement representing the characterised waveguide tested using the VNA and the CS [10].

To perform comparable measurements, the CS reference plane and reference path gain were shifted to match the calibrated VNA values. Here, we present an example at the WR03 band, which was selected as the CS reference measurement, providing

Fig. 19.3 Reference waveguide PDP measurement result consisting of two attenuators and two lines. The measurement was using the VNA and the CS devices at WR03 band

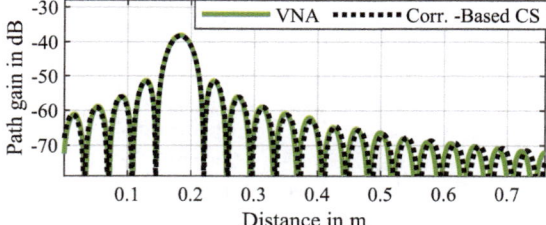

the characterised phase and magnitude measured from the VNA and plotted in terms of the power delay profile (PDP). The reference plane and path gain corrections were performed using a back-to-back connection with a reference waveguide cascade of two short lines and two attenuators of 10 dB and 20 dB measured using the CS and the VNA, respectively [11].

The reference waveguide measurement using the VNA and CS shifted plane and path gain provided a 38.26 dB loss and an electrical length of 18.29 cm. To improve distance accuracy, an interpolation factor of 40 was applied to both measurements as shown in Fig. 19.3.

19.3   Principle of OTA Measurement Verification Artefacts

When an OTA measurement artefact is employed to model a channel, the object's shape, size, material, the antenna radiation patterns, and their positions are the defining parameters to form that artefact scenario. The OTA artefacts need to be tested for repeatability, as their complexity increases when the scenarios become larger. However, they provide higher accuracy results when measurements are performed in a controlled environment [10, 12].

The two correlation-based channel sounding systems and the two sets of VNA extension modules were gathered to perform the measurement campaign for OTA artefacts. The measurements were divided into two frequency bands based on the CS operation frequencies. The VNA linear frequency sweep was selected accordingly.

Data acquisition was carried out in a climate-controlled anechoic chamber at Physikalisch-Technische Bundesanstalt (PTB), ensuring stable environmental conditions throughout the measurements [12]. The time zero and several back-to-back calibrations were recorded before starting the measurement. The VNA and CS calibrations were repeated on a daily basis, and a VNA drift measurement was performed at 1 m distance for 12 hours to check the drift influence on the measurement.

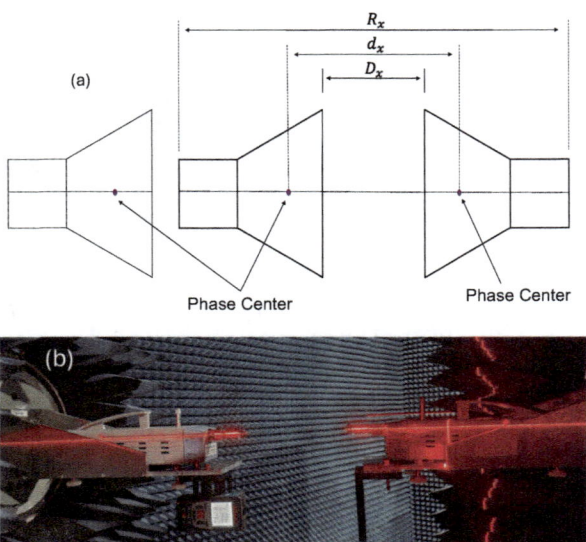

Fig. 19.4 Far-field antenna distance measurement setup: (**a**) Schematic showing the antenna separations and phase centre positions. (**b**) Photograph of the measurement setup (Adapted from [10])

19.3.1 LoS Distance Measurement

This measurement aims to investigate the path loss (PL) of a LoS at different separations using an automated control system that changes the distance between the transmitter and receiver over the range of 10–350 cm at 10 cm intervals (Fig. 19.4). A calibrated laser distance measuring device was used to ensure separation accuracy. Some similar setups focusing on VNA measurements were reported in [12–15]. The measurement was repeated for each CS and VNA frequency extension module. The measurement helped us adjust the reference plane influenced by antenna elements and correct the PL shift to match the Friis loss equation [10].

19.3.1.1 Antenna Calibration

This investigation examined two identical horn antennas for each frequency band that operate within the WR05 and WR03 waveguide standards. Each frequency extension module was fitted with a horn antenna, corresponding to the two ports of the VNA, thereby extending the measurement frequency range [10, 16].

The antenna dimensions provided in Table 19.1 correspond to the labelled parameters A–F in Fig. 19.5. The PDP magnitude was selected to determine the phase center (PC) by computing the root mean square (RMS) deviation between the

Table 19.1 Horn antenna dimensions (Adapted from [10, 16])

Dimensions (mm)	WR05	WR03
Total length (A)	17.6	37.7
Aperture width (B)	6.7	7.9
Aperture height (C)	4.6	5.9
Waveguide width (D)	2.217	0.864
Waveguide height (E)	1.469	0.432
Flare length (F)	12.5	34.7

Fig. 19.5 Horn antenna dimensions (Adapted from [10, 16])

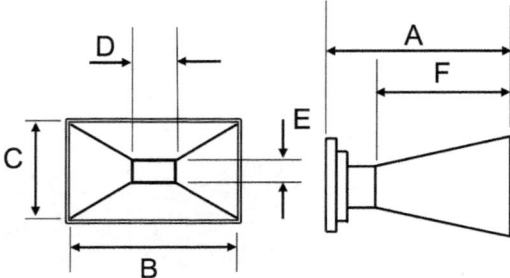

measured data and the adjusted Friis loss calculations [10]. This approach enables the identification of the optimal separation distance that minimises the discrepancy between the measured data and Friis loss calculations. The PDP facilitates the examination of the bandwidth's influence on the PC, supporting the investigation of frequency and time resolution of each measurement.

Even when the full bandwidth of the waveguide is used, the resolution achieved by PDP alone exceeds multiple wavelengths. To refine the estimation of the peak position and magnitude PDP, this investigation adopts an interpolation factor of 40 [10, 16].

The PDP for the forward transmission signal S_{21} is based on the definition, with an additional frequency-domain windowing function [17]:

$$PDP = |IFFT (S_{21} (f) w (f))|^2 , \qquad (19.2)$$

where $w (f)$ is a frequency-domain window. As the antenna gain is not influenced by the measured far-field distance, the combined antenna gain for our horn antennas $G (f)_{dBi}$ can be calculated which represents an average value derived from the Friis equation, resulting in [10, 12]

$$G (f)_{dBi} = \frac{1}{n} \sum_{x=1}^{n} S_{21} (f, P_x)_{dB} - PL_{corr} (f, P_x) - \Delta P_{dB}, \qquad (19.3)$$

where $S_{21} (f, P_x)_{dB}$ and $PL_{corr} (f, P_x) - \Delta P_{dB}$ are the VNA transmission results and the calculated PL at each of the n antenna separation values P_x, corrected for the reference plane offset ΔP, respectively [16].

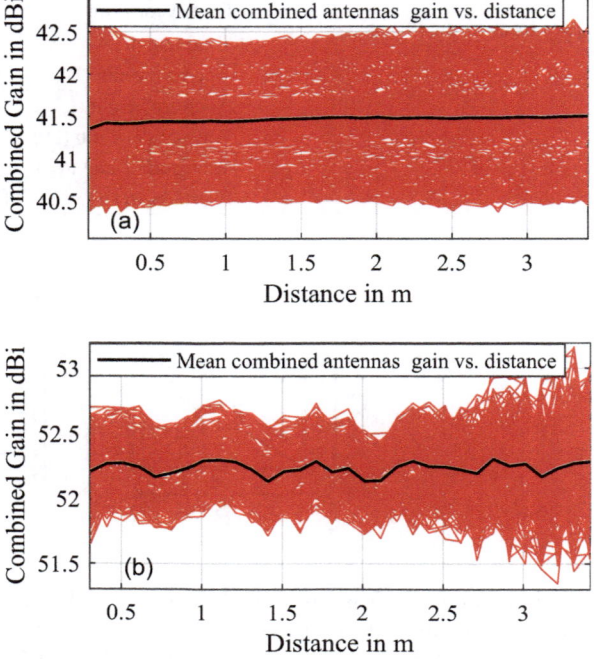

Fig. 19.6 Combined antenna gain vs. separation distances. The combined gain at each frequency point (red) and the mean gain (black). (**a**) WR05 and (**b**) WR03

The antenna gain, representing the common variable at each measured distance, is analysed to estimate the antenna gain behaviour across the entire measurement bandwidth. From this dataset, the mid-frequency combined antenna gain for each horn antenna was calculated to be (20.735 ± 0.025) and (26.125 ± 0.045) dB at 95 % confidence for the WR05 and WR03 horn antennas, respectively.

The combined antenna gain achieved at different measurement distances between 40 cm and 340 cm for the horn antennas is shown in Fig. 19.6.

In Fig. 19.7, we present an example of the RMS difference between the measured and calculated PL based on the PDP calculation for the WR03 band. The measurements were performed for the frequencies between 300 GHz and 310 GHz. The combined PC calculations show a lower RMS difference of 0.15 dB for the measured magnitude, compared to 0.06 dB for the measured PDP, both at 12.6 mm for the WR03 combined antennas. This translates 6.3 mm PC for each WR03 antenna. A complete comparison among different measurements and used techniques based on PDP peak delay and magnitude to investigate the antenna PC using OTA measurements was found in [16].

Fig. 19.7 RMS difference in dB between PDP and measurement distance compared to successive combined PC offset of the combined WR03 horn antennas

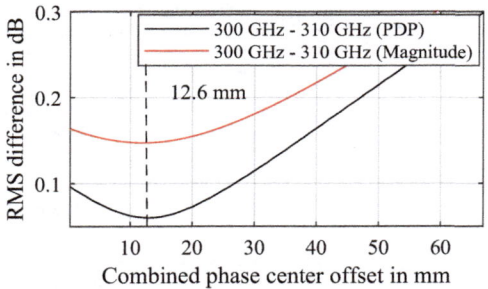

19.3.1.2 VNA Non-Linearity and Cable Movement

The VNA operates as a linear network, establishing a direct relationship between the forward and backward waves detected by the receivers. Non-linearity in the higher frequency range is typically assessed using a fixed attenuation device with known characteristics. However, since such devices are not available for our sub-THz frequencies, we instead evaluated non-linearity by comparing the measurements to the Friis PL as a reference.

To characterise the VNA non-linearity, measurements are taken across the frequency range under test. The S-parameters, particularly S_{21}, are then estimated using error correction. Incremental attenuations for these corrected measurements are calculated at each attenuation state ($j = 1, 2, 3, \ldots$). The maximum differences between the measured and reference incremental attenuations are given by $L_j = \max\left(|S_{21}(j)| - |S_{21}|_{\text{ref}}(j)\right)$ [10]. These differences are bounded by an envelope that spans the entire attenuation range. The differences are then converted to linear units and used as uncertainties in our evaluation [7].

Measurement uncertainties are considered to be a combination of system non-linearity and the effects of cable movement. Figure 19.8 shows the non-linearity difference between the measured $|S_{21}|_{\text{Meas}}$ and the theoretical Friis loss $|S_{21}|_{\text{theor}}$ for the WR05 and WR03 bands. The peak-to-peak differences were 0.07–0.24 dB for the WR05 and WR03 bands, respectively [10].

19.3.1.3 VNA Drift of Transmission Tracking

The stability of a test system can be significantly impacted by the local environment. VNA measurement drift, often due to temperature-induced changes affecting signal path lengths and component performance, can accumulate over time. In our study, we performed our measurement for several hours before performing a new VNA calibration. Therefore, we calculated the measurement drift over a period of 12 hours, which is longer than the interval between recalibrations. Measurements were taken at 15-minute intervals.

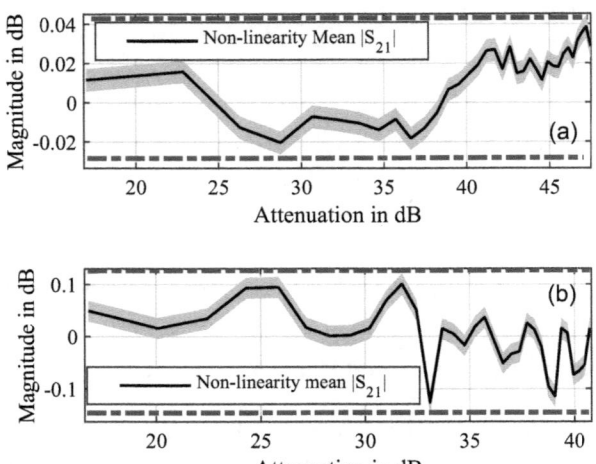

Fig. 19.8 Non-linearity difference between the measured and theoretical Friis loss at (**a**) WR05 and (**b**) WR03 bands. The dashed red lines represent the possible envelope covering the non-linear region

Drift measurements were carried out to evaluate the influence of environmental conditions, using an OTA LoS setup with a fixed 1 m spacing between the two VNA extension modules. The experiment employed our calibrated horn antennas for the WR05 and WR03 bands [10].

Figure 19.9 illustrates the drift of S_{21} relative to the initial measurement at $t = 0$, with 95 % confidence intervals for the extended uncertainties in magnitude and phase for each frequency step.

The drift uncertainty $u(D)$ is calculated as

$$u(D) = k \cdot \frac{\sqrt{\frac{1}{n-1}\sum_{i=1}^{n}\left(D_{j,i} - \hat{D}_{\text{mean}}\right)^2}}{\sqrt{n}} \tag{19.4}$$

where $\hat{D}_{\text{mean}}$ is the mean vector of $S_{21,j,t}$, and $D_{j,t} = S_{21,j,t} - S_{21,j,0}$. The peak-to-peak drift magnitude for all frequencies was 0.24 dB for the WR05 band and 0.41 dB for the WR03 band. The peak-to-peak drift phase was 21° for WR05 and 60° for WR03. Figure 19.9 shows the effect of drift measurement on the measured magnitude and phase.

19.3.1.4 CS Total Spectrum

The accuracy of distance measurements can be used to validate the CS's frequency step size by comparing the achieved measurement delay in the PDP. To ensure the

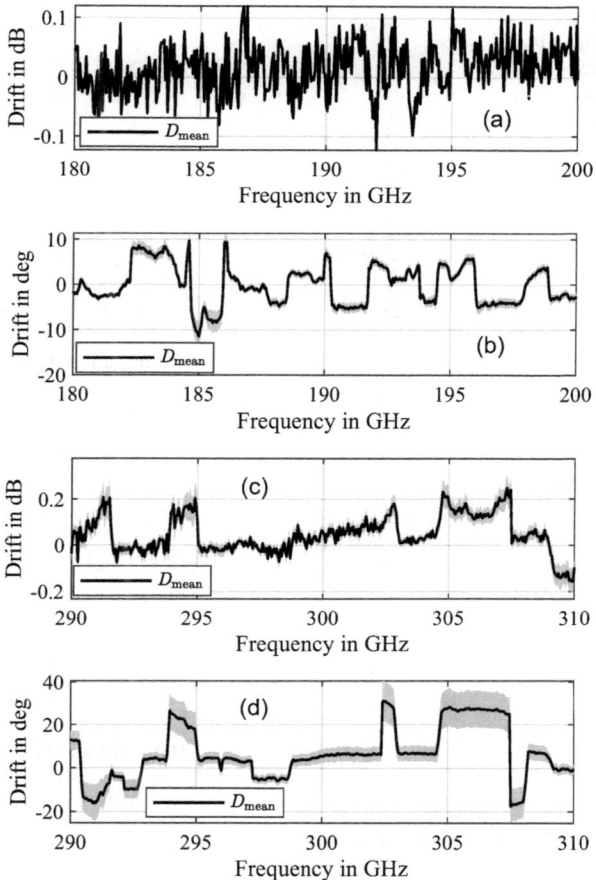

Fig. 19.9 The S_{21} magnitude and phase differences due to drift. Magnitude measurements at (**a**) WR05 and (**c**) WR03 and phase measurements at (**b**) WR05 and (**d**) WR03

frequency step size is correct, the measured delay is compared to the expected delays for different distances. A discrepancy in the frequency step size would result in a shift in the delay value, either increasing or decreasing relative to the reference measurement distance selected for back-to-back calibration.

Figure 19.10 illustrates the difference in PDP delay measured by the CS compared to the delays measured by the VNA across different frequency steps within the total CS spectrum. The measurement indicates the minimal peak-to-peak delay deviation with a total spectrum of 7.5–9.2 GHz for the WR05 and WR03 channel sounders, respectively [10].

Fig. 19.10 Total spectrum of
CS systems based on the
combined peak-to-peak delay
deviation of all measured
separations (**a**) WR05 (**b**)
WR03

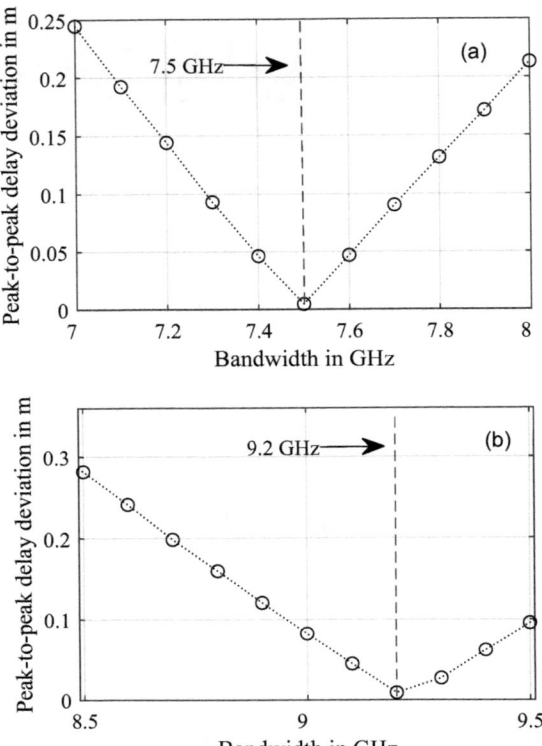

19.3.1.5 CS Reference Measurement Selection

Measurement distances are determined by characterising a reference VNA measurement alongside the OTA separation between antenna elements, based on their PC values. This process allows any measurement to serve as a reference for performing back-to-back deconvolution. Careful characterisation of the reference measurement is critical to evaluating uncertainties linked to using alternative references. By ensuring that the selection of a different reference does not cause significant PL deviations, the reliability of the measurements is improved, and any underlying issues in the procedure can be identified more effectively.

Figure 19.11 demonstrates how PL varies across measurement distances from 10 to 90 cm when compared to a 100 cm calibration reference. To standardise the scale, the PL at 10 cm was normalised to zero. For the WR05 CS, using different references reveals a trend of increasing PDP PL deviation as the measurement distance grows. At distances below 1 m, deviations range from 0.06 to 0.09 dB, increasing to 0.14 dB between 2.3 m and 2.7 m and peaking at 0.18 dB for distances greater than 3 m. In comparison, WR03 CS reference measurements show significantly better consistency, with deviations remaining within a peak-to-peak range of 0.06 dB across all distances [10].

Fig. 19.11 PDP PL difference obtained using measurement distances between 10 cm and 90 cm as a reference measurement compared to the 100 cm measurement reference for CS at (**a**) WR05 and (**b**) WR03

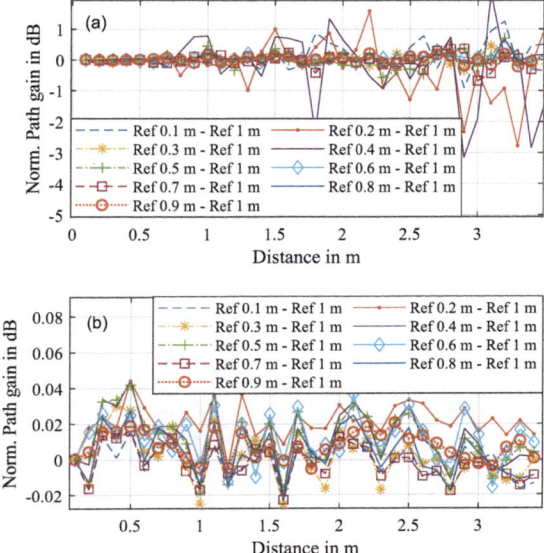

19.3.1.6   CS Bandwidth Feasibility

Determining the feasible measurement bandwidth of a CS is crucial to achieving the best system linearity and resolution. To this end, we tested the bandwidth of both CS systems in terms of PDP PL and delay. We calculated the combined delay deviation compared to VNA and laser metre measurements for each channel sounding system using 500 MHz steps taking the antenna PC calculations into consideration.

Figure 19.12 displays the characterisation measurement results. The WR05 sounder exhibited the lowest delay deviation at 4.5 GHz, with a deviation of 4.66 mm. For the WR03 sounder, the lowest delay deviation was observed in a 5 GHz bandwidth, with a peak-to-peak deviation of 3.18 mm. The bandwidth feasibility based on the magnitude deviation of PDP peaks was presented in [10].

19.3.1.7   VNA and CS Combined Corrected Measurement Uncertainties

The VNA measurements discussed earlier resulted in several uncertainties, including drift u_{drift}, antenna gain u_{ant}, and non-linearity $u_{\text{n-linear}}$. The combined extended uncertainty for each frequency band was calculated using the root-sum-square method:

$$u_{\text{Combin}} = \sqrt{u_{\text{drift}}^2 + u_{\text{ant}}^2 + u_{\text{n-linear}}^2} \tag{19.5}$$

234 M. D. Al-Dabbagh et al.

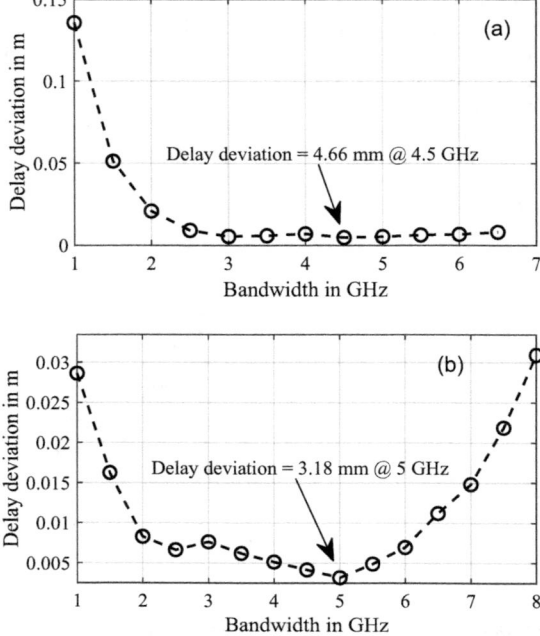

Fig. 19.12 CS peak-to-peak magnitude deviation at different measurement bandwidths: (**a**) WR05 and (**b**) WR03

Delay deviation = 4.66 mm @ 4.5 GHz

Delay deviation = 3.18 mm @ 5 GHz

Table 19.2 Magnitude uncertainties of VNA PDP in the WR05 and WR03 bands

Parameter	WR05 (dB)	WR03 (dB)
Drift	0.2	0.45
Antenna gain	0.05	0.1
Non-linearity	0.06	0.22

The extended uncertainties associated with these terms are summarised in Table 19.2. Figure 19.13a depicts the path gain variations in the WR05 band, comparing VNA and CS measurements across the 185–190 GHz frequency range. The VNA measurements demonstrated deviations with a standard deviation of 0.01 dB and combined uncertainties of ±0.21 dB, whereas the CS measurements exhibited a standard deviation of 0.4 dB [10].

Similarly, Fig. 19.13b illustrates the path gain variations in the WR03 band, comparing VNA and CS measurements over the 300–305 GHz range. The VNA measurements had a standard deviation of 0.04 dB with combined uncertainties of ±0.51 dB. In contrast, the WR03 CS measurements, taken over a 5 GHz bandwidth, exhibited standard deviations of 1.4 dB. Further investigations related to the VNA error terms are presented in [10].

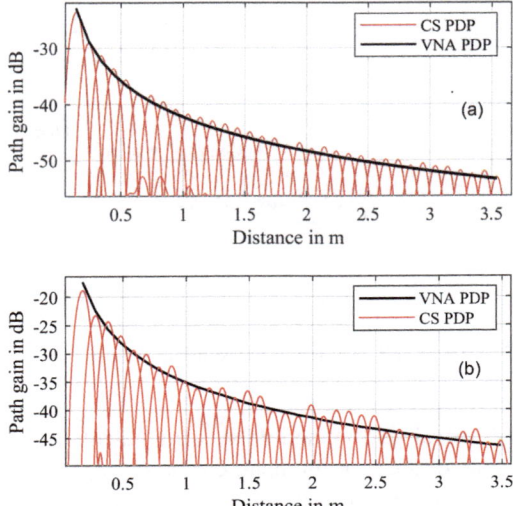

Fig. 19.13 Distance measurement deviation from the theoretical PL at (**a**) WR05 and (**b**) WR03

19.3.2 Multipath Measurements

19.3.2.1 Plane Reflector

This measurement was designed to evaluate antenna performance under controlled conditions. To perform the multipath measurement, an automated system was used to control the rotation of the transmitting antenna. The rotation occurred around a central axis located at the edge of the antenna's aperture [11]. The distance between the transmitting and receiving antennas was fixed at 2 m. At the transmitting port, a horn antenna with a high gain of 26 dBi and a narrow beam was employed, while the receiving port was equipped with a horn antenna featuring a wider beam and a lower gain of 15 dBi, connected to a short waveguide line. Antenna gains were evaluated over a frequency range spanning 300–305 GHz with 500 discrete frequency steps. Measurements were repeated using both types of measurement equipment as the rotational angle varied from 0° to 90°. Since the results depend on the antennas' radiation patterns, identical pairs of antennas were used on the transmitter and receiver sides for both the CS and VNA setups. A diagram of the measurement arrangement is provided in Fig. 19.14.

The measurement setup featured a Rohde & Schwarz ZVA24™ VNA, paired with ZVA-Z325™ frequency converters and calibrated using the unknown-through, offset-short, short, match (UOSM) method for precise operation within the WR3 waveguide band. Measurements were conducted over a narrower bandwidth of 300–305 GHz, utilising a linear frequency sweep with a 101 points and an IF bandwidth of 10 Hz.

A reflector was placed at a distance between the two antennas, and the measurement was repeated for both horizontal and vertical polarisations. The PL, direction-

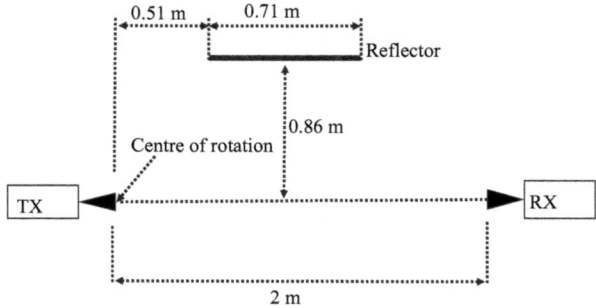

Fig. 19.14 Plane reflector measurement setup

of-arrival (DoA), and delay were calculated to be compared with the geometrical setup of the measurement. Given the dependence on the antenna's radiation pattern, identical antenna combinations at the transmitter and receiver were used with the CS and VNA measurements. Different numbers of points and bandwidths were selected to assess the influence of antenna gain and the change in resolution on delay accuracy and PL for the two paths. Figure 19.15 illustrates the measured LoS and reflected paths at various rotational angles, highlighting the relationship between distance and path loss for both the CS and VNA. Figure 19.15a, b depicts these results, respectively.

The separation between the physical apertures of the transmitting and receiving antennas was measured as 2 m for the LoS path and 2.64 m for the path reflected off the metallic sheet. To achieve consistent radiation patterns, gain, and reference plane shifts, the same antenna combination was used at both transmitting and receiving ports with the VNA and CS. The shift in the reference plane was evident in both LoS and reflected path measurements. The RMS combined reference plane shift for both antennas was 9.60 cm, with ±0.5 mm difference for different rotation angles for both LoS and reflected paths, representing the difference between the measured and the physical aperture distance separation, as shown in Fig. 19.15c.

The tested measurement bandwidth of 5 GHz at both the VNA and CS yielded a distance deviation of ±2.2 mm and ±6.6 mm for VNA and CS measurements, respectively. The measured LoS and reflection at different rotational angles with respect to the distance and PL for the VNA and CS are shown in Fig. 19.15d.

19.3.2.2 Sphere and Wire Measurements

Electromagnetic scattering from a sphere and thin wires has long been discussed in the literature [18]. In this study, these objects were used as artefacts for CS verification. The aim was to calculate the metrological properties of 2-dimensional free-space multipath propagation at THz frequency bands using spherical and

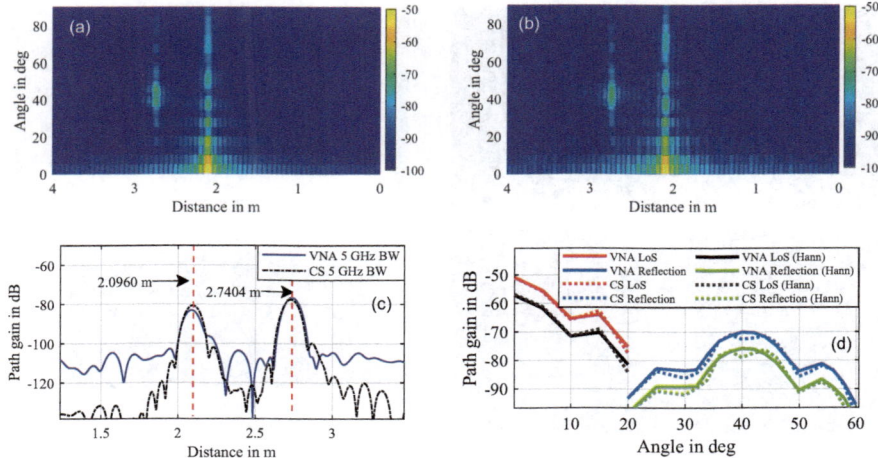

Fig. 19.15 Measurements of PL and delay at various angles (**a**) CS LoS and reflection at different angles (**b**) VNA LoS and reflection at different angles (**c**) PDP at 40° rotational angle showing LoS and reflection peaks (**d**) VNA and CS measured PL based on PDP LoS and reflection peaks

thin wire reflective stainless-steel objects under varying illumination angles. The reflective PL, represented by the PDP, was compared with theoretical values.

Different sphere sizes ranging from 10 mm to 6 mm in diameter and a 1.5 mm diameter thin wire of 10 mm length were tested [19]. The calculable radiation patterns of these reflectors, along with the known antenna radiation patterns, determine the PL. Measurements were compared to the corresponding theoretical values to assess measurement uncertainties. This measurement serves as an indicator of range resolution, where the motion of the circular reflector results in position shifts ranging from approximately 0.05–1.7 mm per measurement, depending on the rotation angle. Additionally, range resolution is analysed when all paths are combined, facilitating the optimisation of windowing, interpolation, and the trade-off between the number of points and bandwidth to enhance measurement accuracy [10]. The experiments were conducted using the CS systems and VNA extension modules at both frequency bands. Some photographs of the measurement setup and results are shown in Fig. 19.16. Figure 19.16a, c shows photographs of the measurement setup featuring single and multiple line reflectors. Meanwhile, Fig. 19.16b, d illustrates the PDP path gain obtained through single and multiple reflections at various rotational stages and distances, respectively. Further results and discussions based on this measurement setup are presented in [10].

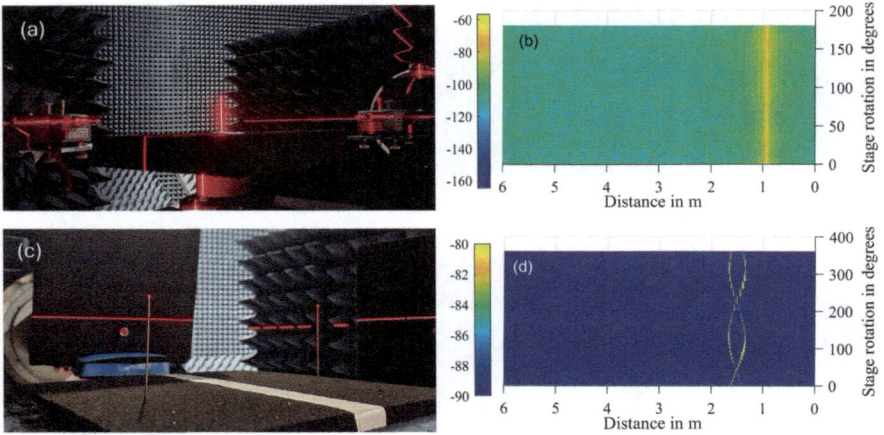

Fig. 19.16 Multipath measurement scenario using a single and multiple reflectors: (**a**) measurement setup photograph, where a single wire reflector was placed at the centre of rotation of the VNA frequency extension modules (**b**) measured PDP showing the distance vs. the stage rotational angles. (**c**) Photograph of the measurement setup of two reflecting wires placed on the rotational stage, where the CS antennas are directed towards them (**d**) measured PDP showing the the reflections originated from changing the rotational angles vs. the changing distance

19.4   Conclusion

In this chapter, we presented some characterisations of two correlation-based sub-THz CS systems measurement scenarios using a VNA at frequencies around 200–300 GHz.

We discussed waveguide-based and OTA measurement setups. Reference measurements were characterised using waveguide cascades and line waveguides fabricated as part of this work.

The VNA calibration involved conducting OTA drift measurements over a 12-hour period for a 1 m separation between two-port antennas. The peak PDP PL drift reached 0.2 dB for WR05 and 0.45 dB for WR03. Furthermore, phase drift caused noticeable shifts in peak delay, with WR05 drifting by 1.5 mm and WR03 by 2.3 mm.

Distance characterisation provided insight into the behaviour of VNA-interpolated PDP peaks, highlighting peak-to-peak non-linearity differences of 0.05 dB for WR05 and 0.22 dB for WR03. Comparisons with theoretical Friis loss enabled the calculation of antenna PC RMS differences and gains, contributing to a deeper understanding of system performance.

The CS measurements across varying separation distances offered valuable insight into frequency step size, bandwidth feasibility, and PL deviations across different reference conditions. WR05 exhibited sensitivity to reference variations, with PL deviations starting at 0.06 dB below 1 m and increasing to 0.18 dB beyond

3 m. In contrast, WR03 showed greater stability, with a peak-to-peak deviation of 0.06 dB across distances.

Measurement uncertainties were also investigated. WR05 VNA measurements exhibited combined uncertainties of ±0.2 dB, while CS measurements had a standard deviation of 0.39 dB. For WR03, the combined PL uncertainty reached ±0.5 dB, with CS standard deviations of 1.4 dB.

We extended the study to multipath scenarios by testing various reflectors at different rotational positions between transmitting and receiving antennas. The multipath components were analysed by varying the PDP delays across different rotational angles, and the VNA and CS systems provided consistent results.

In this chapter we showed the importance of precise calibration and accurate OTA alignment for ensuring consistent results across different measurement systems operating at sub-THz frequencies. PDP has proven to be an effective measurement metric for optimising high-frequency CS systems by serving as a spatial filter, mitigating the influence of unintended multipath reflections that become significant at longer distances.

References

1. S.-R. Moon et al., 6G indoor network enabled by photonics-and electronics-based sub-THz technology. J. Lightwave Technol. **40**(2), 499–510 (2022)
2. H. Elayan, O. Amin, B. Shihada, R.M. Shubair, and M.-S. Alouini, Terahertz band: The last piece of RF spectrum puzzle for communication systems. IEEE Open J. Commun. Soc. **1**, 1–32 (2019)
3. S. Tripathi, N.V. Sabu, A.K. Gupta, H.S. Dhillon, Millimeter-wave and terahertz spectrum for 6G wireless, in *6G Mobile Wireless Networks* (Springer, Berlin, 2021), pp. 83–121
4. X. Cai, X. Cheng, F. Tufvesson, Toward 6G with terahertz communications: Understanding the propagation channels. IEEE Commun. Mag. **62**(2), 32–38 (2024)
5. A. Ghosh, M. Kim, THz channel sounding and modeling techniques: an overview. IEEE Access. **11**, 17823–17856 (2023)
6. S. Rey, J.M. Eckhardt, B. Peng, K. Guan, T. Kürner, Channel sounding techniques for applications in THz communications: a first correlation-based channel sounder for ultra-wideband dynamic channel measurements at 300 GHz, in *2017 9th International Congress on Ultra Modern Telecommunications and Control Systems and Workshops (ICUMT)* (IEEE, Piscataway, 2017), pp. 449–453
7. M. Zeier, D. Allal, R. Judaschke, *Euramet Calibration Guide No. 12: Guidelines on the Evaluation of Vector Network Analysers (VNA)*, vol. 3 (European Association of National Metrology Institutes, Braunschweig, 2018)
8. J.P. Dunsmore, *Handbook of Microwave Component Measurements: With Advanced VNA Techniques* (John Wiley & Sons, London, 2020)
9. J. Campion, J. Oberhammer, Silicon micromachined waveguide calibration standards for terahertz metrology. IEEE Trans. Microwave Theory Tech. **69**(8), 3927–3942 (2021)
10. M.D. Al-Dabbagh, D. Ulm, T. Doeker, D. Dupleich, A. Ebert, R.S. Thomä, T. Kürner, D.A. Humphreys, T. Kleine-Ostmann, Characterization of sub-THz channel sounding systems in OTA measurement scenarios using a vector network analyzer. IEEE Trans. Antennas Propag. **73**(6), pp. 3943–3958 (2025)

11. M.D. Al-Dabbagh, T. Doeker, T. Kleine-Ostmann, T. Kürner, D. Humphreys, THz channel sounder and VNA verification measurement based over-the-air multipath artifact, in *2022 47th International Conference on Infrared, Millimeter and Terahertz Waves (IRMMW-THz)* (IEEE, 2022), pp. 1–2

12. M.D. Al-Dabbagh, T. Kleine-Ostmann, D. Humphreys, Radiative reference plane estimation and uncertainty for THz path loss measurements, in *2021 46th International Conference on Infrared, Millimeter and Terahertz Waves (IRMMW-THz)* (IEEE, 2021), pp. 1–2

13. K. Tekbıyık et al., Statistical channel modeling for short range line–of–sight terahertz communication, in *2019 IEEE 30th Annual International Symposium on Personal, Indoor and Mobile Radio Communications (PIMRC)* (IEEE, 2019), pp. 1–5

14. N.A. Abbasi, A. Hariharan, A.M. Nair, A.F. Molisch, Channel measurements and path loss modeling for indoor THz communication, in *2020 14th European Conference on Antennas and Propagation (EuCAP)* (IEEE, 2020), pp. 1–5

15. K. Tekbıyık, A. R. Ekti, A. Görçin, G. K. Kurt, "On the advances of terahertz communication for 5G and beyond wireless networks," in Flexible and Cognitive Radio Access Technologies for 5G and Beyond, H. Arslan and E. Başar, Eds., U.K.: IET, 2020

16. M.D. Al-Dabbagh, D. Ulm, T. Kleine-Ostmann, D. Humphreys, Horn antenna phase center position influence on sub-THz measurements uncertainties, in *2024 18th European Conference on Antennas and Propagation (EuCAP)* (IEEE, 2024), pp. 1–5

17. J. Quimby et al. *Channel Sounder Measurement Verification: Conducted Tests* (US Department of Commerce, National Institute of Standards and Technology, Gaithersburg, 2020)

18. M. Döbereiner et al., Joint high-resolution delay-doppler estimation for bi-static radar measurements, in *2019 16th European Radar Conference (EuRAD)* (IEEE, 2019), pp. 145–148

19. D. Dupleich, S. Semper, M.D. Al-Dabbagh, A. Ebert, T. Kleine-Ostmann, R. Thomä, Verification of THz channel sounder and delay estimation with over-the-air multipath artifact, in *2022 16th European Conference on Antennas and Propagation (EuCAP)* (IEEE, 2022), pp. 1–5

Chapter 20
Integrated Photonic-Assisted Signal Processing and Thermal Crosstalk

Souvaraj De, Younus Mandalawi, Ranjan Das, and Maxim Weizel

Abstract Integrated photonic-assisted signal processing has multiple applications such as signal amplification, multiplexing, and high-Q filtering in optical communication systems, optical sensing systems, and also microwave photonics. We will review recent works on integrated photonic-assisted signal processing for sinc-shaped Nyquist pulse generation, high-bandwidth Nyquist signal detection with low bandwidth devices, arbitrary waveform generation and measurement, and on-chip photonic frequency decoding. However, in such photonic integrated circuits (PICs), the photonic components are placed very close to each other on the chip, resulting in thermal crosstalk which degrades the system performance. Air-filled oxide and deep trench designs have proven to be very effective in mitigating the thermal crosstalk for various frequently deployed photonic devices like Mach-Zehnder modulators (MZMs), ring resonators, optical switches, and photodetectors designed on a standard silicon-on-insulator (SOI) platform. In this chapter, we will additionally review the basics of optical signal processing and some results for such trench-enhanced thermal crosstalk resilient circuits.

20.1 Photonics Integrated Samplers

Photonic integrated signal processing [1, 2] can be a viable solution to the ever-increasing global demands for high bandwidth systems coupled with low latency and reduced power consumption. When it comes to sampling, which is defined

S. De (✉)
Physikalisch-Technische Bundesanstalt (PTB), Braunschweig, Germany
e-mail: de.souvaraj@ptb.de

Y. Mandalawi · R. Das
Technische Universität Braunschweig, THz-Photonics Group, Braunschweig, Germany
e-mail: younus.mandalawi@ihf.tu-bs.de; ranjan.das@ieee.org

M. Weizel
Universität Paderborn, Schaltungstechnik (SCT)/Heinz Nixdorf Institut, Paderborn, Germany
e-mail: maxim.weizel@hni.uni-paderborn.de

© The Author(s) 2026   241
T. Kürner et al. (eds.), *Metrology for THz Communications*, Springer Series
in Optical Sciences 256, https://doi.org/10.1007/978-3-032-01986-8_20

in the time domain as the transformation of a continuous signal amplitude into equally spaced, discrete-time values, optical sampling is a more viable option compared to sampling in the electrical domain [3] as the former sampling method has higher operational bandwidths at lower power consumption and immunity to electromagnetic interference.

As per the sampling theorem, any bandwidth-limited signal can be represented as the superposition of time-shifted orthogonal sinc pulses weighted with the sampling points. To retrieve a single sampling point, the signal has to be multiplied with a sinc pulse with the right bandwidth and time shift. Nevertheless, due to the limited spectrum of such pulses in the frequency domain, they are unlimited in the time domain. However, the same bandwidth-limited signal can as well be represented by a superposition of sinc pulse sequences (SPSs) [4] that correlate to a rectangular, phase-locked frequency comb in the frequency domain and can be generated with intensity modulators like integrated MZMs by driving them with sinusoidal radio frequencies (RFs). In this context, a compact integrated photonic sampler based on ring modulators [5] is also a viable option.

However, especially for high-bandwidth signals the limited bandwidth of electronic signal processing is a severe bottleneck [6]. These systems have very complex architectures and extensive electronics and computational hardware requirements. Thus, time-frequency coherence-based all-optical sampling with SPS [7] may open a route to achieve much higher bandwidths with present-day technologies. Such a sampling technique is suitable for any integrated photonic platforms, including silicon photonics and lithium niobate on insulators. Using this method, an experimental demonstration for detecting high-bandwidth Nyquist signals with low-bandwidth silicon photonics was carried out [8]. Additionally, by the same method, high-speed arbitrary waveforms can as well be generated with low-speed electronics [9]. Besides sampling and signal generation lots of different signal processing techniques can be integrated into photonic platforms. A fully reconfigurable and adaptable compact photonic frequency decoder-PIC (PFD-PIC) with low system complexity, power consumption, and footprint was successfully illustrated, for instance [10–12]. Nonetheless, when a PIC is tightly packed with all the photonic components, the thermal crosstalk is a major problem leading to significant system performance reduction. Useful techniques to mitigate such a crosstalk will be investigated in the following sections as well.

20.1.1   Integrated Sinc-Shaped Nyquist Pulse Generation

Optical sinc-shaped Nyquist pulse sequences have numerous advantages owing to the rectangular spectrum in the frequency domain [13] like multiplexing the signal in the wavelength domain, high-speed data transmission, and all-optical signal processing [4].

Additionally, such pulses are very advantageous for high-bandwidth integrated optical communication systems on account of their temporal and spectral characteristics like mutable periodicity and operability at multiple wavelengths. Generation of the optical sinc-shaped Nyquist pulse sequences using MZMs has been discussed in detail with the experimental setup in the Chap. 23.

Silicon modulator technologies [14] are compatible with silicon photonic foundries which allow high-volume production and multichannel integration on a single chip. Therefore, we generate the Nyquist pulses in an integrated platform with an on-chip silicon MZM [15]. This modulator is designed with photonic BiCMOS technology as shown in Fig. 20.1 with photonic components like multi-mode interferometers and high-speed germanium photodiodes for observing the MZM bias points and phase shifters along with electronic bipolar transistors having transit frequencies up to 220 GHz on the same silicon substrate [16]. The on-chip amplifier has a linear two-stage driver designed with push-pull topology; the first one provides amplification and gives a 50 Ω interface and the second one is a distributed driver along the transmission line with driver-phase-shifter segments. The 3-dB bandwidth was evaluated to be 18 GHz with the modulator AC response and the DC extinction ratio was 40 dB as shown in Fig. 20.1. The V_{p_i} of the modulator is calculated to be 420 mV with the transmitter requiring a power of 1.8 W from the two DC supplies, giving 3.5–5 V for the on-chip drivers. With this on-chip modulator, optical sinc-shaped Nyquist pulses have been generated with a very low root mean square error (RMSE) of 0.7%.

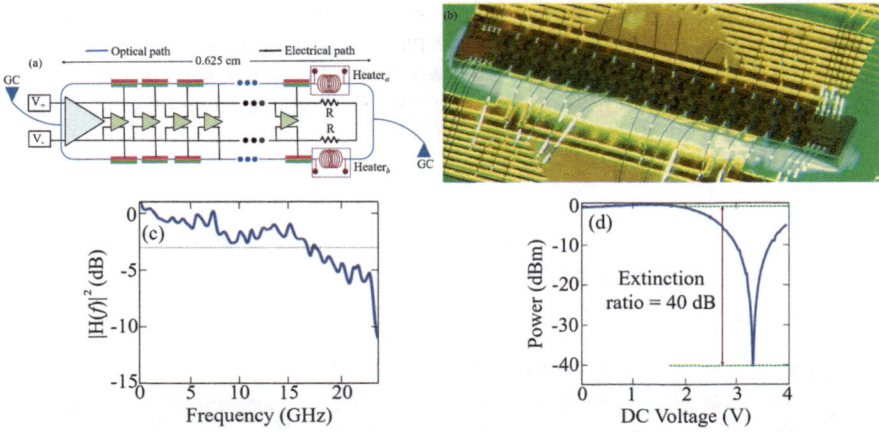

Fig. 20.1 (**a**) Block diagram of the modulator chip. (**b**) Fabrication of the on-chip MZM on PCB. (**c**) Modulator AC response showing a −3-dB bandwidth of 18 GHz. (**d**) Modulator DC response

20.1.2 Integrated Photonics for High-Bandwidth Nyquist Signal Detection

Detections of high-bandwidth signals in low-bandwidth electronics assisted with photonics can be achieved with orthogonal sampling [8]. A high-bandwidth signal can be split into lower-bandwidth signals (N equal to three lower-bandwidth signals) in real time through a linear sampling with sinc-pulse sequences, as discussed in Chap. 22. This can be done by integrated modulators, like electronic-photonic co-integrated dual-drive silicon MZMs, which offer high modulation efficiency (external $V = 420\,mV$), low power consumption, and a high DC extinction ratio (40 dB) [8]. If combined with wavelength division multiplexing, such a design facilitates transceivers for signals with bandwidths up to around 700 GHz [17].

A laser source was modulated with a Nyquist high-bandwidth signal (B) containing three different low-bandwidth Nyquist data ($B/3$) streams that are orthogonally multiplexed with sinc-pulse sequences (2 zero-crossings) in the time domain. After transmission through a standard single-mode fibre (SMF), the high-bandwidth broadband Nyquist signal (dotted curve of 24 GBd QPSK signal in Fig. 20.2c was subjected to a silicon electronic-photonic co-integrated MZM, which is shown in Fig. 20.2a, b, for demultiplexing. This was done through an optical 3-line frequency comb, which is a 2 zero-crossing sinc-pulse sequence in the time domain. The flat frequency comb is generated in the same integrated MZM by modulating a radio-frequency oscillator with DC bias. The $B/3$ bandwidth demultiplexed signal (highlighted solid line with 8-GHz passband signal in Fig. 20.2c) was then coherently detected and processed with low-bandwidth electronics. The eye diagram for the detected 8-GBd QPSK (1/3 of the 24 GBd Nyquist channel) using 4-GHz

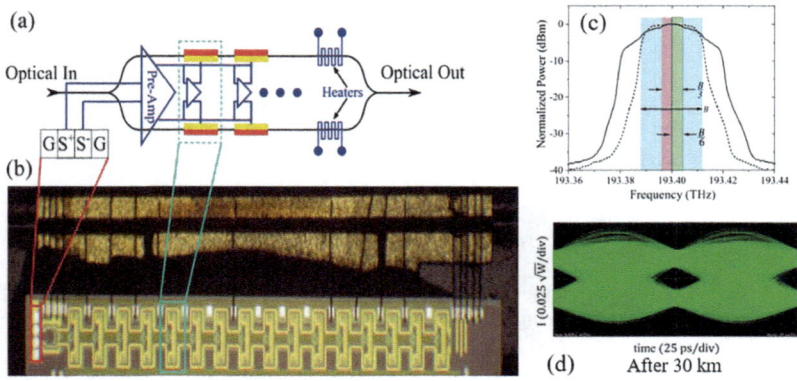

Fig. 20.2 (a) Component diagram of the fabricated MZM, featuring a pre-amplifier and a distributed driver that provides complementary signal inputs to the segmented phase shifters in both arms. (b) Chip image showing bond wires that connect the chip pads to the printed circuit board lines for DC connections. (c) The measured optical spectrum before demultiplexing shown in dashed curves and after demultiplexing shown in solid curves. (d) Eye diagram for 1/3 of the 24-GBd Nyquist channel after 30 km of SMF (Reprinted with permission from [8] © Optical Society of America)

devices after 30 km of SMF is displayed in Fig. 20.2d with 16.22-dB Q-factor and 4.85E-11 bit-error-rate (BER).

20.1.3   Arbitrary Waveform Generator with Segmented IQ-Modulator

One very promising concept for a photonic-assisted DAC is the implementation of orthogonal sampling with sinc-pulse sequences [9, 18]. The concept allows for the generation of high-bandwidth, high-sampling rate arbitrary waveforms from low-bandwidth, low-sampling rate electronics [9, 18]. Another innovative idea is the aggregation of two-level signals like non-return-to-zero (NRZ) into higher levels and higher data rate signals using an integrated segmented MZM modulator [19]. Utilising different segment lengths (electro-optic phase shifters) results in a difference in the relative V_π values, which is the voltage required to achieve 180° phase shift. Thus, applying two NRZ with the same peak-to-peak voltage to two segments, where one of the segments has double the length relative to the other, produces a four-level signal. By merging the two mentioned concepts, an integrated segmented IQ modulator is designed to generate a high-bandwidth 16-QAM signal from low-bandwidth NRZ signals.

Figure 20.3a displays the schematic diagram of the envisioned integrated modulator. It is set up in an IQ configuration. Each arm of the modulator includes

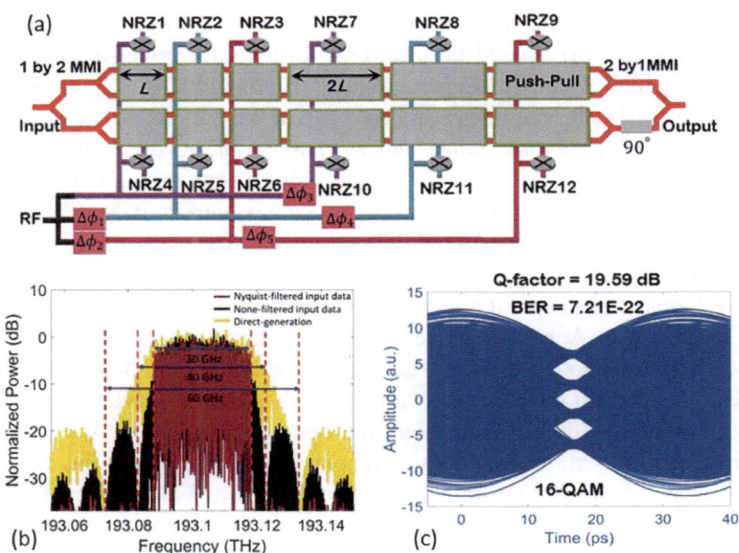

Fig. 20.3 (a) Schematic of the integrated IQ-segmented optical modulator. (b) Optical spectrum of the generated 16-QAM 30-GBd Nyquist signal from 5-GHz NRZs, and (c) eye diagram

six-phase shifter segments. All segments are silicon PN phase shifters operating in reverse bias carrier depletion mode and a series push-pull configuration. For the orthogonal sampling, in the basic setup with $N = 3$ (featuring two zero crossings), three NRZ low-bandwidth signals (first three segments) with a baseband bandwidth of $B/6$ and a repetition rate of $B/3$ (Nyquist limited to 5 GHz) are each multiplied by a single radio frequency from a 10-GHz oscillator. Each segment is assigned a different phase shift ($\Delta\phi_1 = 120°$, $\Delta\phi_1 = 240°$), as illustrated in Fig. 20.3a. The last three segments are twice the length of the first three. Therefore, applying different NRZs for the last three segments gives a 16-QAM signal of full bandwidth B (30 GBd). Figure 20.3b shows the optical spectrum of the generated 30-GBd 16-QAM with Nyquist-filtered input NRZs (red), none-filtered input NRZs (black), and direct-generation (yellow). The eye diagram of the generated 30-GBd Nyqusit 16-QAM is depicted in Fig. 20.3c with a Q-factor of 19.5 dB [9].

20.2 Photonic-Based Frequency Decomposer

Real-time spectral sensing involves the continuous scanning of the radio-frequency spectrum to identify and detect unused white spaces for the effective use and distribution of communication bands. This on-fly signal analysis in both the temporal and spectral domains is particularly significant for broadband low-latency applications such as radar, Lidar, and 5G, as well as the forthcoming 6G communication systems. To provide dependable, seamless, and low-latency services, a dynamic real-time spectrum sensing framework necessitates real-time scanning and detection across a broad-frequency spectrum. This may be achievable through high-speed digital signal processing (DSP) based on electronic systems for lower-frequency bands. However, for higher carrier frequencies and bandwidths, this approach suffers considerable computational power and costs. Furthermore, electronic DSP systems have a restricted operational range and are susceptible to electromagnetic interference (EMI). In contrast, photonic-based spectrum sensing can deliver extensive bandwidth, high measurement precision and accuracy, immunity to electromagnetic interference, and low-latency performance for various delay-based signal processing applications [20, 21].

The schematic concept of the photonic radiowave frequency decomposer is shown in Fig. 20.4. A multi-tone test RF signal under test (SUT) consisting of unknown frequencies (f_1, f_2, and f_3), as presented in Fig. 20.4a, b, is convoluted by a Gaussian-shaped pulse to resolve them at the output delayed signals. Such a test signal subsequently passes through the optically engineered group delay unit (on-chip or fibre based) to decompose temporarily as illustrated in Fig. 20.4.

The concepts were demonstrated both theoretically and experimentally for mm-wave signals in [10–12]. Integrated photonic-based ring resonators were used to tune group delays for different frequency bands and designing the PIC-based group delay unit (GDU) as shown in Fig. 20.5. A multi-tone mm-wave signal was successfully tested by leveraging such as a PIC-based compact design approach as depicted in

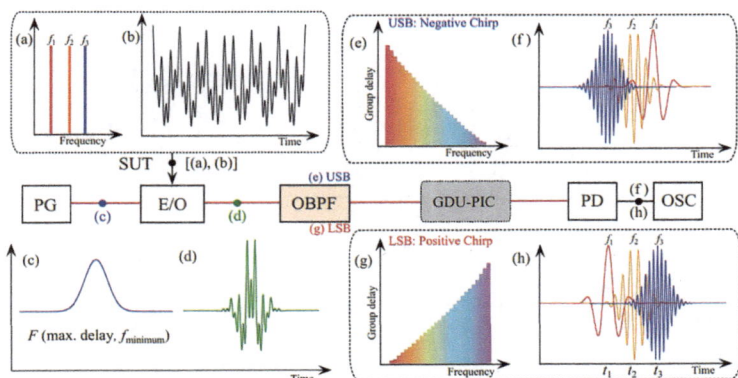

Fig. 20.4 Conceptual block diagram presentation of the proposed photonic frequency decoder. The signal under test (SUT) containing the undetermined frequency components f_1, f_2, and f_3 is displayed in (**a**), and the corresponding time-domain signal is illustrated in (**b**). Test signal limited in time by a pulse modulation (**c**). The test input signal to the photonic frequency decomposer (PFD-PIC) is highlighted in green (**d**). Optical band pass filter (OBPF) deployed to filter out any sideband and, thus, the negative or positive chirp can be processed for time-domain filtering. A different group delay encounter for each signal frequency in the test multi-tone signal (USB, (**e**), (**f**)) and (LSB, (**g**), (**h**)). PG, pulse generator; PD, photodiode; OSC, oscilloscope; E/O, electrical to optical conversion (Adapted with permission from [11] © Optical Society of America)

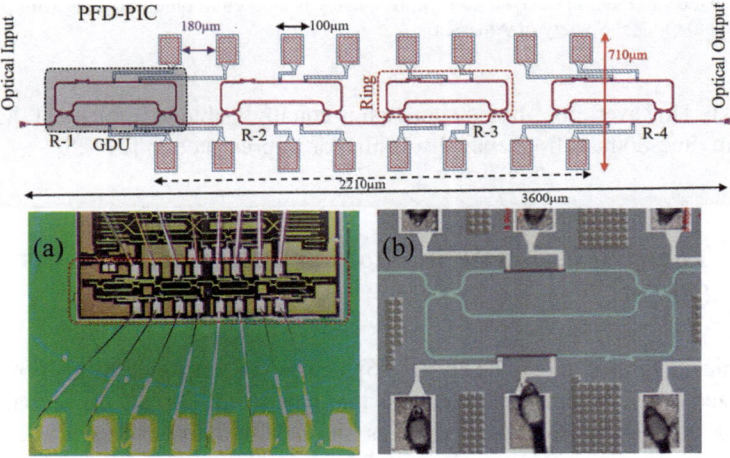

Fig. 20.5 Design mask layout of the fabricated photonics chip on a SOI platform with cascaded four-ring resonators as a group delay unit (GDU) embedded by a Mach-Zehnder interferometer (MZI). The DC connections are used for tuning the thermal phase shifters of the MZI and the ring for delay and resonance tuning purposes. (**a**) Zoomed image of the bonded photonic chip (red box) on a PCB and (**b**) zoomed-in view of a single micro-ring resonator (Adapted with permission from [11] © Optical Society of America)

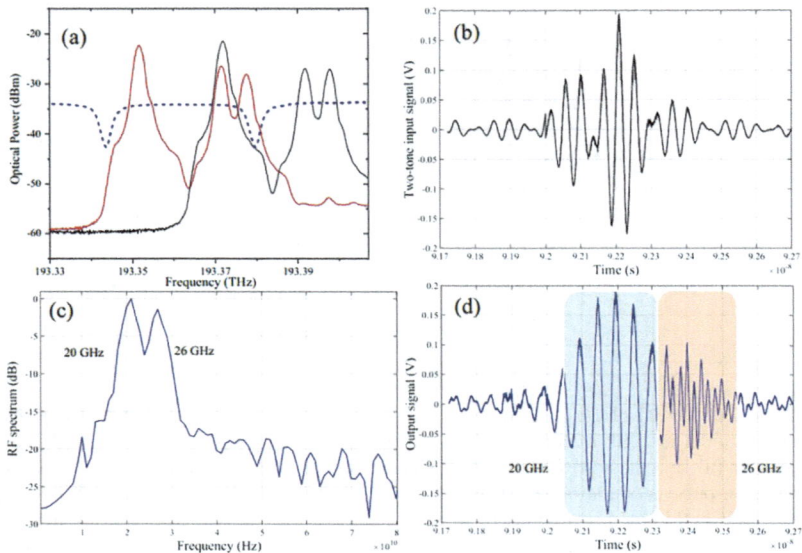

Fig. 20.6 (**a**) Spectrum of the input two-tone test signal detuned from resonance (black) and the spectrum after adjusting the laser wavelength (red). (**b**) Gaussian-shaped pulse with the test signal in time domain. (**c**) The spectrum of the 26-GHz component, aligned close to the delayed resonance. (**d**) Output signal after processed by the chip. The 26-GHz component is shown by the orange shade and the non-delayed 20-GHz frequency is marked in blue (Adapted with permission from [11] © Optical Society of America)

Fig. 20.6. However, fibre-based approach is equally viable to implement the similar off-chip time-domain frequency discriminator as presented in [22–25].

20.3 Thermal Crosstalk in Photonic Integrated Circuits (PICs)

Photonic integrated circuits (PICs) [26] consist of very tightly packed discrete integrated and tunable optical elements that enable ultra-broad analog bandwidths that can complement digital electronics and achieve throughputs in the range of petabits per second. They are useful for optical communication, optical signal processing, and even quantum computing. The most common platform for the PICs is silicon-on-insulator (SOI) which is also compatible with the complementary metal oxide semiconductor (CMOS) technology [27].

However, the chip constituents like silicon and silicon dioxide have a finite thermal conductance resulting in thermal crosstalk, which, in densely packed PICs with co-integrated photonic and electronics functionalities, will severely impact the overall system performance. In this section, we study the effect of crosstalk on the photonic components in such circuits and techniques to overcome thermal crosstalk.

20.3.1 Crosstalk in Thermally Controlled Ring Resonators

Silicon photonic-integrated ring resonators are used in a multitude of applications in the broadest sense as optical filters. The resonance frequency and free spectral range (FSR) of a ring resonator is determined by its diameter and the effective refractive index. A single ring resonator has usually a very high frequency selectivity, and thus, series or parallel cascading is used to create more complex transfer functions. Here, we want to focus on cascading multiple rings in series in a so-called coupled resonator optical waveguide (CROW). By carefully adjusting the number (corresponds to order), diameter, and coupling sections, higher-order filters with desired frequency response in terms of bandwidth, free spectral range, and out-of-band rejection can be created. In [28] the authors show a design strategy using the Butterworth and Chebyshev response as an example. These can be used in a wavelength division multiplexed (WDM) system, where a flat top, steep roll-off characteristic is desired to filter out a certain band and direct it to the according receiver. While in theory a fixed design with precisely aligned rings is sufficient to achieve the required functionality in practical applications due to fabrication tolerances and temperature fluctuations, a tuneable design is preferred. Other applications like ring modulators and optical switching make the necessity to tune the resonance frequency more obvious.

20.3.1.1 Hourglass-Shaped Resonators

Some techniques to mitigate thermal crosstalk are described in Sect. 20.3.2, but these are not always available in the technology at hand. As an example study, the authors in [29] implemented 3rd-order CROW filters with three cascaded and individually tunable rings. To increase the tuning efficiency the rings are divided into a coupling section and a phase tuning section. In the tuning section the two opposite waveguides are routed together as close as possible without optically coupling them. Overall, this leads to an hourglass shape of the individual rings that can be seen in the inset of Fig. 20.7.

Due to the close proximity of the rings, it becomes crucial to consider the thermal crosstalk. Müller et al. [29] evaluates two different thermal heating concepts: N-doping of the optical waveguide along the resonator circumference (see Fig. 20.7) and, secondly, meandering the aluminium metal layer closest to the waveguide. For the n-doped CROWs the distance between the heating sections is $36\,\mu m$ and the thermal crosstalk is found to be 1.8% with an insertion loss of 3 dB. For the metal heating CROWs the distance between the heaters was $9.7\,\mu m$, with 7.3% of thermal crosstalk and 1.5 dB insertion loss. Overall, the hourglass shape increased the phase

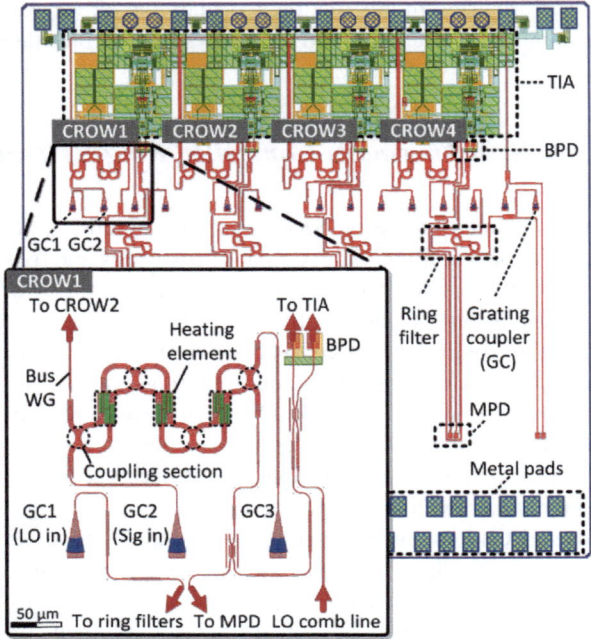

Fig. 20.7 Layout of a photonically assisted ADC implemented in an electronic photonic integrated circuit (EPIC) SiGe BiCMOS technology. Four CROW filters slice the incoming optical signal into equally spaced and partially overlapping segments, directing the light to the corresponding trans-impedance amplifier (TIA). Inset: layout of a 3rd-order coupled resonator optical waveguide (CROW) filter with hourglass shape for reduced thermal crosstalk. Heating elements (doped sections) are shown in green (Reprinted with permission from [31] © 2021 IEEE)

tuning efficiency compared to its circular counterpart by up to 30%, which in turn means that the power is reduced and thus the thermal crosstalk.

The functionality of the hourglass-shaped 3rd-order CROW filters is demonstrated in a photonically assisted ADC experiment on a silicon photonics platform in [30, 31]. Figure 20.7 shows the full EPIC chip with four CROWs, four 1st-order ring-filters, and four TIAs as heterodyne receivers. The spacing between the channels is 450 μm such that channel to channel thermal crosstalk is negligible.

20.3.2 Techniques to Mitigate Thermal Crosstalk

20.3.2.1 Oxide Trench

In PICs, implementing trench designs can help reduce the impact of thermal crosstalk. The effect of an air-filled oxide trench is investigated for a basic design on an SOI platform consisting of a doped heater, buried oxide (BOX) layer, and

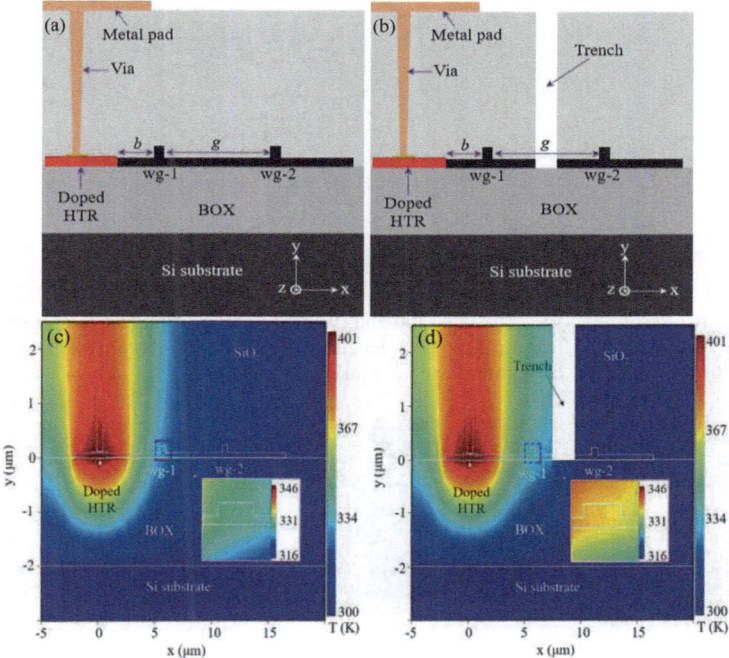

Fig. 20.8 Rib waveguides for a N^{++}-doped silicon heater with metal contacts: (**a**) without trench and (**b**) with an air-filled trench. (**c**)–(**d**) Corresponding temperature profiles for a gap (**g**) of 5 µm and a heater power of 150 mW with a magnified view of temperatures at the main waveguide (wg-1) (the figures have been taken from [32])

two rib waveguides with standard parameters [32]. Practically, such trenches can be implemented by oxide etching with the removal of the cladding region.

The trench assists in thermally shielding the adjacent integrated components as air has a lower thermal conductivity of 0.026 W/(m·K). Full-wave thermal and optical simulations indicate that the adjacent waveguide remains around the ambient temperature of 300 K as depicted in Fig. 20.8. For evaluation of the undesired phase change in the second waveguide (wg-2), the phase crosstalk ratio, which is the ratio of the phase shift in the second waveguide (wg-2) to the phase shift in the main waveguide (wg-1), is calculated for conventional and trench scenarios as shown in Fig. 20.9. For the trench design, the heat is localised within the main waveguide (wg-1) resulting in a sixfold reduction in the phase crosstalk for a gap (g) of 5 µm between the waveguides. Consequently, the same phase shift is achieved for a lower heater power, as shown by the phase change ratio which is the ratio of phase shifts in the main waveguide (wg-1) for conventional and trench designs in Fig. 20.9.

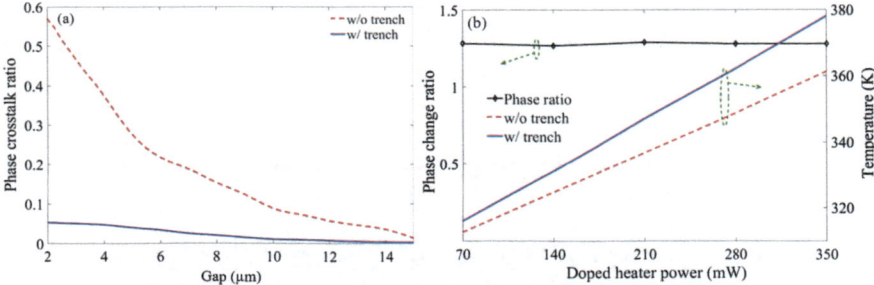

Fig. 20.9 (**a**) Resulting phase crosstalk ratio and (**b**) phase change ratio and temperature plot of the main waveguide (wg-1 in Fig. 20.8) for a length of 500 μm considering without and with an air-filled trench scenarios (the figures have been taken from [32])

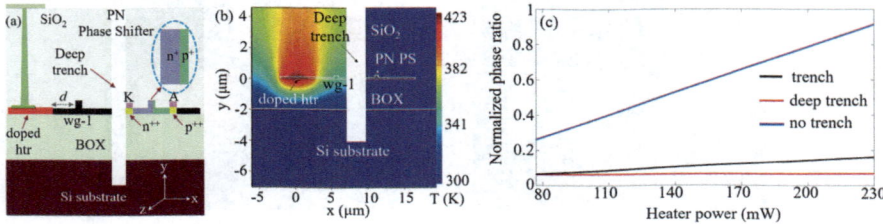

Fig. 20.10 (**a**) Cross-sectional view of a deep trench assisted PN phase shifter design with a thermally tuned waveguide (wg-1) and (**b**) the respective temperature profile at a separation of 5 μm with a heater power of 211 mW. (**c**) Normalised phase ratio comparison for with a trench [32] and with a deep trench [33] and for no trench. The normalised phase ratio is defined as the ratio of the phase shift in the PN phase shifter to the phase shift (at maximum heater power) in the main waveguide (wg-1)

20.3.2.2 Deep Trench

Implementing deep trench designs can further reduce the thermal crosstalk, especially in densely packed PICs, where a large number of photonic components are placed very close to each other. In this chapter, the thermal crosstalk performance of two common photonic devices, namely, PN phase shifters and Mach-Zehnder odulators (MZMs), is thoroughly analysed in terms of simulations.

The design of the PN phase shifter with an air-filled deep trench [33] starting at 2 μm inside the silicon substrate with standard dimensions and doping concentrations is shown in Fig. 20.10. From the temperature profile, it can be observed that the deep trench enables maximum thermal shielding of the PN phase shifter waveguide. Therefore, the phase is unchanged for a fixed bias voltage, shown by the red curve, regardless of the power applied across the adjacent doped heater which eventually leads to a better mitigation of thermal crosstalk in PICs.

A CMOS-compatible traveling-wave MZM with appropriate dimensions and doping concentrations [34, 35] and a deep trench in an SOI platform is designed

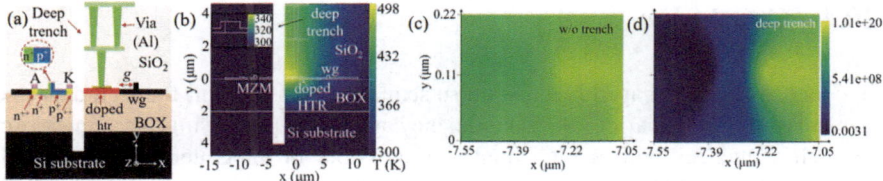

Fig. 20.11 (**a**) Cross-sectional view of the TW-MZM and the thermally tuned waveguide with a deep trench and (**b**) the respective temperature distribution at a heater power of 250 mW. The charge profile in the active region of the TW-MZM waveguide can be seen for (**c**) no trench and (**d**) deep trench at $V_{bias} = 1.5$ V. The colour bar shows the charge concentration per cm^3 (Reprinted with permission from [34] © Optical Society of America)

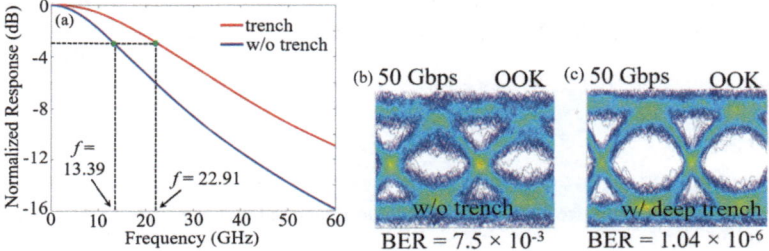

Fig. 20.12 (**a**) Bandwidth comparison of the TW-MZM with 500 μm length, with and without a trench. Eye diagrams of the TW-MZM at 50 Gbps without trench (**b**) and with a deep trench (**c**) (Reprinted with permission from [34] © Optical Society of America)

as per Fig. 20.11. From the temperature profile, it is clear that the deep trench provides thermal isolation for the TW-MZM from the adjacent doped heater. Also, due to the thermal insulation with a deep trench, the amount of p$^+$ and n$^+$ charge carriers diffusing into the PN active region is drastically reduced as compared to the conventional design. This leads to a significant reduction in the junction capacitance and resistance which eventually results in a 71% enhancement in the electro-optic (EO) bandwidth up to 22.91 GHz, without any pre-compensation or pre-emphasis data processing [34] as shown in Fig. 20.12. In terms of data transmission at higher data rates of 50 Gbps for an on-off keying (OOK) modulation scheme, the bit error rate (BER) is lower and the eye opening is wider for a deep trench scenario. Such deep trench designs have proven to be effective in crosstalk mitigation for numerous photonic devices like dual-drive MZMs [36], microring switches [37], ring modulators [38], photodetectors [39–42], optical attenuators [43], quantum well lasers [44], and even amorphous silicon platforms [45].

20.4 Conclusion

In conclusion, sinc-shaped Nyquist pulse sequences can account for high data rate transmission with superior spectral efficiency and signal processing. The generation of such sequences with a high quality is possible in integrated platforms with on-chip silicon MZMs. The generation of high-speed arbitrary waveforms with low-speed electronics via integrated segmented IQ-modulator was discussed. A sampling rate of above 300 GS/s might be achievable in the near future with high-bandwidth integrated modulators together with appropriate electronics to provide the driving signals. Nonetheless, in integrated platforms, particularly in PICs, there is thermal crosstalk owing to the densely packed photonic components. The alleviation of crosstalk with deep trenches is demonstrated by simulation results of photonic devices like MZM and PN phase shifters.

References

1. W. Liu et al., A fully reconfigurable photonic integrated signal processor. Nat. Photon. **10**, 190–195 (2016)
2. B. Bai et al., Microcomb-based integrated photonic processing unit. Nat. Commun. **14**, 66 (2023)
3. D. Fang et al., 320 GHz analog-to-digital converter exploiting Kerr soliton combs and photonic-electronic spectral stitching, in *Proceedings of the European Conference on Optical Communication* (2021), pp. 1–4
4. J. Meier et al., High-bandwidth arbitrary signal detection using low-speed electronics. IEEE Photon. J. **14**(2), 1–7 (2022)
5. M.I. Hosni et al., Low power, compact integrated photonic sampler based on a silicon ring modulator. IEEE Photon. J. **14**(4), 1–6 (2022)
6. E.P. da Silva et al., Combined optical and electrical spectrum shaping for high-baud-rate nyquist-wdm transceivers, IEEE Photon. J. **8**(1), 1–11 (2016)
7. A. Misra et al., Integrated source-free all optical sampling with a sampling rate of up to three times the RF bandwidth of silicon photonic MZM. Opt. Express **27**(21), 29972–29984 (2019)
8. A. Misra et al., Reconfigurable and real-time high-bandwidth nyquist signal detection with low-bandwidth in silicon photonics. Opt. Express **30**, 13776–13789 (2022)
9. Y. Mandalawi et al., Integrated segmented iq-modulator for orthogonal sampling and multi-level high-bandwidth signal generation. Opt. Lett. **49**(8), 2193 (2024)
10. R. Das, T. Schneider, Integrated group delay units for real-time reconfigurable spectrum sensing of mm-wave signals. Opt. Lett. **45**(17), 4778–4781 (2020)
11. K. Singh, R. Das, A. Venugopalan, S. De, M.I. Hosni, L. Zhou, T. Schneider, Real-time reconfigurable on-chip photonic frequency decoder. Opt. Express **31**(19), 30160–30170 (2023)
12. K. Singh, R. Das, A. Venugopalan, S. De, M.I. Hosni, L. Zhou, T. Schneider, Reconfigurable RF frequency sniffer using tunable micro ring resonator, in *CLEO 2023* (2023)
13. M.A. Soto et al., Optical sinc-shaped nyquist pulses of exceptional quality. Nat. Commun. **4**(1), 2898 (2013)
14. X. Zhou et al., Silicon photonics for high-speed communications and photonic signal processing. NPJ Nanophoton. **1**, 27 (2024)
15. S. De et al., Roll-off factor analysis of optical nyquist pulses generated by an on-chip Mach-Zehnder modulator. IEEE Photon. Technol. Lett. **33**(21), 1189–1192 (2021)

16. C. Kress et al., High modulation efficiency segmented Mach-Zehnder modulator monolithically integrated with linear driver in 0.25 μm bicmos technology, in *Advanced Photonics 2021 (IPR, NOMA, Sensors, Networks, SPPCom, SOF), OSA Technical Digest* (2021)
17. A. Venugopalan et al., Filterless reception of terabit, faster than nyquist superchannels with 4 GHz electronics. J. Light. Technol. **42**(15), 5121–5127 (2024)
18. Y. Mandalawi et al., Segmented Mach-Zehnder modulator for orthogonal sampling based high bandwidth thz transmitters, in *2024 15th German Microwave Conference (GeMiC)*, pp. 45–48 (2024)
19. Z. Zheng et al., Transmission of 120 Gbaud QAM with an all-silicon segmented modulator. J. Light. Technol. **40**(16), 5457–5466 (2022)
20. R. Das et al., Gain-enabled optical delay readout unit using cmos-compatible avalanche photodetectors. Photon. Res. **10**(10), 2422–2433 (2022)
21. H. Frankis et al., Application of the TDFA window in true optical time delay systems. Opt. Express **30**(17), 30164–30175 (2022)
22. J.E. Kadum et al., Temporal disentanglement of wireless signal carriers based on quasi-light-storage. J. Lightwave Technol. **40**(20), 6762–6768 (2022)
23. J.E. Kadum et al., Brillouin-scattering-induced transparency enabled reconfigurable sensing of RF signals. Photon. Res. **9**(8), 1486–1492 (2021)
24. J.E. Kadum et al., Quasi-light storage enabled cognitive RF sensing, in *Conference on Lasers and Electro-Optics* (2022)
25. J.E. Kadum et al., Slow light enabled temporal frequency discriminator, in *OSA Advanced Photonics Congress 2021* (2021)
26. X. Xu et al., Self-calibrating programmable photonic integrated circuits. Nat. Photon. **16**, 595–602 (2022)
27. M.F. Gonzalez-Zalba et al., Scaling silicon-based quantum computing using cmos technology. Nat. Electron. **4**, 872–884 (2021)
28. A. Melloni, M. Martinelli, Synthesis of direct-coupled-resonators bandpass filters for wdm systems. J. Lightwave Technol. **20**(2), 296–303 (2002)
29. J. Müller et al., Optimized hourglass-shaped resonators for efficient thermal tuning of crow filters with reduced crosstalk, in *2021 IEEE 17th International Conference on Group IV Photonics (GFP)*, pp. 1–2 (2021)
30. A. Zazzi et al., Optically enabled adcs and application to optical communications. IEEE Open J. Solid-State Circuits Soc. **1**, 209–221 (2021)
31. D. Fang et al., Optical arbitrary waveform measurement (oawm) using silicon photonic slicing filters. J. Lightwave Technol. **40**(6), 1705–1717 (2022)
32. S. De et al., Design and simulation of thermo-optic phase shifters with low thermal crosstalk for dense photonic integration. IEEE Access **8**, 141632–141640 (2020)
33. S. De et al., Cmos-compatible photonic phase shifters with extremely low thermal crosstalk performance. J. Light. Technol. **39**(7), 2113–2122 (2021)
34. S. De et al., Athermal travelling wave Mach-Zehnder modulators for optical interconnects, in *OSA Advanced Photonics Congress 2021* (2021)
35. S. De et al., Effect of thermal crosstalk on travelling-wave Mach-Zehnder modulator, in *2021 Conference on Lasers and Electro-Optics Europe and European Quantum Electronics Conference (CLEO/Europe-EQEC)* (2021), pp. 1–1
36. S. De et al., Thermal crosstalk mitigation in a dual-drive Mach-Zehnder modulator, in *2022 International Conference on Numerical Simulation of Optoelectronic Devices (NUSOD)*, pp. 27–28 (2022)
37. R.C. Heim et al., Thermal crosstalk alleviated silicon microring switches, in *2023 Conference on Lasers and Electro-Optics Europe and European Quantum Electronics Conference (CLEO/Europe-EQEC)*, pp. 1–1 (2023)
38. S. De et al., Thermal insulation of silicon ring modulators in densely-packed photonic integrated circuits, in *Proceedings of SPIE*, vol. 12575 (2023)

39. S. De et al., Study of thermal crosstalk in avalanche photodetectors in thz domain, in *47th International Conference on Infrared, Millimeter and Terahertz Waves (IRMMW-THz)* (2022), pp. 1–2

40. S. De et al., Temperature insensitive avalanche photodetectors, in *Optica Advanced Photonics Congress 2022* (2022)

41. S. De et al., Crosstalk resistant integrated uni-traveling carrier photodetector, in *2023 48th International Conference on Infrared, Millimeter, and Terahertz Waves (IRMMW-THz)* (2023), pp. 1–2

42. S. De et al., Crosstalk immune high-speed photonic transmitter-receiver system, in *2024 15th German Microwave Conference (GeMiC)* (2024), pp. 25–28

43. S. De et al., Thermal crosstalk resilient integrated optical attenuator, in *Frontiers in Optics/Laser Science* (2020)

44. S. De et al., Thermal-aware multi-quantum well laser, in *Frontiers in Optics + Laser Science 2022 (FIO, LS)* (2022)

45. S. De et al., Amorphous silicon based crosstalk resilient photonic phase shifters, in *2023 Conference on Lasers and Electro-Optics Europe and European Quantum Electronics Conference (CLEO/Europe-EQEC)* (2023), pp. 1–1

Chapter 21
Error Metrics for Photonics-Assisted Signal Processing

Souvaraj De, Younus Mandalawi, and Ranjan Das

Abstract For the next generation of wireless communications (6G and beyond), peak data rates of several Tbit/s are foreseen. To keep up with such ever-rising demands, photonics-assisted signal processing has been considered a viable solution where conventional copper-based electronics suffers from excess power dissipation issue. This is also attributed to its high bandwidth, energy efficiency, CMOS compatibility, and compact design. Photonics-assisted analog-to-digital converters (PADCs) have been regarded as an alternative to bandwidth and signal integrity-limited electronic analog-to-digital converter (EADC). For the comparison of the quality, different error metrics such as SINAD, ENOB, and root mean square error (RMSE) are used. In this chapter, we begin by describing and discussing these various metrics, followed by using them to evaluate the performance of PADC and compare it with EADC. We address different error sources including time jitter and laser relative intensity noise (RIN). As we further validate by simulation and experiment, PADC may provide better signal-to-noise and distortion ratio (SINAD), effective number of bits (ENOB), and RMSE than state-of-the-art EADC.

21.1 Introduction

The effective resolution of any digital-to-analog converter (DAC) or analog-to-digital converter (ADC) can be represented by the equivalent effective number of bits (ENOB) [1–3]. The value of ENOB can be derived by the ratio of signal-to-

S. De (✉)
Physikalisch-Technische Bundesanstalt (PTB), Braunschweig, Germany
e-mail: de.souvaraj@ptb.de

Y. Madalawi · R. Das
Technische Universtität Braunschweig, THz-Photonics Group, Braunschweig, Germany
e-mail: younus.mandalawi@ihf.tu-bs.de; ranjan.das@ieee.org

© The Author(s) 2026
T. Kürner et al. (eds.), *Metrology for THz Communications*, Springer Series in Optical Sciences 256, https://doi.org/10.1007/978-3-032-01986-8_21

noise and distortion ratio (SINAD) for the sampled and digitized signal under test
by [3–5]

$$ENOB = \frac{SINAD - 1.76}{6.02}. \qquad (21.1)$$

The value of the SINAD can be measured from the frequency spectrum of all
sampled test points after the time-domain trace is transformed to the frequency
domain by a fast Fourier transform (FFT). Then, the ratio of the signal component
under test to all other unwanted components including noise floor and distortions
is calculated. To follow the IEEE standards for ADC characterization, a sine wave
signal has to be considered as the signal under test or signal to be sampled. The
SINAD can also be measured using the root mean square error (RMSE) of the
sampled points in the time domain compared to the fitted signal. Then, the SINAD
can be calculated as [3, 6]

$$SINAD = 20 \log_{10}(\frac{A}{\sqrt{2} \cdot RMSE}), \qquad (21.2)$$

where A is the full amplitude of the signal, which is considered to be 2 in the case
of a normalized amplitude. One of the main limitations for the SINAD in electronic
ADC (EADC) is time jitter [3–5, 7]. When ignoring other noise sources such as shot
and thermal noise, the SINAD can be calculated from the time jitter as

$$SINAD = 10 \log_{10}(1/(2\pi f \sigma_{EADC})^2), \qquad (21.3)$$

where σ_{EADC} is the time jitter of the EADC and f is the input frequency. There
are two sources of jitters that arise in an EADC. The first is the clock jitter which
is caused by the clock source used in the system. Sampling with EADC is based
on sample-and-hold circuits, which disconnect the hold capacitor from the input
buffer amplifier [7, 8]. Variations in switching from sample to sample, caused by
noise and switching uncertainty, result in what is referred to as aperture jitter, which
contributes to the jitter from the clock.

21.2 Resolution Improvement and Bandwidth Reduction Using Photonics-Assisted Analog-to-Digital Converters

The effect of input frequency at a certain time jitter (Δt_{jitter}) on the voltage error
Δv is shown in Fig. 21.1. The introduction of Δv results in an uncertainly for
the represented values of the sampling points, which degrades the overall SINAD
and ENOB values of the sampling information points after the conversion in an
EADC. Figure 21.1 shows that a low input frequency (a) should exhibit lower
(Δv) and hence higher signal resolution compared to a higher input frequency

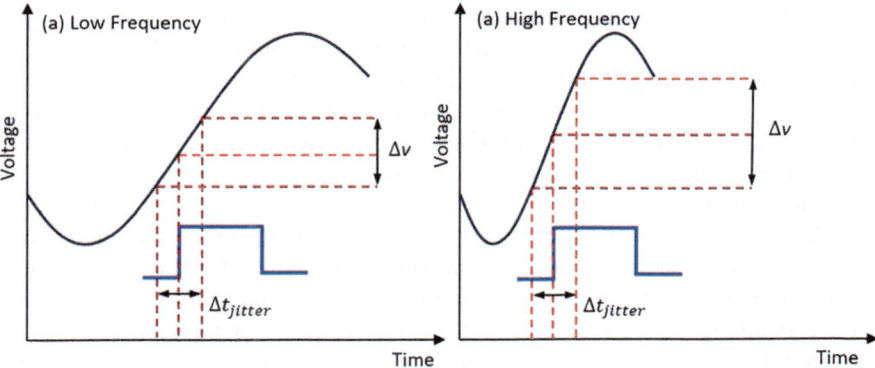

Fig. 21.1 Sampling error as a function of the analog input frequency and time jitter

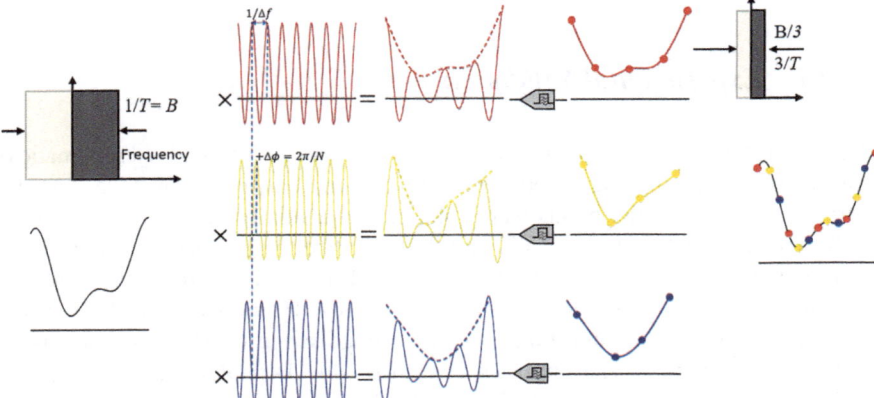

Fig. 21.2 Schematic illustration of the concept of photonics-assisted ADC with orthogonal sampling in three branches

(b) at the same value of Δt_{jitter}. This can also be concluded from Eq. (21.3). Therefore, a high-quality down-conversion of a high-bandwidth signal into parallel low-bandwidth sub-signals is advantageous in achieving higher signal integrity and effective resolution [1, 5, 9, 10]. It also allows for ultra-wideband wireless and optical signal reception while using low-bandwidth electronics.

Figure 21.2 presents a concept of a PADC with orthogonal sampling in a three-branch configuration [5, 11–15]. Please see as well in Chap. 22 for details. The system offers the reception of a high-bandwidth signal beyond the capability i of today's state-of-the-art EADC. It consists of two stages of sampling, a high-speed optical sampling along with a low-speed electrical sampling and digitization. By converting the high-bandwidth signal to be sampled into low-bandwidth parallel sub-signals, the system lowers the input frequency to the EADC, which results in improved SINAD and ENOB values. This is achieved while sampling a wide bandwidth signal using low-bandwidth back-end devices. A wireless or THz signal to be sampled has first to be converted to the optical domain by an optical modulator. The signal to be sampled with bandwidth B is split into a number of N branches ($N = 3$), before multiplication and sampling with a sinc-pulse sequence that has $N - 1$ zero-crossings (2 zero-crossings) in the time domain. This is equal to a convolution with an N line flat frequency comb at a spacing of Δf.

Unlike the traditional sample-and-hold CMOS-based circuits, no aperture jitter is added [5, 7]. This is due to the multiplication nature of the first optical sampling stage. The down-converted sub-signal carries one-third ($1/N$) of the total information at the Nyquist sampling limit. In the other parallel branches, the rest of the sampling information points are collected by shifting the same sinc-pulse sequence to the next zero-crossing with electrical phase shifts of the radio-frequency oscillator (RFO) (120° at the second branch, 240° at the third branch) achieving orthogonal sampling [5, 11–15].

21.2.1 Experimental Validation

A proof-of-concept experiment was carried out. Figure 21.3 provides a schematic of the experimental setup. Due to equipment limitations, only one branch was tested at a time, and the sampling points were collected sequentially by adjusting the phase shift $\Delta \phi$ of the RFO to 0°, 120°, and 240°. A fiber laser operating at 1550 nm served as the optical source, with the first MZM having a 20 GHz bandwidth and a relative intensity noise (RIN) level of -135 dBc/Hz at 10 MHz. A 14.5 GHz sine wave, biased at the null point for full voltage swing testing, was used as the signal to be sampled. The optical spectrum of this 14.5 GHz signal is depicted in Fig. 21.3a. An EDFA amplified the optical signal to a constant output power of 12 dBm to ensure equal power input to the CD and EADC when comparing direct measurements with those from the PADC. To minimize the EDFA's amplified spontaneous emission (ASE) noise, a 1 nm optical band-pass filter (O-BPF) was used. Polarization controllers (PC) were used to adjust the polarization before the modulators. The MZM in the sampling stage was driven by a 10 GHz sinusoidal frequency, with appropriate DC biasing and RF power for orthogonal sampling. The oscillator's measured jitter was approximately 75 fs.

The optical spectrum after sampling is shown in Fig. 21.3b, with the spectral copies from the convolution [5]. The sampled signal was amplified by a second EDFA to 12 dBm and then passed through another 1 nm OBPF. The power loss from splitting the input signal, which amounted to 4.8 dB due to the three branches, was not factored into the experiment since it could have been offset by a 4.8 dB higher gain from the EDFA. To eliminate the complexity of digital signal processing related to the carrier frequency and phase recovery algorithms in coherent detection, the signal was homodynously combined with the carrier laser wave using a 90° hybrid and a 43 GHz balanced detector. As a low-bandwidth detector was unavailable, a 5 GHz electrical low-pass filter was applied immediately after detection to reduce

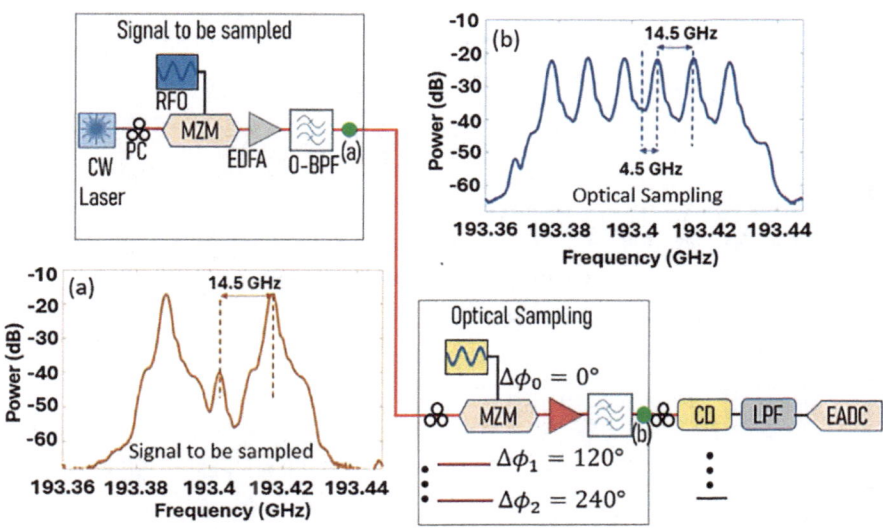

Fig. 21.3 Diagram of the experimental arrangement used to evaluate the performance of the proposed PADC. The carrier laser wave, utilized for homodyne coherent detection, is not depicted. Panels (**a**) and (**b**) display the spectrum of the 14.5 GHz input signal and the corresponding spectrum obtained after convolution with a 30 GHz three-line comb, respectively. The place at which the two spectra (insets (**a**) and (**b**)) are measured is represented by the green circles in the block diagram. CW, continuous wave; PC, polarization controller; RFO, single tone radio-frequency oscillator; MZM, Mach–Zehnder modulator; EDFA, erbium-doped fiber amplifier; O-BPF, optical band-pass filter; CD, coherent detection; LPF, electrical low-pass filter; EADC, electronic analog-to-digital converter

the bandwidth. In a setup with a 5 GHz bandwidth coherent detector (CD), this filter would not be necessary. Finally, the three low-bandwidth signals, each of $B/(2N)$, were digitized into discrete values by an EADC with a jitter of 1 ps in the second sampling stage. The digitized sampling points were interleaved through offline processing.

Figure 21.4a displays the measured sampling points (circles) after sampling the 14.5 GHz signal with the three-branch PADC, along with the fitted 14.5 GHz signal (curve). The frequency spectrum is shown in Fig. 21.4b. The measured SINAD and ENOB are 29.30 dB and 4.57 bit, respectively. The calculated RMSE for the captured and interleaved sampling points was 0.048. To evaluate the performance against conventional EADCs, the 14.5 GHz signal was received, sampled, and directly digitized using an EADC, as illustrated in Fig. 21.4c, d. The input power to the EADC was kept constant in both tests to ensure a fair comparison. The resulting SINAD and ENOB were 20.95 dB and 3.18 bit, respectively. The sampled points had an RMSE of 0.14 compared to the fitted curve. The spectrum of the reconstructed signal, shown in Fig.21.4b, exhibits slight distortions between 3 and 6 GHz. These distortions may be due to the asynchronous measurement of the three branches,

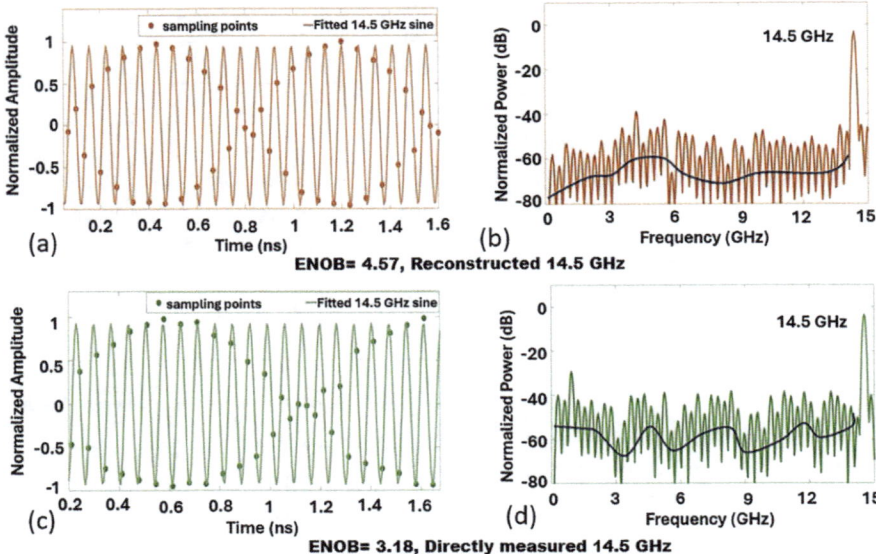

Fig. 21.4 (a) Time-domain and (b) frequency-domain representations of the 14.5 GHz signal sampled and reconstructed using the two-stage PADC, compared to direct measurements shown in (c) and (d) obtained with a high-bandwidth EADC. The frequency spectrum was derived from approximately 1000 sample points

possibly causing a small time mismatch and explaining the minor discrepancies with the simulation results.

21.2.2 Simulation Setup and Results

Figure 21.5 illustrates the simulation setup, with the left block showing the signal generation process and the right block depicting the sampling configuration. We simulated systems with three and nine branches. In the nine-branch system, a single modulator is capable of being driven by four equidistant radio frequencies. The simulation assumes the presence of an additional MZM in each branch, which further reduces the bandwidth requirements of the MZM. This second MZM must operate at three times the frequency of the first, achieved using either an extra source or a frequency tripler. The extension for the nine-line system is represented by dashed lines in Fig. 21.5.

To adhere to IEEE standards for ADC characterization, we used a sine wave as the sampling signal. This sine wave was generated by an oscillator and converted to the optical domain via the first MZM. A continuous wave (CW) laser served as the laser source. The signal intended for sampling was amplified by an EDFA and then passed through a band-pass filter to eliminate ASE noise. Subsequently, the

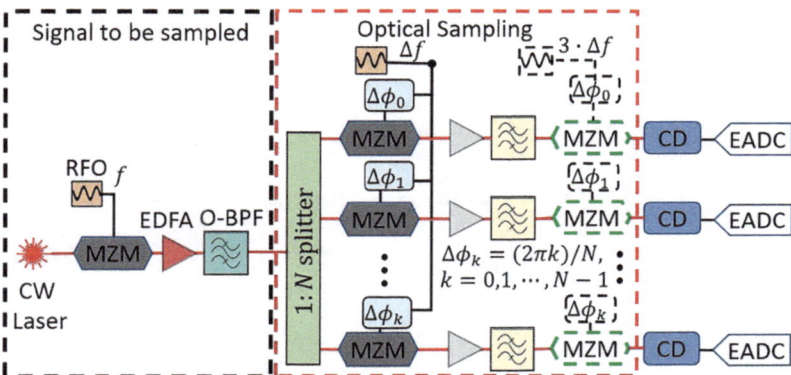

Fig. 21.5 Simulation configuration for the proposed PADC. Dashed lines represent additional MZMs and an extra source for a nine-branch PADC, with the option of using a frequency tripler as an alternative. CW, continuous wave; RFO, single tone radio-frequency oscillator; MZM, Mach–Zehnder modulator; EDFA, erbium-doped fiber amplifier; O-BPF, optical band-pass filter; CD, coherent detection; EADC, electronic analog-to-digital converter

signal was split into $N = 3$ or $N = 9$ branches using a power splitter. In the three-branch system, subsampling occurs in three parallel branches, with each branch using an additional MZM driven at a sinusoidal frequency of 10 GHz and phase shifts of 0°, 120°, and 240°, respectively, achieving a total sampling rate of 30 GS/s. For the nine-branch system, two cascaded MZMs in each branch were driven at 14 GHz for the first and 42 GHz (3×14) for the second, with appropriate phase shifts, as indicated by the dashed lines in Fig. 21.5. This configuration achieves a total sampling rate of 126 GS/s. An additional EDFA compensates for the power loss from the power splitter and MZM insertion loss. In all setups, the EDFA output power is adjusted to ensure that the input power to the electronic ADC matches that of a direct measurement performed with a high-bandwidth EADC, enabling a fair comparison. Following amplification, the signal is detected by a low-bandwidth CD and then sampled and analyzed using a low-bandwidth EADC and digital signal processing (DSP). For the 14.5 GHz signal, three branches operate at 5 GHz, while for the 62.5 GHz signal, nine branches operate at 7 GHz.

Initially, to ensure our simulation aligns with the experiment described in Sect. 21.2.1, we configure the simulation parameters to replicate the experimental conditions. All noise parameters were matched to that of the experiment. Unless otherwise stated, other simulation parameters were maintained at their default settings in the Optisystem software. The CW laser was given a linewidth of 100 kHz, and the RIN was set to −135 dBc/Hz. The time jitter for the RFO was set at 75 fs, while the EADC had a jitter of 1 ps, as per the datasheet. We assumed the EDFA had a noise figure of 4 dB and the CD had a dark current of 5 nA. The measured noise floor in the simulation, consisting solely of shot noise and thermal noise, was around −160 dBc/Hz.

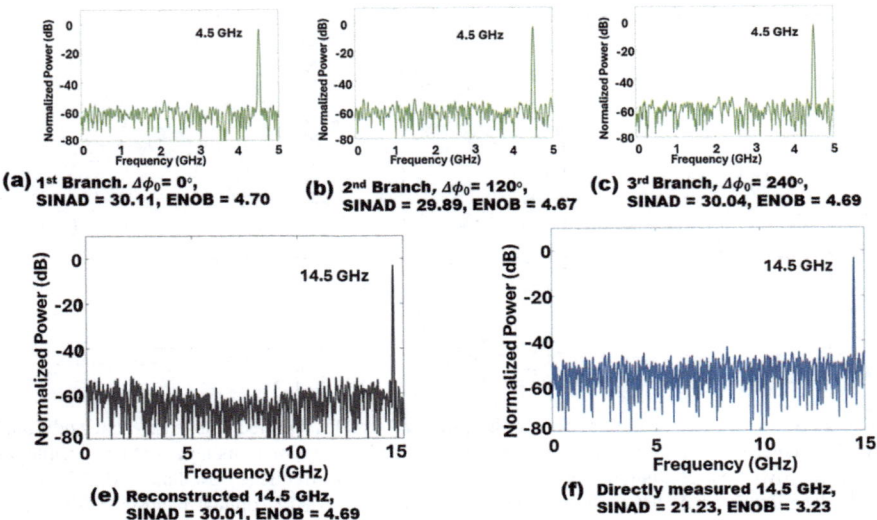

Fig. 21.6 Simulation results of the normalized spectrum for a three-branch system in (**a**)–(**c**), illustrating the down-conversion of a 14.5 GHz signal into three 4.5 GHz signals. The normalized spectrum of the reconstructed signal is shown in (**e**), with the directly detected 14.5 GHz signal provided for comparison in (**f**). The spectra were obtained using the FFT of 4095 sample points for (**a**)–(**c**) and 8190 sample points for (**e**) and (**f**)

The results for down-converting a 14.5 GHz signal into three 4.5 GHz sub-signals and measuring them with a 5 GHz EADC are shown in Fig. 21.6a, c. Each branch's modulator was driven at a frequency of 10 GHz with phase shifts of 0°, 120°, and 240° for the first, second, and third branches, respectively. Each branch demonstrated a SINAD of approximately 30.01 dB and an ENOB of 4.69 bit. We ensured correct ADC characterization by employing a linearly biased MZM and coherent detection, using the full voltage swing. The sampling points from all three branches were time-interleaved to reconstruct the 14.5 GHz signal, as shown in Fig. 21.6e. This reconstructed signal also had a SINAD of 30.01 dB and an ENOB of 4.69 bit. For comparison, a simulation of the direct measurement of the 14.5 GHz signal with a 15 GHz EADC is shown in Fig. 21.6f, where the SINAD and ENOB were 8.78 dB and 1.46 bit worse, respectively. These results match well with the experiment presented in Sect. 21.2.1.

To explore the capability and potential of the method, further simulations were conducted using parameters aligned with the best-in-class components currently available on the market. The oscillator's time jitter was set to 10 fs, as provided by the Keysight E8257D UNY. For the laser, a RIN of −160 dBc/Hz was assumed, similar to that of models like the ULN15PC and CWL-Pxx-WLxx-168-1-FC, with a linewidth of 100 kHz. The time jitter for the EADC was assumed to be 100 fs [16]. A signal under test of 14.5 GHz was sampled with three-branch PADC, which resulted in an ENOB of 7.80 bit. Compared to the EADC, the PADC has a 1.5

Table 21.1 Result summary

	14.5 GHz signal to be sampled			62.5 GHz signal to be sampled
	Experiment	Simulation	Best-in-class simulation	Best-in-class simulation
Sampled with	EADC			EADC
SINAD, ENOB	20.95 dB, 3.18 bit	21.23 dB, 3.23 bit	39.68 dB, 6.3 bit	27.04 dB, 4.20 bit
Sampled with	PADC with three branches			PADC with nine branches
SINAD, ENOB	29.30 dB, 4.57 bit	30.01 dB, 4.69 bit	48.73 dB, 7.80 bit	43.41 dB, 6.92 bit
Improvement	1.4 bit	1.46 bit	1.5 bit	2.72 bit

bit ENOB and 9.05 dB SINAD improvement. A high-frequency 62.5 GHz signal was tested with a nine-branch PADC system at 126 Gs/s, which showed an ENOB of 6.92 bit. Comparing this value to an EADC, 2.72 bit ENOB and 16.07 dB SINAD improvements were achieved. Table 21.1 presents a summary of the results from the experiments, simulations, and simulations using best-in-class component parameters.

21.3 Analysis of PADC Performance

The performance of the PADC system was evaluated through additional simulations, examining key parameters including the used laser's RIN (Sect. 21.3.1), the time jitter of both the optical sampling stage and the electronic ADC (Sect. 21.3.2), and the number of branches (N) in the PADC (Sect. 21.3.3). For analysis and metrology purposes, the best-in-class parameters are implemented here which were mentioned in Sect. 21.2.2.

21.3.1 Laser Relative Intensity Noise

For a wireless or electrical signal to be sampled, it must be converted to the optical domain by modulating it on an optical carrier via an optical modulator as depicted in the left block of Fig. 21.5. In such cases, the relative intensity noise (RIN) of the laser becomes an important factor that should be investigated. In the laser simulation model, the RIN is assumed to be perfect white noise, meaning it presents a constant noise level across all considered frequencies. The mean-square optical intensity fluctuation, $< \Delta P^2 >$, was estimated from the RIN with the equation

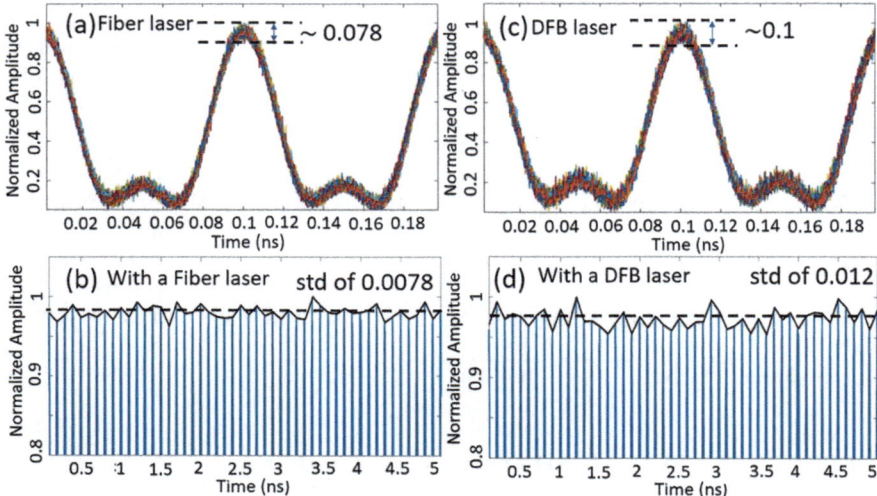

Fig. 21.7 (**a**) Sinc-pulse sequence recorded for a three-line frequency comb with 10 GHz spacing generated by a fiber laser. (**b**) Displays the variation of the normalized amplitude over time. Similarly, (**c**) and (**d**) illustrate the results for a DFB laser. std, standard deviation (The figures have been taken from [6])

$RIN = \frac{<\Delta P^2>}{P_m^2}$, where P_m^2 is the measured power, set to 5 dBm, and the RIN represents the average relative intensity noise in dBc/Hz.

To analyze the RIN effect, we examined the generation of sinc-pulse sequences using both a distributed feedback (DFB) laser and a more stable fiber laser. The findings are presented in Fig. 21.7. The data reveals that the laser source's noise impacts the quality of the initial sampling stage. Despite the fiber laser having a standard deviation an order of magnitude better than that of the DFB laser (see Fig. 21.7b, d), the amplitude noise of the generated sequences remains nearly identical for both lasers (see Fig. 21.7a, c).

For wireless and electrical signals, the RIN influences the quality of sampling in the first stage, significantly affecting the SINAD and ENOB as depicted in Fig. 21.8. Performance degrades sharply when the RIN exceeds −145 dBc/Hz. Although lasers such as the CWL-Pxx-WLxx-168-1-FC and the fiber laser used in the experiment exhibit RIN levels around −135 dBc/Hz, demonstrating good performance at approximately 10 MHz.

21.3.2 Time Jitter

The discussed PADC system, whose design was shown previously in Sect. 21.2, involves two distinct sampling stages: optical orthogonal sampling in the first stage

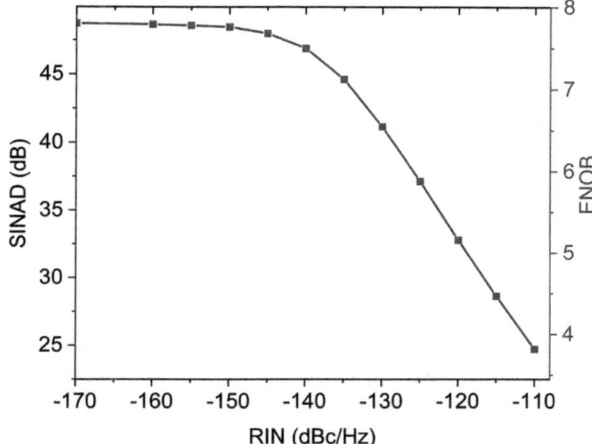

Fig. 21.8 Simulated impact of RIN on the SINAD and ENOB for sampling a 14.5 GHz signal across three branches (ripple of 0.01 dB, with jitter of 10 fs in the first stage and 100 fs in the second stage)

and electronic sampling in the second stage with a conventional low-bandwidth EADC. Both stages are subject to time jitter; however, as noted, the optical sampling relies on multiplication rather than switching. Hence, the optical sampling jitter only comes from the clock jitter of the RFO used to generate the optical sinc-pulse sequences, while the EADC suffers from an aperture jitter in addition to the clock jitter. Consequently, the PADC eliminates aperture jitter during the initial sampling stage, allowing the low jitter of modern oscillators to be directly utilized in sampling high-bandwidth input signals. Even on the integrated platform, a low jitter RFO down to 20 fs was shown [17]. Additionally, microwave oscillators that utilize stabilized mode-locked lasers can achieve extremely low jitter values in the zeptosecond range [18, 19].

Figure 21.9 shows the experimentally measured phase noise and time jitter for the implemented electrical RFO (blue curve) along with, the optical domain sinc-pulse sequences (orange curve), and after optical sampling (blue curve) (sinc-pulse sequence modulated by the sampling points at the frequency of the repetition rate). As can be seen, no additional jitter is added, and the clock jitter of the RFO is directly transmitted to the optical sinc-pulse sequence and optical sampling stage. Further evaluation of the impact of jitter is discussed in more detail in Chap. 22. The jitter values for both the first and second stages were varied from 10 to 100 fs. Even when the jitter values of the EADC and the first sampling stage are matched, there is still room for improvement.

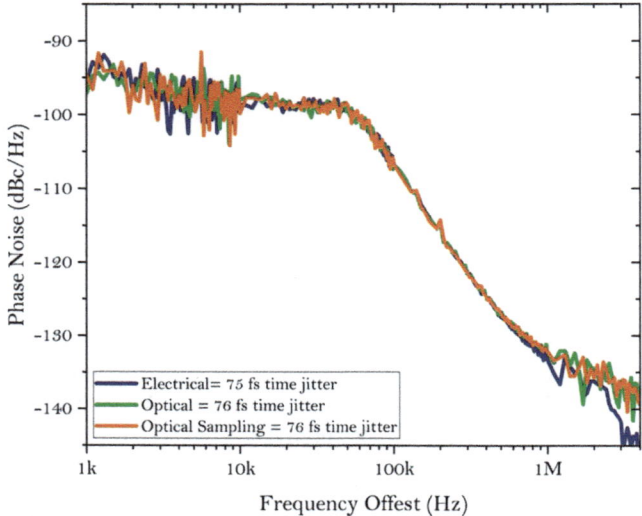

Fig. 21.9 Measured phase noise and jitter for the radio-frequency oscillator (electrical, blue), for generating sinc-pulse sequences with two zero-crossings in an MZM (green), and for sampling a 14.5 GHz sinusoidal signal (orange)

21.3.3 Number of PADC Parallel Branches

The impact of the number of parallel branches (N) in the PADC system on the bandwidth requirement of the back-end devices (e.g., CD, EADC, and DSP bandwidths) and achievable ENOB and SINAD have been analyzed. A 62 GHz signal was selected as signal to be sampled (signal under test), which was then sampled and interleaved by a varying number of branches (N) and a 126 GHz bandwidth frequency comb with N number of lines. Figure 21.10a illustrates the relationship between the number of comb lines (N) and the required bandwidth for both the detector and electronic signal processing necessary to sample the 62 GHz signal. For direct measurement with EADC, the bandwidth matches the input signal at 62 GHz. However, with the signal down-conversion (frequency convolution) in the PADC, this bandwidth is reduced to 20.66 GHz with a three-line comb (three-branch PADC) and further to just 6.88 GHz with a nine-line comb (nine-branch PADC).

The enhancements in SINAD and ENOB with different numbers of comb lines are depicted in Fig. 21.10b. As explained in the Sects. 21.1 and 21.2, the signal effective resolution at the EADC is highly dependent on the input frequency. Thus, by reducing the input frequency from 62 GHz with the direct measurement to 6.88 GHz with a nine-branch PADC, the ENOB value jumps from around 4.25 bit to around 7 bit, providing around 2.75 bit ENOB improvement. It must be noted that the complexity and size of the system scale up with the number of parallel

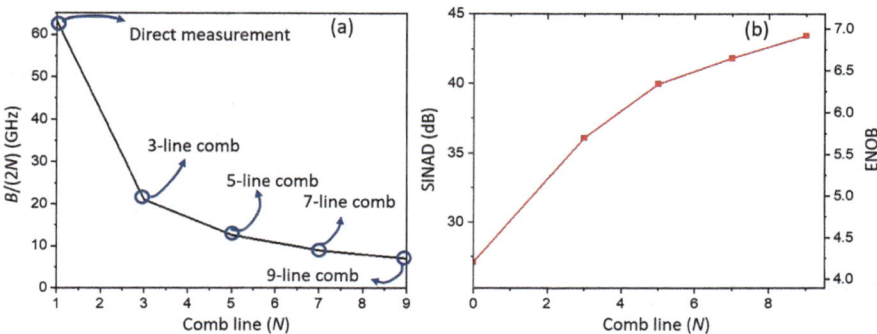

Fig. 21.10 (a) Required bandwidth for the detector and EADC as a function of the number of comb lines (N-branch). (b) SINAD and ENOB plotted against the number of comb lines N (the figures have been taken from [6])

branches. However, especially for integrated devices, this may not be a big issue. If the modulator in the optical sampling stage is powered by a single-tone RFO, the signal's bandwidth will be reduced by a factor of three. The signal's bandwidth can exceed the modulator's RF bandwidth requirement by 50% and potentially even by 300% [14, 20]. As a result, employing a 100 GHz integrated Mach–Zehnder modulator (MZM) [21, 22] enables the sampling of signals with analog bandwidths reaching 150 GHz or even extending to 300 GHz.

21.4   Conclusion

For the upcoming generation of wireless communications, such as 6G and beyond, data rates of up to 1 Tbit/s are anticipated. To meet these escalating requirements, PADC has emerged as a promising solution for high-bandwidth signal reception as an alternative for the bandwidth and effective resolution limited EADC. Its advantages include high bandwidth, energy efficiency, compatibility with CMOS technology, and a compact form factor. In this chapter, we outlined various quality metrics used for performance comparison, including SINAD, ENOB, and RMSE. We then applied these metrics to evaluate the performance of a PADC and compared it with a direct measurement using an EADC. The system testing was done by simulations and proof-of-concept experiments.

The presented PADC utilizes optical sinc-pulse sequences with $N - 1$ zero-crossing to down-convert a high-bandwidth input signal into a low-bandwidth sub-signal in a first-stage optical sub-Nyquist sampling. The sinc-pulse sequences in their simplest form (2 zero-crossing) can be generated from an RFO and a DC bias to modulate an optical carrier with an optical intensity modulator. A 62.5 GHz signal to be-sampled was sampled with a nine-branch PADC and only 7 GHz bandwidth detectors and EADCs. Compared to the direct measurement with

EADC, a 16.07 dB SINAD improvement was calculated along with a 2.72 bit ENOB improvement. We have addressed various sources of error, such as time jitter and laser RIN. Additionally, we showed bandwidth reduction and effective resolution improvements with different numbers of system branches in the PADC. Both simulation and experimental results confirm that PADCs achieve superior SINAD, ENOB, and RMSE compared to the latest EADC technology.

References

1. T. Schneider, Toward terabit receivers for optical and wireless communications. IEEE Commun. Mag. **61**, 169–174 (2023)
2. C. Schmidt, H. Yamazaki, G. Raybon, et al., Data converter interleaving: current trends and future perspectives. IEEE Commun. Mag. **58**, 19–25 (2020)
3. A. Khilo, S.J. Spector, M.E. Grein, et al., Photonic ADC: overcoming the bottleneck of electronic jitter. Opt. Express **20**, 4454 (2012)
4. R.H. Walden, Analog-to-digital converter survey and analysis. IEEE J. Sel. Areas Commun. **17**, 539–550 (1999)
5. Y. Mandalawi, J. Meier, K. Singh, et al., Analysis of bandwidth reduction and resolution improvement for photonics-assisted ADC. J. Lightwave Technol. **41**, 6225–6234 (2023)
6. Y. Mandalawi, J. Meier, M.I. Hosni, et al., Photonics assisted analog-to-digital conversion of wide-bandwidth signals by orthogonal sampling, in *2023 53rd European Microwave Conference (EuMC)* (IEEE, Berlin, 2023), pp. 464–467
7. W. Kester, Aperture time, aperture jitter, aperture delay time (2008). MT-007 Tutor
8. W. Kester, Understand SINAD, ENOB, SNR, THD, THD+N, and SFDR so you don't get lost in the noise floor (2009). MT-003, Rev A, pp. 2–9
9. S. Varughese, J. Langston, V.A. Thomas, et al., Frequency dependent enob requirements for m-qam optical links: an analysis using an improved digital to analog converter model. J. Lightwave Technol. **36**, 4082–4089 (2018)
10. S. Varughese, D. Lippiatt, S. Tibuleac, S.E. Ralph, Frequency dependent ENoB requirements for 400G/600G/800G optical links. J. Lightwave Technol. **38**, 5008–5016 (2020)
11. M.A. Soto, M. Alem, M.A. Shoaie, et al., Generation of nyquist sinc pulses using intensity modulators, in *2013 Conference on Lasers and Electro-Optics (CLEO)* (2013), pp. 3–4
12. M.A. Soto, M. Alem, M. Amin Shoaie, et al., Optical sinc-shaped nyquist pulses of exceptional quality. Nat. Commun. **4**, 1–11 (2013)
13. S. Preußler, G. Raoof Mehrpoor, T. Schneider, Frequency-time coherence for all-optical sampling without optical pulse source. Sci. Rep. **6**, 1–10 (2016)
14. A. Misra, C. Kress, K. Singh, et al., Integrated source-free all optical sampling with a sampling rate of up to three times the RF bandwidth of silicon photonic MZM. Opt. Express **27**, 29972 (2019)
15. J. Meier, K. Singh, A. Misra, et al., High-bandwidth arbitrary signal detection using low-speed electronics. IEEE Photon. J. **14**, 1–7 (2022)
16. W. Shi, Y. Tian, A. Gervais, Scaling capacity of fiber-optic transmission systems via silicon photonics. Nanophotonics **9**(16), 4629–4663 (2020)
17. S. Levantino, Recent advances in high-performance frequency synthesizer design, in *Proceedings of the Custom Integrated Circuits Conference* (IEEE, Newport Beach, 2022), pp. 1–7
18. X. Xie, R. Bouchand, D. Nicolodi, et al., Photonic microwave signals with zeptosecond-level absolute timing noise. Nat. Photon. **11**, 44–47 (2017)
19. M. Hyun, C.G. Jeon, J. Kim, Ultralow-noise microwave extraction from optical frequency combs using photocurrent pulse shaping with balanced photodetection. Sci. Rep. **11**, 1–7 (2021)

20. A. Misra, C. Kress, K. Singh, et al., Reconfigurable and real-time high-bandwidth nyquist signal detection with low-bandwidth in silicon photonics. Opt. Express **30**, 13776 (2022)
21. B. Pan, H. Liu, H. Xu, et al., Ultra-compact lithium niobate microcavity electro-optic modulator beyond 110 GHz. Chip **100029** (2022)
22. Y. Xue, R. Gan, K. Chen, et al., Breaking the bandwidth limit of a high-quality-factor ring modulator based on thin-film lithium niobate. Optica **9**(10), 1131–1137 (2022)

Chapter 22
Photonics-Assisted Signal Processing Systems Using Parallelization in Time and Frequency Domain

Thomas Schneider

Abstract Data hungry applications like streaming and gaming but as well new services like the Internet of Things, autonomous driving, and Industry X.0 lead to a steady increase in the worldwide data rates. Higher data rates require the processing of higher-bandwidth signals. Standard CMOS-based digital signal processors (DSP), analog-to-digital converters (ADC), and digital-to-analog converters (DAC) are approaching their limits. Due to the increasing losses for higher frequencies, especially the jitter, the power consumption and the accompanied heat dissipation are a severe problem for the processing of high-bandwidth signals. The higher losses lead to a heating of the chip which cannot be dissipated for very small integrated structures. Photonics instead can handle much higher bandwidths. A dense wavelength division multiplexed (DWDM) channel has a typical bandwidth of 40 GHz, and integrated modulators with bandwidths above 100 GHz are commercially available. Therefore, photonics may help to keep pace with extremely high data rates even for future applications. The basic idea of photonics assisted signal processing is to use photonics for the down-conversion of high-bandwidth signals into parallel low-bandwidth ones and to use standard electronic CMOS based ADC, DAC, or DSP with a much lower bandwidth to process these signals in parallel branches.

22.1 Introduction

The basic idea behind time- and frequency-domain parallelization is shown in Fig. 22.1. In (a), the input signal is down-converted into parallel low-bandwidth time-domain signals. This down-conversion can be made with mode-locked lasers, in integrated ring modulators or even with time lenses, or time stretching and will be discussed in detail in Sect. 22.3 of this chapter.

T. Schneider (✉)
Technische Universität Braunschweig, THz-Photonics, Braunschweig, Germany
e-mail: thomas.schneider@tu-braunschweig.de

© The Author(s) 2026
T. Kürner et al. (eds.), *Metrology for THz Communications*, Springer Series in Optical Sciences 256, https://doi.org/10.1007/978-3-032-01986-8_22

273

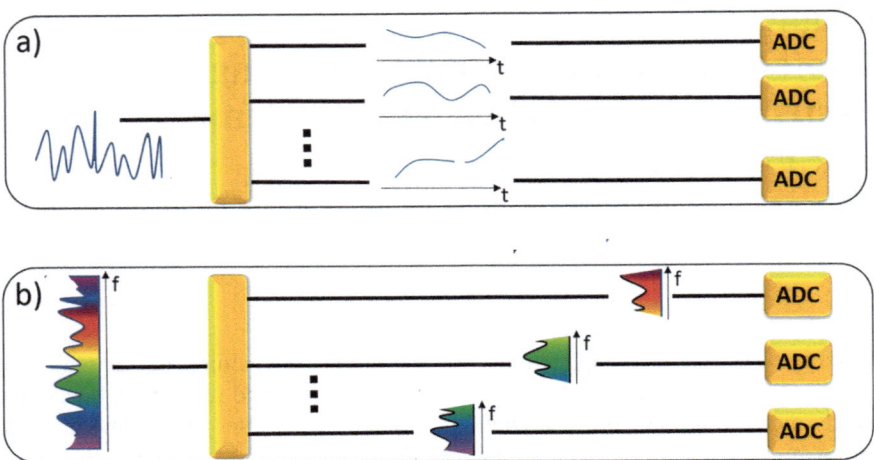

Fig. 22.1 Time- (**a**) and frequency-domain (**b**) parallelization of broadband signals into parallel, low-bandwidth sub-signals, after Ref. [1]

The frequency-domain parallelization is depicted in Fig. 22.1b. This can be directly achieved in the frequency domain by optical filters, for instance, and will be discussed in Sect. 22.2 of this chapter.

22.2 Frequency-Domain Parallelization

The basic idea behind frequency-domain parallelization, or spectral slicing, is straightforward and depicted in Fig. 22.2a. The broadband input spectrum will be divided into parallel low-bandwidth sub-spectra by a bank of overlapping optical filters. These sub-spectra are then processed with low-bandwidth electronics. Since for the reconstruction of the high-bandwidth signal by post-processing the whole information is required (amplitude and phase), the electronic detection must be carried out by a coherent detector (CD). Normal photodiodes, like every optical instrument, can only detect the intensity, which is the squared field information. Therefore, the phase information is lost. A coherent detector instead translates the phase information of the input signal-to-be-measured into an intensity information by interference with a local oscillator wave. This intensity, which depends on the phase information, as relative phase to that of the local oscillator, can then be measured with photodiodes.

In Fig. 22.2a, the local oscillator signal for each sub-spectrum is a single line of an optical frequency comb. Such optical frequency combs can be generated from a single laser line by the nonlinear effect of four-wave mixing in an integrated ring modulator, for instance. Integrated frequency comb generators are commercially available. For the single sub-spectrum, the local oscillator wave should be close

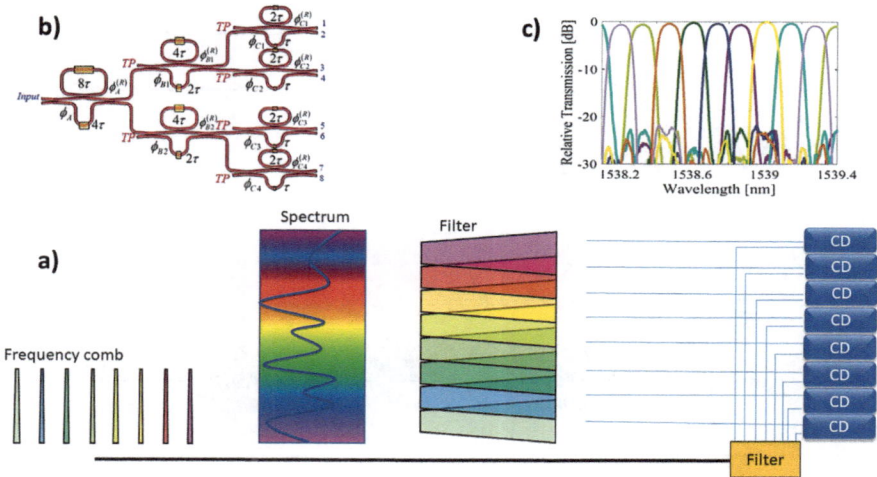

Fig. 22.2 Frequency-domain parallelization with a bank of overlapping rectangular optical filters and a frequency comb, known as spectral slicing [2]. In (**b**), the layout of a silicon-photonic eight-channel bank of filters with a filter spacing of 0.133 nm (17 GHz) is shown. It consists of seven unbalanced Mach-Zehnder interferometers with a nested ring resonator element in a cascaded tree topology [3]. The measured transfer function of the bank of filters, after adjusting the transfer function of each single filter, is shown in (**c**). CD, coherent detector

to the center of the spectrum. Therefore, the frequency spacing of the comb must coincide with the frequency spacing between the filters in the bank of filters, and the single local oscillator line must be filtered out, with another bank of filters, for instance.

The layout and measured transfer function of such an integrated bank of filters is shown in Fig. 22.2b, c [3]. As can be seen, the adjacent filter curves overlap each other on both sides. This overlapping is necessary to stitch the different sub-spectra together in the electronic post-processing. The post-processing of the signals is rather complex, and for this post-processing, the filter curve of the optical filter has to be known.

22.3   Time-Domain Parallelization

22.3.1   Down Sampling

For down sampling, usually mode-locked lasers (MLL) can be used [4]. Such MLL produce a sequence of very short pulses, which can be seen as a Dirac-Delta sequence. For sampling, the signal to be sampled is multiplied with the Dirac-Delta sequence of the MLL. If the intensities are low, only linear optical effects

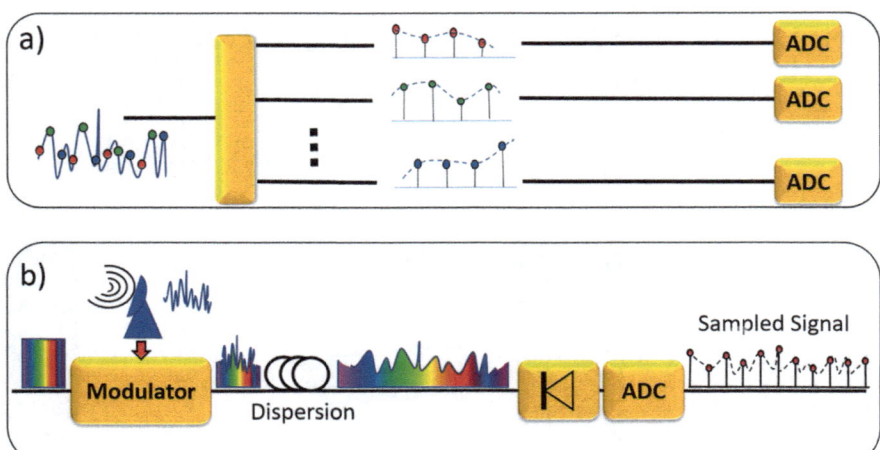

Fig. 22.3 Time-domain parallelization of high-bandwidth signals with a mode-locked laser (**a**) and with a time stretcher or a time lens (**b**), after Ref. [1]

are happening, and the interaction between the different optical waves in a material follows the superposition principle. Therefore, the waves can add up, they can interfere, but no nonlinear interaction, like a multiplication, can take place. For the alteration of the signal by the sequence, high intensities and a nonlinearity are needed [1]. This nonlinearity can be four-wave mixing in a nonlinear crystal or in a highly nonlinear fiber, for instance. Another possibility is to use the nonlinear function of a photodetector.

The main advantages of MLL are their very low pulse durations, which can be a few femtoseconds, and the extremely low jitter values which can be achieved with stabilized lasers. These jitter values can even be in the zeptosecond range [18]. This has made optical sampling with MLL a standard measurement technique for metrology applications. However, the basic idea behind real-time sampling of a high-bandwidth signal is shown in Fig. 22.3a. The high-bandwidth signal is sampled with a sampling rate much lower than twice the highest bandwidth in the signal (Nyquist rate) in each single branch. From branch to branch, the position of this subsampling is slightly shifted, so that by bringing all branches together, the signal is sampled with at least the Nyquist rate, and the whole information is retrieved from the signal.

Mode-locked lasers with a very low jitter require a precise stabilization of the cavity length and the carrier envelope phase (CEP). In MLL, typically operating at 800 nm (Titanium Sapphire) or 1550 nm (optical fiber), the pulses can become so short that only a few wavelengths of the carrier fit under the pulse envelope. So, the phase between the envelope of the pulse and the carrier might change from pulse to pulse. This can be stabilized very precisely by measuring the beat note between the first and last frequency in a comb that spans over an octave. The cavity length can be stabilized by several mechanical means. However, all of this requires space, making the resonator length quite long since the longer the resonator, the lower is the repetition rate of the MLL. So, typically stabilized MLL have repetition rates of a few hundred MHz. Therefore, for the real-time sampling of a 40 GHz signal with 100 MHz MLL, 400 branches would be required. Additionally, the time shift between adjacent branches has to be adjusted very precisely.

The aforementioned properties make MLL very interesting for metrology and for the measurement of repetitive signals, but very impractical for the real-time measurement of arbitrary high-bandwidth signals. Recently, integrated frequency comb sources with repetition rates in the GHz range have been shown [5], and in the meanwhile, they are commercially available. However, the very precise jitter stabilization of these devices has to be shown.

22.3.2   Time Lenses and Time Stretching

The basic idea behind time stretching and time lenses is shown in Fig. 22.3b. First, the high-bandwidth signal is sliced into time windows. One of these windows for a wireless or THz signal is shown in the figure as the input from a wireless antenna. This time window is then transferred to the optical domain by modulating its time function onto a frequency spreaded optical carrier. This carrier can be a short pulse, for instance. Due to the time-frequency principles, the shorter the pulse, the broader is its spectrum. This is shown with the different colors in Fig. 22.3b. These short pulses can be stretched in time by a dispersive optical element [6]. The dispersion of optical devices is a result of the frequency dependence of the refractive index, i.e., different frequencies or wavelengths see slightly different refractive indices. The phase velocity of the optical wave is inversely proportional to the refractive index. Therefore, different frequencies in the short pulse may propagate with different velocities, which eventually leads to a broadening of the pulse.

Once the pulse is as broad as the time window, it can be modulated onto that pulse, as can be seen after the modulator in Fig. 22.3b. Now, another dispersion can be used to stretch the pulse even further. Since the time window is modulated on the pulse, this stretching leads as well to the stretching of the window and the information modulated on it. Finally, the stretched time window can be converted back to the electrical domain, and the information, which is now much longer in time, can be sampled with low-bandwidth electronics.

For the stretching only, the first-order dispersion is required but, usually this first-order dispersion is accompanied by higher-order dispersion terms, which may lead to a distortion of the signal. A time stretching without dispersion can be seen in Fig. 22.4

A short pulse (Fig. 22.4a) enters a ring resonator. The pulse propagates in the ring, and each time it crosses the outcoupling stage, a copy of the pulse is coupled out (Fig. 22.4c). This corresponds to a multiplication of the pulse spectrum with a frequency comb, as depicted in Fig. 22.4b. In the next step, these pulses are

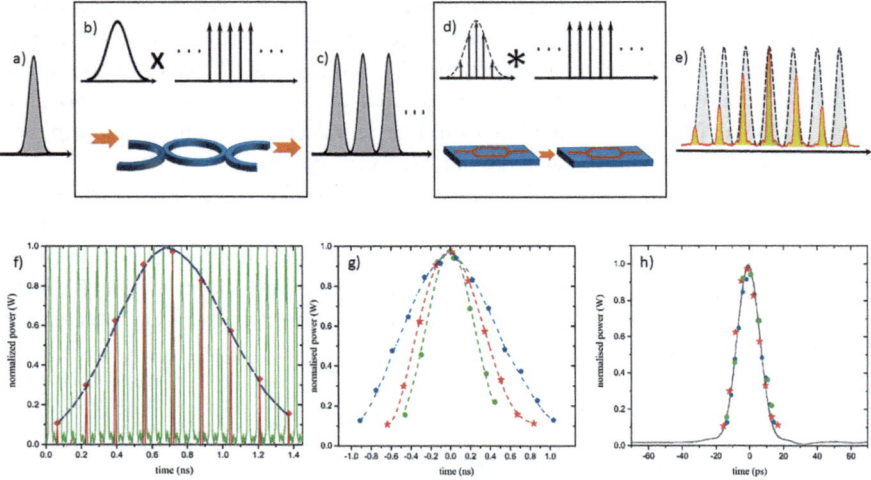

Fig. 22.4 Dispersionless time magnifier with some measured results for short pulses [7]

sampled by orthogonal sampling (which will be described in detail in Sect. 22.4). The sampling rate is a little bit different compared to the pulse repetition rate. Therefore, different copies of the pulse are sampled at different positions, as can be seen with the measured results in Fig. 22.4f, g, and h. The result is a broadened version of the input pulse. Since the broadening depends on the difference between the repetition rate of the pulse copies and the sampling rate, which are both known, the exact pulse shape can be reconstructed, as presented in Fig. 22.4h.

22.4   Orthogonal Sampling

Orthogonal sampling is a kind of time-domain parallelization with sinc-pulse sequences (SPS) [8, 9, 19, 20]. As stated by the sampling theorem, every signal limited in bandwidth corresponds to the superposition of single sinc pulses $sin(x)/x$ which are time-shifted to each other and weighted with the sampling points, as shown in Fig. 22.5a.

But a single sinc pulse cannot be generated in practice, since it is unlimited in time. A sinc-pulse sequence (SPS) instead is the unlimited superposition of single, time-shifted sinc pulses and corresponds to a rectangular frequency comb in the frequency domain. Therefore, SPS can be generated very easily but their main advantage is that the bandwidth limited signal can as well be seen as a superposition of SPS, each of which weighted with periodical sampling points, as shown in Fig. 22.5b.

As single sinc pulses, SPS are orthogonal if they are time-shifted against each other, so that all maxima of each single SPS are in the zero-crossings of all other

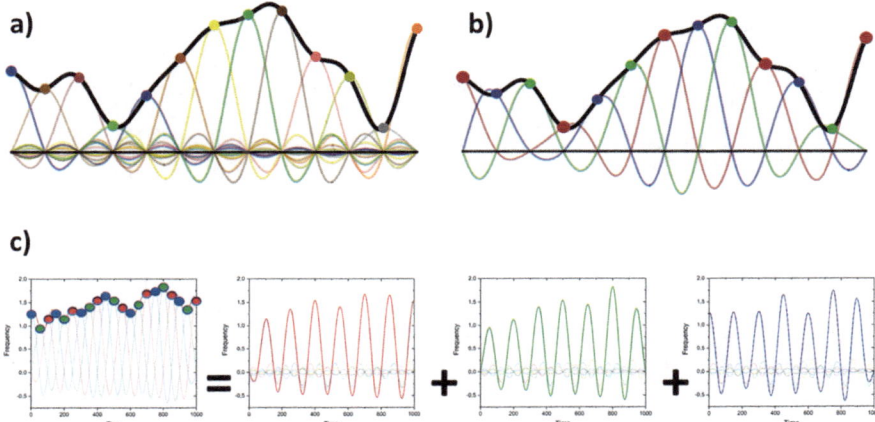

Fig. 22.5 Construction of a bandwidth limited signal by single sinc pulses (**a**) and by sinc-pulse sequences (SPS) in (**b**). The bandwidth limited signal in (**c**) is the superposition of the three SPS (red, green, and blue), each of which weighted with the respective sampling values and time-shifted to each other so that the green SPS is in the first zero-crossings of the red one, whereas the blue SPS is in the first zero-crossing of the green and in the second of the red SPS

SPS. The main difference between describing the signal with SPS or with single sinc pulses is that for the latter each sampling point requires a single sinc pulse, whereas SPS encode periodical sampling points.

The generation of a bandwidth limited signal as the superposition of three SPS is shown in Fig. 22.5c. As depicted, each of the three SPS (red, green, and blue) encodes periodical sampling points. The three SPS are time-shifted against each other, so that the green one is in the first zero-crossings of the red and the blue in its second. As can be shown mathematically, the description of the signal as a superposition of weighted SPS is error-free [10, 19].

22.4.1 Sinc-Pulse Sequences

The similarities and differences between single sinc pulses (top) and sinc-pulse sequences (bottom) with two zero-crossings are shown in Fig. 22.6. A single sinc pulse has a restricted, perfectly rectangular spectrum. Therefore, in the time domain, it is unlimited. The bandwidth of the sinc pulse is defined as the inverse of the time to the first zero-crossing $1/T$. Under the rectangular envelope of the single sinc pulse, all frequencies are present.

An SPS, as shown in the bottom of Fig. 22.6, corresponds to single frequencies under the rectangular envelope. Here, an SPS with two zero-crossings is shown. In the time domain, this simply corresponds to a direct current (DC)-shifted sinusoidal signal. Consequently, in the baseband, it is the DC at a frequency of zero and the

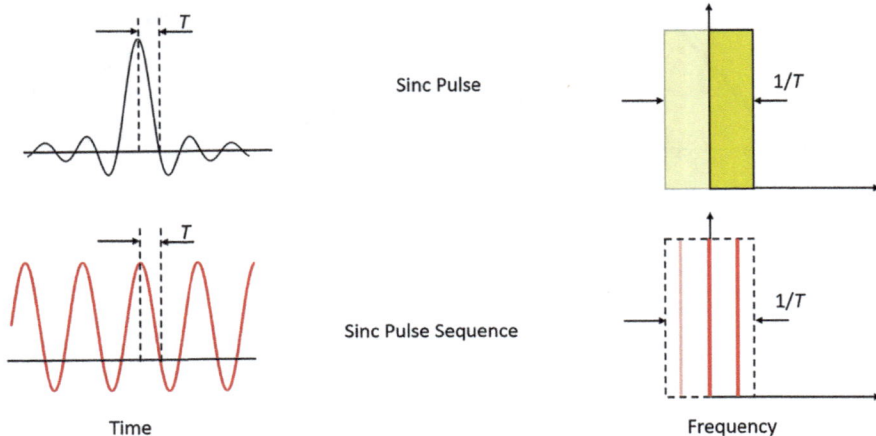

Fig. 22.6 Similarities and differences between single sinc pulses (top) and sinc-pulse sequences (bottom) in the time (left) and frequency domain (right). The transparent parts of the spectrum show the negative mirror frequencies which appear below the carrier after modulation

sinusoidal. After modulation onto a carrier, the left sideband appears as a lower frequency sideband below the carrier.

Of course, an SPS can consist of more than just a DC and a single sinusoidal; this is shown in Fig. 22.7. As can be seen on the left side of Fig. 22.7a, the SPS shown on the right is the unlimited superposition of single sinc pulses, time-shifted to each other. Each single sinc pulse is unlimited in time as well. This superposition is only possible since the single sinc pulses are orthogonal to each other. Without orthogonality, the single functions would interfere with each other, and the result would depend on the phase relationship between them. In communications, that phenomenon is called inter-symbol interference (ISI). However, due to the orthogonality, there is no ISI, and each single sinc pulse is completely independent of all other sinc pulses in the sequence.

In the frequency domain, the SPS is a rectangular frequency comb as shown in Fig. 22.7b. The number of frequency lines N defines the number of zero-crossings $N - 1$ in between two pulses. In each of the zero-crossings, the SPS is orthogonal to another SPS, time-shifted so that its maxima lie in the zero-crossings. Therefore, to fill all zero-crossings, an SPS with N comb lines requires N time-shifted SPS. If all these single SPS are weighted with the sampling points of the signal, the superposition can define the signal without any error. This can be used for the generation and detection of very high-bandwidth signals with low-bandwidth electronics, as will be shown in the Sects. 22.4.2 and 22.4.3.

Of course, an unlimited superposition of unlimited functions is practically as impossible as the generation of a single unlimited function. However, in the experiments, the single frequency lines are generated by oscillators running for

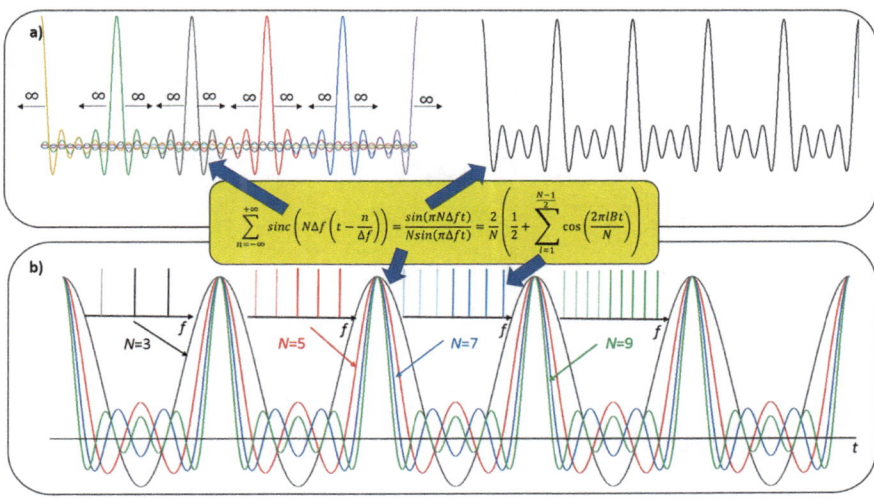

Fig. 22.7 Construction of a bandwidth limited signal by single sinc pulses (**a**) and by sinc-pulse sequences (SPS) in (**b**). The bandwidth limited signal in (**c**) is the superposition of the three SPS (red, green, and blue), each of which weighted with the respective sampling values and time-shifted to each other so that the green SPS is in the first zero-crossings of the red one, whereas the blue SPS is in the first zero-crossing of the green and in the second of the red SPS

hours or days. Compared to the single pulse with a duration of a few picoseconds ($1\,\mathrm{ps} = 10^{-12}\,\mathrm{s}$), this can be seen as practically unlimited.

22.4.2 Orthogonal Sampling for DAC

How the orthogonal sampling can be used to generate high-bandwidth signals from low-bandwidth electronics is shown in Fig. 22.8. In principle, it follows the basic idea already presented in Fig. 22.5c; three SPS each of which weighted with 1/3rd of the sampling values are superimposed to generate the high-bandwidth signal. Of course, any number of N can be used. N defines the bandwidth and sampling rate reduction, but N defines as well the number of required branches. Therefore, with N, the requirements on the electronics decrease, but the complexity increases. However, especially for integrated devices, this complexity is much less severe than the bandwidth requirements for high-bandwidth electronics.

The device in Fig. 22.8 is a digital-to-analog converter (DAC) or an arbitrary signal generator (AWG). The digital values for the generation of the high-bandwidth signal are first divided by N (here $N = 3$). In each of the N branches, an electronic DAC with 1/Nth of the sampling rate and bandwidth is fed with 1/Nth of the sampling values and generates an analog signal with 1/Nth of the bandwidth.

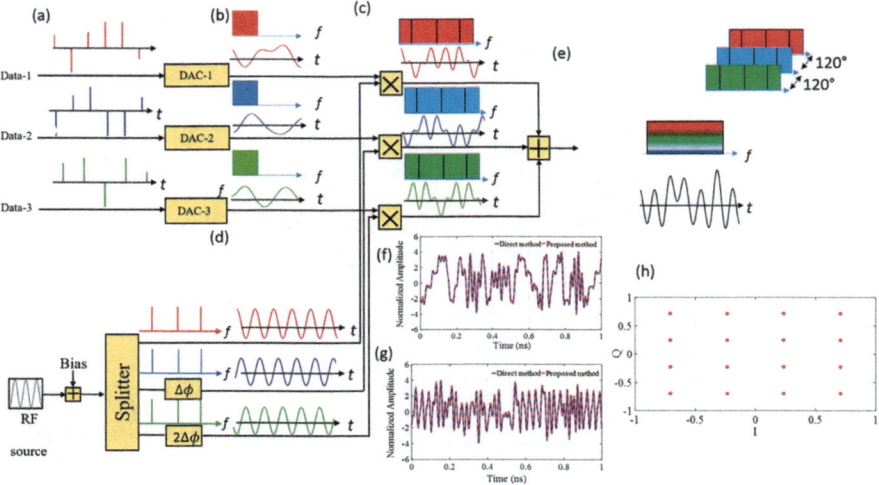

Fig. 22.8 Arbitrary waveform generation by orthogonal sampling. The subsampling points (**a**) are used to generate analog sub-signals (**b**) with low-bandwidth electronic digital-to-analog converters DAC (**d**). In each branch, the analog sub-signal is multiplied with a time-shifted SPS with two zero-crossings (**c**), generated by a radio frequency oscillator (RF) and an additional DC bias. Lastly, the high-bandwidth signal (**e**) is generated by adding up the signals from all three branches. $I(f)$ and $Q(g)$ results achieved by this method for a QAM-16 signal limited to its Nyquist bandwidth of 60 GHz with a bitrate of 480 Gbit/s are shown in (**f**), (**g**), and (**h**). The direct generation of the signal is represented by the blue trace shows, whereas the signal generation with three 20 GHz sub-DACs is shown by the red, and the corresponding constellation diagram is shown in (**h**) [20]

For the orthogonal sampling, SPS are required; for $N = 3$, they can be generated by a sinusoidal signal from a radio frequency generator (RF) together with a DC bias. To provide three time-shifted SPS, the input SPS is power-splitted into three branches and time-shifted against each other. This time shift is just a phase shift of this sinusoidal. So, in the first branch, the phase shift $\Delta\Phi$ is zero, in the second, it is $\Delta\Phi = 120°$, and in the third, it follows $\Delta\Phi = 240°$. This phase change shifts the SPS so that the blue one is in the first zero-crossings of the red and the green one in its second.

In the next step, the three time-shifted SPS are multiplied with the low-bandwidth signals. The results are the three weighted SPS shown in (c) in the time and frequency domain. If we assume a Nyquist signal, the bandwidth of the single low-bandwidth sub-signal is rectangular, as depicted in Fig. 22.8b. The multiplication in the time domain is a convolution in the frequency domain. Therefore, in each branch, the rectangle is convoluted with a rectangular three-line frequency comb. This results in a spreading of the spectrum of the low-bandwidth sub-signal to that of the high-bandwidth signal. In the last step, the sub-signals from all N branches are added up, forming the high-bandwidth signal.

Simulation results for the generation of a QAM-16, 480 Gbit/s signal, limited to its Nyquist bandwidth of 60 GHz, by three branches are shown in Fig. 22.8f, g, and

h. The I and Q traces of the signal can be seen in (f) and (g), respectively. The direct generation of the 60 GHz signal with a 60 GHz electronic DAC is shown by the blue trace, and the generation of the signal with three 20 GHz sub-DACs is depicted by the red one. In the simulation, the DACs were assumed as ideal; therefore, the red and blue trace are almost identical. However, due to jitter, higher frequency DAC have a very low effective number of bit (ENOB), and they show problems with the linearity for increasing bandwidths. Therefore, it can be assumed that the presented orthogonal sampling method for the generation of high-bandwidth signals enables the generation of higher-quality signals.

22.4.3 Orthogonal Sampling for ADC

The sampling points can be retrieved by multiplying the high-bandwidth signal with a sinc-pulse sequence with the right bandwidth and time shift. The required bandwidth of the SPS is at least twice that of the signal to be sampled in the baseband (or the same as the optical bandwidth). For the sampling, the next SPS has to be in the first zero-crossings of the previous one. Since this is the inverse of the bandwidth, the sampling rate is twice the highest bandwidth in the signal, as required by the sampling theorem.

For high-bandwidth optical signal processing, the multiplication can be carried out in a modulator. As modulators, Mach-Zehnder modulators [11], or for integrated, low-footprint, low-power consuming devices, ring modulators [12] can be used. The required radio frequency bandwidth of the modulator can be three [8] or even nine times [14] lower than the bandwidth of the signal. Therefore, with integrated modulators with bandwidths exceeding 100 GHz [15], up to 900 GHz signals can be processed.

The basic block diagram for an orthogonal sampling based ADC is shown in Fig. 22.9. The multiplication with the SPS retrieves the sampling points of the high-bandwidth signal (Fig. 22.9a). The sampled signal in one of the N branches can be seen in Fig. 22.9b. As depicted, the red sampling points of the high-bandwidth input signal are conserved in the still optical signal. These red sampling points represent the low-bandwidth dashed function in Fig. 22.9b. A coherent detector (CD) converts the signal to the electrical domain to retrieve these sampling points. A CD is required since the whole information of the signal (amplitude and phase) is needed. Therefore, the CD has to be driven with a local oscillator signal with a wavelength which corresponds to the center of the signal.

To convert the still high-bandwidth optical signal into a low-bandwidth electrical one, the CD should have a bandwidth which corresponds to the dashed sub-signal in Fig. 22.9b, which is $B/(2N)$, if B is assumed as the optical bandwidth of the input signal. Therefore, the bandwidth (for the baseband signal $B/2$) and at the same time sampling rate of the electronics is reduced by N. To retrieve the blue and green sampling points, two additional branches are required, as shown in Fig. 22.9d, e.

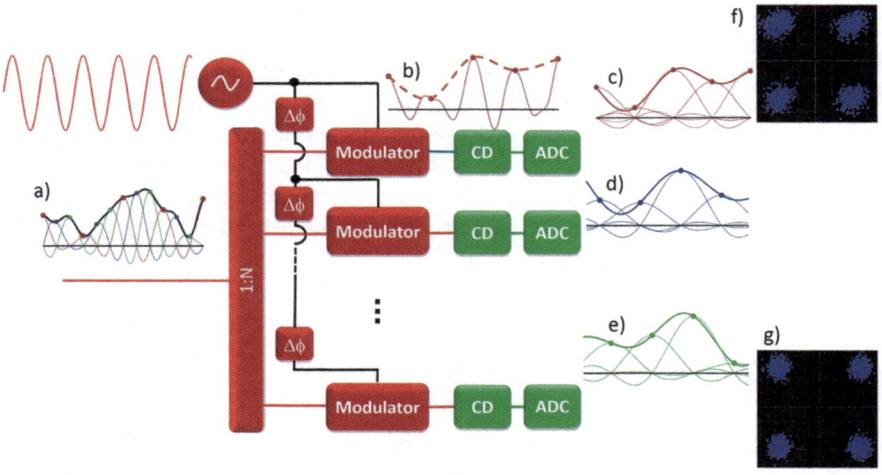

Fig. 22.9 Analog-to-digital conversion (ADC) with orthogonal sampling. In each of the *N* branches, the high-bandwidth signal (**a**) is multiplied with an SPS; the result is a sampling and time-domain interleaving of *N* sub-signals (**c**), (**d**), and (**e**). These sub-signals can now be detected and processed with low-bandwidth electronic devices, after Ref. [1]. The insets (**f**) and (**g**) show the experimental detection of a 24 GBd Nyquist signal (24 GHz bandwidth) with a 24 GHz ADC (**f**) and with three 8 GHz ADC by orthogonal sampling (**g**). Due to orthogonal sampling, a Q-factor improvement of more than 2 dB has been achieved [13]. Red blocks and lines, optical and green electrical functionalities and connections; CD, coherent detector

The insets Fig. 22.9f, g show experimental results for the detection of a 24 GHz, 24 Gbd Nyquist signal with a 24 GHz electronic device (f) or with a three-branch ADC with 8 GHz bandwidth for each detector and ADC in the branch (g). As can be seen, for the low-bandwidth ADC, the noise is lower, and the quality of the received signal is correspondingly higher. This quality enhancement is mainly a result of the lower jitter of the lower-bandwidth electronic ADCs, as already discussed in chapter: Photonics-Assisted Signal Processing.

The jitter of electronic ADC does not come from the clock signals. As discussed, very precise clock sources are available. The jitter is instead a result of the so-called aperture jitter of the sample and hold stage normally used in integrated ADC. This sample and hold can be seen as a kind of switch. After the clock signal has triggered the sample and hold, it is uncertain when it really switches. For state-of-the-art electronic ADC, this uncertainty is in the range of around 100 fs. Due to that jitter, the ENOB and SINAD reduce with increased bandwidth of the signal to be sampled.

Orthogonal sampling instead can be seen as a two-stage sampling. In the first stage, the high-bandwidth signal is down-converted into parallel low-bandwidth sub-signals, which are then measured by lower-bandwidth ADC. The first stage is the multiplication of the high-bandwidth signal with the sinc-pulse sequence. This multiplication is not a switching, and the jitter of low-bandwidth clock signals can be directly transferred to the down-sampled signal [16, 17, 21]. Therefore, the first

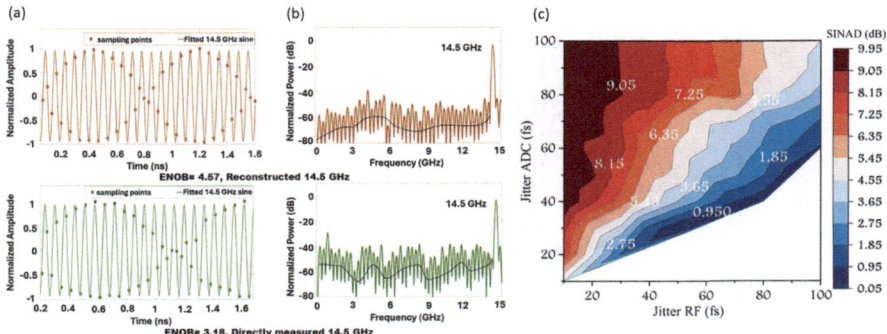

Fig. 22.10 Effective number of bit (ENOB) and signal-to-noise and distortion ratio (SINAD) investigation for orthogonal sampling based ADC [21]. In the upper part of (**a**), a 14.5 GHz sinusoidal signal measured by three-branch orthogonal sampling is shown. The signal Fourier transform is presented in (**b**), which is compared to the broadband direct signal measurement in the bottom of (**a**) and (**b**). In (**c**), the SINAD improvement for the sampling of a 62 GHz sinusoidal signal in a three-branch orthogonal sampling system, compared to a direct measurement of the high-bandwidth signal with a detector with a jitter value of "Jitter ADC," is shown. "Jitter RF" is the radio frequency jitter of the clock signal and "Jitter ADC" is the jitter of the electronic ADC used to measure the signal [21]

stage does not add jitter to the sampled signal, and the measurement quality can be increased.

Figure 22.10 shows a comparison for the measurement of a 62 GHz sinusoidal signal by three-branch orthogonal sampling (top) and with a conventional broadband oscilloscope. In (a), the reconstructed signal is shown, and the fast Fourier transform of the measured signal is presented in (b). As shown, the three-branch orthogonal system enables an ENOB improvement of about 1.5 dB. Consequently, the SINAD improvement is almost 10 dB. The quality improvement may lead to lower requirements for the digital signal processing in the detector. Therefore, the improved signal detection with orthogonal sampling may result in longer distances, higher data rates, or a lower energy consumption of the detection systems.

The ENOB and SINAD improvement is a result of the better jitter of the clock signal compared to the aperture jitter of the electronic devices. In Fig. 22.10c, the SINAD improvement for a three-stage orthogonal sampling system is shown. Here, the "Jitter RF" is the jitter of the clock source which is transferred to the sampling in the first stage, whereas "Jitter ADC" is the jitter of the electronic ADCs. Even integrated clock sources with jitter values of a few fs are available, whereas state-of-the-art electronic ADC have jitter values of around 100 fs. Therefore, for a three-stage orthogonal sampling system, a SINAD improvement of 10 dB can be expected. The improvement increases with the number of stages, as shown by Eq. (22.1) [21]:

$$\text{SINAD} = 10\text{Log}_{10}\left(\frac{1}{\left((\pi f \sigma_{RF})^2 + (2\pi \frac{f}{N}\sigma_{ADC})^2\right)}\right) \quad (22.1)$$

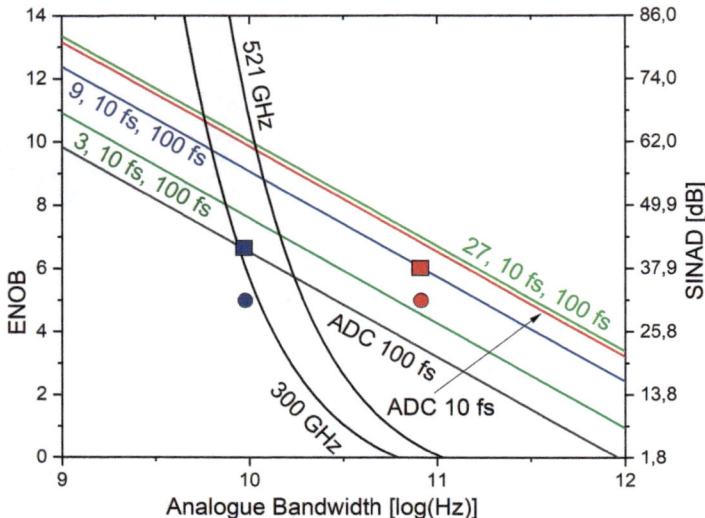

Fig. 22.11 ENOB and SINAD vs. the analog bandwidth for electronic ADCs with a time jitter of 10 and 100 fs and for orthogonal sampling based ADC with 3, 9, and 27 branches (first number). The second number is the jitter of the clock and the third the jitter of the electronic ADC. The requirements for the detection of a 1 Tbit/s signal with 16-QAM are included by the red dot, and the red square represents the resolution of an orthogonal sampling system with nine branches and 80 GHz bandwidth. Such an orthogonal sampling system requires 9 ADC with a bandwidth of only 9 GHz (blue square), and the blue dot depicts the minimum requirements on these ADCs for the reception of a 16-QAM signal [1]

with σ_{RF} and σ_{ADC} as the jitter values for the first and second sampling stage, f is the frequency of the signal to be sampled and N the number of branches. The ENOB values achievable by orthogonal sampling for a different number of jitter values and branches are shown in Fig. 22.11. Impairments due to insertion losses and power splitting have been included. However, not included is a possible further ENOB and SINAD improvement by lower-bandwidth transimpedance amplifiers for the lower-bandwidth photodetectors.

The black and red lines show conventional 100 and 10 fs ADC, and the minimum requirements on a 16-QAM 1 Tbit/s receiver are shown by the red dot. As shown, an orthogonal sampling ADC with 27 branches, 100 fs for the electronic ADCs, and 10 fs jitter for the RF clock can outperform an electronic ADC with 10 fs jitter. But for the reception of a 1 Tbit/s, 16-QAM modulated signal, a nine-branch orthogonal sampling (red square) is sufficient. For this device, modulators with 1/3rd or even 1/6th (14 GHz) of the bandwidth are sufficient, and the electronics (CD and ADC) require a bandwidth of only 9 GHz for the detection of the 80 GHz signal. The resolution of these 9 GHz ADCs is shown with the blue square. As can be seen, for the reception of a 16-QAM signal (blue dot), the offered resolution is higher.

Orthogonal sampling can be straightforwardly integrated into any photonic platform and may offer a solution to address the demands of future Terahertz communition systems with cost-effective, low-power, and low-footprint integrated devices.

References

1. T. Schneider, Toward terabit receivers for optical and wireless communications. IEEE Commun. Mag. **61**(8), 169–174 (2023)
2. N. Fontaine et al., Real-time full-field arbitrary optical waveform measurement. Nat. Photon. **4**, 248–254 (2010)
3. D. Munk et al., Eight-channel silicon-photonic wavelength division multiplexer with 17 GHz spacing. IEEE J. Sel. Top. Quantum Electron. **25**(5), 1–10 (2019)
4. A. Khilo et al., Photonic ADC: overcoming the bottleneck of electronic jitter. Opt. Express **20**(4), 4454 (2012)
5. P. Del'Haye, A. Schliesser, O. Arcizet et al., Optical frequency comb generation from a monolithic microresonator. Nature **450**, 1214–1217 (2007)
6. Y. Zhou et al., A unified framework for photonic time-stretch systems. Laser Photon. Rev. **16**(2100524) (2022). https://doi.org/10.1002/lpor.202100524
7. A. Misra, S. Preußler, L. Zhou et al., Nonlinearity- and dispersion-less integrated optical time magnifier based on a high-Q SiN microring resonator. Sci. Rep. **9**, 14277 (2019)
8. K. Singh, J. Meier, A. Misra, S. Preußler, J.C. Scheytt, T. Schneider, Photonic arbitrary waveform generation with three times the sampling rate of the modulator bandwidth. IEEE Photon. Technol. Lett. **32**(24), 1544–1547 (2020)
9. A. Misra, J. Meier, S. Preußler, K. Singh, T. Schneider, Agnostic sampling transceiver. Opt. Express **29**, 14828–14840 (2021)
10. J. Meier, A. Misra, S. Preußler, T. Schneider, Optical convolution with a rectangular frequency comb for almost ideal sampling, in *Proceedings of SPIE 10947, Next-Generation Optical Communication: Components, Sub-Systems, and Systems VIII* (2019), p. 109470I
11. A. Misra, K. Singh, J. Meier, C. Kress, T. Schwabe, S. Preußler, J.C. Scheytt, T. Schneider, Flexible time-domain de-multiplexing of nyquist OTDM channels by orthogonal sampling in silicon photonics, in *Conference on Lasers and Electro-Optics, Technical Digest Series* (Optica Publishing Group, 2022)
12. M.I. Hosni, A. Vilson, K. Singh, J. Meier, Y. Mandalawi, T. Schneider, Ring modulator based high-sampling rate integrated photonic sampler, in *2023 Conference on Lasers and Electro-Optics Europe, European Quantum Electronics Conference (CLEO/Europe-EQEC)* (2023)
13. K. Singh, J. Meier, M.I. Hosni, Y. Mandalawi, T. Schneider, Enhanced performance high-bandwidth signal detection by time-frequency coherence sampling, in *Frontiers in Optics + Laser Science 2022 (FIO, LS), Technical Digest Series* (Optica Publishing Group, 2022)
14. M. Soto, M. Alem, M.A. Shoaie et al., Optical sinc-shaped Nyquist pulses of exceptional quality. Nat. Commun. **4**, 2898 (2013)
15. P.O. Weigel, J. Zhao, K. Fang et al., Bonded thin film lithium niobate modulator on a silicon photonics platform exceeding 100 GHz 3-dB electrical modulation bandwidth. Opt. Express **26**, 23728–23739 (2018)
16. Y. Mandalawi et al., Photonics assisted analog-to-digital conversion of wide-bandwidth signals by orthogonal sampling, in *2023 53rd European Microwave Conference (EuMC)* (2023), pp. 464–467
17. Y. Mandalawi, M.I. Hosni, J. Meier, K. Singh, S. De, R. Das, T. Schneider, Integrated segmented IQ-modulator for orthogonal sampling and multi-level high-bandwidth signal generation. Opt. Lett. **49**, 2193–2196 (2024)
18. X. Xie et al., Photonic microwave signals with zeptosecond-level absolute timing noise. Nat. Photon. **11**, 44–47 (2017)

19. J. Meier, A. Misra, S. Preußler, T. Schneider, Orthogonal full-field optical sampling. IEEE Photon. J. **11**(2), 1–9 (2019)
20. M.I. Hosni et al., Orthogonal sampling-based broad-band signal generation with low-bandwidth electronics. IEEE Open J. Commun. Soc. **4**, 2930–2938 (2023)
21. Y. Mandalawi, J. Meier, K. Singh, M.I. Hosni, S. De, T. Schneider, Analysis of bandwidth reduction and resolution improvement for photonics-assisted ADC. J. Lightwave Technol. **41**(19), 6225–6234 (2023)

Chapter 23
Metrological Analysis of Non-idealities for Photonics-Assisted Signal Processing

Souvaraj De, Younus Mandalawi, Ranjan Das, and Thomas Schneider

Abstract Waveform metrology provides the key to understand and analyze the waveform behavior of different systems and environments. To ensure reliable measurements with high precision, this is particularly crucial for calibrating and characterizing electrical systems, like high-speed sampling oscilloscopes, spectrum analyzers, waveform generators, etc. Additionally, full waveform metrology techniques make it possible to derive parameters from devices operating in the time domain like oscilloscopes, as well as devices operating in the frequency domain such as vector network analyzers. This ensures a comprehensive system analysis with a complete transfer function with the full measurement uncertainty represented by a covariance matrix. For high-bandwidth signals, especially for Terahertz (THz) or sub-THz communications, photonics-assisted signal processing is advantageous as it offers a much wider bandwidth compared to its electrical counterpart with no electromagnetic interference. As such, it is important to determine the transfer function (TF) of a complete optical sampling system adapting techniques from waveform metrology. Therefore, in this chapter, we perform the metrological evaluation of the optical system in terms of non-idealities like amplitude ripple, sideband suppression ratio (SSR), and spectrum roll-off factor β and study parameters like the root mean square error (RMSE), signal-to-noise and distortion ratio (SINAD), and the effective number of bits (ENOB).

S. De (✉)
Physikalisch-Technische Bundesanstalt (PTB), Braunschweig, Germany
e-mail: de.souvaraj@ptb.de

Y. Mandalawi · R. Das · T. Schneider
Technische Universität Braunschweig, THz-Photonics Group, Braunschweig, Germany
e-mail: younus.mandalawi@ihf.tu-bs.de; ranjan.das@ieee.org;
thomas.schneider@tu-braunschweig.de

© The Author(s) 2026
T. Kürner et al. (eds.), *Metrology for THz Communications*, Springer Series
in Optical Sciences 256, https://doi.org/10.1007/978-3-032-01986-8_23

23.1   Introduction

The principles of waveform metrology have been discussed in Part 1 "Fundamentals." In general, the generation and accurate measurement of the waveforms are invaluable for high-end oscilloscopes and arbitrary waveform generators [1]. Moreover, the metrological assessment of the waveforms has a variety of applications in wireless, wired, optical communication, and remote sensing. In order to characterize modern communication systems, it is essential to measure and characterize the waveforms with error sources that are correlated in time and frequency. Previous works in the field of uncertainty estimation [2–7] and calibration of sampling oscilloscopes [8–10] have paved the way to support the waveform traceability for the photonics and high-frequency electronic industry.

For a complete analysis of the system, it is necessary to perform the "full waveform metrology" [11] which takes into account the full waveforms as a function of time and frequency as target measurements and deriving the conventional parametric descriptions. It includes time domain measurements for devices like oscilloscopes and frequency domain measurements for devices like vector network analyzers, including the effects of time-base corrections [4] and finite bandwidth effects. Therefore, to have a consistent mapping from time to frequency domain and vice versa, it is important to establish traceability and correlated uncertainty analysis that can be propagated through each step of the measurement process.

In recent years, there has been a tremendous surge in the global demand for high-data-rate communication owing to the increasing number of internet users worldwide and the emergence of new technologies demanding high data rates as well as low latencies. As a result, the current network infrastructure is pushed to its limits, especially for high data rates that require high carrier frequencies like terahertz (THz) communication [12] with peak data rates of up to 1 Tbps [13] to facilitate the sixth-generation (6G) and beyond communications. Additionally, compared to low-frequency signals, THz signals can convey more information, resulting in faster data transfer rates and enabling new applications like virtual reality (VR), augmented reality (AR), high-resolution imaging, and sensing as well as further communication applications like centimeter-level localization, imaging, and precise positioning.

However, it is becoming increasingly difficult to process such high-bandwidth signals with conventional electrical methods. In this context, photonics-assisted signal processing is more advantageous as it provides a wider bandwidth, relatively simpler implementation, no electromagnetic interference, and more flexibility than the electrical equivalent.

23.2 Characterization of Optical Sampling

23.2.1 Metrology of Sinc-Shaped Nyquist Pulse Generation

For data transmission with the highest possible symbol rate in an optical communication system, sinc-shaped Nyquist pulses [14] are best suited as they meet the Nyquist criterion of zero inter-symbol interference (ISI) allowing error-free transmission of overlapping pulses, encoding the data with minimum spectral bandwidth and robust in terms of fiber dispersion and channel non-idealities. Additionally, sinc-pulse sequences (SPS) can be used for signal processing in analog-to-digital (ADC) and digital-to-analog converters (DAC), as discussed in the Chap. 22. The sinusoidal modulation of cascaded Mach-Zehnder modulators (MZMs) can produce a phase-locked, rectangular optical frequency comb (OFC) [14] which results in high-quality Nyquist pulse sequences in the time domain as shown in Fig. 23.1.

Nevertheless, the presence of non-idealities in the system can have a drastic impact on the quality of the Nyquist pulses. In this chapter, we demonstrate the metrological analysis of three major non-idealities, namely, ripple, sideband suppression ratio (SSR), and roll-off factor in terms of the RMSE.

23.2.1.1 Effect of OFC Positive Ripple

For Nyquist pulse generation with modulators as shown in Fig. 23.2, ripples can occur in the OFCs due to an improper calibration of biasing and RF power. For the experiments, the OFCs are analyzed in the frequency and time domain for different positive ripple values, where the carrier is higher than the sidebands, by changing the bias voltage of the MZM. Increasing the ripple ensues a further divergence

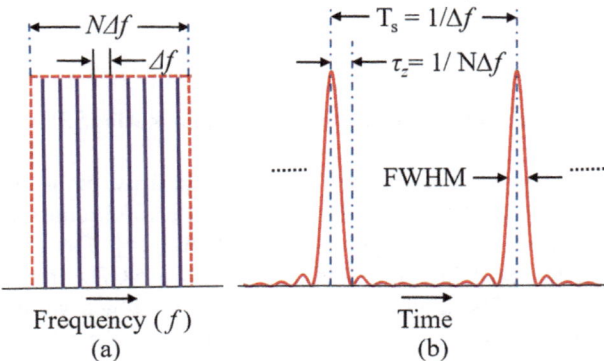

Fig. 23.1 Schematic representation of an N-line: (**a**) a frequency comb of rectangular shape and phases locked linearly and (**b**) the respective pulse sequence in time domain determined by the specific parameters (The figures have been taken from [15])

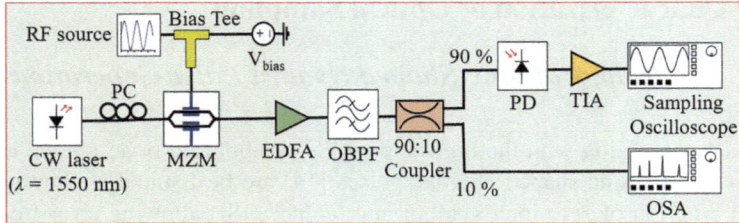

Fig. 23.2 Experimental setup for a comb generation with three lines (The figure has been taken from [15])

from the ideal scenario with a reduction in the duration of zero crossing (τ_z) and full width at half-maximum (FWHM) while having an identical repetition rate. For metrological analysis, the root mean square error (RMSE) is evaluated with the following expression:

$$RMSE = \sqrt{\frac{1}{N}\sum_{k=1}^{N}(M_k - I_k)^2} \qquad (23.1)$$

Here, M_k and I_k represent the kth instances of the measured and the ideal value, and N is the total number of sampling points. Considering a three-line OFC with a frequency spacing of 5 GHz, for a higher positive ripple of 3 dB, as in Fig. 23.3a, b, the main lobe of the measured and ideal sinc-shaped Nyquist pulse highly deviate from one another resulting in a higher RMSE [15]. The ripple analysis for five-line and nine-line OFCs with 50 GHz frequency spacing is carried out as well in terms of the RMSE (in %) and plotted with the results of the three-line OFC as shown in Fig. 23.3c. The best performance with the least RMSE is observed for the nine-line OFC. However, when properly adjusted, even integrated MZMs show ripple values of below 1 dB.

23.2.1.2   Effect of the OFC Sideband Suppression Ratio (SSR)

The unwanted sidebands may result from a nonlinearity of the modulator and increase with increasing RF power. However, by a careful adjustment of the RF power, the sidebands can be suppressed by 30 dB [14]. For the experimental investigation of the SSR in a three-line OFC with a 5 GHz frequency spacing, we consider only the first higher-order sideband [15], which we have artificially generated by another sinusoidal source synchronized to the primary RF source. As observed from Fig. 23.4a, b, increasing the SSR improves the pulse duration and FWHM and reduces the waveform distortion, and the obtained pulse is more congruent to the ideal pulse owing to the reduced influence of the higher-order sidebands. The RMSE (in %) is evaluated for a five- and nine-line OFC with 50 GHz

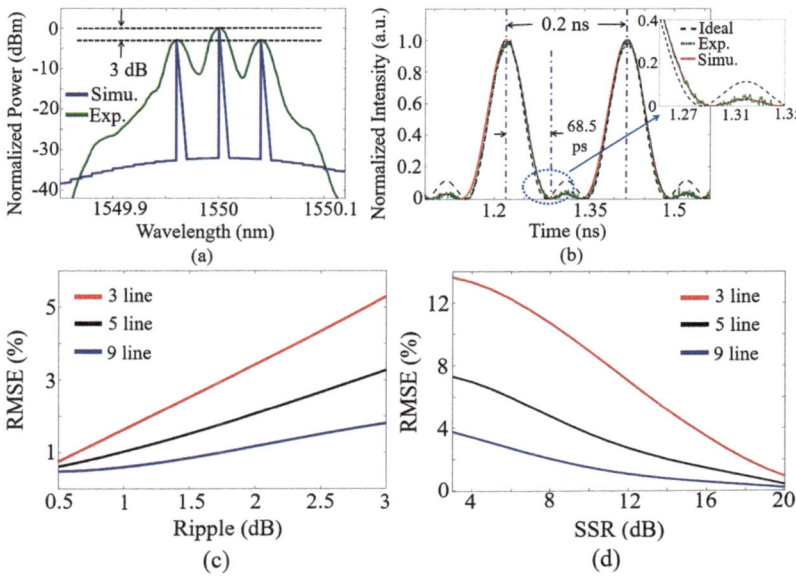

Fig. 23.3 The three-line frequency comb with the corresponding sinc-shaped Nyquist pulse sequence exhibiting a 3 dB positive ripple (**a, b**). The ideal sinc-shaped Nyquist pulse train, without ripple and with a spectral suppression ratio (SSR) of approximately 35 dB, is shown in black, while the simulated pulse with the respective ripple is depicted in orange. Comparison of the RMSE (in %) for a three-, five-, and nine-line comb for (**c**) different ripple values (in dB) and (**d**) for different SSR (in dB) (The figures have been taken from [15])

frequency spacing as presented in Fig. 23.3d. It is noticed that an increase in the number of comb lines results in a narrower pulse width and an increased number of lobes which ensues in the reduction of RMSE for a higher number of comb lines.

23.2.1.3   Effect of the Roll-Off Factor β

When SPS are generated by the abovementioned method with single or cascaded MZM, the sideband supression ratio is 30 dB, and an optical filter is not needed. However, another method for sinc-pulse sequence generation is the optical filtering of the frequency comb from a mode-locked laser [16]. However, as there is no such thing as an ideal rectangular filter, the filtered pulses have a roll-off factor β which defines the excess pulse bandwidth as compared to the ideal sinc-pulse ($\beta = 0$), resulting in higher-order sidebands in the frequency domain [15].

Taking β as the roll-off factor and τ_z as the zero crossing duration, a single sinc-shaped Nyquist pulse in the time domain can be formulated as

$$p(t) = sinc\left(\frac{t}{\tau_z}\right) \cdot \frac{\cos(\beta\pi t/\tau_z)}{1 - (2\beta t/\tau_z)^2} \tag{23.2}$$

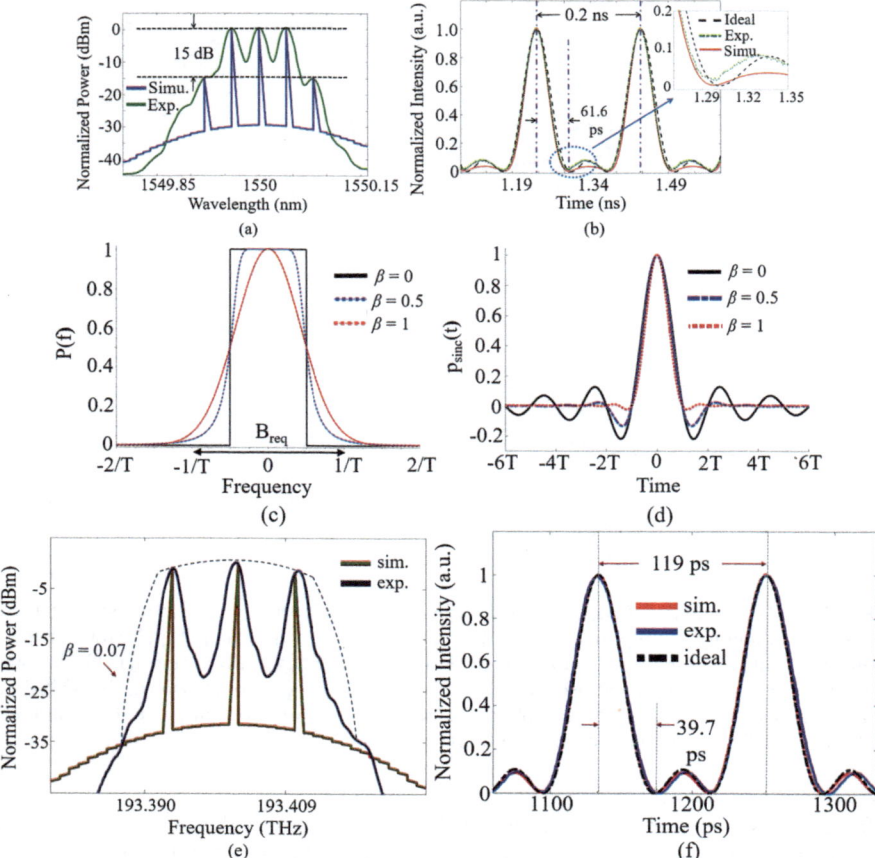

Fig. 23.4 Three-line OFCs with corresponding sinc-shaped Nyquist pulses for 15 dB SSR (**a, b**). In the frequency domain, the simulated spectrum is displayed in blue, while the experimental spectrum appears in green. In the time domain, the ideal sinc-shaped Nyquist pulse with an SSR of approximately 35 dB is shown in black, while the experimental and simulated pulses are represented in green and orange, respectively. Sinc-shaped Nyquist pulses for $\beta = 0, 0.5$, and 1 in (**c**) the frequency and (**d**) the time domain. The required bandwidth (B) is given by (3). Three-line OFCs with respective Nyquist pulses for a roll-off factor of 0.07 (**e, f**). In the time domain, the ideal optical Nyquist pulses with $\beta = 0$ are represented as dashed black lines, and the simulated pulses are shown in red (The figures have been taken from [15])

The Nyquist pulse bandwidth in the frequency domain and the pulse shape in the time domain for different instances of β are visualised in Fig. 23.4c, d where the corresponding bandwidth of the sinc-shaped Nyquist pulses is expressed as

$$B = \frac{1 + \beta}{\tau_z} \qquad (23.3)$$

For the experimental analysis, a three-line OFC with a frequency spacing of 8.4 GHz was generated with an on-chip silicon MZM [17], and the different roll-off factors (β) were generated by changing the parameters of a programmable optical filter (Finisar Waveshaper 1000A) as shown in Fig. 23.4e, f. The optical filter bandwidth was taken as 25.2 GHz, corresponding to the pulse bandwidth. Decreasing the value of β from 1 to 0 results in a filter shape that follows closely to the ideal rectangular shape, reduces the power of unwanted higher-order sidebands, and increases the FWHM and the pulse duration, resulting in the ensuing pulses being congruent to the ideal pulses as explained in Fig. 23.4c, d. For a three-line OFC, the effect of the higher-order sidebands is noticeable for β higher than 0.33. The minimum possible RMSE was calculated to be 0.7 % for $\beta = 0.07$ [17].

23.2.2 Metrology of a Direct Optical Sampling System

Communication at very high data rates of up to 1 Tbit/s [13], as needed for 6G and beyond wireless systems, requires high bandwidths for the transmitter, receiver, and channel. Traditional electronic methods have limited capacity for high-bandwidth signal processing with limited power efficiency. But it is possible to down-convert such high-bandwidth signals into parallel low-bandwidth signals and then process the signals with parallel low-bandwidth electronics (please see the Chap. 22 of this book). Electrical samplers may realize that down-conversion. Nevertheless, owing to the higher clock jitter in electronic devices, the performance significantly decays at higher bandwidths. An alternative is offered by orthogonal sampling with a flat frequency comb generated by a Mach-Zehnder modulator (MZM) [14]. For ideal devices, this method samples the signal error-free, and there is no aperture jitter since the down-conversion technique relies solely on the signal multiplication with a sinc-pulse sequence in the time domain. Hence, the possible low jitter of the oscillators can be directly transferred to the down-converted signal, increasing the effective number of bits (ENOB) (please see the Chap. 21 of this book).

Nonetheless, error sources, like nonlinearities and noise, can deteriorate the signal processing. For metrological evaluation of the complete optical sampling system, it is necessary to characterize the response of the system to such non-idealities, especially for direct detection (DD) systems, which are widely used for data centers and applications based on handheld devices. Consequently, the effect of the positive ripple on the performance of orthogonal sampling is investigated for a three-line OFC system in terms of metrics like RMSE, SINAD, and ENOB [18].

23.2.2.1 Experimental Setup

Orthogonal sampling is useful in detecting a high-bandwidth signal with low-bandwidth electronics [19]. Such sampling systems have been discussed in the Chap. 22. Coherent detectors, which enable the detection of the amplitude and

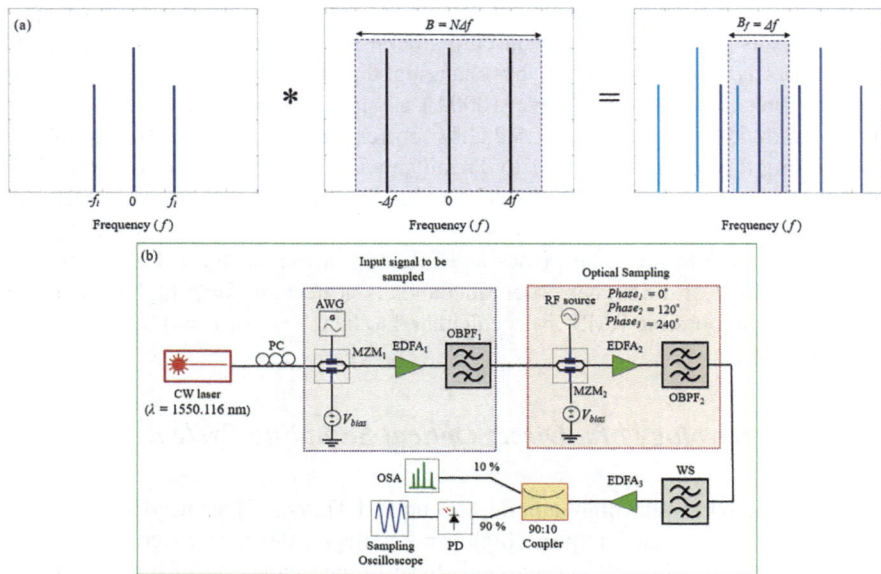

Fig. 23.5 (**a**) Convolution of the signal to be sampled with a frequency f_i with a three-line frequency comb having a spacing of Δf. (**b**) Experimental setup for orthogonally sampling the generated sinusoidal signal with the Mach-Zehnder modulator (MZM-2) with a filter/waveshaper (WS) bandwidth equal to the frequency spacing (Δf) of the comb lines produced by the MZM-2

phase, have been used in this chapter. However, for many applications, like the access network or data centers, a direct detection with a single photodiode would be advantageous. But, in this case, the phase information of the optical field gets lost, and the detected field is squared. However, by employing an optical filter before the detector, orthogonal sampling can as well be used for a direct detection with a simple photodiode. Using this technique for direct detection, the input sinusoidal signal of frequency f_i is convolved with a three-line comb that has a frequency spacing of Δf and bandwidth of $B = N\Delta f$. This signal is then filtered to the bandwidth $B/N = \Delta f$ in the optical domain as shown in Fig. 23.5a.

The proof of concept for the direct detection (DD) [18] is verified by the experimental setup in Fig. 23.5b where the input signal is 12 GHz and the RF phase is changed to 0°, 120°, and 240° to sample the complete signal. The three-line OFC of 16 GHz frequency spacing with a positive ripple of 0.1 dB is generated with the MZM$_2$ and the 12 GHz wave in blue is reconstructed from the sampling points of the red, orange, and gray sinc-pulse sequences as depicted in Fig. 23.6.

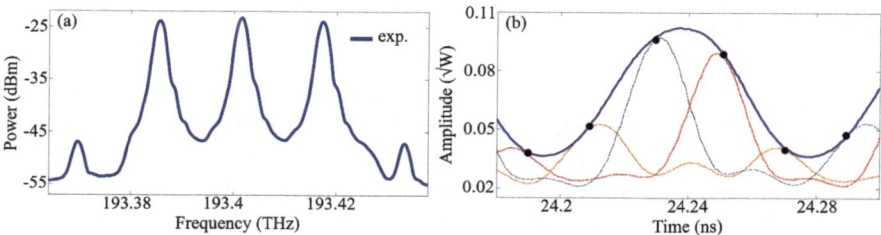

Fig. 23.6 (**a**) Three-line optical frequency comb with 16 GHz spacing generated with MZM$_2$ for optical sampling with a positive ripple of 0.1 dB. (**b**) Experimental results showing the sine wave in blue offline reconstructed from the sampling points of the red, orange, and gray sinc-pulse sequences

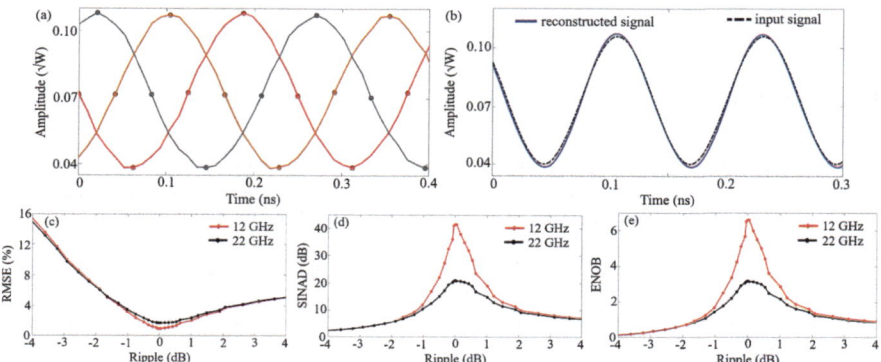

Fig. 23.7 (**a**) The waveforms at the ADC outputs for three different phase occurrences, i.e., 0°, 120°, and 240°, respectively. Here, in each case, the black dotted outlines represent the sampling instances. (**b**) Comparison between the ideal signal with the final 12 GHz reconstructed signal. (**c**) Comparison of the RMSE for different positive and negative ripple values (in dB) along with the corresponding SINAD (**d**) and ENOB (**e**). In each case, we consider sinusoidal input signals with frequencies of 12 GHz and 22 GHz and a system ADC jitter of 100 fs

23.2.2.2   Simulation Results

Considering a ripple of 0.5 dB and an input signal of 12 GHz, the reconstructed waveforms for the three different phases of 0°, 120°, and 240° after the analog-to-digital converter (ADC) with a jitter of 100 fs and the final reconstructed signal are compared to the input signal as shown in Fig. 23.7a, b. With the formulas listed in the Chap. 21, the RMSE, ENOB, and SINAD are calculated for different values of positive and negative ripple at input frequencies of 12 GHz and 22 GHz as demonstrated in Fig. 23.7c–e [18]. Here, the ADC jitter is considered to be 100 fs and the optical filter has a rectangular bandwidth. It is observed that increasing the ripple positively or negatively has a negative impact on the quality of the sampled signal in the directly detected sampling system.

23.3 Conclusion

A complete characterization of the optical sampling system with the transfer function using waveform metrology is of utmost importance to keep pace with the rising global demand for high data rate communication. As a first step, the non-idealities for sinc-shaped Nyquist pulse generation, namely, ripple, SSR, and the roll-off factor (β), were examined for three-, five-, and nine-line OFCs, and it was determined that the nine-line OFCs have the lowest RMSE and hence the best performance. Different metrological parameters like RMSE, SINAD, and ENOB were studied for a direct detection (DD)-based optical sampling system considering the positive and negative ripple, and it was determined that ripple drastically influences the quality of sampling.

References

1. Arbitrary waveform generator market size, share and trends analysis report by product, by technology, by application (telecommunications, electronics, healthcare, education, industrial), by region, and segment forecasts, 2019 - 2025. Report ID: GVR-3-68038-488-8, Industry: Semiconductors and Electronics
2. P.D. Hale, C.M.J. Wang, Calculation of pulse parameters and propagation of uncertainty. IEEE Trans. Instrum. Measure. **58**, 639–648 (2009)
3. S. Eichstaedt, et al., On challenges in the uncertainty evaluation for time-dependent measurements. Metrologia **53**(4), S125–S135 (2016)
4. D.F.W.C.M. Wang, P.D. Hale, Uncertainty of timebase corrections. IEEE Trans. Instrum. Measure. **58**, 3468–3472 (2009)
5. P.D. Hale, et al., Traceable waveform calibration with a covariance-based uncertainty analysis. IEEE Trans. Instrum. Measure. **58**, 3554–3568 (2009)
6. K.A. Remley, et al., Millimeter-wave modulated-signal and error-vector-magnitude measurement with uncertainty. IEEE Trans. Microwave Theory Tech. **63**, 1710–1720 (2015)
7. A. Lewandowski, et al., Covariance-based vector-network-analyzer uncertainty analysis for time- and frequency-domain measurements. IEEE Trans. Microwave Theory Tech. **58**, 1877–1886 (2010)
8. A. Dienstfrey, et al., Minimum-phase calibration of sampling oscilloscopes. IEEE Trans. Microwave Theory Tech. **54**, 3197–3208 (2006)
9. T.S. Clement, et al., Calibration of sampling oscilloscopes with high-speed photodiodes. IEEE Trans. Microwave Theory Tech. **54**, 3173–3181 (2006)
10. P.D. Hale, et al., Compensation of random and systematic timing errors in sampling oscilloscopes. IEEE Trans. Instrum. Measure. **55**, 2146–2154 (2006)
11. P.D. Hale, D. Williams, A. Dienstfrey, Waveform metrology: Signal measurements in a modulated world. Metrologia **55**(5), S135–S151 (2018)
12. T.N.T. Kuerner, D. Mittleman (eds.), *THz Communications: Paving the Way Towards Wireless Tbps* (Springer International Publishing, New York, 2021)
13. T. Schneider et al., Link budget analysis for terahertz fixed wireless links. IEEE Trans. Terahertz Sci. Tech. **2**, 250–256 (2012)
14. M.A. Soto et al., Optical sinc-shaped nyquist pulses of exceptional quality. Nat. Commun. **4**, 2898 (2013)
15. S. De et al., Analysis of non-idealities in the generation of reconfigurable sinc-shaped optical nyquist pulses. IEEE Access **9**, 76286–76295 (2021)

16. T.H.M. Nakazawa, M. Yoshida, Experiments on an fm mode-locked laser as an arbitrary optical function generator. IEEE J. Quant. Electron. **58**(3), 1–16 (2022)
17. S. De et al., Roll-off factor analysis of optical nyquist pulses generated by an on-chip mach-zehnder modulator. IEEE Photon. Technol. Lett. **33**(21), 1189–1192 (2021)
18. S. De et al., Comb flatness dependence for orthogonally sampled high bandwidth signals, in *Proceedings of the SPIE 12994, Terahertz Photonics III*, vol. 1299407 (2024), pp. 1–8
19. J. Meier et al., High-bandwidth arbitrary signal detection using low-speed electronics. IEEE Photonics J. **14**(2), 1–7 (2022)

Part IV
Metrology for RF Components and Propagation

Chapter 24
Metrology of mmW- and THz-Frequency- and Time-Domain Waveforms

Dominik Wrana, Benjamin Schoch, Ingmar Kallfass, and David A. Humphreys

Abstract The characterization of ultra-broadband THz communication links and their corresponding building blocks requires versatile and sophisticated measurement instrumentation. Aside from classical frequency-domain characterization under continuous-wave conditions, including small- and large-signal analysis, the time-domain device evaluation under broadband complex modulated signals is crucial to enable accurate link-level performance prediction and precise system-level modelling of the frontend components. In this chapter, different measurement approaches and setups are discussed, including classical frequency-domain characterization based on vector network or spectrum analyzers as well as time-domain approaches using arbitrary waveform generators and real-time oscilloscopes. Additionally, the CrossLink measurement platform introduces novel capabilities for simultaneous time- and frequency-domain characterization.

24.1 Frequency-Domain Characterization

When it comes to the characterization of millimeter wave and terahertz frontends and their building blocks, frequency-domain characterization is widely used to determine characteristic performance metrics, such as conversion gain, 1-dB compression points, third-order intercept points, or vector metrics like S-parameters.

D. Wrana · B. Schoch (✉) · I. Kallfass
Universität Stuttgart, Institut für Robuste Leistungshalbleitersysteme, Stuttgart, Germany
e-mail: dominik.wrana@ilh.uni-stuttgart.de; benjamin.schoch@ilh.uni-stuttgart.de; ingmar.kallfass@ilh.uni-stuttgart.de

D. A. Humphreys
NPL, UK (Retired), University of Bristol, Ascot, UK
e-mail: david.a.humphreys@ieee.org

T. Kürner et al. (eds.), *Metrology for THz Communications*, Springer Series in Optical Sciences 256, https://doi.org/10.1007/978-3-032-01986-8_24

303

24.1.1   Vector Network Analyzer

Vector network analyzers (VNAs) are a widely used and already well-known instrumentation for circuit characterization. Often used as the first measurement tool after wafer fabrication to investigate process monitors and perform an initial wafer mapping, there is a variety of possible measurements to be performed with a VNA. With its integrated sources and receivers, measurements are performed under the condition of a known input signal. This enables relative measurements, observing the incident as well as the reflected wave, typically resulting in the corresponding S-parameters, also including the transmission from port to port. Not only small-signal measurements are possible but also large-signal measurements like frequency selective power sweeps or the extraction of the so-called X-parameters [1]. VNAs support the characterization of devices with more than just two ports, like mixers, and couplers. Measurements with up to four ports can be considered as an industry standard.

In terms of covered frequency range, the VNA base units typically range from low frequencies in the MHz range up to 110 GHz, but there are single-frequency sweep units working up to 220 GHz. More commonly, a base unit is combined with external frequency range extenders, especially when targeting the THz frequency range around 300 GHz. There, the port interconnects are no longer based on coaxial connectors but rather on rectangular waveguide interfaces such as the WR-3.4 for H-band (220–330 GHz) frequencies.

Among measurement instruments, VNAs have the highest accuracy considering the outstanding calibration and de-embedding capabilities that come with vectorial network analysis. However, the VNA measurement principle is narrowband by design, thus making it not suitable for performing circuit analysis using broadband-modulated signals, at least at the first glance.

24.1.2   Spectrum Analyzer

In contrast to the VNA, a spectrum analyzer is only equipped with a receiver and inherently measures unknown signal sources. Either the actual frequency source is part of the device under test (DUT), e.g., an oscillator, or the DUT is driven by external sources, for instance, laboratory-grade frequency synthesizers, as it is exemplarily shown in Fig. 24.1. There, a 300-GHz superheterodyne transmitter module is placed as a DUT in order to analyze the spectral contributions to the RF spectrum arising from the LO generation. Analog to VNAs, spectrum analyzers are available up to 110 GHz with coaxial connection capabilities. For even more elevated frequencies, external frequency extension units, mostly working as passive down-conversion mixers, have to be used.

Typical measurements along with spectrum observation and evaluation (in and out of band of the DUT) include intermodulation distortion (IMD), adjacent channel

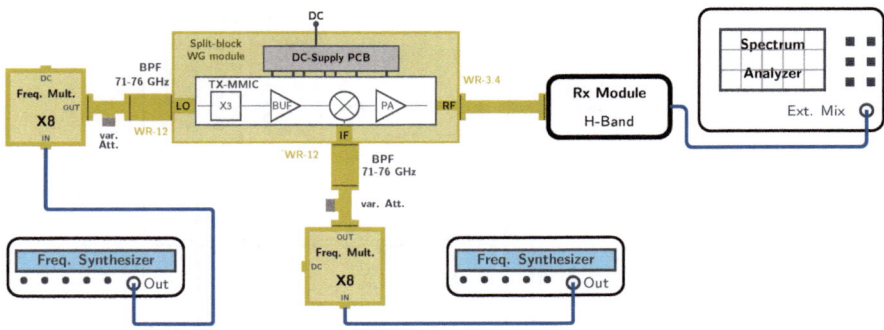

Fig. 24.1 Measurement block diagram for the RF spectrum evaluation of a superheterodyne 300-GHz transmitter module using frequency synthesizers to inject LO and IF continuous-wave test signals and a spectrum analyzer in conjunction with a H-band frequency extension module to spectrally resolved evaluate RF contributions such as spurs and LO leakage (Adapted from [2])

power ratio (ACPR), and others. Some spectrum analyzers offer a small real-time analysis bandwidth in the MHz to low GHz range, enabling, together with vector signal analysis software, the reception and demodulation of complex modulated signals. However, it is not suitable to support ultra-broadband signals with tens of gigahertz of bandwidth.

24.2 CrossLink Measurement Platform

The most versatile method of characterizing the transceiver impairments is accomplished by applying pseudorandom bit sequences (PRBS), using arbitrarily modulated signals of varying modulation formats, symbol rates, and power levels. Transceiver impairments like phase noise, intermodulation distortion, and quadrature channel imbalance can be mapped to a distorted constellation diagram which is quantified by EVM.

The EVM is typically derived from a sampled time-domain signal which is usually obtained from a real-time or sampling oscilloscope, described in Sect. 24.3. But it can also be obtained from its corresponding frequency-domain representation. A frequency-domain spectrum can be measured by the VNA using the standard built-in hardware. The VNA is designed to operate with a narrow IF bandwidth, and in comparison to a spectrum analyzer, there is typically no embedded preselector filter. Measuring the narrow IF bandwidth while sweeping the observation window across the desired bandwidth of interest, the sections can be stitched together in the frequency domain. This requires the continuous repetition of the waveform which is commonly used in RF active component testing according to standardized waveforms like 5G NR OFDM frames or QAM-modulated PRBS. If the capturing is performed across N periods of the waveform, it is possible to reconstruct the

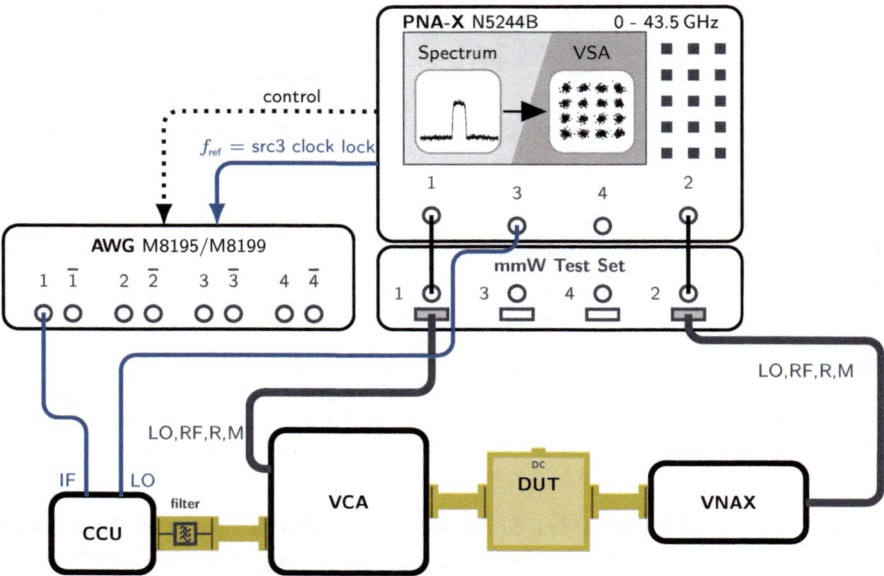

Fig. 24.2 Block diagram of the CrossLink measurement setup containing a compact up-converting mixer (CCU) used to generate wideband modulated signals in desired waveguide bands driven by an arbitrary waveform generator (AWG). A VCA module at P1 and a standard VNAX module at P2 were used for vectorial network analysis as well as receiving wideband modulated signals

complex frequency spectrum. With frequency stitching and FFT processing, a nearly infinite bandwidth can be analyzed and further processed by the vector signal analyzer (VSA) software [3].

The fundamental element of the instrumentation configuration is a four-port 43.5 GHz vector network analyzer (VNA) as it is shown in Fig. 24.2. The four ports P1 through P4 are driven by two internal frequency sources Src1 and Src2. A third RF source (Src3) up to 13.5 GHz with a built-in direct digital synthesis (DDS) synthesizer is available at the rear panel.

While the module at P2 is a standard vector network analyzer extender (VNAX), the module at P1 is a novel vector component analyzer (VCA) module. This module is capable of feeding and amplifying broadband signals pass through the receivers which initially capture the incident and reflected waves while performing narrowband frequency sweeps to calculate the S-parameters. Accordingly, the VNA is capable of analyzing not only the S-parameters but also the broadband signal injected at the wideband input port of the VCA module [4]. A simplified block diagram of the VCA module is shown in Fig. 24.3.

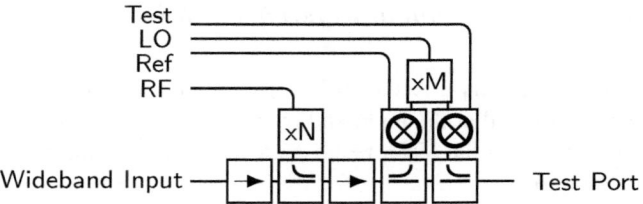

Fig. 24.3 Simplified block diagram of the VCA module illustrates the module's ability to offer standard narrowband vectorial network analysis capabilities in conjunction with the analysis of ultra-wideband complex modulated signals

24.2.1 Data Generation

The VNA is utilized for the computation of the wideband multitone waveform, which is then transmitted to the arbitrary waveform generator (AWG) for repetitive playback. The AWG incorporates digital-to-analog converters (DACs) with 8-bit resolution and a maximum sample rate of 65 GSa/s, facilitating analog bandwidths of up to 25 GHz. To enhance the signal quality captured by the VNA, an increased reference clock frequency operating at approximately 2 GHz driven by Src3 is utilized. To guarantee coherence, all frequencies within the measurement setup were derived from the internal clock of the VNA. This is a prerequisite step in order to enhance the performance of time-domain analysis via the frequency stitching procedure [5]. A variety of complex modulated signals can be calculated in terms of baud rate and spectral efficiency.

The quality of the wideband modulated signal is directly related to the signal generation, here the AWG, and the path to the analyzing receiver at a reference plane. This can be overcome using the spectral digital pre-distortion method, also called source correction [6]. The measured signal spectrum is subjected to an analysis of the distortion content, with the objective of computing a pre-distortion waveform. Subsequently, the waveform is transmitted to the signal generation, thereby initiating the next iteration of the correction process. The aforementioned correction procedure allows for the attainment of an exceptionally high signal quality under the specified measurement conditions. That means the source correction must be repeated when modifying the waveform or any environmental parameter.

The continuous repetition of the waveform in conjunction with coherent measurement enables the application of a vector averaging technique, whereby the averaging is performed with phase consistency. A principal advantage is the reduction of the noise floor, which directly enhances the SNR in demodulation. This methodology allows the impact of instrumentation noise to be substantially diminished while simultaneously enhancing the effective dynamic range of the test system [7].

24.2.2 Source Correction at Reference Planes

Considering a test setup as it is shown in Fig. 24.2 and used in [8], two reference planes can be determined at the input and output of the DUT. It is possible to shift the reference plane from the VNA receivers to the DUT interface by means of S-parameter calibration.

To demonstrate the efficacy of the source correction, a GaAs-based power amplifier is utilized as the DUT, operating within the partial W-band (68 to 90 GHz). A center frequency of 77.5 GHz was selected to apply two different symbol rates, specifically 1 and 4 GBd, respectively. This setting resulted in an analog RF bandwidth of 1.35 and 5.4 GHz, respectively, when a root-raised cosine filter with a roll-off factor of 0.35 was applied. The modulation format is varied from a basic QPSK format with a spectral efficiency of 2 bit/s/Hz to a more sophisticated 256-QAM format with an efficiency of 8 bit/s/Hz.

In the vicinity of the center frequency, the amplifier exhibits a small-signal gain of 15 dB, a saturated output power of 20 dBm, and an output-related 1-dB compression point of 17 dBm.

Table 24.1 presents a comparison between the uncorrected and corrected waveforms measured by the receiver at port 1. The received waveform is directly processed by the VSA software to display the constellation diagrams, as well as the EVM for QPSK and 256-QAM signal. The calibration process has been observed to enhance the EVM by approximately one order of magnitude. In the absence of a discernible demodulation of the uncorrected waveform, the source correction is demonstrated to be capable of recuperating the signal, as evidenced by the 256-QAM 1 GBd waveform.

Table 24.2 shows the modulated signal measured at port 2 while applying the source correction calculated from port 1. This would be a typical measurement to characterize the influence of the DUT on the constellation diagram and the EVM. A reduction in signal quality can be observed since the DUT introduces a frequency and phase behavior as well as starting to be nonlinear at these power levels.

Since the setup can perform the source correction procedure at any receiver, the waveform can be corrected at the output of the DUT. Following the implementation of the aforementioned corrections, the signal at the output of the DUT, which incorporates it into the measurement setup, is now nearly perfect. A next DUT, i.e., from a multistage transceiver, can be measured. Furthermore, it is noteworthy to state that due to an increase in signal level, which allows for a more efficient source correction, the signal at the amplifier output exhibits an enhanced quality.

24.3 Time-Domain Characterization

So far, the most common way of performing transmission experiments using broadband modulated data is a pure time-domain (TD) characterization setup.

Table 24.1 Measured EVM performance uncorrected and corrected at interface ① that means receiver at port 1 in the W-band

	Interface ① uncorrected	Interface ① corrected
1 GBd, $\alpha = 0.35$		
	EVM 13%	EVM 1.3%
	-8 dBm	-8 dBm
4 GBd, $\alpha = 0.35$		
	EVM 27%	EVM 3.8%
	-4.5 dBm	-5 dBm
1 GBd, $\alpha = 0.35$		
	no sync	
	no sync	EVM 0.6%
	-8 dBm	-8 dBm
4 GBd, $\alpha = 0.35$		
	no sync	
	no sync	EVM 1.2%
	-4.5 dBm	-5 dBm

Figure 24.1 shows such a setup, used for the characterization of a simplex wireless 300-GHz link, including superheterodyne transmit and receive frontends [9]. The I/Q data signals are generated using an AWG, in this specific case a Keysight M8195A with a maximum sampling rate of 64 GSa/s. The data is injected into the DUT, transmitted wirelessly over a distance of a few meters up to multiple hundred meters, down-converted by the receiver, and ultimately captured with a real-time oscilloscope, in this specific case a Keysight DSOZ204A with a bandwidth of 20 GHz and a sampling rate of 80 GSa/s. The sampled data is then processed using offline vector signal analysis software. The ability to use this setup when transmitter and receiver are far apart from each other already sets it apart from VNA-

Table 24.2 Measured EVM performance at the interface ② that means receiver at port 2, with the correction from interface ① and separate correction at port 2

	Interface ② **cal from** ①	Interface ② **cal**
1 GBd, $\alpha = 0.35$		
	EVM 4.5%	EVM 0.6%
	6 dBm	6 dBm
4 GBd, $\alpha = 0.35$		
	EVM 14%	EVM 1.2%
	9 dBm	9 dBm
1 GBd, $\alpha = 0.35$		
	EVM 3%	EVM 0.3%
	6 dBm	6 dBm
4 GBd, $\alpha = 0.35$		
	no sync	
	no sync	EVM 0.9%
	9 dBm	8 dBm

based setups. With those, it is impossible to operate with such long link distances. However, one drawback of the TD measurement scenario is the limitations in moving the calibration plane directly to the DUT. As shown in Fig. 24.1, there might be some auxiliary up- and down-conversion needed to reach the frequency ranges required at the DUT interfaces, especially for frequencies as high as 300 GHz. This auxiliary hardware also comes with a frequency response but even more importantly with nonlinearities and other distortion mechanisms. Those will already significantly reduce the signal quality, making it extremely difficult to assign a quantitative value of signal distortion induced by the actual DUT (Fig. 24.4).

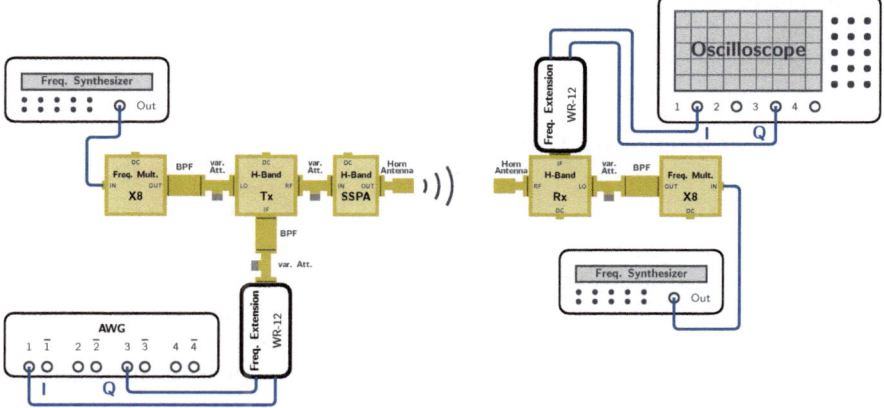

Fig. 24.4 Measurement block diagram of the link-level time-domain characterization setup of a 300-GHz superheterodyne wireless link using an arbitrary waveform generator for broadband modulated signal generation as well as a real-time oscilloscope for capturing of the received waveform (Adapted from [9])

24.3.1 Waveform Generation

The waveform generation is done either by tools directly provided with the hardware, such as IQTools, or by third-party processing tools, e.g., Matlab. Generally speaking, there is no limit to the type of waveform, considering single-carrier, multi-carrier, or OFDM waveforms and with respect to the applied coding, possibly integrating any communication standard. Often, a first characterization of the hardware in form of a data transmission experiment is done using a simple single-carrier waveform based on a pseudorandom bit sequence (PRBS) with an equally distributed symbol probability. Typical modulation formats include quadrature-amplitude modulation (16-QAM, 32-QAM, 64-QAM, etc.) or phase-shift keying (BPSK, QPSK, 8-PSK, etc.) schemes. The PRBS is mapped to the symbols of the chosen modulation scheme and it is pulse shaped afterward using a root-raised cosine (RRC) filter. Applying the matched filter principle, the second RRC filtering is applied during demodulation. The waveform length in terms of symbols depends on the sampling rate of the AWG as well as the bitrate and thus the bandwidth of the waveform. Together with the available memory size of the AWG, the maximum number of symbols can be calculated. Considering a looped playback of the waveform by the AWG, it is important to include a so-called wraparound to avoid any amplitude or phase discontinuities when jumping from the last sample back to the first sample in the memory. To enhance the constellation synchronization at the receiver, a synchronization pattern or header can be added to the waveform and searched for in the demodulator. This way, the transmission is still not data aided, where the exact waveform needs to be known by the receiver, but a correct orientation of the received constellation is eased, especially in the presence of signal distortion.

24.3.2 Signal Quality Limitations

To accurately quantify even small-signal distortions induced by the DUT, the initial quality of the injected modulated test waveform has to be as ideal as possible. However, the signal quality in terms of signal-to-noise ratio (SNR) and error vector magnitude (EVM) is limited by the AWG hardware, e.g., by the quantization noise of the D/A converter. Some hardware non-idealities like channel skew between different output channels, e.g., used for I and Q data signals, or the frequency response of the individual AWG channels can be calibrated out.

Therefore, an in-system calibration is performed beforehand and applied to the generated waveform, which also accounts for and compensates for the frequency response of the cables connecting the AWG to the DUT. This way, the setup is calibrated up to the DUT input or in case of the back-to-back characterization up to the real-time oscilloscope port. In contrast to the VNA-based CrossLink setup in 24.2, the initial signal quality cannot be further improved by vector averaging but is limited by the noise floor and dynamic range of the arbitrary waveform generator and real-time oscilloscope, respectively. Figure 24.5 exemplarily evaluates the signal quality in terms of EVM_{rms} and SNR as a function of the AWG output

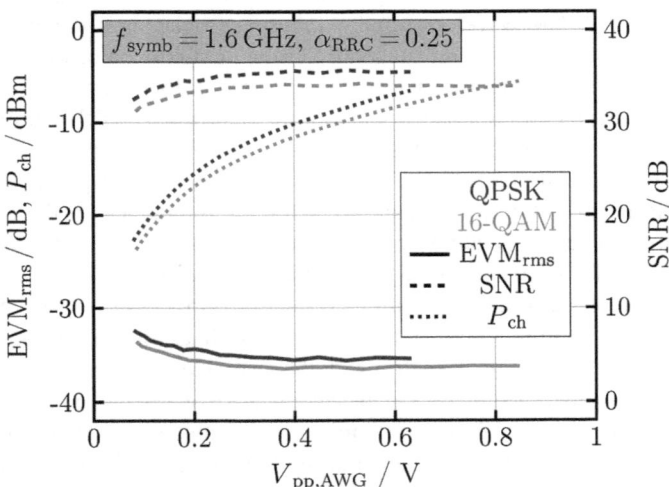

Fig. 24.5 Measured EVM, SNR, and corresponding channel power for QPSK and 16-QAM modulated signals with a symbol rate of 1.6 GHz as a function of peak-to-peak voltage using a Keysight M8195A arbitrary waveform generator and a Keysight DSOZ204A real-time oscilloscope connected in back-to-back configuration

peak-to-peak voltage, which in conjunction with the applied pulse shaping using a root-raised cosine filter with a roll-off of 0.25 translates to a channel power P_{ch}. For the shown measurement, an internal frequency offset of 4 GHz is applied during signal generation and a single channel is evaluated. For very low amplitudes, the signal quality is limited by the noise floor and spurious-free dynamic range of the AWG. Above 0.3 $V_{pp,AWG}$, the EVM converges -35.5 and -36.3 dB for QPSK and 16-QAM, respectively. The corresponding SNR lies at 35.5 dB for QPSK and 33.9 dB for 16-QAM. Due to the higher peak-to-average power ratio of the 16-QAM signal, the channel power is reduced compared to a QPSK modulation for the same amplitude setting, indicating the decreasing dynamic range with increasing modulation order.

24.4 Conclusion

The characterization of THz frontends and their building blocks for ultra-broadband communication links is carried out using time-domain as well as frequency-domain measurement setups. In this chapter, different setups for small- and large-signal as well as CW and broadband modulated characterization of frontend components have been described. The VNA-based CrossLink measurement platform enables novel measurement possibilities with dedicated signal correction capabilities at the input of the DUT, even at THz frequencies. This approach offers the highest initial measurement signal quality compared to classical time-domain analysis based on AWG and oscilloscope and enables the characterization of even minor distortions due to DUT non-idealities.

References

1. D.E. Root, J. Verspecht, D. Sharrit, J. Wood, A. Cognata, Broad-band poly-harmonic distortion (PHD) behavioral models from fast automated simulations and large-signal vectorial network measurements. IEEE Trans. Microwave Theory Tech. **53**(11), 3656–3664 (2005). https://doi.org/10.1109/TMTT.2005.855728
2. D. Wrana, S. Haussmann, B. Schoch, L. John, A. Tessmann, I. Kallfass, Effects of harmonics from frequency-multiplicative carrier generation in a superheterodyne 300 GHz transmit frontend, in *2023 53rd European Microwave Conference (EuMC)*, Berlin (2023), pp. 138–141. https://doi.org/10.23919/EuMC58039.2023.10290717
3. J.-P. Teyssier, J. Dunsmore, J. Verspecht, J. Kerr, Coherent multi-tone stimulus-response measurements with a VNA, in *2017 89th ARFTG Microwave Measurement Conference (ARFTG)*, Honolulu, HI (2017), pp. 1–3. https://doi.org/10.1109/ARFTG.2017.8000824
4. J. Verspecht, A. Stav, T. Nielsen, S. Kusano, The vector component analyzer: a new way to characterize distortions of modulated signals in high-frequency active devices. IEEE Microwave Mag. **23**(12), 86–96 (2022). https://doi.org/10.1109/MMM.2022.3203939

5. J.P. Teyssier, N. Messaoudi, J. Dunsmore, J. Verspecht, Deterministic detection and vector band stitching for the measurement of 6G wideband test signals, in *2023 53rd European Microwave Conference (EuMC)*, Berlin (2023), pp. 766–769. https://doi.org/10.23919/EuMC58039.2023.10290328
6. S. Kusano, A. Stav, T. Li, J. Verspecht, Correcting nonlinear distortion of wideband modulated signals using new frequency domain methods, in *2022 98th ARFTG Microwave Measurement Conference (ARFTG)*, Las Vegas, NV (2022), pp. 1–3. https://doi.org/10.1109/ARFTG52954.2022.9844123
7. J.P. Dunsmore, J.-P. Teyssier, The evolution of vector network analyzers to provide precision vector spectrum analysis for 6G applications: VNA evolves for 6G EVM signals. IEEE Microwave Mag. **25**(12), 77–90 (2024). https://doi.org/10.1109/MMM.2024.3438711
8. I. Kallfass et al., Instrumentation for the time and frequency domain characterization of terahertz communication transceivers and their building blocks, in *2023 IEEE/MTT-S International Microwave Symposium - IMS 2023*, San Diego, CA (2023), pp. 1030–1033. https://doi.org/10.1109/IMS37964.2023.10188006
9. D. Wrana, L. John, B. Schoch, S. Wagner, I. Kallfass, Short range wireless transmission using a 295–315 GHz superheterodyne link targeting IEEE802.15.3d applications, in *2021 51st European Microwave Conference (EuMC)*, London (2022), pp. 205–208. https://doi.org/10.23919/EuMC50147.2022.9784217

Chapter 25
Behavioral Characterization of Nonlinear Distortion

Dominik Wrana, Simon Haussmann, Jonas Gedschold, Reiner Thomä, and Ingmar Kallfass

Abstract Components of wireless communication systems often do not operate completely within their linear operating range. This applies in particular to power amplifiers in the sub-THz and THz frequency range. While the best possible linear approximation of the transfer characteristic can still be found in the weakly nonlinear case, the resulting nonlinear induced distortion depends on the input signal. Therefore, a behavioral characterization is performed that describes the transmission characteristics for a specific operating mode. Here, we pursue two approaches. The first one is the EVM (error vector magnitude) approach, which reveals the nonlinear induced deviation of the modulation symbols from a reference in terms of magnitude and phase (AM-AM and AM-PM distortion). The other approach is more statistical. It is reminiscent of the NPR (noise power ratio) test, which can be implemented quite effectively for typical communication signals such as OFDM and SC-FDM. The key issue is that the test signal should reproduce the application signal with regard to the relevant statistical characteristics.

25.1 Introduction

Transmission of signals in communication systems very often undergoes nonlinear distortions. Examples are saturation of power amplifiers or mixers where nonlinearities are deliberately exploited. Therefore, nonlinearities need to be thoroughly characterized in detail and considered for design on the circuit level. In this chapter, we follow another approach that addresses the overall input/output (I/O) behavior of some transmission systems (or subsystems) for a certain class of applications.

D. Wrana · S. Haussmann (✉) · I. Kallfass
Universität Stuttgart, Institut für Robuste Leistungshalbleitersysteme, Stuttgart, Germany
e-mail: dominik.wrana@ilh.uni-stuttgart.de; simon.haussmann@ilh.uni-stuttgart.de; ingmar.kallfass@ilh.uni-stuttgart.de

J. Gedschold · R. Thomä
Technische Universität Ilmenau, Institute of Information Technology, Ilmenau, Germany
e-mail: jonas.gedschold@tu-ilmenau.de; reiner.thomae@tu-ilmenau.de

T. Kürner et al. (eds.), *Metrology for THz Communications*, Springer Series in Optical Sciences 256, https://doi.org/10.1007/978-3-032-01986-8_25

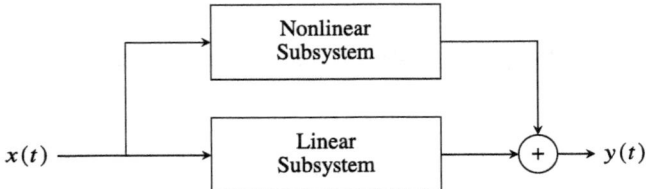

Fig. 25.1 Modeling a weakly nonlinear system by two subsystems (compare [1, Ch. 3])

Linear systems can be identified by various input signals, including swept sinusoids, short pulses, multicarrier waveforms, or other broadband pseudo-noise signals. The reason behind this is the superposition principle that makes sure that the system response does not depend on the input waveform. The resulting I/O transfer functions in the time or frequency domain (i.e., the impulse or frequency response) can be applied in a generic way for calculating the system response for many applications. However, this no longer holds if the system under test undergoes nonlinear distortions since the superposition principle is no longer valid. In this case, the system response depends to a certain extent on the input waveform.

Although there are some general I/O models for nonlinear dynamic systems such as Volterra series or generalized memory polynomials, these models are not easy to handle and identify. In addition, the effect of the nonlinear distortion is not immediately recognizable. Therefore, we follow a behavioral approach that represents a specific application. This approach refers to characterizing the I/O relation for a test signal that emulates an application-specific input signal. At the output, the desired part has to be separated from the nonlinear signal distortions. For communication systems, it can be exploited that they can be sufficiently characterized as *weakly* nonlinear. In this case, the system can be modeled by two subsystems as in Fig. 25.1.

One is a linear system and the other one is a nonlinear system producing distortion signals orthogonal to the linearly transformed output. The term *weakly nonlinear* refers to the assumption that the linear subsystem is sufficiently independent of the input signal. This may change as the input power increases. Then, the underlying linear system may be better described as the best linear approximation of the system under test.

In this chapter, we pursue two approaches that are most important for the behavioral characterization of weakly nonlinear systems such as power amplifiers with saturation behavior. One approach is the Error-Vector-Magnitude (EVM) technique which reflects the difference between the vector representing the demodulated symbol and the ideal vector in magnitude and phase. Therefore, the test signal should follow the state diagram of multilevel modulation for a certain operating condition. The difference is determined symbol by symbol by comparison with a certain reference modulator. This shows how much the performance of the modulator is affected by the power amplifier. The other approach is more statistical. Since the nonlinear induced distortion depends on the input signal, we apply waveforms that reproduce the statistical signal characteristics in typical operating situations. Relevant statistics are, e.g., the probability density function (PDF) and the power spectral density (PSD). For wideband signals such as Orthogonal Frequency-Division Multiplex (OFDM) or Single Carrier Frequency-Division Multiplex (SC-FDM), it is desired to characterize the PSD of the interference induced by the nonlinearity which will inevitably overlap with the desired signal spectrum that undergoes a linear transfer from the input. However, the orthogonality of the OFDM carriers together with discrete Fourier transform (DFT) processing offers a simple solution. Removing one of the carriers of the input signal would have a minor influence on the shape of the time-domain PDF of the input signal, and its power can be kept constant. Then, it is possible to measure/estimate the distortion power at the position of the empty carrier. By repeating this procedure for several frequencies/carriers, an estimate of the in-band PSD of the nonlinear distortion is achieved. This procedure is reminiscent of the well-known noise-power ratio (NPR) measurement [2]. Unlike EVM, this method shows how strongly the interference affects other channels or users in a multiuser access scheme as well as in-band as out-of-band.

25.2 Behavioral Characterization Based on Vector Component Analysis

In THz-systems, an in-depth knowledge of the system components is vital. The nonlinearities of power amplifier, in particular, affects the system performance as most THz-transmitters are optimized for maximum output power. The last output stage, also in systems, without state-of-the-art output power, is driven near the compression because of the increased power-added efficiency. So mostly, the mainly present nonlinearity is in the power amplifiers as they are operated at high signal levels. The origins of the nonlinearities are manyfold: transistors are nonlinear devices itself. Nonlinearities occur, when the signal levels reach nonlinear operation points, sometimes deliberately in class-B or class-C amplifiers, sometimes as inevitable result of amplifier compression. But also, other physical effects can cause nonlinearities: gate-trapping effects or nonlinear capacities, for instance. For the de-embedding purpose, a behavioral approach, which does not incorporate the physical origins of the nonlinearity, but rather models the behavior as a mathematical representation of input-to-output relation, is most constructive. The benefit is that behavioral modeling is based on universal models, which can theoretically be arbitrarily accurate. Circuit models on the other hand represent better the physical characteristics, but the model itself is mainly influencing the model accuracy. A disadvantage of behavioral characterization is the dependency on the selected waveforms or signals. A change of average signal power, peak-to-average–power-ratio, or bandwidth needs new modeling. Another important aspect of measuring or simulate nonlinearity is the assessment of the influence of the system. Popular metrices are error vector magnitude, bit error rate, total harmonic distortion, or the spectral distribution of distortion power.

Depending on the application, different behavioral models can be used: in case of more severe nonlinearities, a model-based approach yields good results. If a certain level of abstraction can be allowed, memoryless or quasi-memoryless models are suitable, as shown in Sect. 25.2. Alternatively, if the modeling should represent the reality accurately, Volterra-series or generalized memory polynomials are a well-established approach. In weakly nonlinear systems, a linearization around the operation point can be a valid solution, as outlined in Sect. 25.3.

Thus, unimportant of the use case, a distinct knowledge of the amplifiers' nonlinearities is a crucial aspect in communication and metrology. A central consideration in characterizing the behavior of stand-alone devices, such as PA, is to calibrate the measurement setup. Naturally, this is enabled by calibration functionality for vectorial network analyzers (VNA) like short-open-load-through (SOLT) or through-reflect-load (TRL). Based on this technique, single-tone AM-AM and AM-PM measurement can be conducted, and simple behavioral models like quasi-memoryless polynomial, Rapp, and Saleh models can be extracted, as shown in [3]. For multitone or modulated data, the VNA-Setup has been extended for stimulation and analysis of ultra-broadband single-carrier communication signals with complex modulation format as shown in [4] Chap. 24.

3GPP standardization are proposing a modified Rapp model for amplifiers with operation frequencies above 6 GHz [5]. The output amplitude and the output phase are expressed as

$$F_A(x) = \frac{Gx}{(1 + |\frac{Gx}{V_{sat}}|^{2p})^{\frac{1}{2p}}}, \quad F_P(x) = \frac{Ax^{q_1}}{1 + |\frac{x}{B}|^{q_2}}, \tag{25.1}$$

As the quasi-memoryless model does not include frequency-dependent behavior, the model-parameters can be obtained by measuring the amplitude-to-amplitude modulation (AM-AM) and amplitude-to-phase modulation (AM-PM) with single-tone VNA measurements. As single-tone measurements have infinitely low bandwidth, the measurement is repeated over frequency, in order to capture the deviations across the frequency. As shown in Fig. 25.2, this results in three-dimensional data. This data is used to fit the model, i.e., the Rapp model with a least mean square method.

25.2.1 AM-AM

The AM-AM characteristic of the device under test is performed by sweeping the input power of the device. By capturing the output power of the fundamental frequency at input and output, the gain can be expressed as a function of the input power. The three-dimensional result is shown in Fig. 25.2a. As shown in the

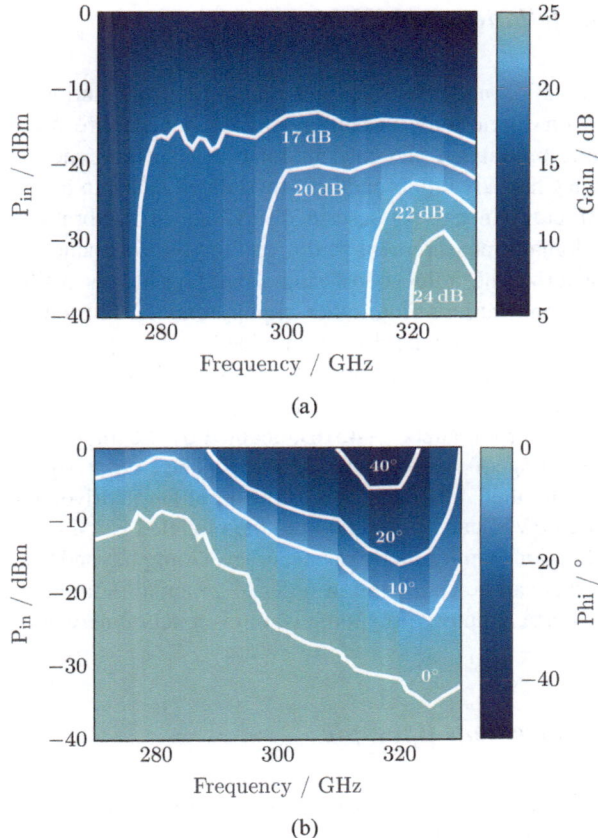

Fig. 25.2 Measured AM-AM (**a**) and AM-PM (**b**) conversion of the 300-GHz power amplifier (Adapted from [3])

measurement data, not only the absolute gain but also the compression curves are varying over the frequency.

25.2.2 AM-PM

Similarly to the AM-AM characterization, the phase shift induced by the nonlinearity can be characterized for single tones, as shown in Fig. 25.2b. As a clear separation, between linear and nonlinear behavior is aspired, the phase shift of the small-signal behavior is deducted from the measurement data.

25.2.3 Constellation and EVM

Single-tone excitation naturally does not reflect intermodulation distortion. Also, models, based on single-tone measurements, do not capture memory effects. In order to include broadband and intermodulation effects in measurements, EVM characterization Chap. 26 with broadband-modulated data can be used to model the behavior of RF components. As described in Chap. 24, calibrated time-domain as well as cross-domain measurement setups can apply broadband data to a DUT in order to characterize the influence of single building blocks while minimizing or even excluding contributions from other impairments. The value of the data-aided EVM is finally a metric for the distortion added to the signal. Figure 25.3 shows a measurement of the input and output EVM of a 300-GHz amplifier in a cross-domain setup. As shown, the EVM of the input signal is almost at a very low, constant level of approximately $-45\,$dB, enabled by the high calibration accuracy and vector averaging of the VNA-based setup. The output signal's EVM has a constant offset, and the EVM increases, as the amplifier is driven into saturation.

The measured EVM can be compared to the modeled EVM of the device which serves as a validation step for modeling accuracy. Going beyond, even the input and output waveforms can be compared in order to generate more complex behavioral models like Volterra, memory polynomials, or even neural networks.

25.2.4 Measurement Example

In this chapter, we want to emphasize the difference between single-tone and broadband measurement and modeling techniques, based on [3]. With the innovative measurement VNA-based measurement approach introduced in Chap. 24, it is possible to accurately characterize the nonlinear distortion of power amplifiers at millimeter and submillimeter wave frequencies. The DUT, a 300-GHz amplifier

Fig. 25.3 Cross-domain measurements of the EVM of a 300-GHz-amplifier, measured at the input and output port (Adapted from [3]. The constellation is measured at a carrier frequency of 280 GHz)

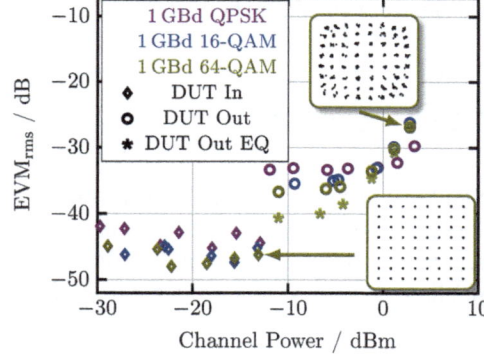

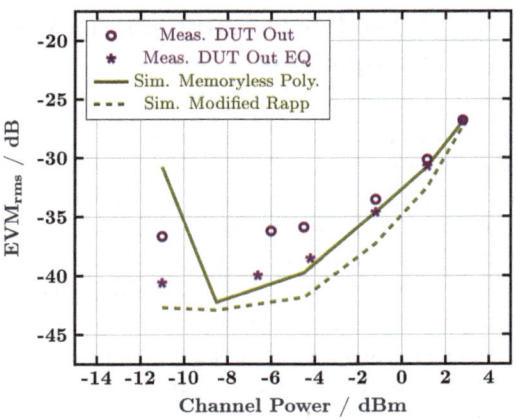

Fig. 25.4 Wideband cross-domain measurements at the output port of the EVM of a 300-GHz amplifier compared to EVM obtained by simulations. The simulated amplifier model is on AM-AM and AM-PM behavior, obtained from single-tone measurements. The applied waveform is 64-QAM with symbol-rate of 1 GBd (Adapted from [3])

device based on a 35 nm mHEMT technology [6], is characterized for the following:

- S-parameters
- CW power sweeps (compression behavior of magnitude and phase) as shown in the AM-AM and AM-PM plots in Fig. 25.2 for center frequencies between 270 and 330 GHz
- Broadband complex-modulated signals using the CrossLink measurement setup, as shown in Fig. 25.3. The graphic shows EVM measurements with a channel bandwidth of 1.35 GHz and a carrier frequency of 280 GHz at different band-power levels

Figure 25.4 shows an evaluation of the modeling accuracy of implemented behavioral models. The measured EVM of a 64-QAM signal is compared to simulations of memoryless polynomial model and Rapp model which are derived from the single-tone characterization. The comparison shows that the degradation of the EVM due to amplifier compression agrees with the simulations. However, the residual EVM floor in linear operation points is worse in measurements than in simulations. This is due to non-flat amplitude frequency response and group-delay distortions that are not covered by memoryless or quasi-memoryless models. A linear equalization of the measured data shows that the results converge, hinting toward the component channel flatness causing the modeling inaccuracy.

The data obtained by measurements is a crucial input for the development and verification of system-level PA models targeting accurate performance prediction in link- or system-level simulations for ultra-broadband wireless communication systems. In conclusion, memoryless behavioral models do not represent an accurate representation of the EVM degradation in system simulations. Memoryless models in conjunction with linear component behavior like S-parameters give acceptable results. Toward ultra-wideband applications, usage of memory models like Volterra or memory polynomials is advised.

25.3   Noise Power Ratio Test for Wideband Radio Components

A multitone or multicarrier sounding signal (comparable to OFDM) allows a straightforward way of detecting and characterizing in-band nonlinear distortions (compare Chap. 16 Sect. 16.2.3 and [7]). Usually, in-band measurements are implemented as noise power ratio tests. A notch filter is applied to the excitation signal to allow the noise floor to be sensed at the suppressed frequencies. For a multitone signal, the notch filter can easily be implemented by deactivating the transmit power for specific subcarriers of the excitation signal as exemplified in Fig. 25.5a. In the case of a linear system response, the sensed power at the suppressed carrier positions corresponds to the measurement noise level. However, if the system responds, nonlinear, intermodulation distortions and harmonics sum up at these frequencies increasing the observed power level [1, Ch. 3.5]. The assumption is that deactivating a small number of carriers has a negligible impact on the amplitude realization of the signal which effectively defines the response of the nonlinear system to this signal. Hence, this method allows a reasonable approximation of the in-band nonlinear distortions. Figure 25.5b shows the relation between in and output power as well as the sensed power level at a deactivated carrier position measured with an amplifier under test (compare [7]). As long as the amplifier operates in its linear operating region, the output power is proportional to the input power. Also, the power sensed at the deactivated carrier corresponds to the noise level. As soon as the amplifier starts operating nonlinearly, the power at the deactivated carrier rises rapidly, indicating the generation of intermodulation distortions between the other carriers. Also, a compression behavior between in and output power is observed.

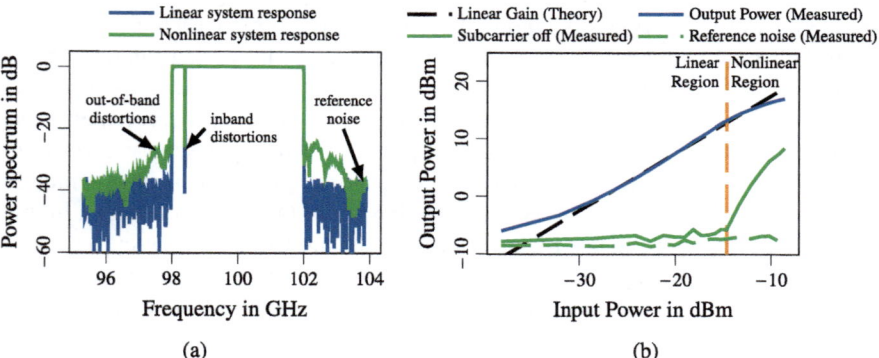

Fig. 25.5 Multitone measurement procedure showing the implementation of the NPR test using a multicarrier waveform (**a**) and the measured output and distortion power for a single carrier for varying input power levels (**b**) (reused from [7])

25.3.1 Noise and Distortion Power Distribution for Communication Signals

The distortion power spectrum as measured by the multitone procedure is analytically related by the physical characteristics of the nonlinear component to the input signal. As such, the distortion components (harmonics and intermodulations) deterministically result from the realization of the system input. However, they have to be considered a random process for varying realizations of the input signal. This gets important, for example, for communication signals, where the signal changes depending on the (random) data bits that have to be transmitted. For such applications, the distortion spectrum can only be characterized statistically, e.g., by estimating the probability distribution of the distortion power. Hence, the characterization process requires a statistically representative set of input signals resembling the characteristics of *real* communication signals. In this way, the nonlinear component can be characterized depending on the modulation type (e.g., OFDM vs. SC-FDM) and the selected power backoff.

25.3.1.1 System Model and Simulations

In this section, we want to exemplify the process of characterizing the distortion power based on two different modulation types, OFDM and SC-FDM. These two modulation types allow DFT processing due to their periodicity (enforced by a cyclic prefix) and, hence, can straightforwardly be used for the multitone measurement procedure. The system model for the simulation is shown in Fig. 25.6. We simulate the data bits drawn from a Bernoulli distribution with equal probability. Afterward, the bits are mapped via a constellation diagram to symbols, in this case via 16-quadrature amplitude modulation (QAM). The difference between OFDM and SC-FDM signal modulation is implemented straightforwardly according to [8]. For OFDM, the symbols are considered complex carrier weights in the frequency

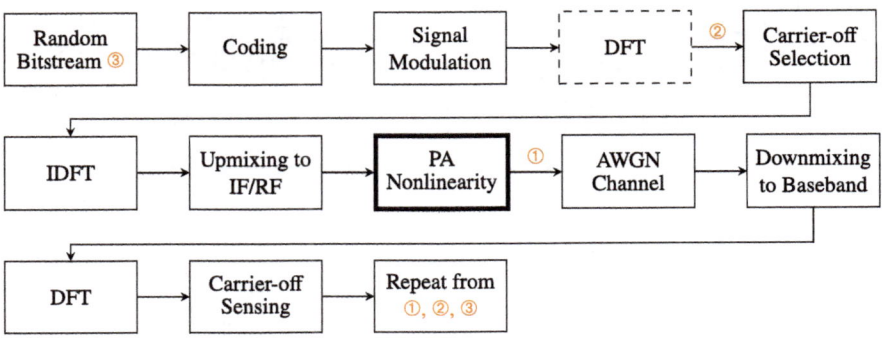

Fig. 25.6 Simulation signal flowchart

domain, while for SC-FDM, they are considered complex time-domain samples. Hence, the processing only differs by a DFT. Subsequently, the power at one frequency bin is set to zero. This can be interpreted for OFDM as sacrificing one carrier for sensing in-band distortions and for SC-FDM as implementing a narrow notch filter in the frequency domain. Afterward, the processing for both modulation types is identical. An inverse discrete Fourier transform (IDFT) transforms to the complex baseband signal, and an IQ mixer mixes up to IF or RF frequencies resulting in a real-valued signal. The power amplifier (PA) is simulated as a simple and memoryless compression curve similar to the observations in Fig. 25.5b as

$$\tilde{x}[n] = \begin{cases} x[n] & x[n] \leq t \\ t + (x[n] - t)/r & x[n] > t \end{cases} \tag{25.2}$$

with time-domain waveform $x[n]$, threshold t, and compression ratio r. For OFDM and SC-FDM, it is well known that they differ heavily in the amplitude distribution of the time-domain waveform and their peak-to-average power ratio (PAPR). For OFDM, the IDFT leads to a (weighted) sum over the carrier weights for each time-domain sample. Consequently, the central limit theorem states that the resulting samples tend to be normally distributed for a sufficiently high number of carriers since their weights are drawn independently from the same distribution (constellation map). On the other hand, the time-domain modulation of SC-FDM leads to a strictly limited amplitude distribution that can achieve very low PAPRs, e.g., in the case of binary phase-shift keying. Hence, for a fair comparison, we normalize both waveforms to their signal power before applying it to (25.2) resulting in selecting the same power backoff relative to the signal power for both modulation types. This is shown in Fig. 25.7. It is apparent that OFDM results in a much higher

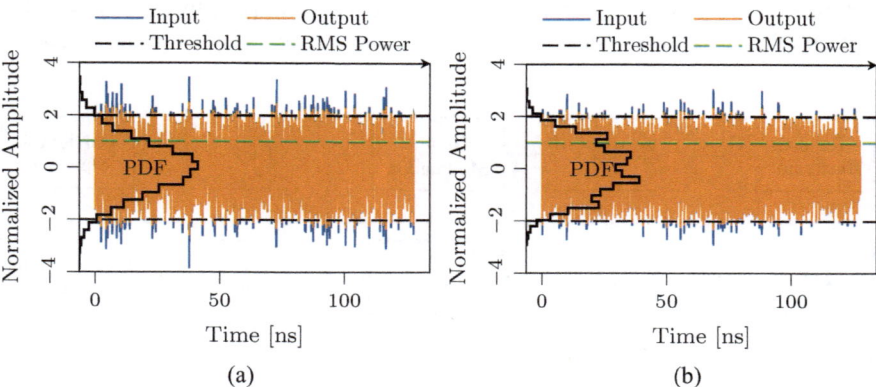

Fig. 25.7 Input and output waveform of the simulated PA nonlinearity for (**a**) an OFDM modulated signal and (**b**) a SC-FDM modulated signal. The PDF denotes the amplitude distribution of the input waveform

PAPR, and hence, the waveform gets more severely distorted compared to the SC-FDM signal. For the sake of simplicity, an additive white Gaussian noise (AWGN) channel is considered after the PA, followed by the receiver blocks responsible for down-mixing to baseband, calculating the DFT and sensing the power at the deactivated carrier position.

Now, several repetitions of the process are required:

① If the distortions are weak, they may be covered by the noise floor. To increase the distortion-to-noise ratio, it can be exploited that the distortions deterministically depend on the transmit signal. Hence, by periodically repeating the same transmit signal, coherent averaging of the sensed power allows a reduction of the noise floor while keeping the distortion power unchanged.

② To ensure that measuring the distortions with deactivated carriers provides a reasonable approximation for signals with a fully occupied frequency spectrum, the number of deactivated carriers should be small. In our case, we just deactivated one carrier at a time. To still get a broadband view on the distortion spectrum, the same symbol has to be sent with a different carrier deactivated each time.

③ As discussed previously, the distortion power has to be considered a random process for varying transmit symbols. Hence, it is required to estimate the distribution of the distortion power as it may occur in an application due to random and varying data streams. We estimate this distribution by performing the simulation several times for different randomly drawn data bits.

The simulation results are shown in Fig. 25.8 and discussed in the next section.

25.3.1.2 Numerical Examples

Figure 25.8a, b shows the simulation result in the frequency domain for linear system behavior and OFDM and SC-FDM modulation, respectively. Hence, the threshold (power backoff) of the PA has been chosen sufficiently high. The blue line shows the mean power over all transmitted symbols for one carrier-off position. The black dots denote the mean of the sensed power at the deactivated carrier positions and the error bars the standard deviation of the power around the mean. In this case, just the AWGN is observed.

Figure 25.8c, d shows the results for nonlinear system behavior as in Fig. 25.7 for a single OFDM and SC-FDM symbol, respectively. The green line shows the mean power for a single carrier-off position. It is observable that the variance for the power measurement is quite small just resulting from the AWGN.

Figure 25.8e, f shows the simulation for a sequence of random symbols. It is interesting to see that the overall shape of the distortion spectrum as well as the variance is comparable between OFDM and SC-FDM, although the time-domain waveforms are completely different. However, the overall distortion power is lower for SC-FDM due to the lower PAPR as expected. A deeper look on the underlying distribution for the distortion power is given in the next section.

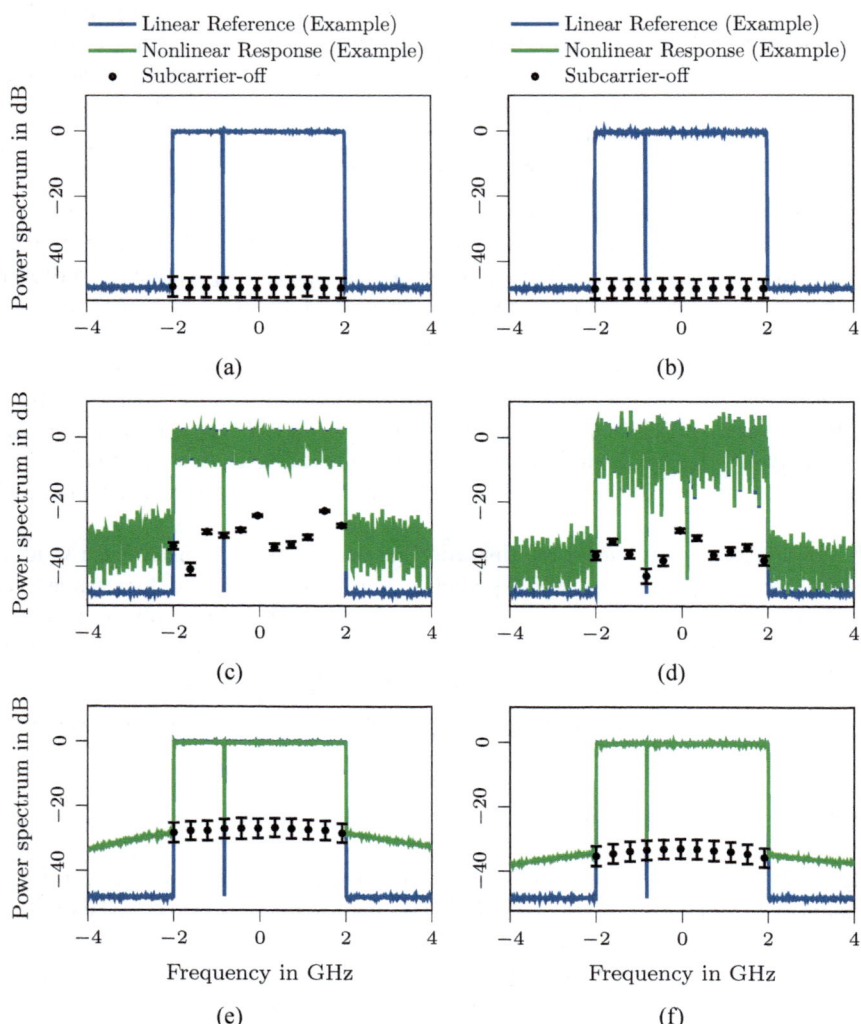

Fig. 25.8 Sensed carrier-off power in the case of (**a, b**) a linear system response, (**c, d**) a nonlinear system response while periodically repeating a single symbol, and (**e, f**) a nonlinear system response for a random symbol sequence. The blue line shows for every case a linear response as a reference for one carrier-off position. The green line shows the nonlinear response for the same carrier-off position. The black dots visualize the mean power at all carrier-off positions and the error bars the standard deviation of the sensed power around the mean. The figures on the left (**a, c, e**) correspond to OFDM symbols, while (**b, d, f**) correspond to SC-FDM symbols

25.3.1.3 Statistical Distribution of the In-Band Distortion Power

Figure 25.9a–c shows the distribution of the sensed power for a single carrier-off position for an OFDM modulated signal where the three cases correspond to the linear operating range and the nonlinear operating range for periodic transmission of a single symbol and the transmission of multiple random symbols (comparable to Fig. 25.8a, c, e). The power for the complex-valued reading at the carrier position is

$$P[f] = X_{real}^2[f] + X_{imag}^2[f] \tag{25.3}$$

If real and imaginary part are drawn from zero-mean Gaussian distributions as in the *linear* case where the sensed value is just composed of AWGN, it is well known that the power follows a χ^2 distribution with two degrees of freedom:

$$\left(\frac{X_{real} - \mu_{real}}{\sigma_{real}}\right)^2 + \left(\frac{X_{imag} - \mu_{imag}}{\sigma_{imag}}\right)^2 \sim \chi^2. \tag{25.4}$$

Real and imaginary part have to be normalized by their variance. This case is shown in Fig. 25.9a.

When periodically transmitting a single symbol, real and imaginary parts are still Gaussian distributed, while the mean is defined by the (deterministic) distortion power and the variance by the AWGN. In this case, the power follows a noncentral χ^2 distribution with two degrees of freedom and noncentrality parameter $\lambda = \mu_{real}^2 + \mu_{imag}^2$:

$$\left(\frac{X_{real}}{\sigma_{real}}\right)^2 + \left(\frac{X_{imag}}{\sigma_{imag}}\right)^2 \sim \chi_{nc}^2. \tag{25.5}$$

This case is shown in Fig. 25.9b.

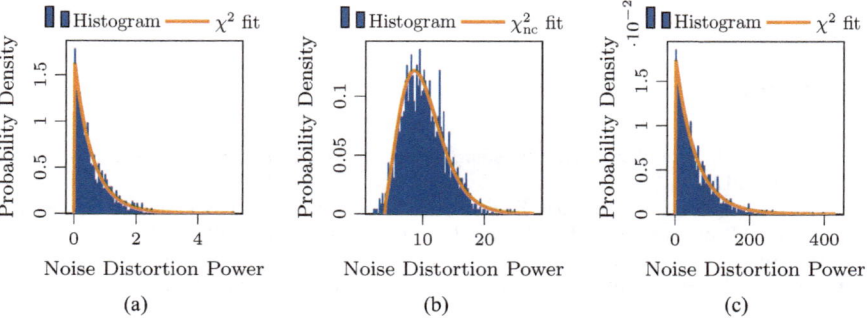

Fig. 25.9 Histogram of the power distribution and distribution fit for a single carrier-off position and an OFDM modulated signal. (**a**) corresponds to the linear operating case, (**b**) to the nonlinear case when periodically repeating a single symbol, and (**c**) to a random symbol stream and nonlinear operation

Interestingly, even in the case of random symbol transmission, real and imaginary parts are Gaussian distributed. Compare, e.g., Theorem 3.9 and 3.11 in [1] to see that stochastic nonlinearities are zero mean and circular complex normally distributed for random multisine excitation signals. Both OFDM and SC-FDM can be considered as belonging to this class of excitation signals when considering the frequency spectrum of the SC-FDM as carrier weights for an OFDM signal. Hence, in this case, the power follows again a χ^2 distribution as shown in Fig. 25.9c. The problem of estimating the distributions can now, in all cases, be formulated as estimating the mean and the variance of the real and imaginary part as measured/simulated at the deactivated carrier position.

25.4 Conclusion

In this chapter, we presented two approaches for behavioral modeling of THz-systems.

First, we extracted the behavior of a Sub-THz PA, based on polynomial and Rapp models. The models were verified by comparing the EVM of system-level simulations to modulated measurements, which are obtained by a VNA-based cross-domain measurement platform. Within the simulation, the compression behavior of the models matches the measurement data toward strong compression. For more accurate operation, the requirement of frequency-dependent PA models is indicated, to suit ultra-broadband systems.

Second, we presented a statistical modeling approach based on the noise-power ratio (NPR) test. This method allows estimating and modeling the in-band power spectral density (PSD) of the distortion power by introducing *gaps* in the frequency spectrum of the measurement waveform. The measurement waveform itself must statistically emulate the waveforms of a desired application. In this chapter, we presented simulation results using OFDM and SC-FDM waveforms. In these cases, the in-band distortion power follows a χ^2 distribution whose parameters can be straightforwardly estimated from a sequence of measurements.

References

1. R. Pintelon, J. Schoukens, *System Identification: A Frequency Domain Approach* (Wiley, Hoboken, 2012)
2. A. Geens, Y. Rolain, W. Van Moer, K. Vanhoenacker, J. Schoukens, Discussion on fundamental issues of NPR measurements. IEEE Trans. Instrum. Measure. **52**(1), 197–202 (2003)
3. S. Haussmann, D. Wrana, L. Samara, T. Zugno, I. Kallfass, Towards system-level modeling of broadband sub-THz amplifiers for wireless communication, in *2025 IEEE Radio Wireless Week* (2025)

4. I. Kallfass, D. Wrana, B. Schoch, J. Hesler, M. Kohler, J.-P. Teyssier, J. Dunsmore, Instrumentation for the time and frequency domain characterization of terahertz communication transceivers and their building blocks, in *2023 IEEE/MTT-S International Microwave Symposium - IMS 2023* (2023), pp. 1030–1033
5. 3GPP, Realistic power amplifier model for the New Radio evaluation, TSG RAN WG4 Meeting 79, R4-163314 (2016)
6. L. John, A. Tessmann, A. Leuther, P. Neininger, T. Merkle, T. Zwick, Broadband 300-GHz power amplifier MMICs in InGaAs mHEMT technology. IEEE Trans. THz Sci. Technol. **10**(3), 309–320 (2020)
7. J. Gedschold, D. Dupleich, S. Semper, M. Döbereiner, A. Ebert, G.D. Galdo, R.S. Thomä, Metrology of multicarrier-based delay-doppler channel sounding for sub-THz frequencies. Open J. Antennas Propag. (2024) [Online]. Available: https://doi.org/10.36227/techrxiv.173337536. 61771937/v1
8. D. Falconer, S. Ariyavisitakul, A. Benyamin-Seeyar, B. Eidson, Frequency domain equalization for single-carrier broadband wireless systems. IEEE Commun. Mag. **40**(4), 58–66 (2002)

Chapter 26
Electronic Signal Generation

Dominik Wrana and Ingmar Kallfass

Abstract This chapter focuses on the electronic signal generation of THz carrier signals at 300 GHz based on electronic frequency multipliers. With the inherent generation of unwanted harmonics posing the risk of interferers in the RF domain, evaluation of the harmonic suppression is a key measurement requirement. A 300-GHz LO chain is characterized and analysed as a measurement example. Substrate-integrated waveguide (SIW) filters present a suitable option to, aside from on-chip filtering, further increase the spectral purity in-between cascaded multiplier microwave monolithic integrated circuits (MMICs) as well as at the desired output frequency range. Additionally, phase noise measurements carried out at 300 GHz are presented, and compared to the theoretical phase noise degradation values.

26.1 Introduction

Local oscillator (LO) signal or carrier generation is one of the key parts of analogue THz communication frontends. In communication frontends, it is driving the up- and down-conversion mixers, the core element for frequency translation of the broadband modulated signal to and from the THz frequency domain. In wireless links but also in measurement instrumentation for component characterization at THz frequencies, the generation of clean test signals, e.g., for vector signal analysis is of great importance.

There are different approaches to either generate carriers directly at THz frequencies or to generate the fundamental signal at a lower frequency and to bring it to the required frequency range using frequency multiplication. While there are also optical/photonics-based approaches, this chapter will focus on purely electronic frequency multiplicative carrier generation. Also, hybrid approaches are possible as done in [1]. There, an optical carrier is generated at around 70 GHz using photo

D. Wrana (✉) · I. Kallfass
Universität Stuttgart, Institut für Robuste Leistungshalbleitersysteme, Stuttgart, Germany
e-mail: dominik.wrana@ilh.uni-stuttgart.de; ingmar.kallfass@ilh.uni-stuttgart.de

© The Author(s) 2026
T. Kürner et al. (eds.), *Metrology for THz Communications*, Springer Series in Optical Sciences 256, https://doi.org/10.1007/978-3-032-01986-8_26

331

mixing in conjunction with a uni-traveling carrier photodiode (UTC-PD). This signal is injected into a superheterodyne frontend where a further multiplication by a factor of three is performed on-chip of the electronic frontend MMIC. The widely used approach in current state-of-the-art wireless link demonstrators is a frequency source at lower frequencies (e.g., at X-band frequencies) delivering the input to a chain of electronic frequency multipliers. One drawback of this approach is the required high multiplication factor to reach the 300-GHz frequency range, thus making the LO generation bulky for full monolithic integration on the transmit and receive frontend MMICs and generating a multitude of harmonics at the RF domain, potentially degrading the link capabilities.

26.2   Electronic Frequency Multipliers

Electronic frequency multipliers are implemented using the nonlinear characteristic of diodes or transistors in class-B or class-C operation mode. While it is possible to design and specifically match the output to suit higher harmonics ($n > 3$) for a single multiplier cell, most common implementations focus on frequency doublers ($n = 2$) or triplers ($n = 3$). This is also due to the reduced available power at higher frequencies, increasing the post-amplification effort. Balanced architectures have the advantage of inherently suppressing odd-order harmonics. The overall required multiplication factor is achieved by a concatenation of multiple multiplier cells, potentially also inserting an intermediate amplification stage to drive the following stage accordingly. Depending on the chosen fundamental input frequency as well as the frontend mixer topology being fundamental or subharmonic, typical multiplication factors vary between six and 36 for H-band (220–330 GHz) wireless links. Targeting the coverage of wide RF frequency ranges in the H-band (220–330 GHz) as proposed, e.g., by the IEEE802.15.3d standard, large single-channel bandwidths are foreseen as well as the aggregation of multiple considerably narrowband channels with bandwidths as low as 2.16 GHz [2]. To be able to address all the corresponding center frequencies, a considerably large tuning range of the LO signal is necessary. This directly translates to the bandwidth requirements of the frequency multipliers and corresponding suppression of harmonics to ensure a spur-free RF spectrum.

26.2.1   Harmonic Suppression

The main design goal in the development of electronic frequency multipliers, other than output power and bandwidth, is the suppression of unwanted harmonics at the output. Those, due to the limited LO isolation of the mixing devices in communication frontends, are partially leaked to the RF domain and therefore pose

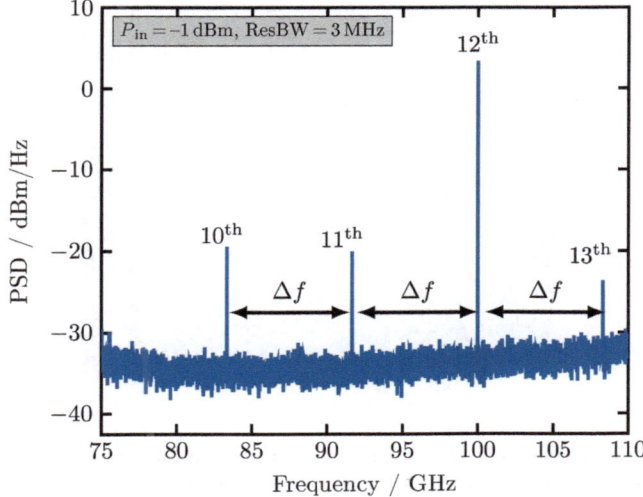

Fig. 26.1 Measured output power spectral density of the X12 electronic frequency multiplier at an input frequency of 8.333 GHz, corresponding to a wanted output signal of 100 GHz at an input power of −1 dBm. The measurement is limited to the W-band (75–110 GHz) frequency range of the spectrum analyzer extension module

the risk of generating in-band interferers for the broadband modulated signals, ultimately degrading the signal quality and reducing the achievable data rate.

In the following paragraph, a 300-GHz LO chain, similar to the one used for the frontends presented in [3] and [4], with an overall multiplication factor of 36, achieved by concatenation of individual X12 and X3 frequency multiplier modules, is evaluated exemplarily with respect to its harmonic suppression as function of input frequency and fundamental input power. First, in Fig. 26.1, the output spectrum of the X12 electronic frequency multiplier, similar to the one presented in [5], is shown for a input frequency f_{in} of 8.333 GHz and corresponding output frequency of 100 GHz. At an input power of − 1 dBm, the output power at the desired output frequency of 100 GHz is around 5 dBm, while the unwanted neighboring harmonics 10, 11, and 13 with equal separation of $\Delta f = f_{in}$ are suppressed by 25 dB or more.

However, the suppression is subject to variation along with frequency as well as input power. Usually, electronic frequency multipliers are operated in saturation to avoid variation of the LO drive level for following mixers and thus potentially varying their IF-to-RF conversion gain.

The evolution of the output power of the desired 12th harmonic is shown in Fig. 26.2 along with the unwanted harmonics 10, 11, 13, and 14 in Fig. 26.3a–d. The saturated output power plateau of at least 4.5 dBm is reached for an input power of −7 dBm or higher within a frequency range of 7.1 to 8.9 GHz. For the other harmonics, there is not necessarily a monotonous increase of the power observed

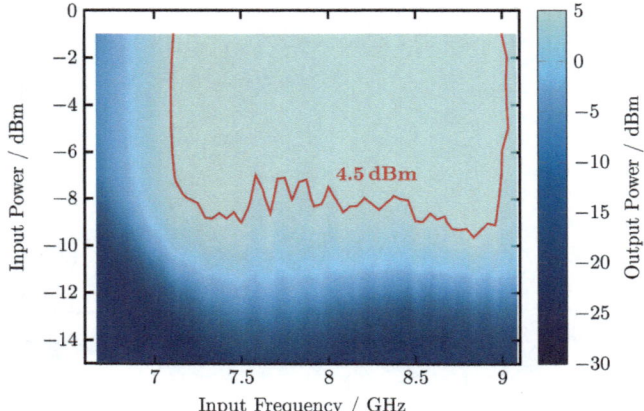

Fig. 26.2 Measured output power over input frequency and input power of the X12 frequency multiplier for the wanted 12th harmonic. The measurement is limited to the W-band (75–110 GHz) frequency range of the spectrum analyzer extension module

but rather selective sweet and sour spots. Especially the 14th harmonic for selective frequencies reaches up to 0 dBm, resulting in less than 5 dB suppression. Overall for a wide range of frequencies, a typical suppression of 20 to 25 dB is achieved.

Analogue to the evaluation of the intermediate interface after the X12 frequency multiplier, the output spectrum of the X3 multiplier, comparable to the ones presented in [6], is characterized in Fig. 26.4 over the frequency range of 250 to 330 GHz, operated in conjunction with the X12 module at the same fundamental f_{in} of 8.333 GHz and fundamental P_{in} of −1 dBm. Along with the desired tone at 300 GHz, a total of seven harmonics associated with the frequency multiplication mechanism is visible, whereas the four closest harmonics 34, 35, 37, and 38 significantly obstruct the RF spectrum, each with a suppression of around 20 dB only.

The evolution of the power contributions of the individual harmonics over the fundamental input frequency and power to the X3 module is depicted in Figs. 26.5 and 26.6. Reaching output power levels of −5 dBm, the worst suppression of 15 dB is achieved at the 37th at lower frequencies. Other than that, typically the power of the harmonics does not significantly exceed −25 dBm, corresponding to 20 dB of suppression.

26.2.2 Off-Chip Harmonic Filtering Using Substrate-Integrated Waveguide Filters

While full monolithic integration of the entire analogue transmit or receive frontend is possible, lower frequency parts of the LO generation are still often implemented

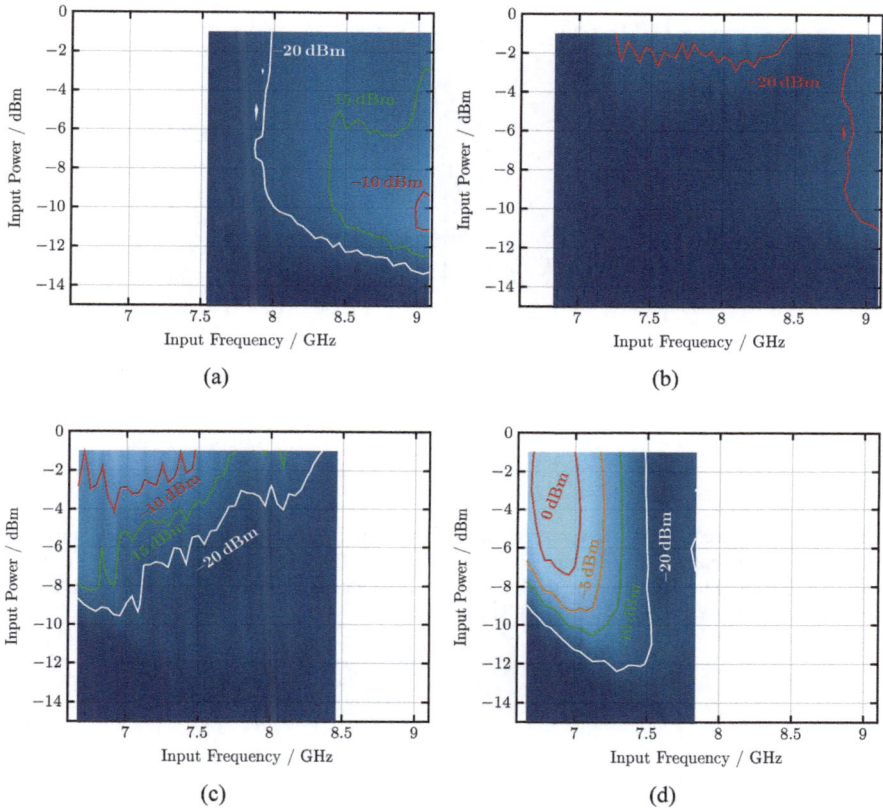

Fig. 26.3 Measured output power over input frequency and input power of the X12 frequency multiplier for the unwanted 10th (**a**), 11th (**b**) 13th (**c**), and 14th (**d**) harmonic. The measurement is limited to the W-band (75–110 GHz) frequency range of the spectrum analyzer extension module

separately on another MMIC, potentially using even different semiconductor technologies than the THz Tx and Rx frontends. This interface offers the possibility to insert dedicated filters for the improvement of harmonic suppression and increasing the spur-free available RF bandwidth. Depending on the MMIC packaging technology, there are several possibilities to implement filtering into the chain. On waveguide level, there are commercially available waveguide filters. Printed circuit board (PCB) based hetero-integration of the multiplier MMICs enables the usage of substrate-integrated waveguide filters (SIWs), like the ones presented in [7], offering a compact and low-cost solution. However, a desired wide LO tuning range potentially interferes with the usage of fixed filters. Tunable filters could be a more flexible solution.

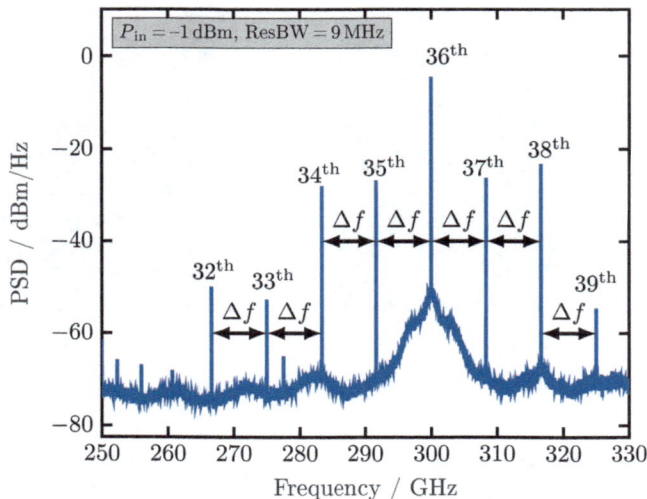

Fig. 26.4 Measured output power spectral density of the X36 electronic frequency multiplier chain at a fundamental input frequency of 8.333 GHz and a fundamental input power of -1 dBm to the X12, resulting in a wanted output signal at 300 GHz along with multiple unwanted harmonics over the frequency range of 250 to 330 GHz

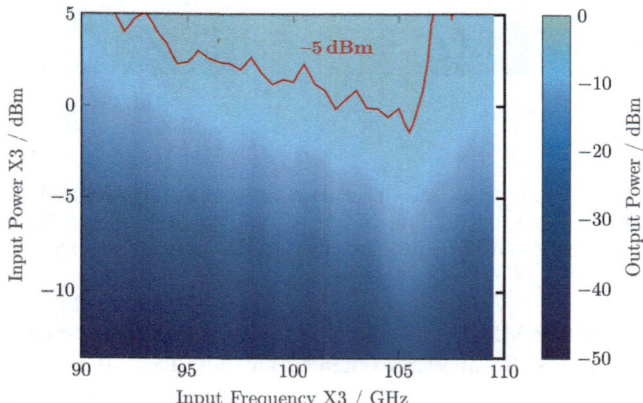

Fig. 26.5 Measured output power over input frequency and fundamental input power of the X3 frequency multiplier for the wanted overall 36th harmonic

26.3   Phase Noise Measurement at THz Frequencies

Apart from unwanted harmonic generation, the phase noise of the initial frequency source and the phase noise degradation introduced by the frequency multiplication is another impairment that needs to be taken into account when characterizing the performance of electronic signal generation for THz applications. Phase noise in

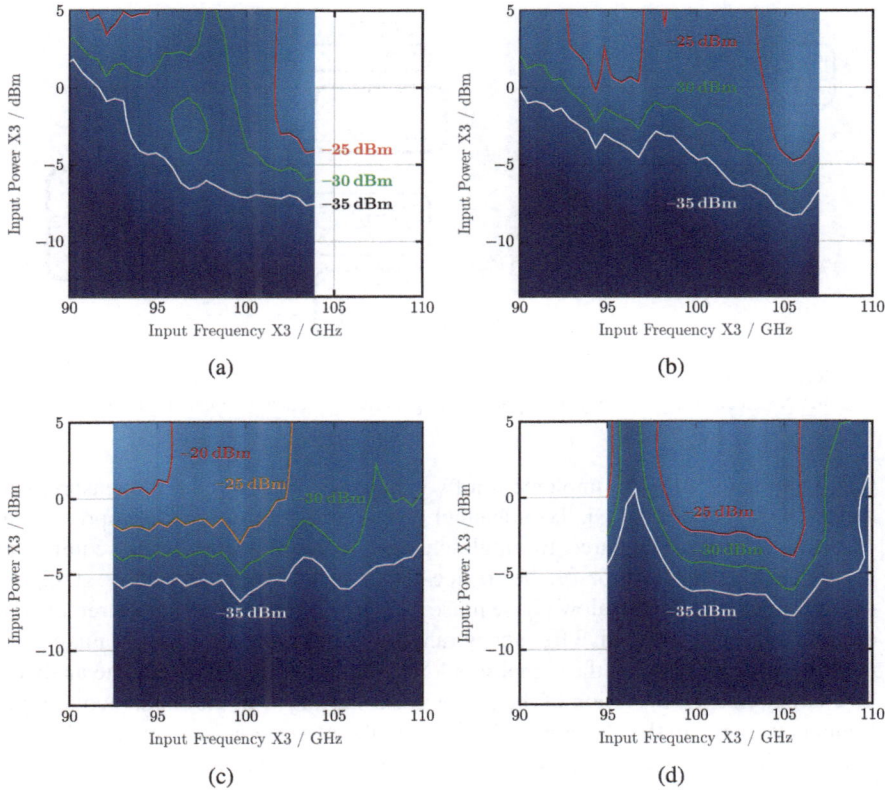

Fig. 26.6 Measured output power over input frequency and fundamental input power of the X3 frequency multiplier for the unwanted 34th (**a**), 35th (**b**), 37th (**c**), and 38th (**d**) harmonic

general is a performance metric obtained by a frequency domain measurement and describes the relative signal power at a certain frequency offset from the carrier frequency in dBc/Hz, commonly plotted as the single-sideband (SSB) phase noise. Depending on the type of frequency source and corresponding noise sources, the phase noise is shaped differently. The phase noise degradation for an ideal frequency multiplication is given as

$$PN_{\text{degradation, dB}} = 20 \cdot log(n). \tag{26.1}$$

26.3.1 Measurement Setup

Spectrum analyzers as well as vector network analyzers typically have built-in capabilities for characterizing phase noise within their frequency range of opera-

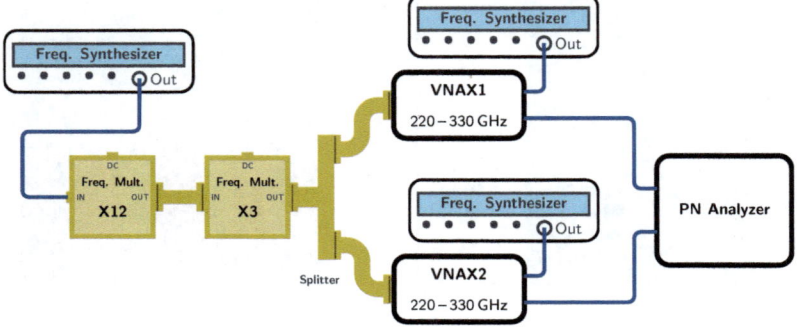

Fig. 26.7 Measurement setup for the characterization of the phase noise degradation through the X12 and X3 electronic frequency multiplier chain generating a carrier signal at 306 GHz

tion. Only some of them implement a two-channel cross-correlation measurement method as explained in [8]. Two-channel cross-correlation offers an improvement of the noise floor compared to single-channel measurements and is required to get to the real noise floor of the device under test (DUT), especially at higher offset frequencies for ultralow phase noise DUTs. For phase noise measurements of carriers exceeding the native frequency range of the instrument, external mixers are required to down-convert the signal to a low IF frequency, suitable for the analysis. To characterize the evolution of the phase noise starting with the fundamental input frequency source in the X-band, and following the corresponding signals in W- and H-band, the setup shown in Fig. 26.7 is used. To precondition the carriers from W- and H-band frequencies, two uncorrelated down-conversion paths to a low IF frequency around 2 GHz are implemented. Both outputs are fed to an Anapico APPH20G phase noise analyzer unit.

26.3.2 Phase Noise of a 300-GHz LO Chain

With the measurement setup presented in Sect. 26.3.1, the LO chain from X- to H-band described in Sect. 26.2.1 is characterized with respect to the evolution of the phase noise degradation, using a laboratory-grade Agilent N5183B frequency synthesizer as a signal source. In the upper plot of Fig. 26.8, the measured single-sideband phase noise for offset frequencies from 10 Hz to 50 MHz is shown for the synthesizer output at 8.5 GHz, after the X12 frequency multiplier at 102 GHz, and finally at the output of the X36 chain at 306 GHz. In the lower plot of Fig. 26.8, the corresponding PN degradation is plotted separately for the X12 and overall X36 multiplication and compared to the theory. According to Eq. 26.1 and assuming negligible additive phase noise of the down-conversion paths, the theoretical degradation of the phase noise is equal to 21.6 dB for the frequency

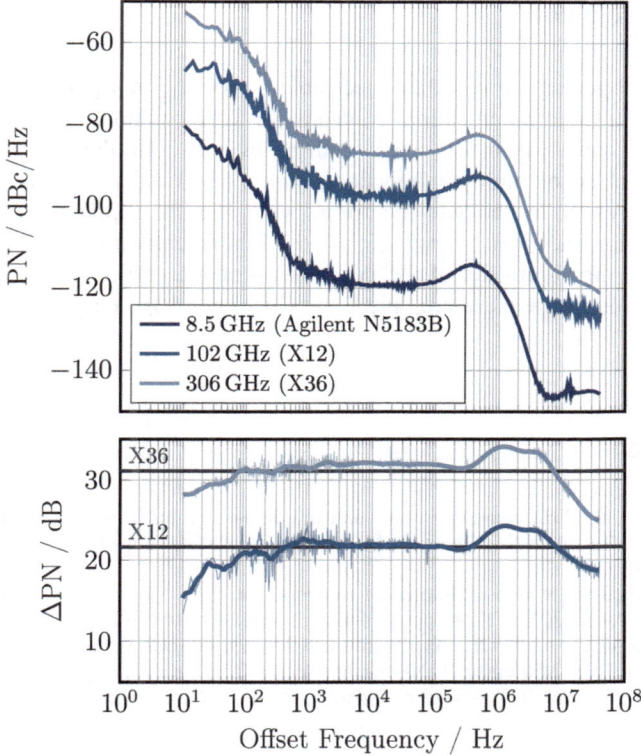

Fig. 26.8 The upper plot shows the evolution of the measured phase noise in an electronic frequency multiplier chain from X- to W-band and from W- to H-band with an overall multiplication factor of 36, targeting fundamental up- and down-conversion of baseband modulated signals to and from the 300 GHz frequency range. The lower plot additionally compares the measured phase noise degradation with the theoretical values obtained from the 20·log(n) rule of thumb

multiplication by a factor of twelve and to 31.1 dB for the total multiplication by a factor of 36. Overall, the measurements show excellent agreement with the theoretical calculation from 50 Hz to 500 kHz, proving the applicability of the formula for the approximation of the PN degradation for THz frequencies generated by LO chains with high multiplication factors. For higher offset frequencies, the measured values drop below the calculated one which can be the result of the actual noise floor for the initial frequency source being lower as it was measured with the setup.

26.4   Conclusion

Carrier signal generation based on electronic frequency multiplication is a widely used approach in communication frontends for THz communication systems. As exemplarily shown for an LO chain with an overall multiplication factor of 36, the concatenation of multiple frequency multipliers results in several unwanted harmonics present in the RF spectrum, with varying suppression along with input frequency and input power level of the fundamental multiplier drive signal. Additionally, measurements of the degradation of the carrier phase noise due to frequency multiplication are presented, showing a good agreement with theoretical values.

References

1. T. Kürner et al., THz communications and the demonstration in the ThoR–Backhaul link. IEEE Trans. Terahertz Sci. Technol. **14**(5), 554–567 (2024). https://doi.org/10.1109/TTHZ. 2024.3415480
2. IEEE Standard for Wireless Multimedia Networks, in *IEEE Std 802.15.3-2023 (Revision of IEEE Std 802.15.3-2016)* (2024), pp. 1–684. https://doi.org/10.1109/IEEESTD.2024.10443750
3. I. Dan, TERAPAN: a 300 GHz fixed wireless link based on InGaAs transmit-receive MMICs, in T. Kürner, D.M. Mittleman, T. Nagatsuma, (eds.) *THz Communications. Springer Series in Optical Sciences*, vol. 234 (Springer, Cham, 2022). https://doi.org/10.1007/978-3-030-73738-2_33
4. I. Kallfass et al., MMIC chipset for 300 GHz indoor wireless communication, in *2015 IEEE International Conference on Microwaves, Communications, Antennas and Electronic Systems (COMCAS)*, Tel Aviv (2015), pp. 1–4. https://doi.org/10.1109/COMCAS.2015.7360494
5. U.J. Lewark et al., Ultra-broadband W-band frequency multiplier-by-twelve MMIC, in *2015 10th European Microwave Integrated Circuits Conference (EuMIC)*, Paris (2015), pp. 5–8. https://doi.org/10.1109/EuMIC.2015.7345054
6. C.M. Groetsch, S. Wagner, A. Leuther, D. Meier, I. Kallfass, Ultra-broadband frequency multiplier MMICs for communication and radar applications, in *2018 13th European Microwave Integrated Circuits Conference (EuMIC)*, Madrid (2018), pp. 113–116. https://doi.org/10.23919/EuMIC.2018.8539865
7. D. Wrana, M. Guenter, S. Haussmann, I. Kallfass, Compact E-band SIW filters targeting PCB-based MMIC hetero-integration, in *2024 15th German Microwave Conference (GeMiC)*, Duisburg (2024), pp. 209–212. https://doi.org/10.23919/GeMiC59120.2024.10485315
8. G. Feldhaus, A. Roth, A 1 MHz to 50 GHz direct down-conversion phase noise analyzer with cross-correlation, in *2016 European Frequency and Time Forum (EFTF)*, York (2016), pp. 1–4. https://doi.org/10.1109/EFTF.2016.7477759

Chapter 27
Electronic Signal Conversion, Signal Amplification, and Distortion Mechanisms

Dominik Wrana, Ingmar Kallfass, and David A. Humphreys

Abstract This chapter focuses on electronic fully integrated transmit and receive frontends and their building blocks, targeting ultra-broadband wireless communication systems operating at 300 GHz. Distinguishing between direct-conversion and heterodyne frontend architectures, several circuit-level non-idealities and signal impairments arising from the LO generation as well as the IF-to-RF conversion are discussed.

27.1 Electronic THz Frontend Architectures

Transmit and receive frontends for wireless THz communication systems operating at a carrier frequency around 300 GHz have been reported in a wide variety of semiconductor technologies, both silicon based, like CMOS and SiGe, and III–IV compound semiconductor based, like InGaAs or InP. Independent of the technology, the same common building blocks are used to realize the microwave monolithic integrated circuits (MMICs) commonly containing a local oscillator (LO) chain driving the up- or down-conversion mixer and, depending on the frontend being a transmitter or receiver, an RF power or low-noise amplifier, respectively. The core building blocks can be separated in frequency-converting blocks, like frequency multipliers or mixers, and non-frequency-converting blocks like amplifiers or passives like couplers and splitters. There are also other building blocks, like power detectors, switches, or phase shifters which enable more and more sophisticated frontend capabilities like pre-distortion, self-calibration, or built-in self-test (BIST). In terms of general frontend architectures, two main variants are widely used in

D. Wrana (✉) · I. Kallfass
Universität Stuttgart, Institut für Robuste Leistungshalbleitersysteme, Stuttgart, Germany
e-mail: dominik.wrana@ilh.uni-stuttgart.de; ingmar.kallfass@ilh.uni-stuttgart.de

D. A. Humphreys
NPL, UK (Retired), University of Bristol, Ascot, UK
e-mail: david.a.humphreys@ieee.org

© The Author(s) 2026
T. Kürner et al. (eds.), *Metrology for THz Communications*, Springer Series in Optical Sciences 256, https://doi.org/10.1007/978-3-032-01986-8_27

343

wireless communication systems, the direct-conversion approach and the (super-)heterodyne approach.

27.1.1 Direct-Conversion Architecture

The direct-conversion architecture typically implements DC-coupled IF interfaces supporting baseband signal injection with up to 30 GHz of bandwidth as in [1]. To enable the transmission of complex modulated signals, the direct-conversion frontend implements separate mixer cells for in-phase (I) and quadrature-phase (Q) baseband signals. The quadrature phase is achieved either by a 90° phase shift of the LO signals of the I and Q mixer cells or in the RF domain by combining the I and Q branches using a 90° hybrid, as done in [2]. With at least two mixer cells (four if the mixer is realized in a balanced fashion), the LO buffer amplifier is required to deliver twice (or four times) the output power compared to the heterodyne approach and comparable fundamental LO frequencies. When injecting a quadrature-modulated signal on a low intermediate frequency to the I and Q branches, the architecture can be used like an image-reject mixer, only providing the upper or lower sideband of the RF spectrum. When used as a standard double-sideband mixer, the LO leakage is always located at the very center of the modulated signal, making it difficult to suppress or filter out the leakage component from the RF signal.

27.1.2 (Super-)Heterodyne Architecture

The characteristic feature of the heterodyne or superheterodyne architecture as shown in Fig. 27.1b is that there is only one frequency-converting mixer cell whose IF interface is usually not DC coupled and supports frequencies from a couple of gigahertz up to some tens of gigahertz or as presented in [3, 4] even operates at E-band (60–90 GHz). The injection of complex modulated signals using quadrature-amplitude modulation (QAM) or phase-shift keying (PSK) formats requires the data signal to be already modulated onto a carrier frequency. This is either done digitally using arbitrary waveform generators which can deliver up to 256 GSa/s and 80 GHz of analogue bandwidth, modems, or FPGAs in conjunction with high-performance DACs/ADCs. Or it is done by using an intermediate analogue frequency conversion unit, as it is done in [4] where the baseband IF signals are converted to E-band (60–90 GHz) using a balanced I/Q transmitter module.

There are several advantages of the superheterodyne frontend, compared to the direct-conversion architecture. One, the frontend does not add I/Q impairments such as gain and phase imbalances, as the quadrature phase is either achieved digitally or by another analogue frontend. Second, the overall circuit complexity is lower as there is no need for I and Q mixer cells and correspondingly also the required LO drive power is less, potentially making the LO buffer amplifier smaller in size. Third,

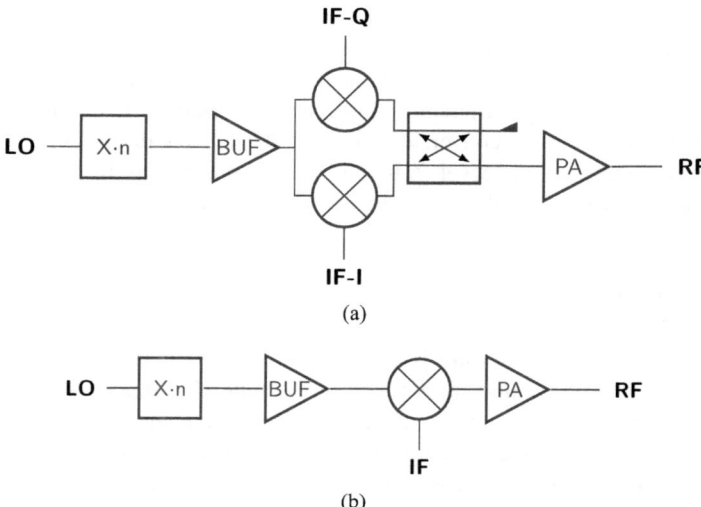

Fig. 27.1 Generic block diagrams of (**a**) a typical direct-conversion and (**b**) a (super)heterodyne transmitter architecture, including frequency multiplier (Xn), buffer (BUF), and power amplifier (PA) building blocks. A local oscillator (LO) signal drives the frequency-converting mixers to translate the intermediate frequency (IF) signals to the radio frequency (RF) domain. For the direct-conversion architecture, a quadrature coupler is used in the RF domain to combine the in-phase (I) and quadrature (Q) signal paths

with considerably high IF carrier frequencies, the multiplication factor required in the LO chain to generate the fundamental LO signal is lower than for a fundamental direct-conversion frontend, considering an identical input frequency.

27.2 Frontend Impairments

Electronic analogue transmit and receive frontends are prone to several circuit-level non-idealities, leading to distortion of the signal and thus degradation of the signal quality. The LO generation based on electronic frequency multipliers provides unwanted harmonics as well as the LO-to-RF isolation of the mixer is limited. Both result in unwanted spectral components in the RF domain. Furthermore, the IF-to-RF conversion is limited by the linearity of the mixer and amplifier building blocks as well as potentially by I/Q impairment arising from quadrature hybrids.

27.2.1 LO Generation

In state-of-the-art fully integrated electronic THz frontends, the LO chain, together with the RF power amplifier, occupies the largest part of the frontend chip area. This is due to the fact that so far the initially injected LO source frequency lies in the range of 10 to 20 GHz which then requires electronic frequency multiplication with a large overall multiplication factor. In between the multiplier stages but mostly at the end of the multiplier chain, large buffer amplifiers are needed that are capable of generating sufficient output power to achieve the required LO drive level of the up- and down-conversion mixers. The LO chain also plays a crucial role when looking at circuit-level frontend impairments which limit the achievable performance regarding broadband modulated data transmission.

27.2.1.1 LO Harmonics

In Chap. 26, it is pointed out that electronic frequency multiplicative carrier generation is prone to limited suppression of unwanted harmonics. Especially with the requirement of a wide tuning range of the LO frequency, it is impossible to implement dedicated filters on-chip at the output of the frequency multiplier to isolate the required LO tone from any other spur. However, when filtering at an intermediate stage in the frequency multiplier chain, at least some spurs can be suppressed as it is demonstrated in [5] for the closest harmonics arising from the external times-eight frequency multiplier. While the problem of fundamental LO leakage described in 27.2.1.2 predominately is a problem in direct-conversion frontends, harmonics in general and particularly LO harmonics have to be considered as well in superheterodyne architectures targeting an ultra-broadband RF frequency range of operation. As elaborated in [5], the fourth harmonic of the on-chip frequency multiplier by three falls into the passband of the LO buffer amplifier but even more importantly the RF frequency band of operation, posing the risk of an in-band interferer. Figure 27.3a, b shows the characteristic behavior of the converted RF tone $f_{RF} = 3 \cdot f_{LO} + f_{IF}$ and the fourth LO harmonic $4 \cdot f_{LO}$, respectively, as a function of the LO input power for different LO frequencies. To obtain the maximum IF-to-RF conversion gain, the minimum required LO input power lies at -1 dBm. The power of the fourth LO harmonic has strong dependency on the LO frequency and reaches its maximum at around -20 dBm for an LO frequency of 70.5 GHz. In this case, the unwanted harmonic contributes as much power to the RF spectrum as the actual RF signal (Fig. 27.2).

Figure 27.3a–d further elaborates on the dependency over IF input power and derive the corresponding carrier-to-interferer power ratio:

$$CIR = 10 \cdot \log_{10}\left(\frac{P_{carrier}}{P_{interferer}}\right) \tag{27.1}$$

from the measured RF spectrum. For a very low IF input power of -18 dBm, the CIR amounts 11.2 dB, and it is almost linearly increasing with the IF input power of -12 dBm and -5 dBm to 17.4 and 24.1 dB, respectively. With the input-related 1-dB compression point of the module in the range of -4 to -2 dBm, the 0 dBm of IF input power in Fig. 27.3d is already in the nonlinear regime. While the spectral component of the fourth LO harmonic is around -35.8 dBm, the intermodulation

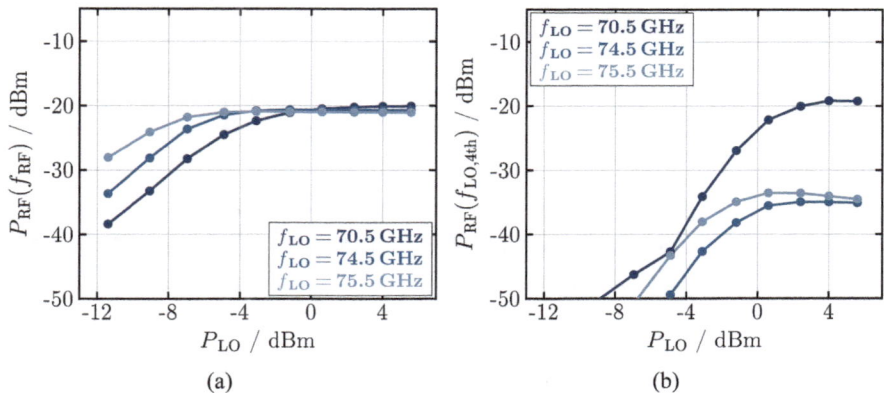

Fig. 27.2 Measured output power as a function of the LO input power for spectral contributions linked to (**a**) the converted RF tone $f_{RF} = 3 \cdot f_{LO} + f_{IF}$ and (**b**) the leaked fourth harmonic $4 \cdot f_{LO}$ (Adapted from [5])

products gain more power and now dominate the CIR of 24.3 dB. Overall, this leads to an optimum CIR of 24.3 dB for IF input power of -5 dBm.

In [6], a real-time full-duplex data transmission experiment is carried out, using the superheterodyne Tx and Rx hardware described in Sect. 27.1.2 as well as this section. It is demonstrated that even commercial modems with sophisticated signal correction and pre-distortion capabilities are data throughput limited in the corresponding transmission channels by the severely degraded CIR.

27.2.1.2 LO-to-RF Leakage

While the leaked fourth harmonic of the on-chip frequency multiplier described in 27.2.1.1 to some extent could also be categorized as LO-to-RF leakage, the term LO leakage or LO feedthrough usually refers to the limited isolation of the mixer regarding the fundamental LO tone driving the mixer. The severeness of this effect is slightly reduced in the presented superheterodyne architecture, as the fundamental tone between 210 and 230 GHz is located outside of the RF band of operation and therefore not affecting the RF amplifier. Generally speaking, in heterodyne architectures, the leaked fundamental LO tone is never in-band of the modulated signal. But with direct-conversion transceivers, LO-to-RF leakage is one of the factors, which limit the achievable signal quality already at the transmitter side. Arising from the finite LO isolation of the mixing devices, the leaked LO power can reach or even surpass the RF power of the converted modulated signal and possibly prematurely compress the RF power amplifier. In [7], a balanced I/Q up-converter for E-band (60–90 GHz) frequencies is characterized with respect to the LO leakage at different LO frequencies. Therefore, the total RF power is measured in Fig. 27.4a for LO frequencies of 73.5, 78.25, and 83.5 GHz with corresponding

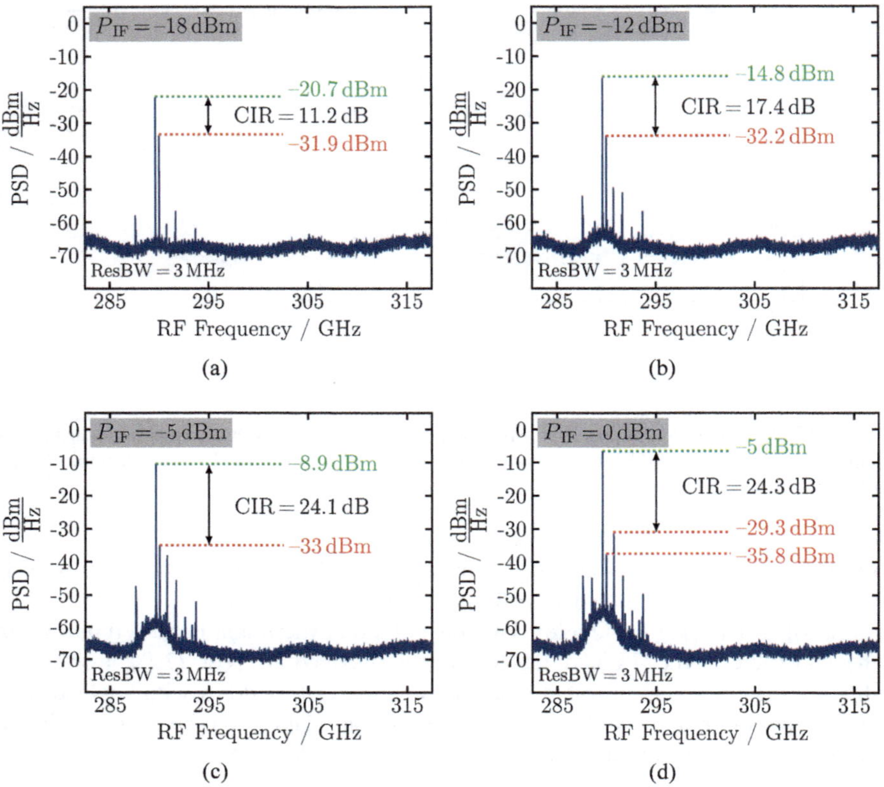

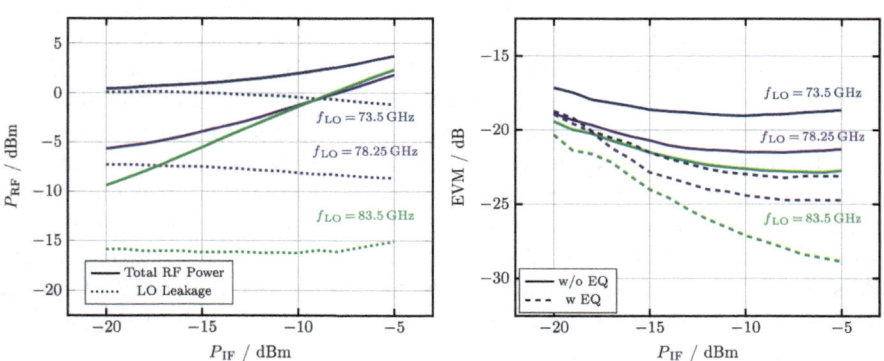

Fig. 27.3 Measured RF output spectrum of the superheterodyne Tx module for different IF input powers, indicating the evolution of the carrier-to-interferer power ratio dominated by a LO harmonic spur (Adapted from [5])

Fig. 27.4 Measured (**a**) total power present at the RF port of the E-band I/Q up-converter module and (**b**) EVM for a 1 GBd QPSK modulated signal for different LO frequencies (Adapted from [7])

power contributions of the LO leakage around 0, -8, and $-16\,$dBm, respectively. Performing an IF input power sweep, for 83.5 GHz, a linear increase of the total RF power is visible, pointing out that the modulated signal is the main contributor to the RF power. For 78.25 GHz, there is still a region above $-10\,$dBm of IF input power, where the total RF power linearly rises with the IF input power. However, for 73.5 GHz, there is already more than 0 dBm present at the RF port even for very low IF input powers, and no linear dependency is visible between IF input and total RF output power, leading to the assumption of the integrated RF power amplifier already operating beyond its linear region.

As described in [8] and [9], the E-band up-converter module possesses variable terminations at the isolated ports of the Lange couplers generating the quadrature phase. That way, the LO isolation can be tuned and optimized for a specific LO frequency. Figure 27.5a shows the leaked LO power at the RF port of the E-band module for two different optimizations. When optimized for 73.5 GHz of LO frequency, the leaked LO power gets as low as $-30\,$dBm when operated at 73.5 GHz but reaches values close to 0 dBm when operated above 82 GHz. In turns, optimized for 83.5 GHz, values of $-18\,$dBm are reached when operated at 83.5 GHz, but for 73.5 GHz the level even exceeds 0 dBm. Overall, the optimization delivered up to 30 dB of LO leakage improvement. It is also important to note that, as shown in Fig. 27.5b, the IF linearity and the conversion gain are unaffected by the tuning.

To illustrate the corresponding impact on the signal quality of modulated signals, Fig. 27.6 shows a comparison of a 16-QAM waveform with symbol rates of 1, 2, and 2.5 GBd for the two optimizations OP 73.5 GHz and OP 83.5 GHz and operated at 73.5 GHz. Even with the equalizer turned on, there is an almost constant offset of 2 to 2.5 dB in error vector magnitude (EVM) between the two setups.

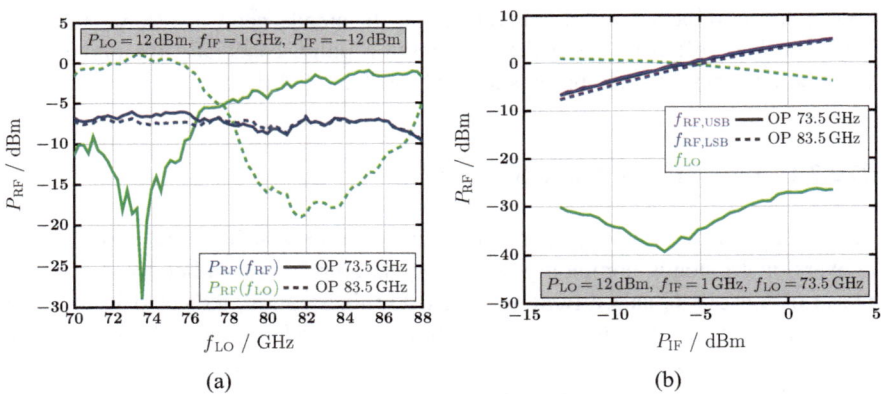

(a)   (b)

Fig. 27.5 Measured RF power contribution of the LO leakage and the converted RF tone of the E-band IQ up-converter module, bias optimized for lowest LO leakage at two different operation points (OP) of 73.5 (solid lines) and 83.5 GHz (dashed lines) of LO frequency, plotted (**a**) as a function of LO frequency and (**b**) as a function of IF input power (Adapted from [9])

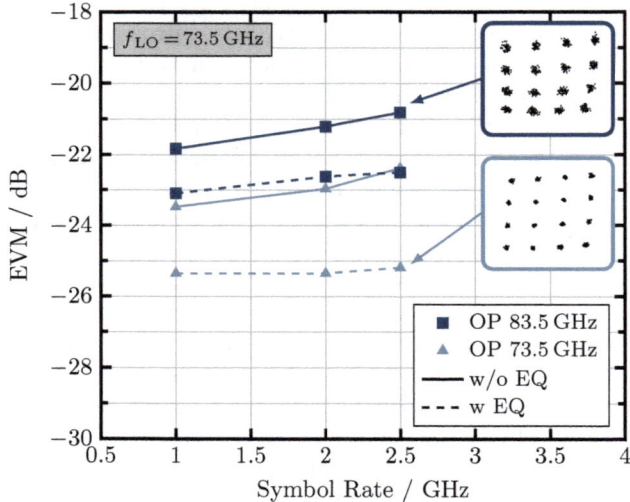

Fig. 27.6 Measured EVM for 16-QAM modulated signals using a LO frequency of 73.5 GHz and an IF input power of -12 dBm for the E-band IQ up-converter (Adapted from [9])

27.2.2 IF-to-RF Conversion

The path of the IF signals to the RF port in the transmitter and vice versa in the receiver is mainly distorted by the limited linearity of the mixer and amplifier devices as well as the I/Q impairments of the quadrature hybrids if a direct-conversion approach is used. Therefore, the frontends need to be operated with sufficient back-off from the 1-dB compression point, especially for higher-order modulation schemes and employ suitable broadband hybrids.

27.2.2.1 I/Q Gain and Phase Imbalances

While (super-)heterodyne frontends inherently do not add IQ impairments, direct-conversion frontends commonly achieve quadrature phase either by shifting the phase of the LO signals driving the I and Q mixers or, as indicated in the block diagram in Fig. 27.1a, using a 90° hybrid in the RF domain. Typical hybrids include Lange, Tandem-X, and branchline couplers. By design, all hybrids exhibit asymmetries with respect to gain and phase over frequency for the I and Q branches and thus do not split or combine signals evenly by magnitude and 90° phase over the entire frequency range. Therefore, targeting a broadband frequency range of operation in the H-band from 220 to 330 GHz, some trade-offs have to be made when choosing the quadrature hybrid. Apart from the I/Q imbalances, the absolute insertion loss and the required chip area might be considered as well as

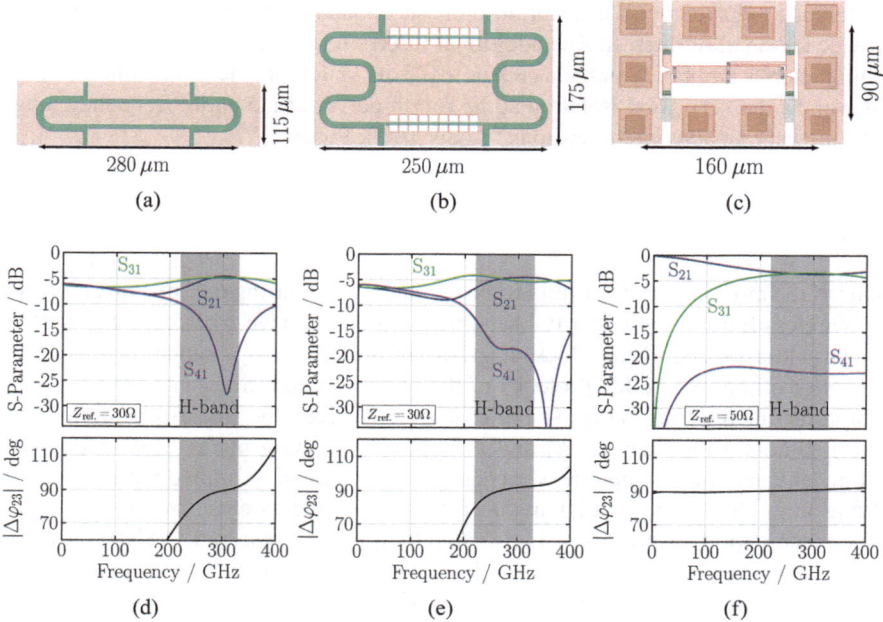

Fig. 27.7 Comparison of size-optimized layouts including core dimensions and nominal simulation results of exemplary designs of 300-GHz quadrature hybrids in a 35 nm InGaAs semiconductor technology. In (**a**), (**d**) a single-box and (**b**), (**e**) a double-box branchline coupler is shown, whereas (**c**), (**e**) show a four-finger Lange coupler (Partially adapted from [6])

the robustness of the performance against process variations, e.g., affecting the characteristic impedance or the coupling factor.

Figure 27.7 shows a comparison of three 90° hybrid designs realized in an InGaAs semiconductor technology, targeting an operation from 220 to 330 GHz. First, a single-stage branchline coupler designed for a reference impedance of 30 Ω is shown in Fig. 27.7a, d. Within the design goals for maximum gain and phase imbalance of 1 dB and $\pm 3°$, respectively, the bandwidth target is not met. The limited bandwidth can be improved by adapting a double-stage branchline design as shown in Fig. 27.7b, e. However, this poses additional challenges with regard to the achievable characteristic line impedances in the technology-specific backend of line (BEOL). Therefore, in this case, a slotted ground plane has to be introduced to achieve the 72 Ω lines. As a trade-off with the increased bandwidth, the required chip area also increases from $280 \times 115\,\mu m$ to $250 \times 175\,\mu m$. Both branchline couplers show an additional insertion loss of around 2 dB, and both show by design a rather narrowband plateau in the phase differences between the quadrature ports. The flatness and therefore bandwidth of this plateau can be traded off at the cost of more amplitude imbalance. In contrast, the Lange coupler designed for a reference impedance of 50 Ω shown in Fig. 27.7c, e inherently exhibits a very flat phase difference behavior with a quadrature phase imbalance below $\pm°$ over the entire frequency band of interest. The additional insertion loss of only around 0.6 dB and simultaneous amplitude imbalance below 0.5 dB undercuts the branchline couplers as well as the very good isolation at the fourth port over the entire frequency range and the very small core dimensions of only $160 \times 90 \, \mu$m.

27.2.2.2 Linearity

One of the most important criteria when it comes to increasing the spectral efficiency of wireless links is the linearity of the transmit and receive frontends. The system linearity, according to the block diagrams in Sect. 27.1, is a superposition of the linearity of the two building blocks of the up-conversion mixer and the RF power amplifier for the transmitter. In the case of a receiver, the linearity is defined by the low-noise amplifier (LNA) and the down-conversion mixer. Due to physical limits, the maximum output power becomes more and more limited with increasing frequency. Therefore, the systems need to be operated as close as possible to the nonlinear region to provide a high enough signal-to-noise ratio (SNR) to cover significant transmission distances. This contradicts with the increasing required back-off from the nonlinear regime for higher-order modulation schemes due to their increased peak-to-average-power ratio (PAPR). For m-QAM modulations with an even order, the

$$\text{PAPR} = \frac{P_{\text{peak}}}{P_{\text{rms}}} = \frac{a_n^2}{\frac{1}{m} \sum_{i=1}^{n} a_i^2 \cdot n_i} \tag{27.2}$$

with the amplitude levels a_i weighted by the number of symbols n_i landing on the respective level and a total number of m symbols in the constellation. This leads to a theoretically required back-off due to the PAPR of 2.55 dB for 16-QAM modulation, 3.7 dB for 64-QAM, and 4.2 dB for 256-QAM. Figure 27.8 shows the EVM curves obtained by a back-to-back measurement of the superheterodyne Tx and Rx modules for 1 GBd modulated signals. With respect to the input-related 1-dB compression point (IP1dB) of the Tx module, which was determined by standard continuous-wave power sweep, the valley points of the EVM curves for QPSK, 8-PSK, and 16-QAM indicate required back-offs of -0.75, -2.5, and -3 dB, respectively, which is in good agreement with Eq. 27.2.

27.3 Conclusion

Electronic signal conversion and amplification for terahertz communication systems are prone to several distortion mechanisms arising in the analogue transmit and receive frontends, which degrade the signal quality and thus limit the achievable data

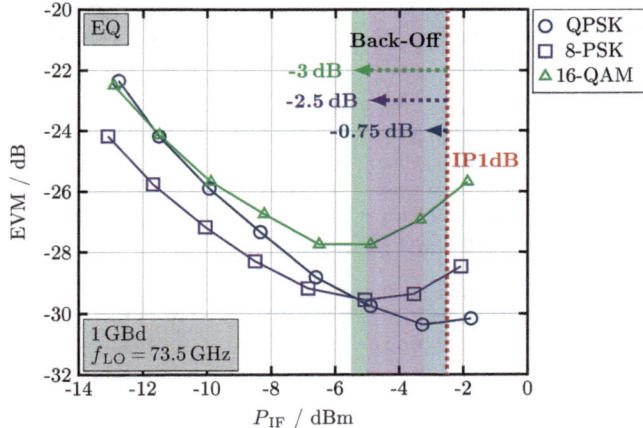

Fig. 27.8 Measured EVM and required back-off for QPSK, 8-PSK, and 16-QAM 1 GBd modulated signals using a LO frequency of 73.5 GHz for the superheterodyne H-band Tx and Rx modules (Adapted from [4])

rate. Depending on the transceiver architecture, the distortion mechanism includes mixer and amplifier nonlinearity as well as non-idealities arising from the LO generation such as LO leakage and spurious tones but also I-Q gain and phase imbalances of quadrature from the hybrids required to enable the transmission of complex modulated signals.

References

1. I. Kallfass et al., MMIC chipset for 300 GHz indoor wireless communication, in *2015 IEEE International Conference on Microwaves, Communications, Antennas and Electronic Systems (COMCAS)*, Tel Aviv (2015), pp. 1–4. https://doi.org/10.1109/COMCAS.2015.7360494
2. D. Lopez-Diaz et al., A subharmonic chipset for gigabit communication around 240 GHz, in *2012 IEEE/MTT-S International Microwave Symposium Digest*, Montreal, QC (2012), pp. 1–3. https://doi.org/10.1109/MWSYM.2012.6258404
3. I. Dan, C. Grötsch, L. John, S. Wagner, A. Tessmann, I. Kallfass, A superheterodyne 300GHz transmit receive chipset for beyond 5G network integration, in *2021 16th European Microwave Integrated Circuits Conference (EuMIC)*, London (2022), pp. 117–120. https://doi.org/10.23919/EuMIC50153.2022.9783947
4. D. Wrana, L. John, B. Schoch, S. Wagner, I. Kallfass, Sensitivity analysis of a 280–312 GHz superheterodyne terahertz link targeting IEEE802.15.3d applications. IEEE Trans. Terahertz Sci. Technol. 12(4), 325–333 (2022). https://doi.org/10.1109/TTHZ.2022.3172008
5. D. Wrana, S. Haussmann, B. Schoch, L. John, A. Tessmann, I. Kallfass, Effects of harmonics from frequency-multiplicative carrier generation in a superheterodyne 300 GHz transmit frontend, in *2023 53rd European Microwave Conference (EuMC)*, Berlin (2023), pp. 138–141. https://doi.org/10.23919/EuMC58039.2023.10290717

6. D. Wrana, Y. Leiba, L. John, B. Schoch, A. Tessmann, I. Kallfass, Short-range full-duplex real-time wireless terahertz link for IEEE802.15.3d applications, in *2022 IEEE Radio and Wireless Symposium (RWS)*, Las Vegas, NV (2022), pp. 94–97. https://doi.org/10.1109/RWS53089.2022.9719954
7. D. Wrana, B. Schoch, L. Manoliu, S. Haußmann, A. Tessmann, I. Kallfass, Investigation of the Influence of LO Leakage in an E-Band Quadrature Transmitter, in *2022 14th German Microwave Conference (GeMiC)*, Ulm (2022), pp. 33–36
8. C. Grötsch, A. Tessmann, S. Wagner, I. Kallfass, On-chip post-production tuning of I/Q frequency converters using adjustable coupler terminations, in *2017 12th European Microwave Integrated Circuits Conference (EuMIC)*, Nuremberg (2017), pp. 273–276. https://doi.org/10.23919/EuMIC.2017.8230712
9. B. Schoch et al., E-Band active upconverter module with tunable LO feedthrough, in *2023 IEEE Radio and Wireless Symposium (RWS)*, Las Vegas, NV (2023), pp. 80–83. https://doi.org/10.1109/RWS55624.2023.10046296

Chapter 28
Frequency Synthesis Based on MLLs

Meysam Bahmanian and J. Christoph Scheytt

Abstract In this chapter, the precision of optical clocks based on mode-locked laser (MLL) is compared with more conventional types of clock sources. It is shown that the phase noise of the optical pulse train from the MLL can be better than other types of clock sources by orders of magnitude. Then, an abstract representation of frequency synthesizer is demonstrated. Different techniques for RF generation using MLL are shown, and their pros and cons are discussed. Finally, a comparison of all these techniques is made with respect to their phase noise and capability to generate RF signal with different frequencies for different applications.

28.1 Timing Precision of Frequency References

Low-jitter frequency references have a wide range of applications from wireless and wireline communications, high-speed analog-to-digital converter (ADC), and digital-to-analog converter (DAC) to fundamental research facilities, such as large array telescope systems and free electron lasers (FELs) [1]. Depending on the operating principle of these frequency references, their output could be either in the electrical or in the optical domain. In the electrical domain, oven-controlled quartz oscillators and surface acoustic wave (SAW) oscillators can offer phase noise levels down to $(-120, -140)$ dBc/Hz at $(1\,\text{kHz}, 100\,\text{kHz})$ offset frequencies normalized to a 10-GHz carrier [2–4]. Sapphire-loaded cavity oscillators (SLCOs) exhibit better phase noise performance by approximately two orders of magnitude but with higher manufacturing cost and size [5, 6]. The so-called optoelectronic oscillators (OEOs) use a continuous wave (CW) laser and a feedback loop in a mixed electro-optical domain and have better phase noise than quartz and SAW oscillators, but their output signal is spurious because of their long cavity [7]. The coupled optoelectronic oscillators (COEOs) replace the CW laser with another

M. Bahmanian (✉) · J. C. Scheytt
Schaltungstechnik (SCT)/Heinz Nixdorf Institut, Universität Paderborn, Paderborn, Germany
e-mail: meysam.bahmanian@uni-paderborn.de; cscheytt@hni.upb.de

© The Author(s) 2026
T. Kürner et al. (eds.), *Metrology for THz Communications*, Springer Series
in Optical Sciences 256, https://doi.org/10.1007/978-3-032-01986-8_28

optical feedback loop consisting of an optical amplifier and an optical filter. The smaller delay of the feedback path in COEO increases the intervals of the spurs in the frequency spectrum of the RF output and enhances the spectral purity [7]. Using this technique, Matsko et al. [8] reported a 10-GHz COEO with a phase noise of $(-125, -145)$ dBc/Hz at $(1 \text{ kHz}, 100 \text{ kHz})$ offset frequencies, and Ly et al. [9] reported a COEO-based millimeter-wave signal generation at 90 GHz with a phase noise of $(-104, -129)$ dBc/Hz at $(1 \text{ kHz}, 100 \text{ kHz})$ offset frequencies. In the optical domain, medium-priced and compact MLLs achieve a phase noise performance better than quartz and SAW oscillators at offset frequencies above 1 kHz [10–12]. Further improvement of MLL phase noise (beyond or comparable with SLCOs) has been achieved using different techniques such as optical frequency division (OFD) in which one of the MLL optical comblines is locked onto an ultrastable CW laser [6, 13]. Recently, Kalubovilage et al. [14] demonstrated a compact monolithic mode-locked laser (MMLL) with an exceptional open-loop phase noise performance comparable to OFD systems. Figure 28.1 compares the phase noise of the state of the art for different types of reference oscillators normalized to 10-GHz carrier frequency. The phase noise scaling is based on an ideal frequency multiplier without any additive phase noise.

In addition to the phase noise performance of these oscillators, temporal and spectral properties of their output signals should be considered. Especially, if such an oscillator is used as a reference oscillator in a phase-locked loop (PLL), a waveform with high harmonic content is desired, as it enables the designer to lock a tunable oscillator on any of these harmonics in order to maximize the output

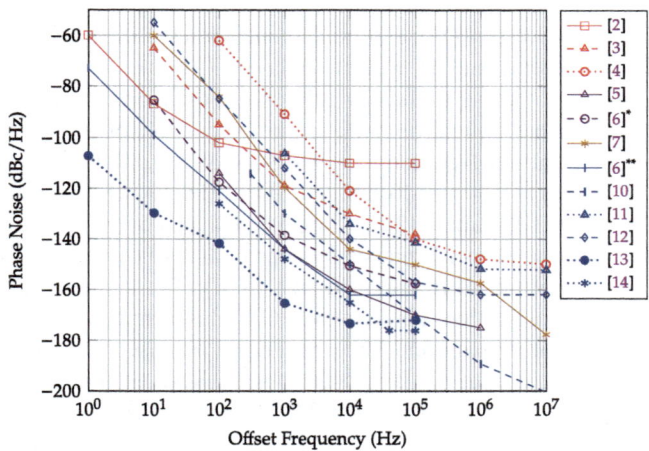

Fig. 28.1 Comparison of the phase noise of the state of the art, normalized to 10-GHz carrier frequency: (red) quartz and SAW oscillators, (brown) coupled optoelectronic oscillator, (violet) sapphire-loaded cavity oscillators, and (blue) optical sources. [6]* and [6]** correspond to the phase noise of the sapphire-loaded cavity oscillator and the mode-locked laser reported in [6], respectively

frequency range of the PLL. Quartz and SAW oscillators have usually sinusoidal output waveforms. The harmonic content of their output signals barely exceeds a few gigahertz because of their narrowband resonator and limited bandwidth of their electronic components. Therefore, higher harmonic content is usually generated using step recovery diodes (SRDs) or nonlinear transmission lines (NLTLs) [15, 16]. The harmonic content of a COEO is limited by the bandwidth of its electronic components and the optical filter used in the feedback loop. Therefore, the spectral width of COEO is in the range of a few nanometers (less than 100 GHz) [8, 9]. In contrast to these electronic and optoelectronic reference oscillators, MLLs achieve subpicosecond pulsewidths and THz-wide optical frequency combs [10, 13, 17]. These frequency combs correspond to a harmonic rich intensity (optical cycle-averaged intensity) waveform and are therefore well suited for microwave signal generation.

28.2 Frequency Synthesizer

The main objective in designing a frequency synthesizer is to translate the frequency of a low-noise reference signal to an RF signal with the desired frequency. This translation is ideally without introducing additional noise. This is illustrated in Fig. 28.2 at an abstract level with the phase noise power spectral densities. The phase noise of the reference is scaled by the frequency multiplication factor. Any additional noise above this level is additive phase noise of the frequency synthesizer.

The simplest approach to translate the reference frequency to the desired frequency is using a nonlinear device that generates harmonics of its input signal, known as the frequency multiplier. The additive phase noise of frequency multipliers is dependent on their technology and generally with proper device selection and design can be below the input signal phase noise. The main drawback of this

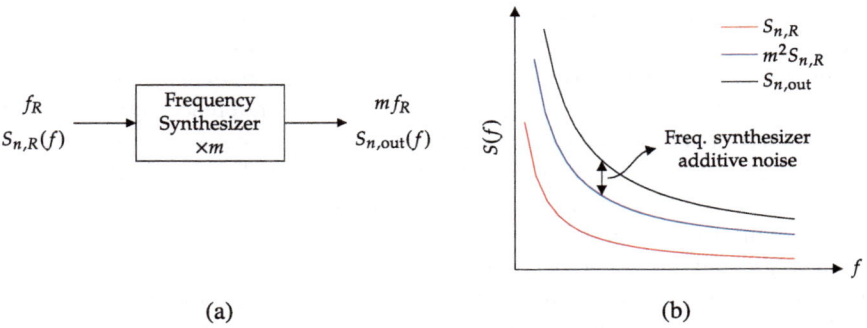

(a)   (b)

Fig. 28.2 (a) Abstract representation of frequency synthesizer and (b) its phase noise plots; f_R, reference frequency; $S_{n,R}(f)$, reference phase noise; m, frequency multiplication factor; $S_{n,\text{out}}(f)$, frequency synthesizer phase noise

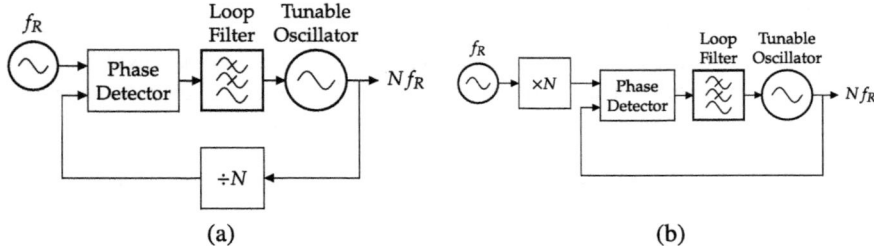

Fig. 28.3 Block diagram of PLL with (**a**) loop frequency divider and (**b**) reference frequency multiplier

technique is that the available frequencies are limited only to the harmonics of the reference signal, and non-integer multiplication factors are not possible. Besides, the output signal of frequency multipliers contains both leakage of the input signal as well as other undesired harmonics. High-quality RF generation would then require post-multiplication filtering and suppression of undesired harmonics. This can be more complicated for wideband frequency synthesizers, as the output signal has a wide frequency range and a tunable filter is required.

The standard approach for wideband frequency synthesis from a low phase noise reference signal is using a PLL. On the one hand, the frequency of the reference signal of a PLL is usually in the megahertz range, and the output frequency is in the gigahertz range. One the other hand, the PLL requires equal frequencies at the input of its phase detector. In order to match these frequencies, there are two fundamental approaches:

1. Reducing the frequency of the output signal, illustrated in Fig. 28.3a
2. Increasing the frequency of the reference signal, illustrated in Fig. 28.3b

The first method, reducing the frequency of output signal, is popular in low-cost frequency synthesizers. By using a digitally programmable fractional frequency divider, the output frequency can be tuned in fine frequency steps [18]. The main drawback is that the additive phase noise of the PLL scales with the loop division factor and surpasses the phase noise of the ultralow-noise frequency references.

The second method, increasing the frequency of the reference signal, does not have a loop divider to reduce the frequency of the RF signal. Therefore, it has lower additive phase noise compared to the first method. The drawback is that the frequency of the reference signal is increased using harmonic generators, and only integer multiples of the reference frequency are available. In order to improve the frequency resolution, the output frequency of the PLL is mixed with the signal from a fine-step secondary PLL, using direct mixing or offset phase-locked loop scheme.

The low-noise optical pulse trains of MLLs have shown a great potential for low-noise RF generation. The phase noise of these optical references can be better than their electronic counterparts by three orders of magnitude. This has led to efforts to generate a low-noise microwave signal from an MLL [19–23]. Among these methods, phase locking of a microwave oscillator onto an MLL using balanced optical microwave phase detector (BOMPD) is a cost-effective and relatively compact solution, as its only electro-optical components are an intensity modulator and one pair of photodetectors [23, 24]. In addition, the RF signal is sampled with the optical reference using an intensity modulator which has a significantly higher bandwidth compared to double-balanced mixers that are typically used in high-performance fully electronic phase-locked loops. The downside of such an optoelectronic phase-locked loop (OEPLL) is the requirement for phase adjustment of its microwave signal paths and relatively sophisticated microwave setup. An alternative topology for a BOMPD proposed by [25] significantly simplifies the architecture and does not require microwave phase shifters, bandpass filters, and balanced mixers by using an electro-optical balanced intensity modulator (BIM). Using this method, Jung et al. [26] locked a dielectric resonator oscillator (DRO) at 8 GHz onto an MLL with a residual rms-jitter of 838 as integrated from 1 Hz to 1 MHz.

The results from [26, 27] demonstrate the potential of OEPLLs for ultralow-phase-noise frequency synthesis. However, the small bandwidth of the DRO makes the approach impractical for broadband frequency synthesis. In addition, the fiber-based Sagnac-modulator is bulky, expensive, and sensitive to mechanical vibrations. Currently, most OEPLL systems operate only at a single frequency or a very limited frequency range. Therefore, more research is needed toward microwave OEPLL frequency synthesizers with a large output frequency range and compact size. Nejadmalayeri et al. [28] replaced the fiber Sagnac-modulator with a lithium niobate (LiNbO3) MZM and the DRO with a microwave instrument. This allowed to reduce the size and increase the output frequency range but [28] did not achieve the performance of [26].

The OEPLLs have the potential for wideband frequency synthesis, since the MHz wide loop bandwidth can filter the optical reference harmonics. The phase noise of the current state-of-the-art frequency synthesizers currently is limited by their quartz and SAW-based reference they use. Figure 28.4 compares the phase noise of the

Fig. 28.4 Phase noise comparison of microwave frequency synthesizers at 10 GHz and frequency references normalized to 10-GHz carrier frequency

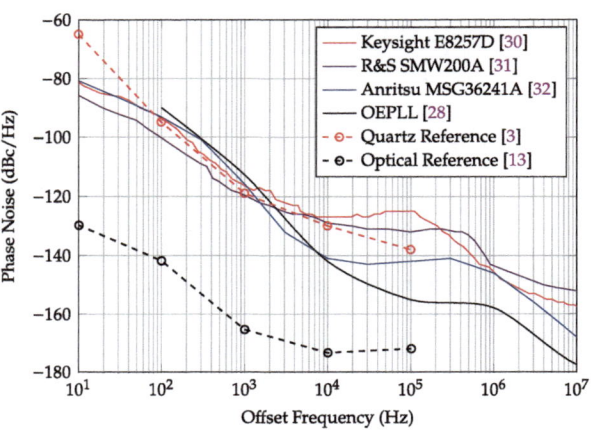

state-of-the art wideband frequency synthesizers (from Keysight, Rohde & Schwarz and Anritsu [29–31]) with the OEPLL reported in [27]. It can be seen that at offset frequencies above approximately 1 kHz the OEPLL outperforms traditional fully electronic frequency synthesizers, thanks to the clean reference signals of MLLs. The phase noise of OEPLLs at close-in offset frequencies below approximately 10 kHz can further be improved if a better optical reference is used. Therefore, the OEPLLs can have better phase noise by approximately three orders of magnitude.

28.3 RF Generation Using Mode-Locked Laser

In the previous sections, we demonstrated that the phase noise of mode-locked lasers (MLLs) can be better than that of traditional frequency references such as quartz and SAWs oscillators. In addition, we discussed that the role of frequency synthesizer is to generate an arbitrary frequency that is phase locked to the low-noise frequency reference. In this section, we discuss state-of-the-art RF frequency generation techniques based on MLL optical reference.

28.3.1 Direct Detection

The simplest approach for MLL-based RF generation is direct detection of MLL pulses using a photodiode, illustrated in Fig. 28.5a. Since the generated photocurrent is harmonic rich, a bandpass filter is required to select to desired harmonic.

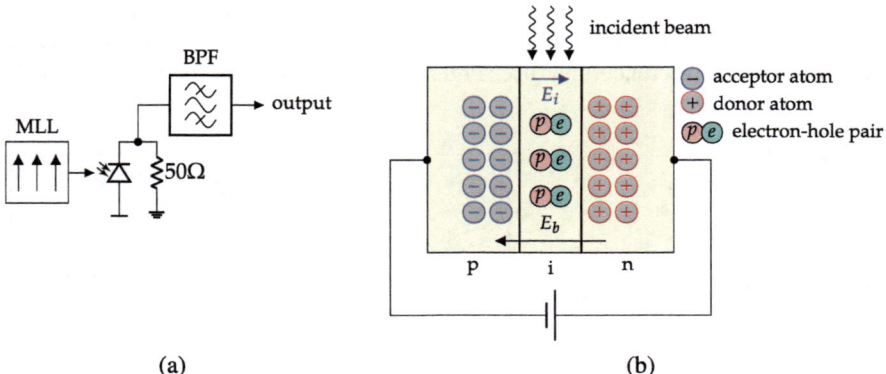

(a)   (b)

Fig. 28.5 (a) Schematic of optical pulse direct detection and (b) illustration of space-charge effect in PIN photodiode

The signal-to-noise ratio (SNR) of this type of RF generator is limited by the thermal noise of output termination load and the maximum achievable RF power from the photodiode. The output power of the photodiode is also limited by high energy MLL pulses mainly due to the nonlinear space-charge effect [32–35]. The main limitation of this technique is the bandwidth of the photodiode. High frequency photodiodes can easily achieve tens of gigahertz bandwidth when stimulated with low-pulse-energy optical beams. However, high peak power optical stimulation of photodiodes, which is required for low-noise RF generation, limits the photodiode bandwidth due to the nonlinear space-charge effects, illustrated in Fig. 28.5b. High energy pulses of MLLs generate many charge carriers (electron-hole pairs) in the active region of photodiode. The Coulomb interaction between these charge carriers creates an electric field, E_i, that counteracts the device built-in electric filed, E_b. This counteracting field reduces the charge carrier velocities and consequently leads to an increase of the transit time of charge carriers and the response time and reduction of the photodiode bandwidth [36]. While the thermal noise floor remains at the same level, the decrease of the photodiode bandwidth leads to reduction of the desired harmonic power and lower SNR. Another undesired phenomenon is the conversion of the optical pulse amplitude noise to photocurrent phase noise, leading to further degradation of phase noise of the RF signal [37].

In order to mitigate the photodiode space-charge effect, uni-traveling carrier (UTC) photodiode and modified uni-traveling carrier (MUTC) photodiode use an undepleted p-layer to absorb light to inject the electrons into the drift region and reduce the transit time and the space-charge effect [38, 39]. The illumination condition has also been optimized through beam shaping using a graded index (GRIN) lens coupling which increases the illumination cross section and reduces the peak magnitude of the optical field [40, 41]. It has been shown that increasing the optical beam diameter leads to a more uniform distribution of mobile carriers in the device and reduction of the counteracting electric field [33, 34, 42].

28.3.2 Pulse Interleaving Rate Multiplier

The space-charge effects can be mitigated by reducing the energy of the optical pulses. This can be realized in a lossless fashion by increasing the repetition rate of the optical pulses using pulse interleaving technique illustrated in Fig. 28.6a for a three-stage ×8 multiplier [6, 13, 22, 43, 45]. The optical pulses are interleaved via three segments, each segment divides the optical beam into two paths, and one of the paths adds a delay equal to half of the period of the pulses and then both paths are combined, increasing the repetition rate by a factor of 2 after of each segment. Figure 28.6b shows the effect of rate multiplication on the generated RF power.

State-of-the-art MLL-based RF generators use this technique to enhance the photodiode output power and increase the SNR [6, 13, 43]. The main drawback of this method is the output frequencies that can be achieved. While the intensity of the output of MLL has harmonics of f_R, where f_R is the repetition rate of the pulses,

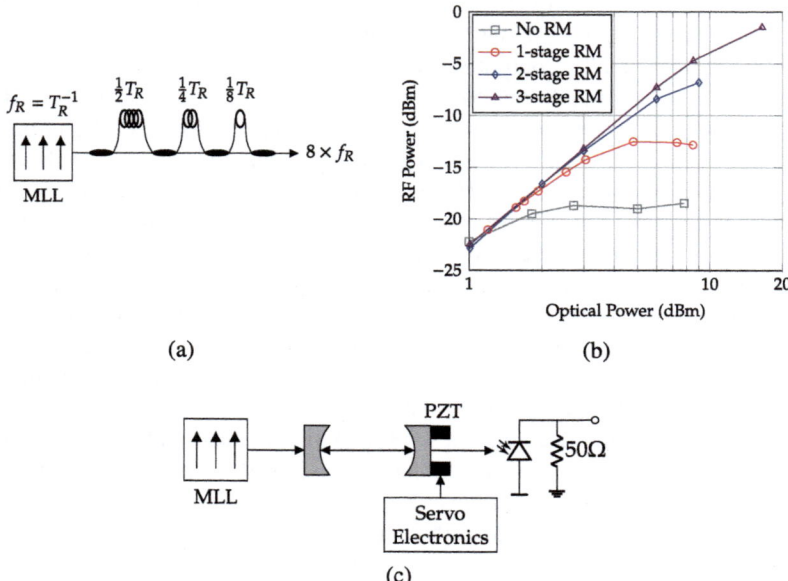

Fig. 28.6 (**a**) Schematic of a three-stage rate multiplier and (**b**) the effect of rate multiplication on generated RF power at 10 GHz characterized with an MLL with a repetition rate of 250 MHz [43]. (**c**) MLL rate multiplier based on combline filtering using Fabry-Perot cavity [44]; RM, rate multiplier

the interleaved pulses have harmonics of Nf_R, where N is the rate multiplication factor. Therefore, using a rate multiplier enhances the SNR and phase noise of the RF signal, but it comes at the cost of higher frequency steps in the RF generator.

28.3.3 Spectral Filtering Rate Multiplier

Another technique to increase the repetition rate of the optical pulses is combline filtering using a Fabry-Perot cavity shown in Fig. 28.6c [43, 44]. Such a cavity has a transmission frequency response with multiple passbands located at integer multiples of the fundamental frequency passband. Therefore, by adjusting the length of the cavity with a Piezo stage, various rate multiplication factors can be achieved. One drawback compared to pulse interleaving technique is that more optical power budget is needed, as the optical power reflected from the cavity is dissipated. In addition, controlling the cavity length makes the setup more complicated compared to static pulse interleaver, and the moving Piezo stage leads to acoustic vibrations and phase noise degradation at close-in offset frequencies below 1 kHz [43].

28.3.4 Heterodyne Mixing and Optical PLL

Rate multipliers can be seen as optical filters with a periodic transfer response that pass comblines that have a certain frequency difference. The periodicity of the transfer function of these filters means the photocurrent generated at the photodiode still has many harmonics. If the periodicity feature of these filters is removed and only two desired comblines are allowed to pass, then the generated photocurrent would be just a single tone. A practical implementation of this idea is shown in Fig. 28.7a, known as heterodyne mixing. First, two comblines with optical frequencies of f_1 and f_2 are selected via comb selection optics tuned at these frequencies. The selected comblines are then combined, and the intensity of the resulting optical field is detected via a photodiode. The frequency of the RF signal at the photodiode output is the frequency difference between the selected comblines.

One approach for combline selection is filtering the two desired wavelengths from the MLL spectral lines [46, 47]. This technique requires extremely narrowband optical filters with high quality factors to filter the undesired neighboring comblines only a few hundreds of megahertz away from the desired combline. Fabrication and

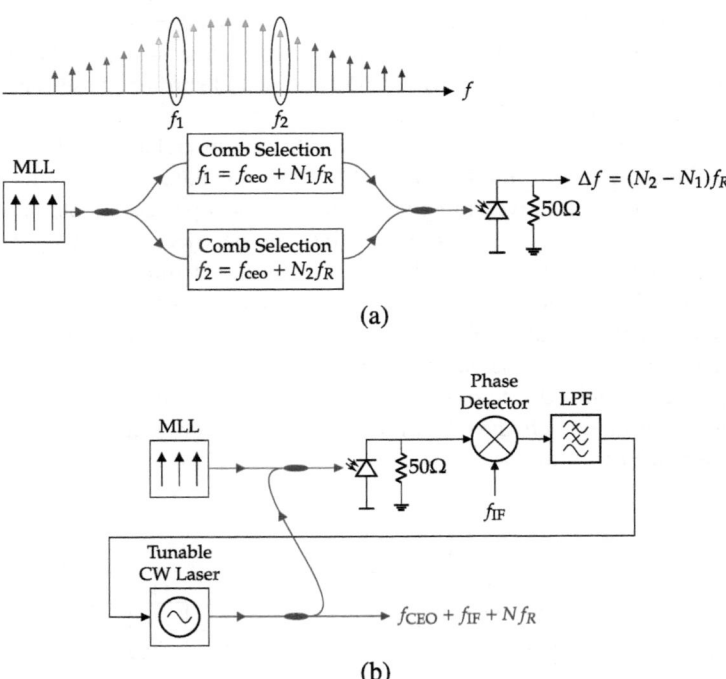

Fig. 28.7 (a) Schematic of RF generator based on heterodyne mixing of two MLL comblines. (b) Combline selection using optical phase-locked loop technique (Optical connections are shown in blue and electrical connections are shown in black)

wavelength tuning of such optical devices is a tedious task. In addition, wideband RF frequency synthesis requires tunability of these filters which makes the realization of a frequency synthesizer using this technique even more complicated.

An alternative approach for optical comb selection would be optical phase-locked loop (OPLL), illustrated in Fig. 28.7b [48–51]. The MLL comblines are combined with the output beam of a tunable CW laser which has an offset frequency, f_{IF}, relative to the target combline. This relative offset frequency is then detected via a photodetector and is stabilized using an optical phase-locked loop. The optical PLL uses a phase detector and a stable intermediate frequency (IF) to generate an error signal which is then filtered and fed back to the tunable CW laser to stabilize it.

28.3.5   Optoelectronic PLL

One problem arising in direct photodetection, whether a rate multiplier is used or not, is the excess noise caused by conversion of amplitude noise of the optical pulses to phase noise [22, 23, 37]. Kim et al. [23] suggested extracting the timing information of the optical pulses in the optical domain, before photodetection, and use this timing information to control a microwave tunable oscillator. This idea led to the development of a new class of electro-optical systems, the balanced optical microwave phase detector (BOMPD), and optoelectronic phase-locked loop (OEPLL), illustrated in Fig. 28.8 [23, 24, 26, 52–54].

In this approach, the optical pulses of the MLL are modulated by the RF signal using a balanced intensity modulator (BIM). The modulated optical pulses carry information about the relative timing between the optical pulses and the microwave signal. This information is then converted into an electrical current by a pair of photodiodes and is used to adjust the microwave signal timing with that of the optical pulse train. It is noteworthy to mention that the OEPLL is similar to its fully electronic counterpart, and the BOMPD operates similar to a balanced mixer that is used in many PLLs as a phase detector. However, the phase detector of the OEPLL has to operate in a mixed electro-optical domain. Since the intensity of the optical pulse train is harmonic rich, the OEPLL can potentially lock on any harmonic of the optical reference repetition rate.

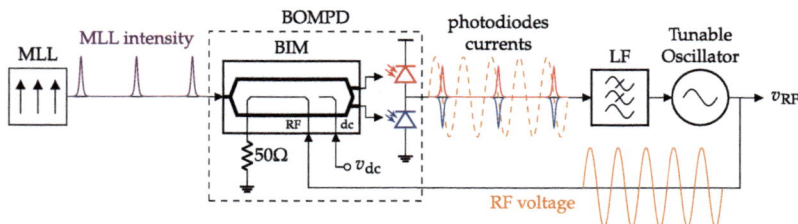

Fig. 28.8 Simplified block diagram of the OEPLL

28.3.6   *Comparison and Conclusions*

Now that we investigated various techniques of RF generation using optical pulses of MLLs, one might ask which method is preferred. The answer to this question mainly lies in the application. Here, we focus on wideband versatile frequency synthesizer and, based on this application, compare these methods.

Figure 28.9 compares the phase noise performance of these methods [13, 26, 27, 43, 46, 48]. Direct detection method can be used to generate any harmonic of the reference repetition rate. However, the RF signal power generated at the photodiode decreases as the pulse energy increases. For a given average optical power, this means higher repetition rates are desired. Although increasing the repetition rate is possible, using a pulse interleaver or a Fabry-Perot cavity, the frequency resolution of the RF generator increases, and fewer frequencies are available at the photodiode output. In addition, for many RF applications, the signal purity is important, and filtering the subharmonics is necessary. This can be difficult from two aspects: firstly, a tunable bandpass filter is required to select the desired harmonic; secondly, sufficient suppression (usually more than 60 dBc) of undesired harmonics is difficult to achieve, as they can be very close to the desired harmonic (for instance, 250 MHz away from a 10 GHz carrier).

The heterodyne mixing approach with filtering the comblines has practical difficulties due to tunability and high quality-factor requirement for the optical filters. For instance, an optical filter at 200 THz with a bandwidth of 200 MHz requires a quality factor of 10,000 which is hard to achieve. The alternative heterodyne-mixing technique, the OPLL, shows a poor phase noise performance and has a high additive phase noise. Possible reasons could be high phase noise of the tunable laser source and failing to sufficiently suppress it in the OPLL. Another possible reason could

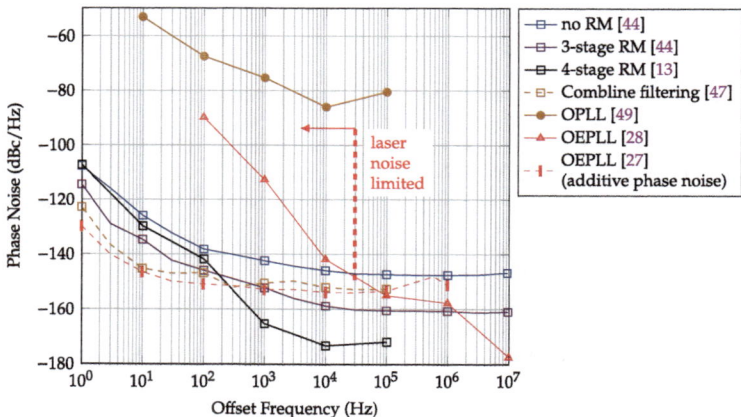

Fig. 28.9 Phase noise comparison of different RF generation methods using MLL optical pulses; RM, rate multiplier

be the high noise of the CEO frequency of the MLL. A simple first-order analysis suggests the CEO noise is cancelled after recombination of the locked optical tones. However, it is possible the CEO noise is not perfectly cancelled due to small delay mismatch in the OPLLs, or the noise can undergo a nonlinear transformation at the baseband components or due to nonlinear characteristic curve of the tunable CW laser. This transformed noise is not cancelled during the photodetection process and shows itself as additive phase noise. These mechanisms include the phase noise of the optical carriers which is orders of magnitude higher than that of microwave signals. Therefore, the leakage of the optical carrier phase noise to the phase of the generated microwave signal can have a strong adverse effect on its phase noise.

The OEPLL has interesting features that make it a viable candidate for low-noise and wideband frequency synthesis. From the phase noise perspective, the OEPLL reported by Jung et al. [27] has a phase noise performance similar to the state-of-the-art multistage repetition rate multipliers. The phase noise at offset frequencies below approximately 30 kHz is limited by the MLL and is expected to improve if an MLL with higher spectral purity is used. The additive phase noise of the same setup has also been reported in [26] and proves that the OEPLL performance is limited by the MLL at those offset frequencies. Although Jung et al. [27] use a narrowband dielectric resonator oscillator (DRO) as the tunable oscillator, there is wide range of microwave oscillators with various bandwidths and phase noise performance available. The balanced intensity modulators (BIMs) with tens of gigahertz bandwidth are standard devices and available. Therefore, the OEPLL does not have the practical limitations of other techniques while having a phase noise performance comparable to state-of-the-art RF generators. Therefore, the OEPLL is well suited for wideband microwave signal generation using the optical pulses of MLLs.

References

1. J.K. Kim, F.X. Kaertner, Attosecond-precision ultrafast photonics. Laser Photon. Rev. **4**, 432–456 (2010)
2. *High Performance (VC)OCXO* (KVG Quartz Crystal Technology GmbH, Neckarbischofsheim, 2017). Datasheet O-40CXXXX-LPN-LGS-LF
3. *Ultra Low Phase Noise Oven Controlled Crystal Oscillator* (Mount Holly Springs, PA, USA, 2016). Datasheet OX-305
4. *Voltage Controlled SAW Oscillator Surface Mount Model* (Synergy Microwave, Paterson, NJ, USA, 2017). Datasheet HFSO1000-5, Rev. B
5. *The Sapphire Loaded Cavity Oscillator (SLCO)* (Poseidon Scientific Instruments, Fremantle, 2004). Datasheet SLCO, Rev. 05/04
6. J.W. Zobel, M. Giunta, A.J. Goers, R.L. Schmid, J. Reeves, R. Holzwarth, E.J. Adles, M.L. Dennis, Comparison of optical frequency comb and sapphire loaded cavity microwave oscillators. IEEE Photon. Technol. Lett. **31**, 1–1 (2019). https://doi.org/10.1109/lpt.2019.2926190
7. L. Maleki, Optoelectronic oscillators for microwave and mm-wave generation, in *2017 18th International Radar Symposium (IRS)* (IEEE, 2017). https://doi.org/10.23919/irs.2017.8008133

8. A.B. Matsko, D. Eliyahu. L. Maleki, Theory of coupled optoelectronic microwave oscillator II: phase noise. J. Opt. Soc. Am. B **30**, 3316 (2013). https://doi.org/10.1364/josab.30.003316
9. A. Ly, V. Auroux, R. Khayatzadeh, N. Gutierrez, A. Fernandez, O. Llopis, Highly spectrally pure 90-GHz signal synthesis using a coupled optoelectronic oscillator. IEEE Photon. Technol. Lett. **30**, 1313–1316 (2018). https://doi.org/10.1109/lpt.2018.2845747
10. T.K. Kim, Y. Song, K. Jung, C. Kim, H. Kim, C.H. Nam, J. Kim, Sub-100-as timing jitter optical pulse trains from mode-locked Er-fiber lasers. Opt. Lett. **36**, 4443 (2011). https://doi.org/10.1364/ol.36.004443
11. *Ultra Low Timing Jitter Performance & Characterization of Origami Femtosecond Laser Series*, Whitepaper P/N 09-001 (Onefive, Berlin, 2009), Rev. 1.1
12. *Menhir Photonics Femtosecond Laser Source* (Menhir Photonics AG, Rümlang, 2019). Datasheet MENHIR-1550
13. X. Xie, R. Bouchand, D. Nicolodi, M. Giunta, W. Hänsel, M. Lezius, A. Joshi, S. Datta, C. Alexandre, M. Lours, P.-A. Tremblin, G. Santarelli, R. Holzwarth, Y.L. Coq, Photonic microwave signals with zeptosecond-level absolute timing noise. Nat. Photon. **11**, 44–47 (2016). https://doi.org/10.1038/nphoton.2016.215
14. M. Kalubovilage, M. Endo, T.R. Schibli, Ultra-low phase noise microwave generation with a free-running monolithic femtosecond laser. Opt. Express **28**, 25400 (2020). https://doi.org/10.1364/oe.399425
15. S. Krakauer, Harmonic generation, rectification, and lifetime evaluation with the step recovery diode. Proc. IRE **50**, 1665–1676 (1962). https://doi.org/10.1109/jrproc.1962.288155
16. E. Afshari, A. Hajimiri, Nonlinear transmission lines for pulse shaping in silicon. IEEE J. Solid-State Circuits **40**, 744–752 (2005). https://doi.org/10.1109/jssc.2005.843639
17. T.D. Shoji, W. Xie, K.L. Silverman, A. Feldman, T. Harvey, R.P. Mirin, T.R. Schibli, Ultra-low-noise monolithic mode-locked solid-state laser. Optica **3**, 995 (2016). https://doi.org/10.1364/optica.3.000995
18. B. Razavi, *RF Microelectronics* (Pearson Education, London, 2011), p. 960. ISBN: 0137134738
19. J. Millo, R. Boudot, M. Lours, P.Y. Bourgeois, A.N. Luiten, Y.L. Coq, Y. Kersalé, G. Santarelli, Ultra-low-noise microwave extraction from fiber-based optical frequency comb. Opt. Lett. **34**, 3707 (2009). https://doi.org/10.1364/ol.34.003707
20. T.M. Fortier, M.S. Kirchner, F. Quinlan, J. Taylor, J.C. Bergquist, T. Rosenband, N. Lemke, A. Ludlow, Y. Jiang, C.W. Oates, S.A. Diddams, Generation of ultrastable microwaves via optical frequency division. Nat. Photon. **5**, 425–429 (2011). https://doi.org/10.1038/nphoton.2011.121
21. F. Quinlan, T.M. Fortier, M.S. Kirchner, J.A. Taylor, M.J. Thorpe, N. Lemke, A.D. Ludlow, Y. Jiang, S.A. Diddams, Ultralow phase noise microwave generation with an Er:fiber-based optical frequency divider. Opt. Lett. **36**, 3260 (2011). https://doi.org/10.1364/ol.36.003260
22. W. Zhang, S. Seidelin, A. Joshi, S. Datta, G. Santarelli, Y.L. Coq, Dual photo-detector system for low phase noise microwave generation with femtosecond lasers. Opt. Lett. **39**, 1204 (2014). https://doi.org/10.1364/ol.39.001204
23. J. Kim, F.X. Kärtner, F. Ludwig, Balanced optical-microwave phase detectors for optoelectronic phase-locked loops. Opt. Lett. **31**, 3659 (2006). https://doi.org/10.1364/ol.31.003659
24. M.Y. Peng, A. Kalaydzhyan, F.X. Kärtner, Balanced optical-microwave phase detector for sub-femtosecond optical-RF synchronization. Opt. Express **22**, 27102 (2014). https://doi.org/10.1364/oe.22.027102
25. J. Kim, F.X. Kärtner, M.H. Perrott, Femtosecond synchronization of radio frequency signals with optical pulse trains. Opt. Lett. **29**, 2076 (2004). https://doi.org/10.1364/ol.29.002076
26. K. Jung, J. Kim, Subfemtosecond synchronization of microwave oscillators with mode-locked Er-fiber lasers. Opt. Lett. **37**, 2958 (2012). https://doi.org/10.1364/ol.37.003758
27. K. Jung, J. Shin, J. Kim, Ultralow phase noise microwave generation from mode-locked Er-fiber lasers with subfemtosecond integrated timing jitter. IEEE Photon. J. **5**, 5500906-5500906 (2013). https://doi.org/10.1109/jphot.2013.2267533

28. A.H. Nejadmalayeri, F.X. Kärtner, Mach-Zehnder based balanced optical microwave phase detector, in *Conference on Lasers and Electro-Optics 2012*, OSA, pp. 1–2 (2012). https://doi.org/10.1364/cleo_si.2012.ctu2a.1
29. *Microwave Analog Signal Generator* (Keysight, Santa Rosa, CA, USA, 2016). Datasheet E8257D
30. *Vector Signal Generator* (Rohde & Schwarz, Munich, 2019). Datasheet SMW200A, Version 11.00
31. *Rubidium RF/Microwave Signal Generator* (Anritsu, Kanagawa, 2021). Datasheet MSG36241A, Rev. A
32. P.-L. Liu, K. Williams, M. Frankel, R. Esman, Saturation characteristics of fast photodetectors. IEEE Trans. Microwave Theory Tech. **47**, 1297–1303 (1999). https://doi.org/10.1109/22.775469
33. K. Williams, R. Esman, M. Dagenais, Effects of high space-charge fields on the response of microwave photodetectors. IEEE Photon. Technol. Lett. **6**, 639–641 (1994). https://doi.org/10.1109/68.285565
34. K. Williams, R. Esman, Design considerations for high-current photodetectors. J. Lightwave Technol. **17**, 1443–1454 (1999). https://doi.org/10.1109/50.779167
35. A. Beling, X. Xie, J.C. Campbell, High-power, high-linearity photodiodes. Optica **3**, 328 (2016). https://doi.org/10.1364/optica.3.000328
36. B. Saleh, M. Teich, *Fundamentals of Photonics (Wiley Series in Pure and Applied Optics)* (Wiley, Hoboken, NJ, USA, 1991). ISBN: 0471839655
37. W. Zhang, T. Li, M. Lours, S. Seidelin, G. Santarelli, Y.L. Coq, Amplitude to phase conversion of InGaAs pin photo-diodes for femtosecond lasers microwave signal generation. Appl. Phys. B **106**, 301–308 (2011). https://doi.org/10.1007/s00340-011-4710-1
38. A. Beling, H. Pan, H. Chen, J.C. Campbell, Linearity of modified uni-traveling carrier photodiodes. J. Lightwave Technol. **26**, 2373–2378 (2008). https://doi.org/10.1109/jlt.2008.927184
39. J. Klamkin, A. Ramaswamy, L.A. Johansson, H.-F. Chou, M.N. Sysak, J.W. Raring, N. Parthasarathy, S.P. DenBaars, J.E. Bowers, L.A. Coldren, High output saturation and high-linearity uni-traveling-carrier waveguide photodiodes. IEEE Photon. Technol. Lett. **19**, 149–151 (2007). https://doi.org/10.1109/lpt.2006.890101
40. A. Joshi, D. Becker, GRIN lens-coupled top-illuminated photodetectors for high-power applications, in *Microwave Photonics, 2007 Interntional Topical Meeting on* (IEEE, 2007). https://doi.org/10.1109/mwp.2007.4378124
41. A. Joshi, S. Datta, D. Becker, GRIN lens coupled top-illuminated highly linear InGaAs photodiodes. IEEE Photon. Technol. Lett. **20**, 1500–1502 (2008). https://doi.org/10.1109/lpt.2008.928532
42. K. Williams, R. Esman, M. Dagenais, Nonlinearities in p-i-n microwave photodetectors. J. Lightwave Technol. **14**, 84–96 (1996). https://doi.org/10.1109/50.476141
43. H. Jiang, J. Taylor, F. Quinlan, T. Fortier, S.A. Diddams, Noise floor reduction of an Er:fiber laser-based photonic microwave generator. IEEE Photon. J. **3**, 1004–1012 (2011). https://doi.org/10.1109/jphot.2011.2171480
44. S.A. Diddams, M. Kirchner, T. Fortier, D. Braje, A.M. Weiner, L. Hollberg, Improved signal-to-noise ratio of 10 GHz microwave signals generated with a mode-filtered femtosecond laser frequency comb. Opt. Express **17**, 3331 (2009). https://doi.org/10.1364/oe.17.003331
45. A. Haboucha, W. Zhang, T. Li, M. Lours, A.N. Luiten, Y.L. Coq, G. Santarelli, Optical-fiber pulse rate multiplier for ultralow phase-noise signal generation. Opt. Lett. **36**, 3654 (2011). https://doi.org/10.1364/ol.36.003654
46. H. Al-Taiy, S. Preußler, S. Brückner, J. Schoebel, T. Schneider, Generation of highly stable millimeter waves with low phase noise and narrow linewidth. IEEE Photon. Technol. Lett. **27**, 1613–1616 (2015). https://doi.org/10.1109/LPT.2015.2432464
47. S. Preussler, T. Schneider, Tunable generation of high quality mm-and THz-waves for wireless communications with carrier frequencies up to several THz, in *WTC 2014; World Telecommunications Congress 2014*, pp. 1–5 (2014)

48. Q. Quraishi, M. Griebel, T. Kleine-Ostmann, R. Bratschitsch, Generation of phase-locked and tunable continuous-wave radiation in the terahertz regime. Opt. Lett. **30**, 3231–3233 (2005). https://doi.org/10.1364/OL.30.003231

49. L. Ponnampalam, R.J. Steed, M.J. Fice, C.C. Renaud, D.C. Rogers, D.G. Moodie, G.D. Maxwell, I.F. Lealman, M.J. Robertson, L. Pavlovic, L. Naglic, M. Vidmar, A.J. Seeds, A compact tunable coherent terahertz source based on an hybrid integrated optical phase-lock loop, in *2010 IEEE International Topical Meeting on Microwave Photonics*, pp. 151–154 (2010). https://doi.org/10.1109/MWP.2010.5664138

50. L. Ponnampalam, M.J. Fice, F. Pozzi, C.C. Renaud, D.C. Rogers, I.F. Lealman, D.G. Moodie, P.J. Cannard, C. Lynch, L. Johnston, M.J. Robertson, R. Cronin, L. Pavlovic, L. Naglic, M. Vidmar, A.J. Seeds, Monolithically integrated photonic heterodyne system. J. Lightwave Technol. **29**, 2229–2234 (2011). https://doi.org/10.1109/JLT.2011.2158186

51. F. Hindle, G. Mouret, S. Eliet, M. Guinet, A. Cuisset, R. Bocquet, T. Yasui, D. Rovera, Widely tunable THz synthesizer. Appl. Phys. B **104**, 763–768 (2011). https://doi.org/10.1007/s00340-011-4690-1

52. M. Bahmanian, J. Tiedau, C. Silberhorn, J.C. Scheytt, Octave-band microwave frequency synthesizer using mode-locked laser as a reference, in *2019 International Topical Meeting on Microwave Photonics (MWP)* (IEEE, 2019), pp. 1–4. https://doi.org/10.1109/MWP.2019.8892046

53. M. Bahmanian, S. Fard, B. Koppelmann, J.C. Scheytt, Ultra low phase noise and ultra wide-band frequency synthesizer using an optical clock source, in *2020 IEEE MTT-S International Microwave Symposium (IMS)* (2020), pp. 1283–1286. https://doi.org/10.1109/IMS30576.2020.9224118

54. M. Bahmanian, J.C. Scheytt, A 2-20-GHz ultralow phase noise signal source using a microwave oscillator locked to a mode-locked laser. IEEE Trans. Microwave Theory Tech. **69**, 1635–1645 (2021). https://doi.org/10.1109/TMTT.2020.3047647

Chapter 29
Integrated Photonically Assisted Samplers

Maxim Weizel, Meysam Bahmanian, and J. Christoph Scheytt

Abstract High-speed ADCs operating in the tens of gigahertz up to potentially terahertz range are largely constrained by the jitter in their clock sources. By incorporating photonically assisted samplers that exploit the ultralow jitter of specific mode-locked lasers (MLLs) as analogue ADC frontends, the performance limits of data converters can be pushed to achieve unprecedented levels of accuracy. Continuous advancements in electronic-photonic integration (silicon photonics) are clearing the path for integrating these systems on a chip scale, thereby leading to increased scalability, as well as reduced cost and power consumption.

29.1 Motivation

Fast samplers as analogue-to-digital converter (ADC) frontends are essential components in all high-speed data acquisition systems, such as optical communication systems or metrology equipment like oscilloscopes. In THz metrology, the performance of the entire measurement system is often limited by the data conversion process. Both the digital-to-analogue converter (DAC) and ADC play pivotal roles in defining the signal bandwidth that can be transmitted or received. However, the expected improvements in complementary metal oxide semiconductor (CMOS) data converters in upcoming technology nodes are insufficient to meet the growing interface data rates required for future communication and measurement systems [1]. To overcome this limitation, parallel (interleaved) data converters have been proposed, and advanced bipolar technologies, such as indium phosphide (InP) or silicon germanium (SiGe), can help to reach the bandwidth and sampling rate requirements. Despite these advances, several challenges remain, from which the timing jitter, respectively, the phase noise, of the utilized clock sources is one of the biggest issues in maintaining a high resolution at high input frequencies [2].

M. Weizel (✉) · M. Bahmanian · J. C. Scheytt
Schaltungstechnik (SCT)/Heinz Nixdorf Institut, Universität Paderborn, Paderborn, Germany
e-mail: maxim.weizel@hni.uni-paderborn.de; meysam.bahmanian@hni.uni-paderborn.de;
cscheytt@hni.uni-paderborn.de

© The Author(s) 2026
T. Kürner et al. (eds.), *Metrology for THz Communications*, Springer Series
in Optical Sciences 256, https://doi.org/10.1007/978-3-032-01986-8_29

Innovative integrated photonically assisted sampling architectures can enable performance beyond the constraints of traditional approaches by leveraging the inherently large bandwidth of optical systems and the ultralow jitter performance of certain mode-locked lasers (MLLs), which can go down to the attosecond level [3, 4]. This allows for higher resolution, bandwidth, and sampling rates [5, 6]. In the following, an overview of some integrated photonically assisted sampling architectures is provided. By addressing the key challenges and outlining potential solutions, this chapter aims to provide a comprehensive understanding of how integrated photonically assisted ADCs can enable the next generation of THz metrology systems.

29.1.1 Integrated Photonically Assisted Samplers

Photonically assisted sampling offers a complementary solution to the limitations of conventional electronic sampling, and the idea of using an electro-optical intensity modulator, like a Mach-Zehnder modulator (MZM), to modulate the amplitude of a pulse train generated by an MLL goes back several decades, though using discrete components, e.g., [7]. Many architectures were proposed thereafter incorporating various time- and frequency-interleaved systems [8]. In the more recent time, huge steps toward an integrated solution were presented in [5, 9, 10].

While most architectures rely on fast MZMs, like the frequency-interleaved sampler presented in Sect. 29.2, another approach is to use an optical clock distribution network [11] and then locally convert the light into an electronic clock signal as in the system presented in Sect. 29.3. The last approach presented in this chapter (Sect. 29.2) relies on ultralow jitter microwave generation based on MLLs.

29.1.2 Challenges and Perspectives

29.1.2.1 Integrated MLL

While still a matter of research, integrated MLLs have made significant progress. Recent advancements in photonic integration have enabled the development of monolithically integrated mode-locked lasers on platforms such as InP and III-V-on-silicon, delivering femtosecond pulses with ultralow timing jitter and high spectral purity. These integrated MLLs are crucial to provide compact and scalable solutions with enhanced noise performance for high-speed applications like terahertz (THz) metrology and photonically assisted analogue-to-digital conversion [12, 13]. Comparably, phase-noise plots of some discrete best-in-class MLLs are shown in [14].

29.1.2.2   Integrated MZM

Increasing the performance of integrated MZMs is an ongoing hot research topic. An ideal modulator would combine parameters like high-EO-bandwidth, low optical insertion loss, short length/device size, and small driving voltages (small V_π) while offering low cost and monolithic co-integration with other advanced photonic building blocks as well as CMOS compatibility.

When it comes to scalability, low cost, and co-integration, the silicon photonic platform with its CMOS compatibility would be the best choice. However, the electro-optic effect in silicon is the plasma dispersion effect. That means the refractive index is modulated by changing the amount of free carriers in the waveguide. To get the highest bandwidth, depletion type phase shifters are preferred over enhancement type, but due to the low modulation efficiency, the MZM arms are in the millimeter-length range leading to large footprints and high optical losses. EO-bandwidths over 50 GHz are still hard to reach [15–17]. Nevertheless, innovative designs are constantly proposed, e.g., the authors in [18] make use of slow light effects to reach an EO-bandwidth of 110 GHz, and recently, in [19], time-frequency equalization was used to reach 110 GHz -1 dB BW. Another all-silicon approach is the micro-ring modulator (MRM). It holds a very compact footprint, and in [20], an EO-bandwidth of 110 GHz is shown, although a disadvantage is its high susceptibility to temperature changes and reduced optical bandwidth.

Furthermore, hybrid approaches exist, building on the silicon photonic platform hybrid modulators can enhance the performance tremendously [21]. Silicon-organic hybrid (SOH) and plasmonic-organic hybrid (POH) modulators with EO-bandwidths of 76 GHz (CC-SOH) [22], 360 GHz (POH) [23], and even 500 GHz (POH) [24] have been proposed. These great results come with some disadvantages. Still organic-hybrid modulators have to prove their long-term stability with advances made in that area [25], but also the fabrication is much more complicated and cost/time intensive.

Another promising candidate is thin-film lithium niobate (TFLN), since the Pockels-based electro-optic effect in lithium niobate (LN) intrinsically happens on femtosecond timescales [26]. Although many modulators over 100 GHz bandwidth with comparably low V_π voltages have been demonstrated, LN remains only partially compatible to today's CMOS or silicon photonic processes. This is mainly due to material contamination issues, including lithium diffusion and etching residues. In addition, LN-modulators typically require a relatively large footprint up to centimeter range [27].

A material which has an even higher Pockels coefficient than lithium niobate is barium titanate (BTO). It offers a path to be integrated into advanced silicon photonic platforms [28] reaching   >200 Gbps/λ transmission [29]. It can be considered as a type of hybrid modulator.

29.2 Frequency-Interleaved Photonically Assisted Sampling

A frequency-interleaved sampling system splits an arbitrary input signal, which covers a bandwidth of BW into N frequency bands with now reduced bandwidth BW/N. N denoting the number of channels. A huge benefit of this approach is that the individual channels can then be built with low bandwidth/low-cost electronics. This well-known principle, which is used in all-electronic data converters, can be translated to a photonically assisted sampler. One possible implementation concept is shown in Fig. 29.1.

29.2.1 Signal Generation and Fiber-to-Chip Coupling

First, an arbitrary electronic signal with bandwidth BW is modulated onto a single frequency (continuous wave) laser by means of an ultra-broadband electro-optical modulator. This broadband modulator can, e.g., be based on organic materials. For example, in [23], the authors present a plasmonic organic hybrid (POH) with 0.36 THz bandwidth. It relies on a silicon photonic platform, making it suitable for monolithic integration with the disadvantage of the organic material not being long-term stable yet. Other now evolving types of fast electro-optic modulators, which can be monolithically integrated, are thin-film lithium niobate MZMs or barium titanate MZMs; see Sect. 29.1.2.

The resulting amplitude-modulated optical signal will span a bandwidth of $2BW$. In order to reconstruct real-valued electronic input signals, only half of the optical spectrum needs to be received and analyzed. Coupling the modulated optical signal

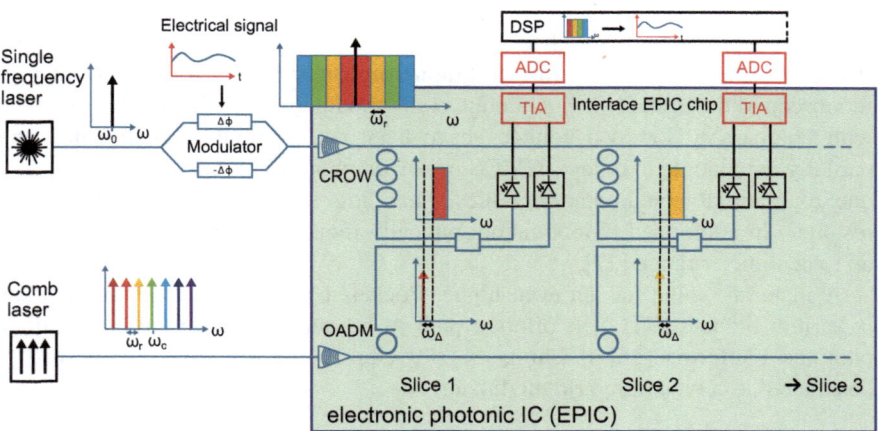

Fig. 29.1 Frequency-interleaved photonic ADC system. (Adapted from [6], used under CC-BY 4.0)

into an electronic photonic integrated circuit (EPIC) can be accomplished using a fiber array and on-chip grating couplers. Another way to couple light into a silicon phonic chip is to use a lensed fiber and an edge coupler. Apart from these two well-known techniques, a rather new and innovative way to realize a fiber to chip connection is photonic wirebonds [30].

29.2.2   Optical Slicing/Filtering

Optical bandpass filters, like ring resonators, monolithically integrated, are used to slice the input signal into N slices. Optimally, the optical filters have a flat passband transfer characteristic and steep roll-offs toward the stopbands. The bandwidth should match that of the co-integrated photodiodes and electronic transimpedance amplifiers (TIAs). Then, the slices are down-converted to baseband by optical mixing with MLL lines serving as local oscillators (LO). Individual comb lines are filtered out by optical add-drop multiplexers (OADM) and directed to their corresponding channel (slice). Mixing can be achieved with directional couplers or multimode interferometers (MMI). The repetition rate of the comb laser is set in such a way that the free spectral range (FSR) matches the bandwidth of the slices. Placing the LO lines with a small offset to the signal slices minimizes the influence of flicker noise generated by the electronic receiver stages at low offset frequencies.

29.2.3   Detection and Post-Processing

Detection in each channel is realized by a balanced receiver with integrated photodiodes and a TIA with reduced bandwidth (BW/N). The following ADCs with high ENOB convert the electronic TIA output to a digital data stream. The original radio frequency (RF) signal is reconstructed by combining all N signal slices using digital signal processing (DSP). The system depicted in Fig. 29.1 uses a so-called heterodyne detection scheme. With modifications, a homodyne detection scheme would also be possible. This idea is discussed further in Sect. 29.2.5.

29.2.4   Experimental Results

The system depicted in Fig. 29.1 is realized and measured in [31, 32]. A die layout view is shown in Fig. 20.7. Four slices are used and the measurement setup can be seen in Fig. 29.2a. Optical bandpass filters, in the form of thermally tuned coupled resonator optical waveguides (CROW), monolithically integrated, are used to slice the input signal. The bandwidth of the CROWs matches that of the co-integrated PDs and electronic TIAs. Due to the complexity of this big system, unfortunately,

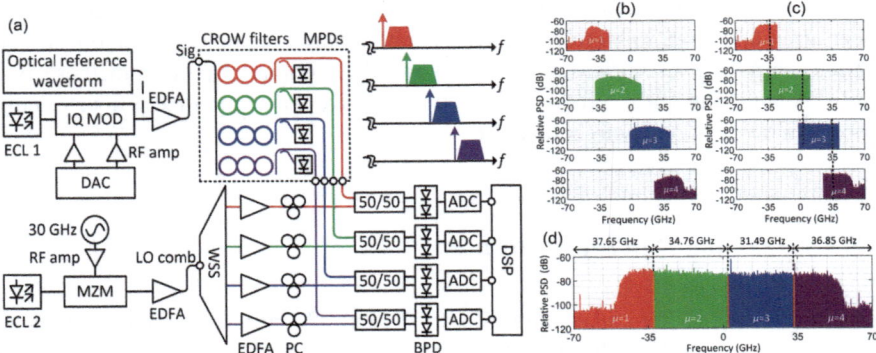

Fig. 29.2 (**a**) Measurement setup of a frequency-interleaved photonic ADC system. (**b**) Raw (unequalized) spectral representation of the received signal per channel. (**c**) Equalized spectrum per channel. (**d**) Final spectrally stitched signal. (**a**) is taken from [31]. (**b**, **c**) and (**d**) reprinted with permission from [32] ©2021 IEEE

not all components were fully functioning, but luckily, the optical CROW filters with their thermal tuning connections (12 chip pads) and monitoring photodiodes (MPDs, four chip pads) were working as specified. Using a fiber array, the optical signal is coupled in through grating couplers into an EPIC, where it is split into four slices. The slices are coupled out of the chip again and are mixed with the local oscillator (LO) comb lines in a 2×2 discrete 50/50 directional coupler. Detection is done with discrete balanced photodiodes and a real-time oscilloscope as ADC. The original radio frequency (RF) signal is reconstructed by combining the signal slices offline using digital signal processing (DSP). More specifically, the reconstruction relies on a one-time system calibration using a femtosecond laser with a known pulse shape and exploits maximum ratio combining (MRC) to spectrally stitch the individual slices with minimum impairments [32]. The received spectrum can be seen in Fig. 29.2b. After applying the calibration to each slice, the corrected spectrum is obtained; see Fig. 29.2c. Finally, stitching the slices together yields the fully reconstructed signal Fig. 29.2d. The experiments showed an overall detection bandwidth of 140 GHz. A 100 GBd QPSK, 16QAM and 64QAM optical data signal was reconstructed with an SNR of 19.8 dB, 19.8 dB, and 19.2 dB respectively.

29.2.5 Discussion and Outlook

Another possible implementation, as mentioned before, would use homodyne instead of heterodyne receivers, but with the drawback of including the low frequency noise. Actually, another drawback is that most practical coherent receivers come with an offset compensation loop having a bandwidth in the low MHz to kHz range. This would filter out any signal components laying in these frequency bands, preventing the system to perform a true arbitrary waveform measurement.

Nevertheless, in a series of experiment, the authors in [33, 34] show a filter-less (slice-less) optical arbitrary waveform measurement system, which relies on homodyne detection. Furthermore, the concept was translated to integrated silicon photonics [35, 36], effectively tackling a drawback of the aforementioned slice-based frequency-interleaved photonically assisted sampler, which required a rather complicated thermal controlling scheme to stabilize the CROW (bandpass) filters.

In future, key components like the comb laser source or the high-frequency modulator are of course envisioned to be monolithically integrated too. Adding the ADCs and a DSP would result in a full system on chip. This would require a large design effort and is also limited by the currently available technology at hand. Especially the integrated pulse lasers are lacking behind in performance compared to their discrete counterparts.

29.3 Time-Interleaved Optically Clocked/Optoelectronic Sampling

29.3.1 Concept

Another possible way of implementing a photonically assisted sampler is the optically clocked electronic sampler or short optoelectronic (OE) sampler. In the simplest form, this sampler type can be implemented by converting the light pulses of an ultralow jitter femtosecond MLL to an ultralow jitter electronic clock signal using a photodiode and a transimpedance amplifier. This signal is then used to trigger a fully electronic sampler/ADC.

To upgrade this concept to an N-channel time-interleaved photonically assisted optically clocked sampler, time-interleaved pulse trains need to be generated without deteriorating the jitter/phase-noise performance of the MLL. One of multiple ways to build such as system is shown in Fig. 29.3. An optical pulse train from an MLL with repetition rate T_s is filtered by an optical bandpass filter and then sent through a dispersive element, which can be a long fiber. The dispersion broadens the pulse in the time domain without changing the power spectral density. N frequency bands can now be filtered out from the broadened pulse and directed through a low loss optical clock distribution network to N channels of optoelectronic samplers as clock signals. On-chip filtering can be achieved by using, e.g., thermally tunable ring resonators [37]. The individual frequency bands are already shifted in time, and by creating enough dispersion, a time-interleaved sampler, with equally spaced samples, is obtained. One of the first experiments with a similar scheme using silicon photonics and comparing it with a discrete setup can be found in [10], where the authors achieved 3.5 effective bits digitizing a 10 GHz signal.

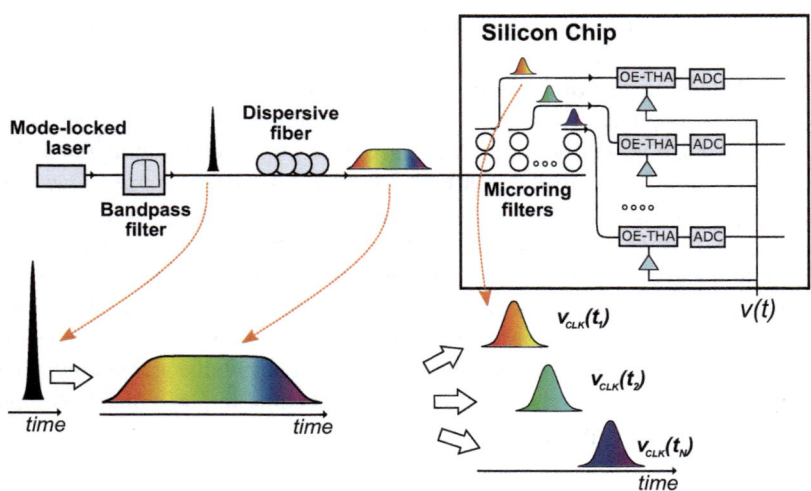

Fig. 29.3 Time-interleaved optically clocked sampling concept

29.3.2 Practical Considerations

On one hand, the advantage of this architecture is its straight forward realization. Already existing conventional electronic sampler designs can be reused, or even existing systems can be upgraded by replacing the electronic oscillators with MLL + optoelectronic conversion systems, e.g., PDs or PDs + TIAs. On the other hand, for a practical system, multiple problems have to be tackled. Firstly, a femtosecond laser, due to its narrow pulse width, provides a very high peak power. This can lead to deterioration in the waveguides, especially in the tiny on-chip silicon waveguides, and increase the jitter [11]. Then, depending on the MLL pulse period, respectively, the desired sampling rate, the type of the TIA has to be chosen accordingly. For low sampling rates below a few GHz, a resistive/passive TIA or a common gate/base TIA [38] can fulfil the speed requirements, but for 10 GS/s or more, a common emitter/source or inverter-based shunt feedback stage with high bandwidth is better suited.

As an example, the Menhir-1550 MLL with a repetition rate of $f_{\text{rep}} = 1/T_{\text{rep}} = 250\,\text{MHz}$ has an average output power of $P_{\text{avg}} > 100\,\text{mW}$ and a full-width-at-half-maximum (FWHM) pulse width of $\tau_{\text{FWHM}} < 250\,\text{fs}$ [39]. Assuming a sech^2 pulse shape as in Eq. (29.1) and with E_p being the pulse energy (29.2), the peak power P_0 can be calculated as seen in Eq. (29.3):

$$P(t) = P_0 \cdot \text{sech}^2 \left(\frac{t}{\tau_p} \right) \tag{29.1}$$

$$E_p = P_{\text{avg}} \cdot T_{\text{rep}} = \int_{-\infty}^{\infty} P(t)\, dt = 2 P_0 \tau_p \tag{29.2}$$

$$P_0 = \frac{1}{2} \cdot \frac{E_p}{\tau_p} \tag{29.3}$$

$$\tau_{\text{FWHM}} = 2 \ln\left(1 + \sqrt{2}\right) \tau_p \approx 1.7627 \tau_p \tag{29.4}$$

For a sech^2 pulse shape, τ_p can be replaced by the more commonly used τ_{FWHM} with the relation in Eq. (29.4) [40, p.72]. It also solves the problem with the high peak power of the MLL pulses, since broadening in time reduces the peak power; see Eq. (29.3).

While for the optically clocked samplers it is possible to upgrade existing systems by replacing the electronic oscillators with an MLL-PD combination, the real advantage of the architecture lies in the silicon photonic integration and in distributing the light pulses to the samplers through on-chip optical waveguides. That is why of course it is envisioned to monolithically integrate the whole system on a silicon photonic chip, but as mentioned in the introduction (Sect. 29.1.2), integrated MLL performance lack behind their discrete counterparts. Furthermore, to create the necessary dispersion, multiple kilometers of fiber are needed, though the length can be significantly reduced by using dispersion compensating fiber. Another possible idea to overcome this issue is to replace the dispersive element with a wavelength-demultiplexer and optical delay lines as in [10]. Wavelength demultiplexing can be achieved with an arrayed waveguide grating (AWG), and delay is created by increasing waveguide lengths per channel.

29.3.3 Implementation of an Optically Clocked THA

An optically clocked track-and-hold amplifier (THA) as a frontend for a high speed photonically assisted ADC is theoretically analyzed and by means of simulation compared to a MZM-based sampler in [41]. The results show that the OE-THA, implemented in a 250 nm Silicon Photonics BiCMOS process with a transit frequency f_t of 220 GHz, can achieve a bandwidth of 78 GHz in post-layout simulations with excellent linearity and therefore no need for additional post-processing. Compared to that, all silicon MZMs are limited in both bandwidth which is approx 40 GHz and linearity resulting from the cos-shaped MZM input-output characteristic.

29.3.3.1 Schematic Description

Encouraged by the simulation results, a chip was designed and fabricated [38]. A microphotograph of the chip from [38] can be seen in Fig. 29.5a. It includes

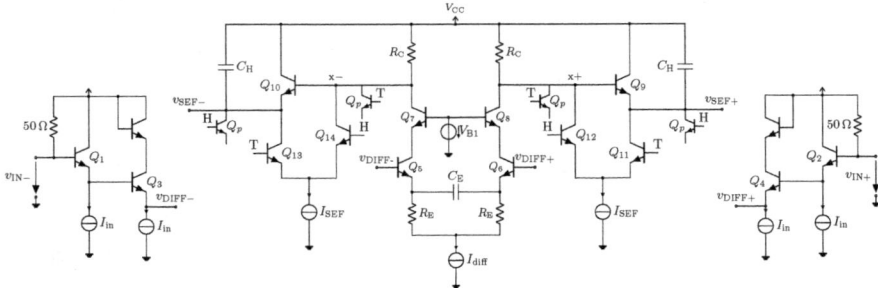

Fig. 29.4 Schematic of the THA with 50 Ohm input buffer, cascode main amplifier, and switched emitter-followers. (Reprinted with permission from [38] ©Optical Society of America)

an advanced differential track-and-hold amplifier in the switched emitter-follower (SEF) architecture. The core schematic is shown in Fig. 29.4. An active feedthrough attenuation circuit and feed-forward capacitors are incorporated (not shown in Fig. 29.4), both reducing the feedthrough of the input signal to the output during hold times significantly. Moreover, to achieve the highest possible bandwidth in the given technology, a main cascode differential amplifier ($Q_5 - Q_8$) with emitter degeneration (R_E) and capacitive peaking (C_E) is used. It should be noted that having a future time-interleaved architecture with multiple channels in mind the THA is designed inductorless. Due to differential clock signals, charge injection from the clock onto the hold capacitors (C_H) through the base-collector capacitance (c_{bc} of the clock-switch transistors ($Q_{11} - Q_{14}$) occurs. It is compensated with transistors Q_p by making use of their parasitic base-collector capacitance. Transistors Q_p are driven by their respective negative differential clock signal, effectively creating a negative capacitance. To convert the light pulses to an electronic clock signal, a single-ended transimpedance amplifier in common-base configuration with a transimpedance of 40 dBΩ is implemented. The counterpart of the single-ended clock signal is an externally controllable reference voltage. This means the applied clock signal is only pseudo-differential.

29.3.3.2   Results

S-parameter measurements show a track-mode small-signal 3-dB bandwidth of 65 GHz and hold-mode isolation of − 30 dB up to 40 GHz. To acquire these measurements, a vector network analyzer (VNA) was used, and the circuit was set to track- or hold-mode through external biasing voltages, respectively. Dynamic performance is measured by connecting a signal generator to the chip via high-frequency probes, and the optical pulses are generated by a solid-state MLL (Menhir-1550) with an FSR of 250 MHz and a FWHM of 250 fs. An external wave shaper introduces an empirically determined dispersion of 6 ps/nm, resulting in a broader pulse FWHM of 60 ps. A standard rectangular-shaped grating coupler and

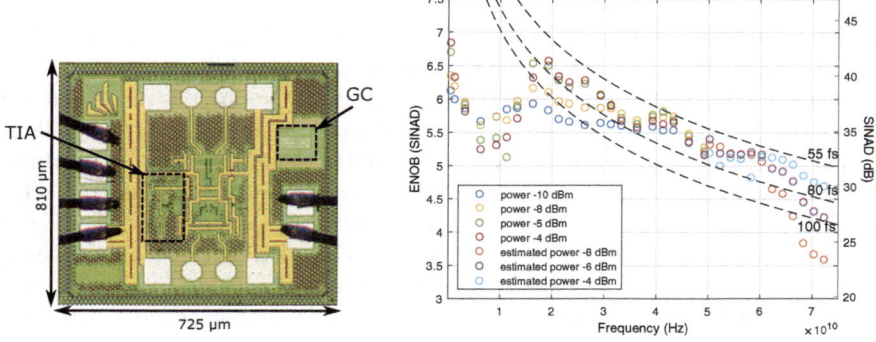

Fig. 29.5 (**a**) Microphotograph of the optoelectronic track-and-hold amplifier. (**b**) ENOB/SINAD measurements depending on frequency and amplitude. Dashed lines show the 55 fs, 80 fs, and 100 fs equivalent jitter limits. (Reprinted with permission from [38] ©Optical Society of America)

long taper is used to couple the light into the chip. Nowadays, focused grating couplers are widely used instead. Signal-to-noise-and-distortion (SINAD) values and the, respectively, calculated effective number of bits (ENOB) can be seen in Fig. 29.5b. The sampler generally shows an ENOB of over 5.5 bits (35 dB SINAD) up to ≈45 GHz. After that, for an estimated input power of − 4 dBm, an ENOB of over 5 bits can be maintained up to 65 GHz. The equivalent jitter ranges from 55 fs to 80 fs up to frequencies over 65 GHz. The best equivalent jitter of 55.8 fs RMS is achieved at 41 GHz using an electrical input power of − 5 dBm.

29.4 Optoelectronic PLL Enabling Photonically Assisted Electronic Sampling

In comparison to the time-interleaved optically clocked THA described in Sect. 29.3 where a mode-locked laser signal is directly converted to an electronic clock by photodiodes and transimpedance amplifiers, the optoelectronic PLL (OEPLL) uses a mode-locked laser as an ultra-stable reference oscillator to lock an electronic oscillator.

29.4.1 Concept

The concept of the OEPLL is described in detail in [14, 42] and will be summarized in the following. For the PLL, an MLL serves as the reference oscillator, while an yttrium iron garnet (YIG) oscillator with a broad 2 GHz–20 GHz tuning bandwidth functions as the tunable oscillator. A balanced optical microwave phase detector (BOMPD) is utilized as phase detector. The BOMPD is made with a dual output Mach-Zehnder modulator with balanced photodetection scheme. In [14], a theoretical framework for an optoelectronic PLL incorporating this phase detector is introduced. Building on this analysis, a broadband frequency synthesizer with an output frequency range spanning from 2 to 20 GHz is subsequently designed and implemented. Phase-noise measurements indicate an integrated RMS jitter (1 kHz–100 MHz) of less than 4 fs within the 5 to 20 GHz range, with a typical value of 4 fs and a minimum of 3 fs. This work presents the first wideband PLL frequency synthesizer that achieves an integrated jitter below 10 fs RMS (measured over 1 kHz–100 MHz) across a 3 to 20 GHz frequency range. A comparison with top-tier laboratory-grade frequency synthesizers in this frequency range demonstrates that this synthesizer exhibits lower phase noise than any electronic frequency synthesizer for offset frequencies above 2 kHz.

Building on these impressive results demonstrated with discrete components, a block diagram of a fully integrated architecture of a 2× time-interleaved photonically assisted track-and-hold amplifier is shown in Fig. 29.6. The integration can, for example, be done in the IHPSG25H5_EPIC process [43]. Considering the better phase-noise performance of discrete pulse lasers, only the MLL is left as a discrete component for now. The MLL signal is coupled into the chip via a grating coupler and then enters the BOMPD that can be implemented as a depletion type dual output MZM modulator [17, 44]. Subsequent balanced detection can be realized with on-chip germanium photodiodes. A lowpass filter and driver circuit generate the control voltage for an on-chip voltage controlled oscillator (VCO) at 64 GHz. A divide by

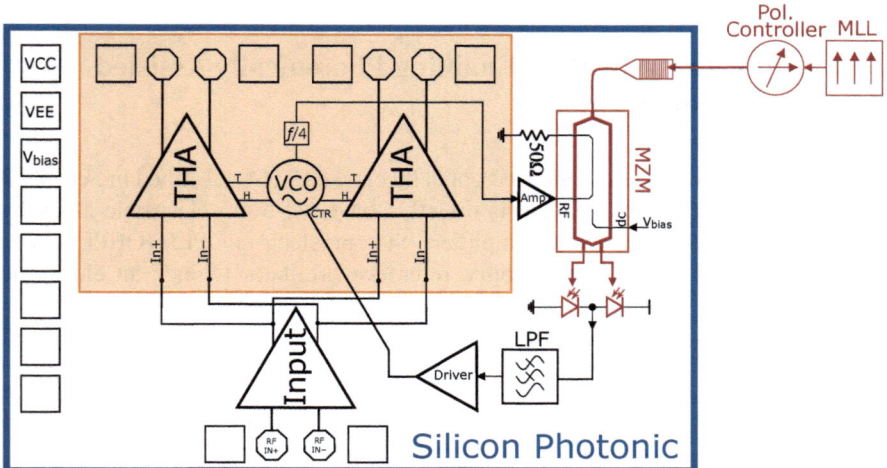

Fig. 29.6 Concept of a 2× time-interleaved track-and-hold amplifier driven by an optoelectronic PLL in a silicon photonic integrated process. All components (VCO, optical phase detector, etc.) besides the MLL are monolithically integrated

four frequency divider creates a signal with a frequency of 16 GHz at its output which goes to the electrical RF-input of the MZM.

29.4.2 Discussion

The advantage of the architecture is that the OE-PLL can be locked to a harmonic of the MLL, and the clock signal up-conversion is performed by the PLL. In comparison to the directly optically clocked sampler, the available MLL repetition rate is not limiting the sampling rate but can be chosen freely. As a disadvantage, an increased phase noise is likely to result from the on-chip VCO compared to a discrete YIG oscillator, requiring a thorough VCO design. Simulations show that from phase-noise perspective, a cross-coupled Colpitts oscillator seems to have a better performance compared to ring-resonator type oscillators in the given technology (IHPSG25H5_EPIC).

29.5 Conclusion

This chapter explores the potential of integrated photonically assisted samplers to overcome the limitations of conventional high-speed ADCs. By leveraging ultralow jitter mode-locked lasers and the inherent large bandwidth of photonic systems, these architectures enable unprecedented resolution, sampling rates, and bandwidths. Key implementations, including frequency-interleaved, optically clocked (time-interleaved), and optoelectronic PLL based sampling systems, have been discussed, highlighting their respective advantages and limitations in THz metrology and high-speed data conversion. While it remains uncertain which architecture will deliver the best results, it is likely to depend on application-specific requirements, requiring further research in all directions.

The integration of photonic and electronic components is central to advancing these systems. Progress in silicon photonics, hybrid modulators, and integrated MLLs will pave the way for scalable, cost-effective, and power-efficient solutions. Fully integrated systems-on-chip, combining optical sources, modulators, ADCs, and DSP, represent a promising future for photonically assisted data acquisition.

References

1. C. Schmidt, H. Yamazaki, G. Raybon, P. Schvan, E. Pincemin, S. J. B. Yoo, D. J. Blumenthal, T. Mizuno, R. Elschner, Data converter interleaving: current trends and future perspectives. IEEE Commun. Mag. **58**(5), 19–25 (2020)

2. R. Walden, Analog-to-digital converter survey and analysis. IEEE J. Sel. Areas Commun. **17**(4), 539–550 (1999)
3. J. Kim, F. Kärtner, Attosecond-precision ultrafast photonics. Laser Photon. Rev. **4**(3), 432–456 (2010)
4. A.J. Benedick, J.G. Fujimoto, F.X. Kärtner, Optical flywheels with attosecond jitter. Nat. Photon. **6**, 97–100 (2012)
5. C.W. Holzwarth, R. Amatya, M. Araghchini, J. Birge, H. Byun, J. Chen, M. Dahlem, N.A. DiLello, F. Gan, J.L. Hoyt, E.P. Ippen, F.X. Kärtner, A. Khilo, J. Kim, M. Kim, A. Motamedi, J.S. Orcutt, M. Park, M. Perrott, M.A. Popovic, R.J. Ram, H.I. Smith, G.R. Zhou, S.J. Spector, T.M. Lyszczarz, M.W. Geis, D.M. Lennon, J.U. Yoon, M.E. Grein, R.T. Schulein, S. Frolov, A. Hanjani, J. Shmulovich, High speed analog-to-digital conversion with silicon photonics, in *Silicon Photonics IV*, ed. by J.A. Kubby, G.T. Reed, vol. 7220 (International Society for Optics and Photonics, SPIE, 2009), p. 72200B
6. A. Zazzi, J. Müller, M. Weizel, J. Koch, D. Fang, A. Moscoso-Mártir, A. Tabatabaei Mashayekh, A.D. Das, D. Drayß, F. Merget, F.X. Kärtner, S. Pachnicke, C. Koos, J.C. Scheytt, J. Witzens, Optically enabled ADCs and application to optical communications. IEEE Open J. Solid-State Circuits Soc. **1**, 209–221 (2021)
7. H.F. Taylor, M.J. Taylor, P.W. Bauer, Electrooptic analog-to-digital conversion using channel waveguide modulators, in *Integrated and Guided Wave Optics* (Optica Publishing Group, 1978), p. TuC1
8. K. E. Al Qubaisi, A. Khilo, Photonic analog-to-digital converters, in *2014 XXXIth URSI General Assembly and Scientific Symposium (URSI GASS)* (2014), pp. 1–3
9. F. X. Kärtner, R. Amatya, M. Araghchini, J. Birge, H. Byun, J. Chen, M. Dahlem, N.A. DiLello, F. Gan, C.W. Holzwarth, J.L. Hoyt, E.P. Ippen, A. Khilo, J. Kim, M. Kim, A. Motamedi, J.S. Orcutt, M. Park, M. Perrott, M.A. Popović, R.J. Ram, H.I. Smith, G.R. Zhou, S.J. Spector, T.M. Lyszczarz, M.W. Geis, D.M. Lennon, J.U. Yoon, M.E. Grein, R.T. Schulein, Photonic analog-to-digital conversion with electronic-photonic integrated circuits, in *Silicon Photonics III*, edm by J.A. Kubby, G.T. Reed, vol. 6898 (International Society for Optics and Photonics, SPIE, 2008)
10. A. Khilo, S.J. Spector, M.E. Grein, A.H. Nejadmalayeri, C.W. Holzwarth, M.Y. Sander, M.S. Dahlem, M.Y. Peng, M.W. Geis, N.A. DiLello, J.U. Yoon, A. Motamedi, J.S. Orcutt, J.P. Wang, C.M. Sorace-Agaskar, M.A. Popović, J. Sun, G.-R. Zhou, H. Byun, J. Chen, J.L. Hoyt, H.I. Smith, R.J. Ram, M. Perrott, T.M. Lyszczarz, E.P. Ippen, F.X. Kärtner, Photonic ADC: overcoming the bottleneck of electronic jitter. Opt. Express **20**, 4454–4469 (2012)
11. E. Krune, K. Jamshidi, K. Voigt, L. Zimmermann, K. Petermann, Jitter analysis of optical clock distribution networks in silicon photonics. J. Lightwave Technol. **32**(22), 4378–4385 (2014)
12. F.X. Kärtner, N. Singh, Integrated CMOS-compatible mode-locked lasers and their optoelectronic applications, in *2019 IEEE BiCMOS and Compound Semiconductor Integrated Circuits and Technology Symposium (BCICTS)* (2019), pp. 1–8
13. K. Van Gasse, S. Uvin, V. Moskalenko, S. Latkowski, G. Roelkens, E. Bente, B. kuyken, Recent advances in the photonic integration of mode-locked laser diodes. IEEE Photon. Technol. Lett. **31**(23), 1870–1873 (2019)
14. M. Bahmanian, J.C. Scheytt, A 2–20-GHz ultralow phase noise signal source using a microwave oscillator locked to a mode-locked laser. IEEE Trans. Microwave Theory Tech. **69**(3), 1635–1645 (2021)
15. D. Patel, A. Samani, V. Veerasubramanian, S. Ghosh, D.V. Plant, Silicon photonic segmented modulator-based electro-optic DAC for 100 Gb/s PAM-4 generation. IEEE Photon. Technol. Lett. **27**(23), 2433–2436 (2015)
16. J. Witzens, High-speed silicon photonics modulators. Proc. IEEE **106**(12), 2158–2182 (2018)
17. C. Kress, T. Schwabe, H. Rhee, J. Christoph Scheytt, Compact, high-speed Mach-Zehnder modulator with on-chip linear drivers in photonic BiCMOS technology. IEEE Access **12**, 64561–64570 (2024)

18. C. Han, M. Jin, Y. Tao, B. Shen, H. Shu, X. Wang, Ultra-compact silicon modulator with 110 GHz bandwidth, in *2022 Optical Fiber Communications Conference and Exhibition (OFC)* (2022), pp. 1–3
19. H. Yue, J. Fu, H. Zhang, B. Xiong, S. Pan, T. Chu, Silicon modulator exceeding 110 GHz using tunable time-frequency equalization (2024)
20. Y. Zhang, H. Zhang, J. Zhang, J. Liu, L. Wang, D. Chen, N. Chi, X. Xiao, S. Yu, 240 Gb/s optical transmission based on an ultrafast silicon microring modulator. Photon. Res. **10**, 1127–1133 (2022)
21. C. Koos, J. Leuthold, W. Freude, M. Kohl, L. Dalton, W. Bogaerts, A.L. Giesecke, M. Lauermann, A. Melikyan, S. Koeber, S. Wolf, C. Weimann, S. Muehlbrandt, K. Koehnle, J. Pfeifle, W. Hartmann, Y. Kutuvantavida, S. Ummethala, R. Palmer, D. Korn, L. Alloatti, P.C. Schindler, D.L. Elder, T. Wahlbrink, J. Bolten, Silicon-organic hybrid (SOH) and plasmonic-organic hybrid (POH) integration. J. Lightwave Technol. **34**(2), 256–268 (2016)
22. S. Ummethala, J.N. Kemal, A.S. Alam, M. Lauermann, A. Kuzmin, Y. Kutuvantavida, S.H. Nandam, L. Hahn, D.L. Elder, L.R. Dalton, T. Zwick, S. Randel, W. Freude, C. Koos, Hybrid electro-optic modulator combining silicon photonic slot waveguides with high-k radiofrequency slotlines. Optica **8**, 511–519 (2021)
23. S. Ummethala, T. Harter, K. Koehnle, Z. Li, S. Muehlbrandt, Y. Kutuvantavida, J. Kemal, P. Marin-Palomo, J. Schaefer, A. Tessmann, S.K. Garlapati, A. Bacher, L. Hahn, M. Walther, T. Zwick, S. Randel, W. Freude, C. Koos, THz-to-optical conversion in wireless communications using an ultra-broadband plasmonic modulator. Nat. Photon. **13**, 519–524 (2019)
24. M. Burla, C. Hoessbacher, W. Heni, C. Haffner, Y. Fedoryshyn, D. Werner, T. Watanabe, H. Massler, D.L. Elder, L.R. Dalton, J. Leuthold, 500 GHz plasmonic Mach-Zehnder modulator enabling sub-THz microwave photonics. APL Photon. **4**, 056106 (2019)
25. A. Schwarzenberger, A. Mertens, H. Kholeif, A. Kotz, C. Eschenbaum, L.E. Johnson, D.L. Elder, S.R. Hammond, K. O'Malley, L. Dalton, S. Randel, W. Freude, C. Koos, First demonstration of a silicon-organic hybrid (SOH) modulator based on a long-term-stable crosslinked electro-optic material, in *49th European Conference on Optical Communications (ECOC 2023)*, vol. 2023 (2023), pp. 859–862
26. Y. Zhang, L. Shao, J. Yang, Z. Chen, K. Zhang, K.-M. Shum, D. Zhu, C.H. Chan, M. Lončar, C. Wang, Systematic investigation of millimeter-wave optic modulation performance in thin-film lithium niobate. Photon. Res. **10**, 2380–2387 (2022)
27. D. Zhu, L. Shao, M. Yu, R. Cheng, B. Desiatov, C.J. Xin, Y. Hu, J. Holzgrafe, S. Ghosh, A. Shams-Ansari, E. Puma, N. Sinclair, C. Reimer, M. Zhang, M. Lončar, Integrated photonics on thin-film lithium niobate. Adv. Opt. Photon. **13**, 242–352 (2021)
28. F. Eltes, C. Mai, D. Caimi, M. Kroh, Y. Popoff, G. Winzer, D. Petousi, S. Lischke, J.E. Ortmann, L. Czornomaz, L. Zimmermann, J. Fompeyrine, S. Abel, A BaTiO3-based electro-optic pockels modulator monolithically integrated on an advanced silicon photonics platform. J. Lightwave Technol. **37**, 1456–1462 (2019)
29. W. Li, F. Eltes, E. Berikaa, M.S. Alam, S. Bernal, C. Minkenberg, S. Abel, D.V. Plant, Thin-film BTO-based MZMs for next-generation IMDD transceivers beyond 200 Gbps/λ. J. Lightwave Technol. **42**(3), 1143–1150 (2024)
30. N. Lindenmann, G. Balthasar, D. Hillerkuss, R. Schmogrow, M. Jordan, J. Leuthold, W. Freude, C. Koos, Photonic wire bonding: a novel concept for chip-scale interconnects. Opt. Express **20**, 17667–17677 (2012)
31. D. Fang, A. Zazzi, J. Müller, D. Drayß, C. Füllner, P. Marin-Palomo, A.T. Mashayekh, A. D. Das, M. Weizel, S. Gudyriev, W. Freude, S. Randel, C. Scheytt, J. Witzens, C. Koos, Optical arbitrary waveform measurement (OAWM) on the silicon photonic platform, in *2021 Optical Fiber Communications Conference and Exhibition (OFC)* (2021), pp. 1–3
32. D. Fang, A. Zazzi, J. Müller, D. Drayss, C. Füllner, P. Marin-Palomo, A. Tabatabaei Mashayekh, A. Dipta Das, M. Weizel, S. Gudyriev, W. Freude, S. Randel, J.C. Scheytt, J. Witzens, C. Koos, Optical arbitrary waveform measurement (OAWM) using silicon photonic slicing filters. J. Lightwave Technol. **40**(6), 1705–1717 (2022)

33. D. Drayss, D. Fang, C. Füllner, G. Likhachev, T. Henauer, Y. Chen, H. Peng, P. Marin-Palomo, T. Zwick, W. Freude, T.J. Kippenberg, S. Randel, C. Koos, Slice-less optical arbitrary waveform measurement (OAWM) in a bandwidth of more than 600 GHz, in *2022 Optical Fiber Communications Conference and Exhibition (OFC)* (2022), pp. 1–3

34. D. Drayss, D. Fang, C. Füllner, G. Lihachev, T. Henauer, Y. Chen, H. Peng, P. Marin-Palomo, T. Zwick, W. Freude, T.J. Kippenberg, S. Randel, C. Koos, Non-sliced optical arbitrary waveform measurement (OAWM) using soliton microcombs. Optica **10**, 888–896 (2023)

35. D. Drayss, D. Fang, C. Füllner, A. Kuzmin, W. Freude, S. Randel, C. Koos, Slice-less optical arbitrary waveform measurement (OAWM) on a silicon photonic chip, in *2022 European Conference on Optical Communication (ECOC)* (2022), pp. 1–4

36. D. Drayss, D. Fang, C. Füllner, W. Freude, S. Randel, C. Koos, Non-sliced optical arbitrary waveform measurement (OAWM) using a silicon photonic receiver chip. J. Lightwave Technol. **42**(14), 4733–4750 (2024)

37. J. Müller, A. Zazzi, G. Vasudevan Rajeswari, A. Moscoso Mártir, A. Tabatabaei Mashayekh, A.D. Das, F. Merget, J. Witzens, Optimized hourglass-shaped resonators for efficient thermal tuning of CROW filters with reduced crosstalk, in *2021 IEEE 17th International Conference on Group IV Photonics (GFP)* (2021), pp. 1–2

38. M. Weizel, J.C. Scheytt, F.X. Kärtner, J. Witzens, Optically clocked switched-emitter-follower THA in a photonic SiGe BiCMOS technology. Opt. Express **29**, 16312–16322 (2021)

39. Menhir Photonics, MENHIR-1550 250MHz. https://menhir-photonics.com/menhir-1550/. Accessed 15 Sept 2024

40. G. Agrawal, *Nonlinear Fiber Optics* (Electronics & Electrical, Academic Press, 2001)

41. M. Weizel, F.X. Kaertner, J. Witzens, J.C. Scheytt, Photonic analog-to-digital-converters – comparison of a MZM-sampler with an optoelectronic switched-emitter-follower sampler, in *Photonic Networks; 21th ITG-Symposium* (2020), pp. 1–6

42. M. Bahmanian, S. Fard, B. Koppelmann, J.C. Scheytt, Wide-band frequency synthesizer with ultra-low phase noise using an optical clock source, in *2020 IEEE/MTT-S International Microwave Symposium (IMS)* (2020), pp. 1283–1286

43. D. Steckler, S. Lischke, A. Peczek, A. Kroh, J. Beyer, L. Zimmermann, Monolithic integration of 80-GHz Ge photodetectors and 100-GHz Ge electro-absorption modulators in a photonic BiCMOS technology. IEEE Trans. Electron Devices **71**(5), 3417–3423 (2024)

44. A. Misra, C. Kress, K. Singh, S. Preußler, J.C. Scheytt, T. Schneider, Integrated source-free all optical sampling with a sampling rate of up to three times the RF bandwidth of silicon photonic MZM. Opt. Express **27**, 29972–29984 (2019)

Chapter 30
Atmospheric Aspects

Enrique Castro-Camus, Fatima Taleb, and Martin Koch

Abstract The terahertz (THz) band has drawn considerable attention for potential telecommunication applications. This band of the electromagnetic spectrum provides greater bandwidth and thus allows for transmission of data at higher speeds. Yet, a limitation for the real-world use of this part of the spectrum is the strong absorption of atmospheric water vapor. The ITU-R 676-13 recommendation from the International Telecommunication Union provides an empirical model that remains the standard reference for the estimation of the attenuation of the atmosphere. This model is widely accepted and has been confirmed by a number of experimental studies of the attenuation of THz signals for a variety of atmospheric conditions, including some data that we present here.

30.1 Introduction

Many studies describing the attenuation of the THz signal under different atmospheric conditions have been published [1–7] including the presence of dust [8], fog [9], rain [10], and snow [11]; these cases will not be addressed in this chapter but are discussed in the references provided.

Short-range links [12], such as those in indoor environments, [13] might be less sensitive to subtle atmospheric condition changes; yet, longer outdoor links are gaining interest in backhaul as one of the first key commercial drivers for the use of frequencies above 100 GHz [14, 15]. While the attenuation of the atmosphere represents important challenges, it also can offer benefits for the use

E. Castro-Camus (✉)
Philipps-Universität Marburg, Physik (Fb13), Marburg, Germany

Centro de Investigaciones en Optica, Lomas del Campestre, Mexico
e-mail: enrique@cio.mx

F. Taleb · M. Koch
Philipps-Universität Marburg, Physik (Fb13), Marburg, Germany
e-mail: fatima.taleb@physik.uni-marburg.de; martin.koch@physik.uni-marburg.de

© The Author(s) 2026
T. Kürner et al. (eds.), *Metrology for THz Communications*, Springer Series
in Optical Sciences 256, https://doi.org/10.1007/978-3-032-01986-8_30

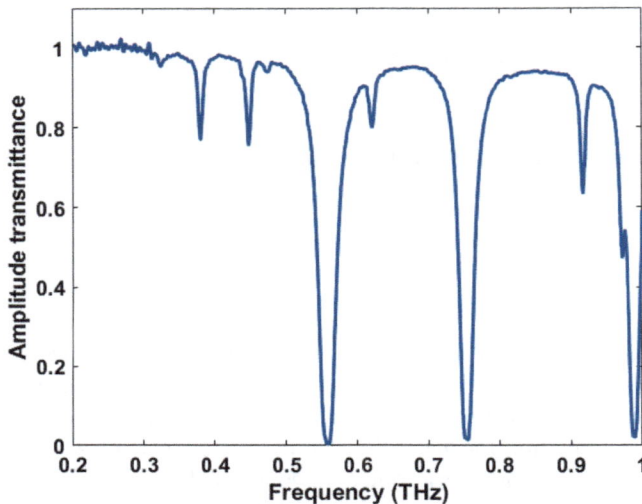

Fig. 30.1 Transmittance of 6.18 m of atmosphere at 21 °C with a relative humidity of 51%. The plot shows that the atmosphere is rich in absorption features in this band, where reasonably transparent spectral windows appear in between absorption peaks that range from weak to extremely deep. (Data obtained from [6])

of terahertz communications that are not available for lower frequency bands. For example, one can use the atmospheric attenuation as a security measure against eavesdropping[16]. Furthermore, the feature-rich absorption spectrum of vapor (Fig. 30.1) might open the possibility of novel encoding strategies for multiple access through hierarchical bandwidth modulation [17]. Therefore, a precise characterization of the temperature and humidity dependent atmospheric properties is fundamental for the future implementation of terahertz telecommunication systems.

The International Telecommunication Union introduced an empirical model as part of their ITU-R 676-13 recommendation. This model allows reasonably accurate calculation of the attenuation of the atmosphere as a function of the temperature and humidity and is probably the most widely used model of this type. This model will be discussed in the next section, and we will also present some experiments as a reference of its accuracy. We constructed a controlled climate chamber for this purpose, allowing long propagation distances and performed terahertz time-domain spectroscopy (THz-TDS) [5, 6, 18] measurements with it. Unlike many previous absorption studies [19–22], we performed all of our measurements at atmospheric pressure.

30.2 The ITU-R 676-13 Model

Probably, the most widely used model to estimate the attenuation of the atmosphere is the one contained in the ITU-R 676-13 recommendation of the International Telecommunication Union [23]. This model, which is completely empirical, is not based on any actual physical model; yet, it provides an analytical expression that works very well, thus, its wide acceptance.

The model considers three main contributions to the attenuation spectrum: firstly, a set of absorption lines associated to oxygen; secondly, a "continuum" featureless and monotonous spectral function; and thirdly, a set of absorption lines associated to water vapor. These tree components are directly superimposed giving an attenuation (in dB/km) of

$$
\gamma(f) = 0.1820 f \left[\sum_{i \in \text{Oxygen peaks}} S_i F_i(f) + N_D(f) + \sum_{i \in \text{Water peaks}} \sigma_i F_i(f) \right],
$$

(30.1)

where

$$
S_i = a_1^i \times 10^{-7} p \theta^3 \exp[a_2^i(1 - \theta)],
$$

(30.2)

$$
\sigma_i = b_1^i \times 10^{-1} e \theta^{3.5} \exp[b_2^i(1 - \theta)],
$$

(30.3)

the a_j^i and b_j^i are fitted parameters, tables containing their values for each individual peak can be found in [23], p is the partial pressure of dry air (in hPa), $\theta = 300/T$, T is the temperature (in K):

$$
e = \frac{\rho T}{216.7},
$$

(30.4)

ρ is the water vapor density, N_D is the continuum component given by

$$
N_D(f) = f p \theta^2 \left[\frac{6.14 \times 10^{-5}}{d \left(1 + \frac{f^2}{d^2}\right)} + \frac{1.4 \times 10^{-12} p \theta^{1.5}}{1 + 1.9 \times 10^{-5} f^{1.5}} \right],
$$

(30.5)

$d = 5.6 \times 10^{-4}(p + e)\theta^{0.8}$,

$$
F_i(f) = \frac{f}{f_i} \left[\frac{\Delta f_i - \delta_i(f_i - f)}{(f_i - f)^2 + \Delta f_i^2} + \frac{\Delta f_i - \delta_i(f_i + f)}{(f_i + f)^2 + \Delta f_i^2} \right],
$$

(30.6)

$$
\Delta f_i = \sqrt{\left[a_3^i \times 10^{-4}(p \theta^{(0.8 - a_4^i)} + 1.1e\theta) \right]^2 + 2.25 \times 10^{-6}}
$$

(30.7)

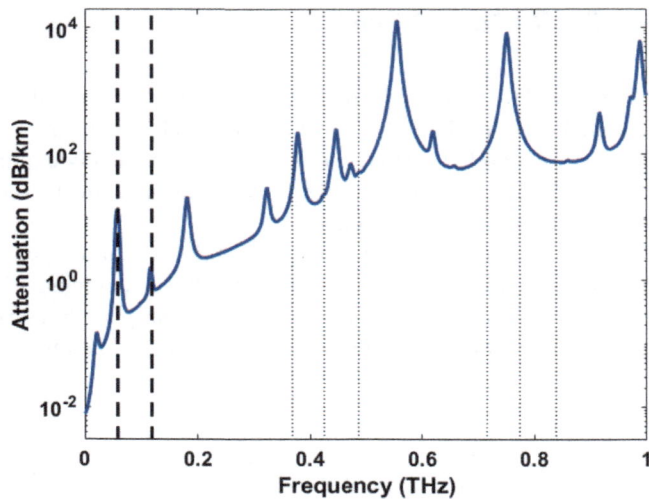

Fig. 30.2 Spectral attenuation calculated with the the empirical model from the ITU-R 676-13 recommendation for a temperature of 15 °C and a humidity of 7.5 g/m³. Two strong lines associated to oxygen are marked by dashed lines, and six more weaker oxygen lines are marked by thinner dotted lines. All other absorption lines are associated to water vapor. (Data taken from [23])

for oxygen

$$\Delta f_i = K + \sqrt{K + \frac{2.1316 \times 10^{-12} f_i^2}{\theta}} \qquad (30.8)$$

for water vapor, with $K = 0.535 b_3^i \times 10^{-4} (p\theta^{b_4^i} + b_5^i e\theta^{b_6^i})$, here $\delta_i = (a_5^i + a_6^i \theta) \times 10^{-4}(p + e)\theta^{0.8}$ for oxygen and $\delta = 0$ for water vapor.

The attenuation spectrum predicted by the ITU model at a temperature of 15 °C and a humidity of 7.5 g/m³ is shown in Fig. 30.2. As mentioned earlier, there is a collection of absorption peaks associated to oxygen, marked with dashed (stronger) and dotted (weaker, not visible in the plot) lines, a monotonously increasing background and a series of other strong peaks related to water vapor. Below, in the results section, present a comparison of the predictions of this model with our measurements.

30.3   Experimental Methods

We will briefly describe the experimental methods that we followed to obtain the temperature- and humidity-dependent properties of air in this section. A much more detailed description of the methods followed, including the construction and characterization of the climate chamber which can be found in our publication [24].

30.3.1   Climate Chamber

A key element of our study mentioned before was the construction of a climate chamber that allowed us to vary the atmospheric conditions[8–10, 16, 25]. This instrument houses a volume of $3\,m \times 0.6\,m \times 0.6\,m$ where the relative humidity (RH) could be controlled over a range roughly between 15% and 95% with an accuracy of $\pm 2\%$; this range corresponds to absolute humidities from $2.7\,g/m^3$ to $36.9\,g/m^3$. The temperature was varied between $5\,°C$ and $45\,°C$ with a precision of $\pm 0.5\,°C$. Probably, one of the unique characteristics of our climate chamber is that it is not hermetically sealed from the external atmosphere; thus, the pressure is always equal to the external atmospheric pressure, which in turn allowed us to carry out our study under atmospheric pressure conditions unlike most similar studies published previously [3, 19–22].

We placed a fiber-coupled photoconductive emitter and detector inside the chamber; the rest of the TDS system remained outside the chamber. In our case, the emitter and detector were gated by an infrared (1550 nm) 80 fs pulsed laser with a 100 MHz repetition rate. A high-density polyethylene lens placed in front of the emitter collimates the THz radiation. The THz signal is split by using a Si wafer at 45 °. The transmitted fraction propagates 6.54 m that we will refer to as the long path, by bouncing off two mirrors before hitting the detector. The reflected fraction of the THz pulse is redirected to the detector by two additional mirrors but propagates only a distance of 0.54 m, which we will refer to as the short path. One of those mirrors is motorized in order to switch the radiation incident on the detector between the two paths. By recording alternately pulses propagating through the long and short paths, we compensated fluctuations of the TDS system.

30.4   Data Processing

30.4.1   THz Time-Domain Data Processing

To calculate the terahertz transmission of the atmosphere, we processed the pulses measured with the TDS in the following manner. The transfer function of the experiment can be defined as $\tilde{H} = \tilde{E}_L / \tilde{E}_S$, where $\tilde{E}_L$ and $\tilde{E}_S$ are the Fourier transformed signals of the pulses after propagation through the long and short paths, respectively. When considering the paths and Fresnel coefficients involved, one can show that

$$\tilde{E}_S = E_0 r_{21} t_{12} t_{21} e^{-2iK\tilde{n}_{Si}d_{Si}} e^{-ik\tilde{n}_{air}d_R}, \tag{30.9}$$

where E_0 is the original pulse from the emitter, r_{21}, t_{12}, and t_{21} are the Fresnel coefficients for the Si-air interfaces at $45°$, and K, k are the wave vectors in silicon and air, respectively. d_{Si} is the thickness of the Si wafer, $\tilde{n}_{Si}$ is the refractive index of Si, and d_R is the short path distance. Likewise, the pulse traveling the long path is given by

$$\tilde{E}_L = E_0 t_{12} t_{21} e^{-iK\tilde{n}_{Si} d_{Si}} e^{-ik\tilde{n}_{air} d_{n+2}},$$ (30.10)

where d_{n+2} is the long path distance. Notice that the transfer function does not depend on E_0 and only depends on the refractive index of air and known physical parameters such as the distances or the refractive index of silicon.

The absorption coefficient of air can be obtained from $|\tilde{H}|$:

$$\alpha(\omega) = \left(\frac{-2}{\Delta d}\right)\left[\ln\left(r_{21}|\tilde{H}|\right) - \frac{1}{2}\alpha_{Si} d_{Si}\right]$$ (30.11)

where α_{Si} is the absorption coefficient of Si. Thus, the attenuation (in dB/m) is given by

$$attenuation_{TDS} = \frac{20\log_{10}(e^{\alpha(\omega)\Delta d/2})}{\Delta d_{TDS}}.$$ (30.12)

30.5   Results and Discussion

30.5.1   Attenuation of THz Radiation

As mentioned earlier, the atmospheric spectrum is composed by a collection of absorption lines; most of these lines are caused by rotational modes of the water molecule. We will center our discussion on two of the absorption lines. Figure 30.3 shows the attenuation for various absolute humidity values at 45 °C. The positions of some of the most prominent water-vapor lines are depicted by gray zones in the plot. We will discuss the attenuation of the lines at 557 GHz and 752 GHz as a function of both the temperature and humidity as an example. The measurements show that the amplitude of the peaks increases linearly with humidity and has hardly any dependence on the temperature. This is consistent with the ITU-R 676-13 model [23].

In Fig. 30.4, we present the attenuation maps for the absorption peaks at 557 GHz (a) and 752 GHz (b), side by side with the ITU model predictions in panels (c) and (d). The lowest (15%) and highest (95%) values of the relative humidity available in our experimental data are indicated by dashed lines. The trend of attenuation with temperature and humidity is similar for both absorption lines. The plots show that the attenuation has only a weak temperature dependence, while it does increase with humidity. The behavior of the experimental measurements and the ITU-R 676-13 model are consistent with each other. This is also true for other absorption lines, in particular some weaker ones such as the 380 GHz and 448 GHz peaks which are frequencies of interest since the absorption can be used to limit the propagation distance of a communication channel serving as a security measure against potential eavesdroppers [16].

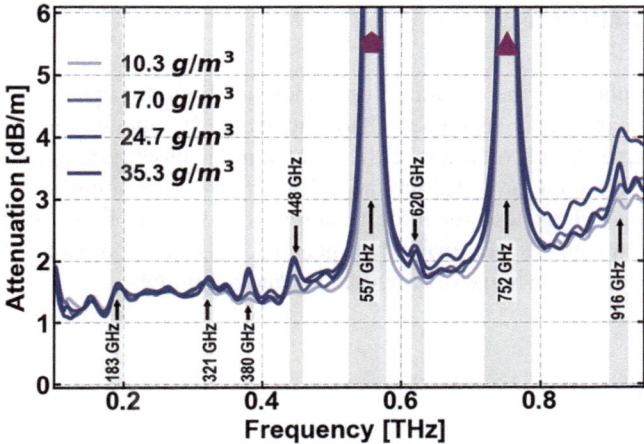

Fig. 30.3 Absorption spectra at 45 ° C corresponding to 10.3 g/m³, 17 g/m³, 24.7 g/m³, 35.3 g/m³ of absolute humidity. Strong resonant water-vapor lines are marked in pale gray. The polygon and triangle symbols mark the analyzed frequencies shown at 557 GHz and 752 GHz. (Reused under the CC-BY license from [24])

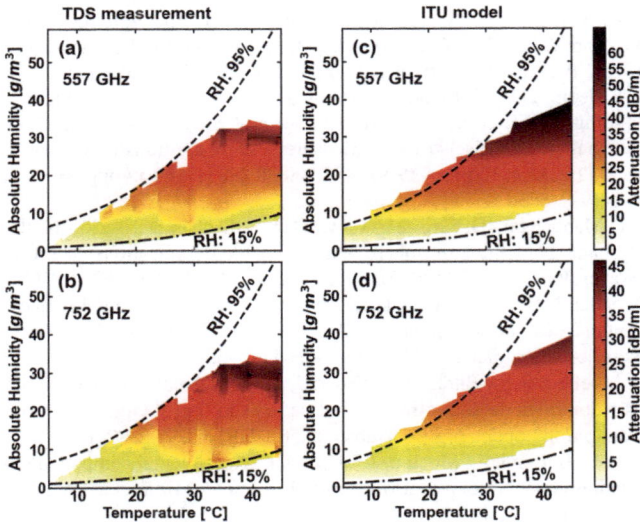

Fig. 30.4 Temperature-humidity attenuation maps for the absorption lines at 557 GHz and 752 GHz as measured with (**a**) and (**b**) THz TDS experiments and (**c**) and (**d**) calculated with the ITU-R 676-13 model. The experimentally obtained attenuation agrees with the theoretical attenuation calculated with the ITU-R 676-13 model. The dashed lines represent the highest (95%) and lowest (15%) measured values of the relative humidity (RH). (Reused under the CC-BY license from [24])

30.6 Conclusions

Since atmosphere is the natural propagation medium for wireless telecommunication channels, it is fundamental to understand its attenuation properties for
the appropriate design of future THz telecommunication systems. We discussed
the empirical model provided in the ITU-R 676-13 recommendation, since it
provides an excellent tool for the estimation of the atmospheric attenuation. We
also presented some of our own experimental results that compare very well with
the model over a wide range of temperatures and humidities.

References

1. Q. Jing, D. Liu, J. Tong, Study on the scattering effect of terahertz waves in near-surface
   atmosphere. IEEE Access **6**, 49007–49018 (2018)
2. X. He, X. Xu, Physics-based prediction of atmospheric transfer characteristics at terahertz
   frequencies. IEEE Trans. Antennas Propagation **67**(4), 2136–2141 (2019)
3. D.M. Slocum, E.J. Slingerland, R.H. Giles, T.M. Goyette, Atmospheric absorption of terahertz
   radiation and water vapor continuum effects. J. Quantit. Spectroscopy and Radiative Transfer
   127, 49–63 (2013). https://www.sciencedirect.com/science/article/pii/S0022407313001702
4. Y. Yang, M. Mandehgar, D.R. Grischkowsky, Broadband THz pulse transmission through the
   atmosphere. IEEE Trans. Terahertz Sci. Technol. **1**(1), 264–273 (2011)
5. E. Moon, T. Jeon, D.R. Grischkowsky, Long-path THz-TDS atmospheric measurements
   between buildings. IEEE Trans. Terahertz Sci. Technol. **5**(5), 742–750 (2015)
6. Y. Yang, A. Shutler, D. Grischkowsky, Measurement of the transmission of the atmosphere
   from 0.2 to 2 THz. Opt. Express **19**(9), 8830–8838 (2011). http://opg.optica.org/oe/abstract.
   cfm?URI=oe-19-9-8830
7. Y. Yang, M. Mandehgar, D. Grischkowsky, THz-TDS characterization of the digital communication channels of the atmosphere and the enabled applications. J. Infrared Millimeter
   Terahertz Waves **36**(2), 97–129 (2015). https://doi.org/10.1007/s10762-014-0099-3
8. K. Su, L. Moeller, R.B. Barat, J.F. Federici, Experimental comparison of terahertz and infrared
   data signal attenuation in dust clouds. J. Opt. Soc. Am. A **29**(11), 2360–2366 (2012). http://
   opg.optica.org/josaa/abstract.cfm?URI=josaa-29-11-2360
9. K. Su, L. Moeller, R.B. Barat, J.F. Federici, Experimental comparison of performance
   degradation from terahertz and infrared wireless links in fog. J. Opt. Soc. Am. A **29**(2), 179–
   184 (2012). http://opg.optica.org/josaa/abstract.cfm?URI=josaa-29-2-179
10. J. Ma, F. Vorrius, L. Lamb, L. Moeller, J.F. Federici, Experimental comparison of terahertz and
    infrared signaling in laboratory-controlled rain. J. Infrared, Millimeter Terahertz Waves **36**(9),
    856–865 (2015). https://doi.org/10.1007/s10762-015-0183-3
11. Y. Amarasinghe, W. Zhang, R. Zhang, D.M. Mittleman, J. Ma, Scattering of terahertz waves by
    snow. J. Infrared Millimeter Terahertz Waves **41**(2), 215–224 (2020). https://doi.org/10.1007/
    s10762-019-00647-4
12. M.H. Rahaman, A. Bandyopadhyay, S. Pal, K.P. Ray, Reviewing the scope of THz communi-
    cation and a technology roadmap for implementation. IETE Tech. Rev. **38**(5), 465–478 (2021).
    https://doi.org/10.1080/02564602.2020.1771221
13. J. Ma, R. Shrestha, L. Moeller, D.M. Mittleman, Invited article: channel performance for indoor
    and outdoor terahertz wireless links. APL Photon. **3**(5), 051601 (2018). https://doi.org/10.
    1063/1.5014037

14. P. Sen, J.V. Siles, N. Thawdar, J.M. Jornet, Multi-kilometre and multi-gigabit-per-second sub-terahertz communications for wireless backhaul applications. Nat. Electron. **6**(2), 164–175 (2023)
15. T. Kürner, T. Kawanishi, Demonstrating 300 GHz wireless backhaul links–the thor approach, in *2022 47th International Conference on Infrared, Millimeter and Terahertz Waves (IRMMW-THz)* (IEEE, 2022), pp. 1–1
16. Z. Fang, H. Guerboukha, R. Shrestha, M. Hornbuckle, Y. Amarasinghe, D.M. Mittleman, Secure communication channels using atmosphere-limited line-of-sight terahertz links. IEEE Trans. Terahertz Sci. Technol. **12**(4), 363–369 (2022)
17. D. Bodet, P. Sen, Z. Hossain, N. Thawdar, J.M. Jornet, Hierarchical bandwidth modulations for ultra-broadband communications in the terahertz band. IEEE Trans. Wirel. Commun. **22**(3), 1931–1947 (2022)
18. G.-R. Kim, T.-I. Jeon, D. Grischkowsky, 910-m propagation of THz ps pulses through the atmosphere. Opt. Express **25**(21), 25422–25434 (2017)
19. D.M. Slocum, R.H. Giles, T.M. Goyette, High-resolution water vapor spectrum and line shape analysis in the terahertz region. J. Quantit. Spectroscopy Radiative Transfer **159**, 69–79 (2015)
20. V.B. Podobedov, D.F. Plusquellic, K.E. Siegrist, G.T. Fraser, Q. Ma, R.H. Tipping, New measurements of the water vapor continuum in the region from 0.3 to 2.7 THz. J. Quantit. Spectroscopy Radiative Transfer **109**(3), 458–467 (2008). https://www.sciencedirect.com/science/article/pii/S0022407307001914
21. V. Podobedov, D.F. Plusquellic, G. Fraser, Investigation of the water-vapor continuum in the THz region using a multipass cell. J. Quantit. Spectroscopy Radiative Transfer **91**(3), 287–295 (2005)
22. M. Koshelev, E. Serov, V. Parshin, M.Y. Tretyakov, Millimeter wave continuum absorption in moist nitrogen at temperatures 261–328 K. J. Quantitative Spectroscopy Radiative Transfer **112**(17), 2704–2712 (2011)
23. S. Series, International Telecommunications Union, *Recommendation ITU-R P.676–13, Attenuation by Atmospheric Gases and Related Effects* (International Telecommunications Union, 2022)
24. F. Taleb, M. Alfaro-Gomez, M.D. Al-Dabbagh, J. Ornik, J. Viana, A. Jäckel, C. Mach, J. Helminiak, T. Kleine-Ostman, T. Kürner et al., Propagation of thz radiation in air over a broad range of atmospheric temperature and humidity conditions. Sci. Rep. **13**(1), 20782 (2023)
25. J. Ma, L. Moeller, J.F. Federici, Experimental comparison of terahertz and infrared signaling in controlled atmospheric turbulence. J. Infrared Millimeter Terahertz Waves **36**, 130–143 (2015)

Chapter 31
Building Materials

Enrique Castro-Camus, Fatima Taleb, and Martin Koch

Abstract The market of wireless communications is driving the technology development toward higher bandwidths and carrier frequencies. Terahertz (THz) frequencies not only open new unused bands for this purpose but also allow higher bandwidths. Yet, atmospheric attenuation above 100 GHz limit communication systems to short distances, and in many cases, non-line-of-sight links, may need to rely on specular or diffuse reflections off-walls, ceilings, or floors around us. We investigated the dielectric properties and scattering behavior of a representative collection of 50 building materials used for construction. The transmission, reflection, and scattering properties of the samples were measured using THz time-domain spectroscopy and their scattering was modelled based on the Fresnel–Rayleigh equations and the Kirchhoff theory. We observed that depending on several characteristics of the materials, the behavior could match the models very well, except for certain characteristics, such as bulk heterogeneity of the material.

31.1 Introduction

As widely discussed throughout this book, the requirement for fast telecommunications systems is the driver for the increase of carrier frequencies in order to be able to allocate more channels, each with broader bandwidth pushing telecommunications to the terahertz band [1–3]. With this increase of frequency, the mm-wave and terahertz bands of the spectrum are expected to be populated with telecommunications signals in the coming years.

E. Castro-Camus (✉)
Philipps-Universität Marburg, Physik (Fb13), Marburg, Germany

Centro de Investigaciones en Optica, Lomas del Campestre, Mexico
e-mail: enrique@cio.mx

F. Taleb · M. Koch
Philipps-Universität Marburg, Physik (Fb13), Marburg, Germany
e-mail: fatima.taleb@physik.uni-marburg.de; martin.koch@physik.uni-marburg.de

© The Author(s) 2026
T. Kürner et al. (eds.), *Metrology for THz Communications*, Springer Series
in Optical Sciences 256, https://doi.org/10.1007/978-3-032-01986-8_31

The use of terahertz waves comes with a unique set challenges that were not an issue in traditional telecommunications at megahertz or gigahertz frequencies [4–7]. One of those challenges is that with shorter wavelengths the dielectric function of many materials is unknown. Their internal structure or superficial roughness in the micron to millimeter regions can no longer be considered smooth or as an effective medium, changing the way in which we have to consider the transmission, reflection, and scattering of walls, floors, ceilings, and even furniture around us [8]. This is an issue that started being discussed in 2005 with the first measurements of such properties [9–12]. This was used in subsequent years for the design of THz communication systems [13–21]. However, it was not, until very recently, that we presented a more systematic study of building materials at terahertz frequencies [22].

In this chapter, some measurements that we performed recently on a relatively large collection of building materials will be discussed. The entire catalog of materials with plots of their dielectric properties can be found in [23]. Furthermore, the frequency-dependent complex refractive indices as well as the angle and frequency-dependent scattering measurements and modelling are available for download [24]. The measurements were performed by terahertz time-domain spectroscopy (THz-TDS), which is described briefly below but in more detail in Chap. 7. The model we adopted for the measurements is a combination of Rayleigh and Kirchhoff scattering.

31.2   Materials and Classification

After studying a collection of 50 building materials, it was found that many with similar functions, such as various types of wood or different kinds of filling foams, present similar dielectric and scattering properties. In Table 31.1, we group materials by their function; in Sect. 31.5, we present the dielectric properties and scattering behavior of one representative material of each group.

31.3   Measurements

31.3.1   Surface Characterization

In order to properly understand the scattering properties of the materials, it is fundamental to measure their roughness profile. A possibility to do this is the use photogrammetry, which is a technique used to reconstruct an object three-dimensionally by combining various photographic images. In our case, each sample was placed on a rotational stage, and fifty photographs were acquired using a digital camera. The 3D reconstruction was performed using 3DF Zephyr software.

Table 31.1 List of materials and their roughness parameters

Type	Name	L (mm)	σ (mm)
Structural	Fire clay brick	2.35	0.024
Structural	Flexfuge cement	2.49	0.029
Structural	Universal cement	1.5	0.024
Structural	Beton concrete	1.99	0.075
Structural	Lime with marble grains	1.05	0.137
Structural	Gypsum with top layer of carton	0.64	0.009
Structural	Gypsum with bottom layer of carton	0.91	0.073
Structural	Gypsum mixed with carton	1	0.025
Structural	Gypsum mixed with stones	1.57	0.017
Structural	Gypsum plaster	0.809	0.128
Structural	Gypsum board	1.01	0.183
Wood	Spruce Fir smooth	1.2	0.014
Wood	Spruce Fir rough	2.25	0.02
Wood	Douglas Fir	1.5	0.019
Wood	Compressed wood	1.78	0.014
Wood	Compressed, multilayered wood	1.22	0.02
Wood	Compressed wood with magnesite	1.78	0.028
Wood	High-density fiberboard	1.016	0.013
Wood	High-density fiberboard with polystyrol	3.03	0.012
Wood	Magnesite-bonded wood	1.335	0.26
Insulation	Bitumen balin membrane	2.45	0.087
Insulation	Acoustic panel	1.175	0.031
Insulation	Mineral insulation board	2.513	0.152
Ceramic and stone	Gray stone tile with rough surface	1.95	0.097
Ceramic and stone	White porcelain tile with smooth surface	1.03	0.006
Ceramic and stone	Brown ceramic tile with rough surface	1.125	0.056
Ceramic and stone	Beige ceramic tile with smooth surface	0.754	0.008
Ceramic and stone	Gray ceramic tile with rough surface	0.5	0.023
Ceramic and stone	Green stone tile with rough surface	1.82	0.087
Ceramic and stone	Marble Murciano	1.34	0.024
Ceramic and stone	Marble graphite	0.48	0.014
Ceramic and stone	Marble mosaic	1.49	0.022
Plastic	Polyurethane coating	0.614	0.062
Plastic	Plexiglass	1	0.005
Glass	White glass	1.5	0.005
Glass	Black glass	1.5	0.005
Caulking compounds	Kitchen silicone	1.97	0.0141
Caulking compounds	Sanitary silicone	2.53	0.034
Caulking compounds	Installation glue	1.42	0.0146
Caulking compounds	Finishing filler	1.97	0.0151
Caulking compounds	Indoor filler	3.52	0.023

(continued)

Table 31.1 (continued)

Type	Name	L (mm)	σ (mm)
Caulking compounds	Acrylic	0.98	0.028
Caulking compounds	Wood putty	2.31	0.013
Caulking compounds	Plaster	4.17	0.022
Foam	Spray foam	0.5	0.1
Foam	Laminate pad	1.57	0.086
Foam	Polystyrene	0.15	0.053
Fabric	Carpet	1.82	0.533
Paint	Colored lacquer	1.51	0.098
Paint	Wall paint	2.76	0.055

The roughness parameters σ, which is the height standard deviation, and L, the correlation length, were obtained processing the 3D models with the open-source program CloudCompare.

31.3.2 Optical Properties

The optical properties of all materials were measured with a fiber-coupled THz-TDS system in transmission configuration. Reference waveforms were recorded in the absence of a sample, and subsequently, samples were placed in the THz beam path. The refractive index was determined as described in [25] by using equations

$$n(\omega) = 1 - \frac{c}{\omega l}(\phi_{\text{sample}} - \phi_{\text{reference}}) \tag{31.1}$$

and

$$\kappa(\omega) = -\frac{c}{\omega l}\ln\left(\frac{(n(\omega) + 1)^2}{4n(\omega)}\frac{|E_{\text{sample}}|}{|E_{\text{reference}}|}\right), \tag{31.2}$$

where ϕ_i and $|E_i|$ are the phase and amplitude of the Fourier transforms for $i =$ sample/reference, c is the speed of light, ω the angular frequency, and l the thickness of the sample.

31.3.3 Reflection Characterization

The time-domain spectrometer was reconfigured, and the emitter and detector were placed on the arms of a goniometer system. With it the THz signal scattered from the sample was recorded as a function of the angle between $\theta_r = 0°$ and $\theta_r = 90°$. The

THz beam was incident at $\theta_i = 30°$. Reference waveforms were recorded replacing the sample by a metal plate. The illuminated area was approximately $4\,\text{cm}^2$.

31.4 Modelling

The scattering patterns of indoor building materials strongly depend on their surface roughness. While smooth, flat, and optically thick materials show a dominantly specular reflection, rough surfaces with a more complex geometry will present a combination of specular and diffuse reflection components. The Rayleigh theory of scattering allows us to analytically calculate the losses, with respect to a perfect specular reflector. It is appropriate for surfaces with mild roughness, typically small compared with the wavelength. The Kirchhoff theory of scattering provides an analytical expression too, to calculate the angular dependent scattering in the case of more rough surfaces.

Since our time-domain spectroscopy experiments span across a very wide range of wavelengths, we decided to generate a combined model that has higher weight from the Rayleigh theory for long wavelengths and a higher weight of the Kirchhoff theory for short wavelengths. The two contributions are weighted by an error function centered at λ_0, which is a wavelength that separates what we can consider a smooth and a rough surface calculated from the Fraunhofer criterion [26]:

$$\sigma = \frac{\lambda_0}{32\cos\theta_i},\tag{31.3}$$

where θ_i is the incident angle with respect to the normal of the surface and σ is the standard deviation of the surface height distribution. In Fig. 31.1, we illustrate how a surface can have a roughness that is large or small with respect to the wavelength; thus, the scattering behavior is expected to have a strong wavelength dependence.

31.4.1 Rayleigh-Fresnel Equations

The reflectivity for a specular reflection is calculated using the Fresnel reflection coefficients. In the case of slightly rough surfaces, the specular reflection is smaller owing to small amounts of diffuse scattering. Hence, in order to account for this loss, the reflectivity in the specular direction for smooth and moderately rough ($\sigma < \lambda_0/32\cos\theta_i$) surfaces, the Fresnel reflection coefficients are multiplied by the Rayleigh roughness factor [10]:

$$\rho = e^{-g/2}\tag{31.4}$$

Fig. 31.1 The figure illustrates how a surface can be rough for a short wavelength (red) and also be smooth for a long wavelength (green) and thus the need to have a combined Rayleigh and Kirchhoff formalism to be able to model a broad spectrum such as the one recorded by time-domain spectroscopy

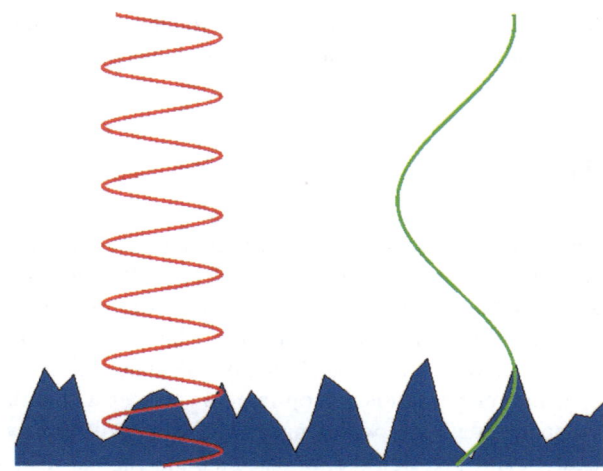

where

$$g = \left(\frac{4\pi \sigma \cos\theta_i}{\lambda_i} \right)^2, \tag{31.5}$$

with θ_i being the angle of incidence, σ the standard deviation of the surface height, and λ_i the free space wavelength of the incident radiation. The modified reflection coefficient in that case is

$$r'_{TM} = \rho \cdot r_{TM}, \tag{31.6}$$

where r_{TM} is the Fresnel reflection coefficient for TM polarized waves.

It is worth noticing that this approximation does not provide an angular dependent function that describes how the radiation lost in the specular direction is redirected in other directions. Yet, if the wavelength is long in terms of the Fraunhofer criterion (Eq. 31.3), the loss would be $\leq 1.5\%$ at an incidence angle of $30°$. Therefore, to first approximation, the only measurable effect is the loss in the specular reflection, and we have thus assumed that the non-specular components are negligible in the Fraunhofer regime.

31.4.1.1 Kirchhoff Theory

The diffuse reflections for rougher surfaces can be described by an extended model of the Kirchhoff theory [27], previously proposed to investigate the scattering properties of typical indoor building materials [11], which states that a wave incident on an optically thick, rough surface at an angle θ_i is scattered at an angle θ_r with respect to the normal and θ_3 away from the incidence plane following

$$\langle \rho\rho^* \rangle = \langle rr^* \rangle \cdot \langle \rho\rho^* \rangle_\infty , \tag{31.7}$$

where $r = (\cos\theta_i - n)/(\cos\theta_i + n)$ is the Fresnel reflection coefficient of the material and $\langle \rho\rho* \rangle_\infty$ is the mean scattering coefficient of a perfectly reflecting material with the same roughness

$$\langle \rho\rho^* \rangle_\infty = e^{-g} \cdot \left(\rho_0^2 + \frac{\pi L^2 F^2}{A} \sum_{m=1}^{\infty} \frac{g^m}{m!m} e^{-(v_x^2 + v_y^2)L^2/4m} \right) \tag{31.8}$$

where

$$\rho_0 = \text{sinc}(v_x l_x) \cdot \text{sinc}(v_y l_y), \tag{31.9}$$

$$v_x = k \cdot (\sin\theta_i - \sin\theta_r \cos(\theta_3)), \tag{31.10}$$

$$v_y = k \cdot (-\sin\theta_r \sin(\theta_3)), \tag{31.11}$$

$$g = k^2 \sigma^2 (\cos\theta_i + \cos\theta_r)^2, \tag{31.12}$$

and

$$F(\theta_i, \theta_r) = \frac{1 + \cos\theta_i \cos\theta_r - \sin\theta_i \sin\theta_r \cos\theta_3}{\cos\theta_i (\cos\theta_i + \cos\theta_r)}. \tag{31.13}$$

In this case, $A = l_x \cdot l_y$ is the illuminated area, k is the free space wave number, and g is given by Eq. 31.12, which measures the relative surface roughness at a given wavelength.

When we combine the two scattering regimes, the scattering coefficient is thus given by

$$s(\lambda) = \text{erf}\left(\frac{\lambda - \lambda_0}{\sigma_{\text{err}}} \right) \cdot \langle \rho\rho^* \rangle + \left[1 - \text{erf}\left(\frac{\lambda - \lambda_0}{\sigma_{\text{err}}} \right) \right] \cdot r'_{\text{TM}} r'^*_{\text{TM}}, \tag{31.14}$$

with λ_0 given by the Fraunhofer criterion in Eq. 31.3, and a standard deviation $\sigma_{\text{err}} = \lambda_0/4$.

31.5 Results

As mentioned earlier, we measured a broad collection of 50 building materials. Presenting the measurements and describing the results for all of them in this chapter would be impractical, and therefore, we will only discuss the measurement and modelling results of some exemplary materials of each type. Yet, the entire collection of plots is available in our publication [22], and the data is freely available for download at [24].

31.5.1 Structural Materials

Many structural materials such as bricks and concrete tend to have relatively smooth surfaces, on the scale of the THz wavelengths, and also a relatively homogeneous internal structure; therefore, it is to be expected to see a strong specular reflection for these kinds of materials, with a relatively small diffuse component. The magnitude of the reflection thus will be very close to the Fresnel reflection which is only dependent on the value of the refractive index. In Fig. 31.2a, the real and imaginary parts of the refractive index of fire clay brick is presented, showing reasonably low losses and almost no dispersion. Figure 31.2b shows the angular dependence of the reflection at 100 GHz. As seen in the plot, the reflection is almost purely specular, with negligible diffuse components. The dashed line on the plot represents the result from the model presented in the previous section fitting the experimental measurement well. Finally, Fig. 31.2c shows the full angular and spectral dependence of the reflection, where similar behavior is seen for all frequencies. This behavior is shared by the majority of the structural materials that we studied, with the amplitude of the specular reflection being larger or smaller depending on the refractive index of each material. It is worth mentioning that gypsum samples tend to have lower reflectivity and a slightly larger scattering, but their behavior is still dominated by the specular reflection.

Fig. 31.2 The figure presents the data for the measurements performed on fire clay brick. (**a**) Real (blue) and imaginary part (red) of the refractive index as measured in transmission. (**b**) Measured (continuous) and modelled (dashed) angular dependence of the scattering at 100 GHz for a beam incident at 30°. (**c**) Full frequency and angle dependent scattering also for a beam incident at 30°

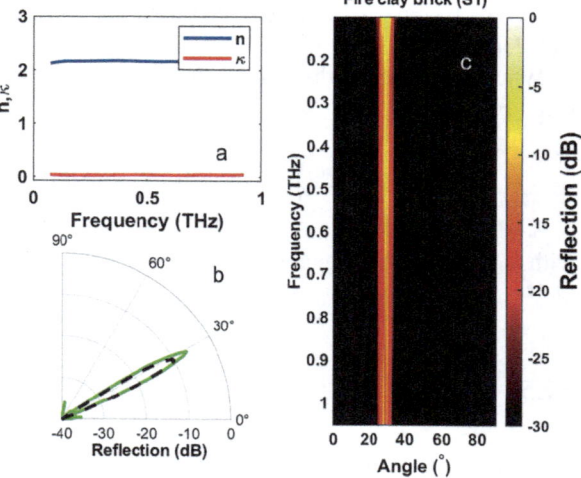

31.5.2 Wood

Wood and wood-based materials are rather important for construction. The densities of natural woods and therefore their refractive index is quite variable but are typically below 1.6 with moderate attenuation. It is worth mentioning that wood shows a bit of birefringence [28], which we will not discuss further here and will also neglect for our calculations. In addition to natural woods, there are a plethora of compressed wood-based materials, which typically show higher refractive indices than natural wood. These materials combine many different types of binders that go from organic glues to inorganic compounds like magnesite and are sometimes rather compact but sometimes with high roughness and complex internal structures. Compressed materials or nicely cut and sanded natural wood tends to have a smooth surface in the scale of the THz wavelengths and therefore shows a predominantly specular behavior. In Fig. 31.3a, the real and imaginary parts of the refractive index of high-density fiberboard with polystyrol are plotted, showing very low losses and almost no dispersion with a refractive index of about 1.5. Figure 31.3b shows the angular dependence of the reflection at 100 GHz; it is possible to see that the reflection is almost purely specular, with negligible diffuse components. The dashed line on the plot represents the result from the model presented in the previous section. The model fits the experimental measurement well. Finally, Fig. 31.3c shows the full angular and spectral dependence of the reflection, where almost perfect specular behavior is seen for all frequencies. This behavior is shared by

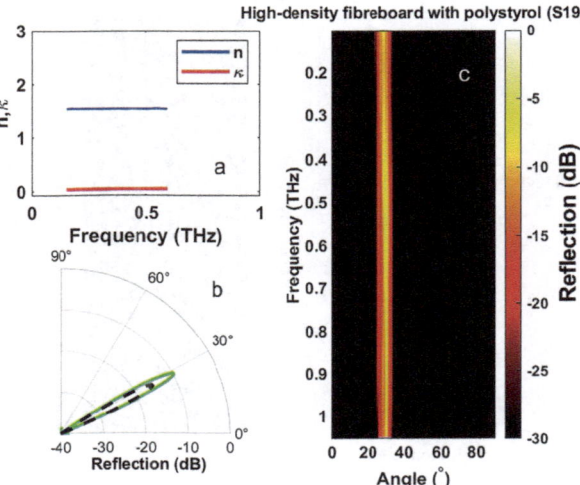

Fig. 31.3 The figure presents the data for the measurements performed on high-density fiberboard with polystyrol. (**a**) Real (blue) and imaginary part (red) of the refractive index as measured in transmission. (**b**) Measured (continuous) and modelled (dashed) angular dependence of the scattering at 100 GHz for a beam incident at 30°. (**c**) Full frequency and angle dependent scattering also for a beam incident at 30°

the majority of the smooth wood-based materials measured, with the amplitude of the specular reflection being larger or smaller depending on the refractive index of each material. It is worth mentioning that materials such as the magnesite-bonded wood sample that we studied have rather large losses and poor reflectivity, especially at higher frequencies, and show more scattering; this is consistent with its comparatively larger roughness.

31.5.3 Insulation Materials

We studied three insulation materials, and in this case, we found a much more noticeable difference in their behavior. While bitumen balin membrane has a refractive index of about 1.57, the refractive index of the acoustic panel is closer to 1.14, and the mineral insulation board is below 1.1, giving them completely different reflection properties. In Fig. 31.4a, the real and imaginary parts of the refractive index of bitumen balin membrane are presented, showing very low losses and almost no dispersion with a refractive index close to 1.57. Figure 31.4b shows the angular dependence of the reflection at 100 GHz. Here, the reflection has a much richer structure than in the previous examples. The scattering shows a wider angular dispersion, the dashed line on the plot represents the result from the model presented in the previous section, and while the fit is not perfect, it does show a broader angular

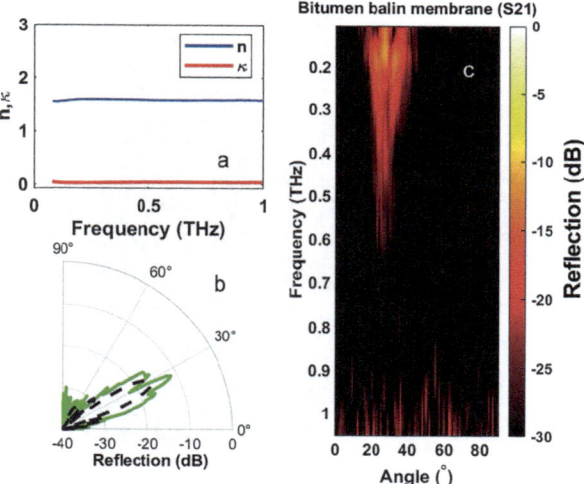

Fig. 31.4 The figure presents the data for the measurements performed on bitumen balin membrane. (**a**) Real (blue) and imaginary part (red) of the refractive index as measured in transmission. (**b**) Measured (continuous) and modelled (dashed) angular dependence of the scattering at 100 GHz for a beam incident at 30°. (**c**) Full frequency and angle dependent scattering also for a beam incident at 30°

dependence than the specular reflection showing lobes that are measurable up to about 10 degrees off the specular direction. Finally, Fig. 31.4c shows the full angular and spectral dependence of the reflection, where a clear angular broadening and a strong spectral dependence is seen.

31.5.4   Ceramics and Stones

We studied nine ceramic and stone materials. We found that this family of materials have refractive indices considerably higher than the ones seen in the previous types, ranging between 1.9 and 3. In Fig. 31.5a, we see the real and imaginary parts of the refractive index of gray stone tile with rough surface which is about 2.84. Figure 31.5b shows the angular dependence of the reflection at 100 GHz. The reflection has a series of lobes that match the results of the model qualitatively well, which is represented by the dashed line. As in previous cases, the model and the measurements match reasonably well when considering the roughness and the refractive index of each of the materials. In addition, Fig. 31.5c shows the full angular and spectral dependence of the reflection, where a rich angular and spectral dependence can be seen.

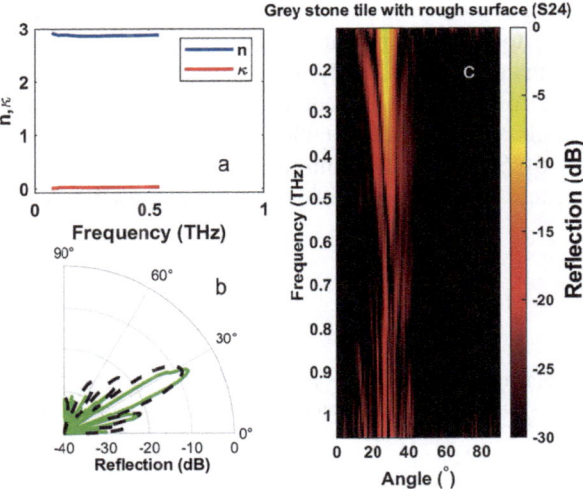

Fig. 31.5 The figure presents the data for the measurements performed on gray stone tile with rough surface. (**a**) Real (blue) and imaginary part (red) of the refractive index as measured in transmission. (**b**) Measured (continuous) and modelled (dashed) angular dependence of the scattering at 100 GHz for a beam incident at 30°. (**c**) Full frequency and angle dependent scattering also for a beam incident at 30°

31.5.5 Other Materials

There are several other materials that we studied. For instance, fabrics and foams tend to have a refractive index very close to 1 and therefore hardly interact with terahertz radiation, so there is almost no scattering or reflection, and their transmission is almost 100%. Furthermore, glasses and plastics usually have surfaces that are smooth even at optical wavelengths, which means that they produce hardly any scattering and show dominantly specular reflections. Paints and other coatings have refractive indices that are from moderate (1.5) to relatively high (~2), but the scattering behavior is strongly dependent on the substrate on which they are applied, since they adopt the roughness of the substrate. Likewise, caulking compounds have a large variation in their refractive indices, and the surfaces they form are strongly dependent on the specific use and also the application procedure.

31.6 Summary

Building materials are all around us; their interaction with terahertz waves is significantly different than with microwaves, currently used for telecommunications, since factors like roughness and scattering are not relevant at long wavelengths. Having a good understanding of the transmission, reflection, and scattering properties of these materials is going to be fundamental for the design of terahertz communication systems for two main reasons. Firstly, the inherent need for directed beams that come with higher frequencies will require beam steering [29–34] and the use of reflections off-walls, ceilings, floors, tables, etc. when direct line-of-sight links are blocked [8, 35, 36]. Secondly, the communication through doors, walls, windows, and so on will require the estimation of losses during these interactions.

References

1. A. Hirata, M. Harada, T. Nagatsuma, 120-GHz wireless link using photonic techniques for generation, modulation, and emission of millimeter-wave signals. J. Lightwave Technol. **21**(10), 2145–2153 (2003)
2. H.-J. Song, K. Ajito, Y. Muramoto, A. Wakatsuki, T. Nagatsuma, N. Kukutsu, 24 gbit/s data transmission in 300 GHz band for future terahertz communications. Electron. Lett. **48**(15), 953–954 (2012)
3. A. Oshiro, N. Nishigami, T. Yamamoto, Y. Nishida, J. Webber, M. Fujita, T. Nagatsuma, Pam4 48-gbit/s wireless communication using a resonant tunneling diode in the 300-GHz band. IEICE Electron. Express **19**(2), 20210494–20210494 (2022)
4. R. Piesiewicz, T. Kleine-Ostmann, N. Krumbholz, D. Mittleman, M. Koch, J. Schoebel, T. Kurner, Short-range ultra-broadband terahertz communications: concepts and perspectives. IEEE Antennas Propagat. Mag. **49**(6), 24–39 (2007)
5. J. Ma, R. Shrestha, L. Moeller, D.M. Mittleman, Invited article: Channel performance for indoor and outdoor terahertz wireless links. APL Photon. **3**(5), 051601 (2018)

6. D. Turchinovich, A. Kammoun, P. Knobloch, T. Dobbertin, M. Koch, Flexible all-plastic mirrors for the THz range. Appl. Phys. A **74**(2), 291–293 (2002)
7. M. Koch, Terahertz communications: a 2020 vision, in *Terahertz Frequency Detection and Identification of Materials and Objects* (Springer, Berlin, 2007), pp. 325–338
8. R. Piesiewicz, M. Jacob, M. Koch, J. Schoebel, T. Kürner, Performance analysis of future multigigabit wireless communication systems at THz frequencies with highly directive antennas in realistic indoor environments. IEEE J. Sel. Top. Quantum Electron. **14**(2), 421–430 (2008)
9. R. Piesiewicz, T. Kleine-Ostmann, N. Krumbholz, D. Mittleman, M. Koch, T. Kürner, Terahertz characterisation of building materials. Electron. Lett. **41**(18), 1002–1004 (2005)
10. R. Piesiewicz, C. Jansen, D. Mittleman, T. Kleine-Ostmann, M. Koch, T. Kurner, Scattering analysis for the modeling of THz communication systems. IEEE Trans. Antennas Propagat. **55**(11), 3002–3009 (2007)
11. C. Jansen, S. Priebe, C. Moller, M. Jacob, H. Dierke, M. Koch, T. Kurner, Diffuse scattering from rough surfaces in THz communication channels. IEEE Trans. Terahertz Sci. Technol. **1**(2), 462–472 (2011)
12. M. Urahashi, A. Hirata, Complex permittivity evaluation of building materials at 200–500 GHz using THz-TDS," in *2020 International Symposium on Antennas and Propagation (ISAP)* (IEEE, 2021), pp. 539–540
13. J. Ma, R. Shrestha, W. Zhang, L. Moeller, D.M. Mittleman, Terahertz wireless links using diffuse scattering from rough surfaces. IEEE Trans. Terahertz Sci. Technol. **9**(5), 463–470 (2019)
14. H.-J. Song, T. Nagatsuma, Present and future of terahertz communications. IEEE Trans. Terahertz Sci. Technol. **1**(1), 256–263 (2011)
15. M. Jacob, S. Priebe, R. Dickhoff, T. Kleine-Ostmann, T. Schrader, T. Kurner, Diffraction in mm and sub-mm wave indoor propagation channels. IEEE Trans. Microwave Theory Tech. **60**(3), 833–844 (2012)
16. T. Schneider, A. Wiatrek, S. Preußler, M. Grigat, R.-P. Braun, Link budget analysis for terahertz fixed wireless links. IEEE Trans. Terahertz Sci. Technol. **2**(2), 250–256 (2012)
17. C. Han, A.O. Bicen, I.F. Akyildiz, Multi-ray channel modeling and wideband characterization for wireless communications in the terahertz band. IEEE Trans. Wirel. Commun. **14**(5), 2402–2412 (2014)
18. Z. Xu, X. Dong, J. Bornemann, Design of a reconfigurable MIMO system for THz communications based on graphene antennas. IEEE Trans. Terahertz Sci. Technol. **4**(5), 609–617 (2014)
19. K. Guan, G. Li, T. Kürner, A. F. Molisch, B. Peng, R. He, B. Hui, J. Kim, Z. Zhong, On millimeter wave and THz mobile radio channel for smart rail mobility. IEEE Trans. Veh. Technol. **66**(7), 5658–5674 (2016)
20. F. Sheikh, Q.H. Abbasi, T. Kaiser, On channels with composite rough surfaces at terahertz frequencies, in *2019 13th European Conference on Antennas and Propagation (EuCAP)* (IEEE, 2019), pp. 1–5
21. M. Wang, Y. Wang, W. Li, J. Ding, C. Bian, X. Wang, C. Wang, C. Li, Z. Zhong, J. Yu, Reflection characteristics measurements of indoor wireless link in D-band. Sensors **22**(18), 6908 (2022)
22. F. Taleb, G.G. Hernandez-Cardoso, E. Castro-Camus, M. Koch, Transmission, reflection, and scattering characterization of building materials for indoor thz communications. IEEE Trans. Terahertz Sci. Technol. **13**(5), 421–430 (2023)
23. Supplementary material to: Transmission, reflection, and scattering characterization of building materials for indoor thz communications. https://doi.org/10.1109/TTHZ.2023.3281773/mm1
24. Data from: Transmission, reflection, and scattering characterization of building materials for indoor thz communications. https://doi.org/10.6084/m9.figshare.21539109.v3
25. W. Withayachumnankul, M. Naftaly, Fundamentals of measurement in terahertz time-domain spectroscopy. J. Infrared Millimeter Terahertz Waves **35**, 610–637 (2014)

26. D. Didascalou, M. Dottling, N. Geng, W. Wiesbeck, An approach to include stochastic rough surface scattering into deterministic ray-optical wave propagation modeling. IEEE Trans. Antennas Propagat. **51**(7), 1508–1515 (2003)
27. P. Beckmann, A. Spizzichino, *The Scattering of Electromagnetic Waves From Rough Surfaces* (Norwood, 1987)
28. M. Reid, R. Fedosejevs, Terahertz birefringence and attenuation properties of wood and paper. Appl. Opt. **45**(12), 2766–2772 (2006)
29. K.-I. Maki, C. Otani, Terahertz beam steering and frequency tuning by using the spatial dispersion of ultrafast laser pulses. Opt. Express **16**(14), 10158–10169 (2008)
30. B. Scherger, M. Reuter, M. Scheller, K. Altmann, N. Vieweg, R. Dabrowski, J.A. Deibel, M. Koch, Discrete terahertz beam steering with an electrically controlled liquid crystal device. J. Infrared Millimeter Terahertz Waves **33**(11), 1117–1122 (2012)
31. Y. Monnai, K. Altmann, C. Jansen, H. Hillmer, M. Koch, H. Shinoda, Terahertz beam steering and variable focusing using programmable diffraction gratings. Opt. Express **21**(2), 2347–2354 (2013)
32. X. Liu, L. Samfaß, K. Kolpatzeck, L. Häring, J.C. Balzer, M. Hoffmann, A. Czylwik, Terahertz beam steering concept based on a mems-reconfigurable reflection grating. Sensors **20**(10), 2874 (2020)
33. J. Wu, Z. Shen, S. Ge, B. Chen, Z. Shen, T. Wang, C. Zhang, W. Hu, K. Fan, W. Padilla et al., Liquid crystal programmable metasurface for terahertz beam steering. Appl. Phys. Lett. **116**(13), 131104 (2020)
34. J.M. Seifert, G.G. Hernandez-Cardoso, M. Koch, E. Castro-Camus, Terahertz beam steering using active diffraction grating fabricated by 3D printing. Opt. Express **28**(15), 21737–21744 (2020)
35. S. Priebe, T. Kurner, Stochastic modeling of THz indoor radio channels. IEEE Trans. Wirel. Commun. **12**(9), 4445–4455 (2013)
36. T. Kürner, D. Mittleman, T. Nagatsuma, *THz Communications: Paving the Way Towards Wireless Tbps* (Springer, Berlin, 2022)

Chapter 32
Interaction of Objects with THz Beams

Enrique Castro-Camus, Felix Gorka, Martin Koch, and Daniel M. Mittleman

Abstract One of the important challenges facing telecommunications at terahertz frequencies is the requirement of highly directional beams to establish the link between two devices. Given that the wavelength in this region ranges from a few millimeters to a few hundreds of microns, it is clear that the geometrical features of many objects are of a size comparable to the wavelength. As a result, many commonly encountered scattering situations cannot be described using either the geometrical limit (where objects are much larger than the wavelength) or the Rayleigh limit (where objects are much smaller than the wavelength). In this chapter, we present measurements and calculations to consider this behavior and discuss the implications for wireless links between two devices.

32.1 Introduction

With the exception of microwave links between two immobile locations, the great majority of current consumer telecommunication devices such as cellular or WiFi networks operate in the spectral range from several hundreds of megahertz to a few gigahertz, which corresponds to wavelengths in the tens of centimeters. Most of the antennas for this regime are either quasi-omnidirectional or have specific emission or detection patterns optimized to cover certain areas [1]. In the cases when

---

E. Castro-Camus (✉)
Philipps-Universität Marburg, Physik (Fb13), Marburg, Germany

Centro de Investigaciones en Optica, Lomas del Campestre, Mexico
e-mail: enrique@cio.mx

F. Gorka · M. Koch
Philipps-Universität Marburg, Physik (Fb13), Marburg, Germany
e-mail: felix.gorka@physik.uni-marburg.de; martin.koch@physik.uni-marburg.de

D. M. Mittleman
School of Engineering, Brown University, Providence, RI, USA
e-mail: daniel_mittleman@brown.edu

© The Author(s) 2026
T. Kürner et al. (eds.), *Metrology for THz Communications*, Springer Series
in Optical Sciences 256, https://doi.org/10.1007/978-3-032-01986-8_32

directional antennas are used, the presence of objects or persons in the line-of-sight between the emitter and detector is generally not considered problematic, because although they can produce scattering losses, full shadowing of the signal is typically not an issue owing to both multipath scattering (which is common at these lower frequencies) and to the fact that such objects do not cast sharp shadows, since the ratio between the size of the scattering features and the wavelength is small. This is clearly illustrated in Fig. 32.1, which contains four panels showing real calculations of the shadow produced at a screen 1 m away from a human hand for frequencies between 10 GHz and 1 THz when illuminated by a plane homogeneous wave.

For systems operating at higher frequency, in the millimeter-wave or terahertz regimes, the situation is quite different. As has been widely discussed [2, 3], such systems will require highly directional beams generated by high-gain antennas, in order to overcome the high free-space path loss associated with even relatively short-range links. Such beams will of course require active steering capabilities, in order to maintain a reliable connection between transmitter and receiver [4]. In this sense, these electromagnetic signals can be treated, at least in some aspects, as quasi-optical beams [5], a term which dates back at least to the emergence of the first high-power coherent millimeter-wave sources in the early 1960s.

In our context, the ideas of quasi-optics are applicable because, although objects that appear in a realistic and dynamic telecommunication scenario such as a book, a ball, or a human subject remain large compared to the wavelength, however, these all contain features that are comparable to the wavelength in size. As a result,

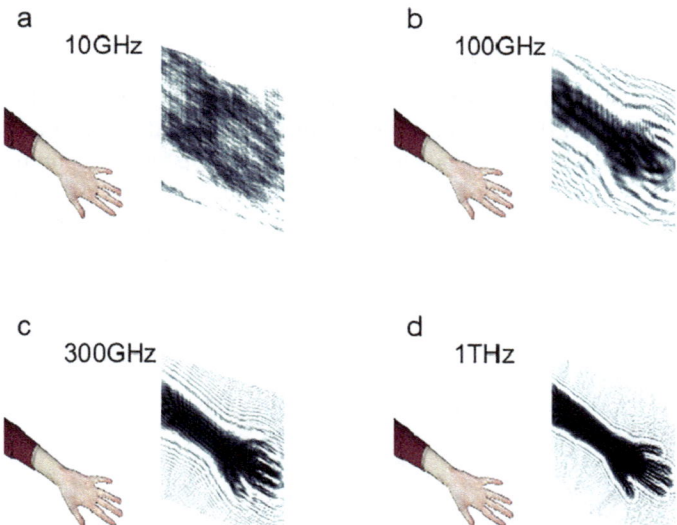

Fig. 32.1 Calculated shades produced by a real-scale human hand at a screen placed 1 m away from the hand when illuminated by a homogeneous plane wave at (**a**) 10 GHz, (**b**) 100 GHz, (**c**) 300 GHz, and (**d**) 1 THz

the relevant scattering phenomena cannot readily be treated using geometric optics as clearly illustrated in Fig. 32.1, where the shadows produced are in most cases not sharp images of the object. The situation is further complicated by the fact that directed signals at terahertz frequencies are unlikely to be well described as collimated beams with a diameter of a few centimeters, but rather as diverging (spherical) waves with a diameter perhaps in the range of tens of centimeters. We must therefore consider the complex and frequency-dependent interaction between such electromagnetic waves and everyday objects in order to develop accurate channel models to account for scattering and shadowing effects in typical environments [6]. Analogous diffraction phenomena at lower frequencies (a few GHz) have been discussed for large structures such as buildings or similarly large objects [7] or as a potential tool for providing coverage around walls and corners [6].

In this chapter, we will discuss measurements of the losses produced by the insertion of some everyday objects with simple geometries. These are compared with theoretical estimates of the purely geometrical shadowing.

32.1.1 Beams and the Challenges That Come with Them

Beam forming and beam steering have been identified as essential parts of future telecommunications systems, particularly as these systems advance toward millimeter-wave (30–300 GHz) and sub-millimeter-wave/THz (300 GHz–1 THz) spectral bands [8]. As discussed in other parts of this book, these frequency ranges are promising for next-generation wireless communication systems such as modern cellular systems and WiFi 8, offering high bandwidth and data rates. However, the shorter wavelengths associated with these bands introduce challenges such as increased propagation losses, limited coverage range, and sensitivity to obstacles as noted in the introduction to this chapter.

Beam forming is used to direct signal energy toward desired users, improving signal-to-noise ratio and mitigating propagation losses in higher frequency bands [9]. By adjusting the phase and amplitude of signals across multiple antenna elements, beam forming creates constructive interference in specific directions, enabling improved performance even in environments with high path losses. For millimeter-wave communication, hybrid analogue-digital beam forming has been identified as a promising approach, providing a trade-off between hardware complexity and beam management capabilities [10]. Reconfigurable and fast beam forming will be particularly important in mobile and dense urban environments where users and devices are constantly in motion. Technologies like phased array antennas and metasurfaces are under active study for electronically controlled beam steering, providing low latency and agile beam adjustments. This requirement is more relevant as the carrier frequency increases, where multipath effects become more sparse and direct line-of-sight paths are preferred for low-loss link budgets [11]. Yet, in a dynamic environment, where the user, as well as people and objects, moves around, the possibility of transient line-of-sight blockage needs to be considered [2]. As noted above, at low frequencies below 6 GHz, the wavelength is relatively long, such that such objects typically do not cast sharp or deep shadows. In contrast, at higher frequencies, a sharp and deep shadow can completely block a signal transmission (see Fig. 32.1d).

Enormous efforts have gone into addressing these issues. Researchers have focused on the development of high-gain antennas, such as dielectric resonator antennas and metal-dielectric hybrid designs [12]. Additionally, reconfigurable intelligent surfaces (RIS) [13] and metasurfaces [14] have been explored for improving beam control, enhancing signal transmission by dynamically reflecting and focusing waves toward receivers and avoiding obstacles in the line-of-sight path [15], and even curving beams around obstacles [16]. Much of this work has been motivated by the challenge of how to deal with blockage events. Yet, there have so far been few studies of the actual impacts of such events [17], including the frequency-dependent losses that arise due to the scattering/shadowing effects of blockage by common objects.

32.2   Beam Propagation

As discussed in the introduction, conventional telecommunication technology has not required careful study of the interaction of the radiation with common objects. Yet, as carrier frequencies approach the terahertz range, the need for reasonably collimated beams and the shorter wavelengths bring us into a quasi-optical regime. It is in these cases that one can benefit from the use of formalisms developed for optics. There are several approaches to predict the way in which electromagnetic waves will interact with objects and the projections (shadows) that they will produce under different circumstances. One reasonable approach relies on the Fresnel diffraction integral, which is an approximation valid for describing the paraxial propagation of waves after interfering with an object; in other words, this approach is applicable when the propagation distance z is large compared to the cross-sectional area of the scattering object projected onto the xy plane. This is often a reasonable approximation for beams with small divergence such as the ones that might be typically encountered in a THz wireless link.

Here, we present some general remarks about the Fresnel diffraction formalism, aiming to provide enough practical information for its use in THz communications. This is not intended to be a detailed derivation or justification of all the steps involved. The Fresnel diffraction formula provides a way to calculate the electric field across a plane, which we will refer to as the screen, after propagating a distance z from another plane, the object plane, where the electric field is known. We define (x, y) as the coordinates at the screen plane and (x', y') as the coordinates at the

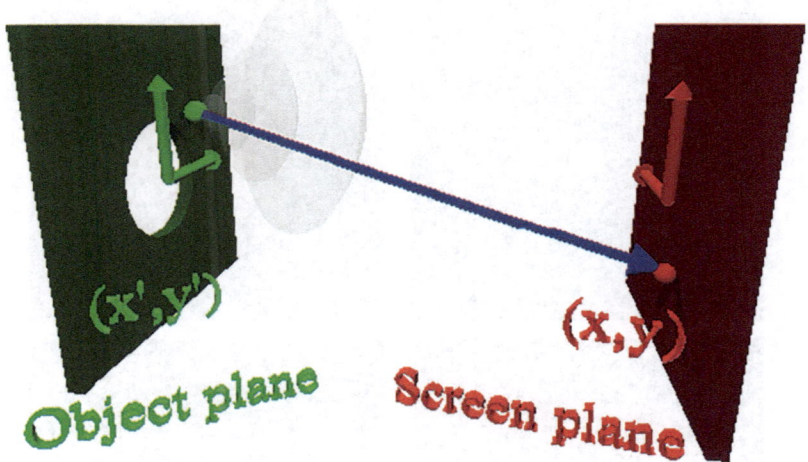

Fig. 32.2 The Fresnel integral allows the calculation of the electric field distribution at the observation (screen) plane (red) by forward propagation of the waves in the object plane (green), by superimposing the contributions from each point in the object plane taking each as a point source

object plane. The electric field in the screen plane is then given by an integral over points in the object plane as

$$E(x, y, z) = \int_{-\infty}^{+\infty} \int_{-\infty}^{+\infty} E(x', y', 0)e^{\frac{ik}{2z}[(x-x')^2+(y-y')^2]}dx'dy'. \tag{32.1}$$

This result arises from the Huygens' principle that states that the electromagnetic wave propagates from the object plane to another subsequent plane, such as the screen, by superimposing the contributions of each point in the object plane as if it produced a spherical wave with the appropriate amplitude and phase. Here, $k = 2\pi/\lambda$ is the wave number and λ is the wavelength (Fig. 32.2).

Many classical problems in optics can be studied with this Fresnel formalism, for instance, the diffraction pattern of a circular or rectangular aperture as shown in Fig. 32.3. For all these cases, the integral can be solved analytically or by using well-known special functions; however, the great majority of realistic problems for objects that we consider will not have simple geometrical forms. Therefore, it is desirable to have a numerical implementation of this integral that is computationally efficient in order to simulate these more complex situations. Firstly, the computation has to be implemented over a finite plane, and secondly, the cross-sectional shape of the object needs to be discretized. This double integral, or in the discrete case double summation, across the entire object plane provides us with the electric field at a single position of the screen; therefore, the integration process has to be repeated for every point in the screen plane, which can result in an enormous computational effort for large models. Yet, a simplification results from realizing that the integral

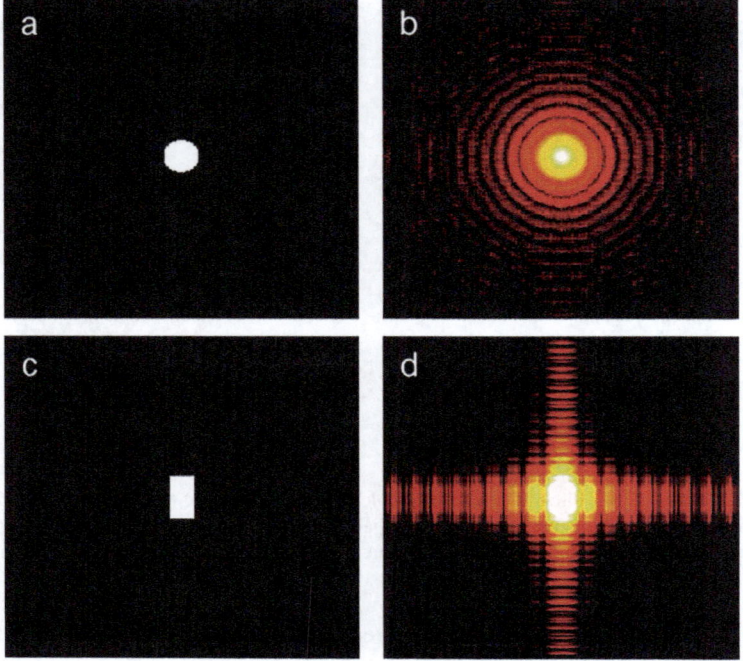

Fig. 32.3 Classical examples of diffraction calculations performed by using the numerical implementation of the Fresnel diffraction integral. (**a**) is the intensity distribution at the object plane for a circular pinhole. (**b**) is the result of the propagation of (**a**). (**c**) is the intensity pattern at the object plane for a rectangular slit, and (**d**) is the resulting diffraction pattern. In both (**b**) and (**d**), the color scale is logarithmic in order to enhance the outer fringe pattern

is a two-dimensional convolution between the function $E(x', y', 0)$ and the function $\exp[\frac{ik}{2z}(x^2 + y^2)]$. By using the convolution theorem, the entire integral can be computed as

$$E(x, y, z) = \mathcal{F}^{-1}\left[\mathcal{F}[E(x', y', 0)] \times \mathcal{F}\left[e^{\frac{ik}{2z}[(x')^2+(y')^2]}\right]\right], \qquad (32.2)$$

where $\mathcal{F}$ represents the two-dimensional Fourier transform. Fortunately, the Fourier transform of the second term can be calculated analytically. Thus, the numerical problem is reduced to the calculation of one product and two two-dimensional Fourier transforms, for which there are highly efficient implementations.

To illustrate the value of this simplification, we compute the shadow cast by a human hand upon illumination by a collimated Gaussian beam with a beam waist of 200 mm, for three different frequencies. A human hand was photographed against an evenly colored background. Image processing was used to remove the background and generate a binary image identifying the pixels that corresponded to the hand and arm as zeros and the pixels that correspond to the background as

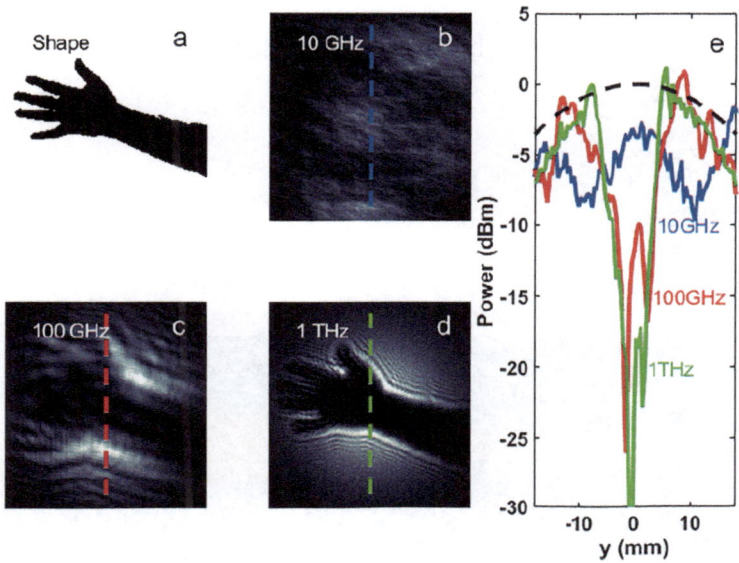

Fig. 32.4 Calculations of the intensity of a collimated Gaussian beam interacting with a human hand after 1 m of propagation. The upper-left panel shows the hand shape (**a**). The subsequent panels show the "shadow" caused by the hand for 10 GHz (**b**), 100 GHz (**c**), and 1 THz (**d**). The dotted lines in panels (**b–d**) illustrate the position of cuts across the center of each image. In panel (**e**), the quantitative position-dependent power distributions across the central cut of the images are shown in dBm for the three frequencies; a dashed line shows the Gaussian beam before being shadowed as a reference

ones. The image was resized in such a way that pixels represented 1 mm × 1 mm squares. Subsequently, the calculation of the propagation for 10 GHz, 100 GHz, and 1 THz was performed for a distance of 1 m resulting in the images presented in Fig. 32.4. The differences between the images are remarkable; at 10 GHz, it is hard to identify a shadow with dark areas, and, while there is some structure, it is possible to see signal more or less across the entire beam waist. At 100 GHz, the diffraction structure is rather rich, and there is an area that is more or less dark in the central part. At 1 THz, the shadow of the hand is clearly identifiable producing an area that is very dark where signal would be virtually impossible to detect as quantitatively shown in Fig. 32.4e.

32.3   Terahertz Time-Domain Spectroscopic Study of Objects Interacting with Large Beams

In this study, we used a terahertz time-domain spectrometer such as the one described in Chap. 7 to study the losses produced by the insertion of everyday

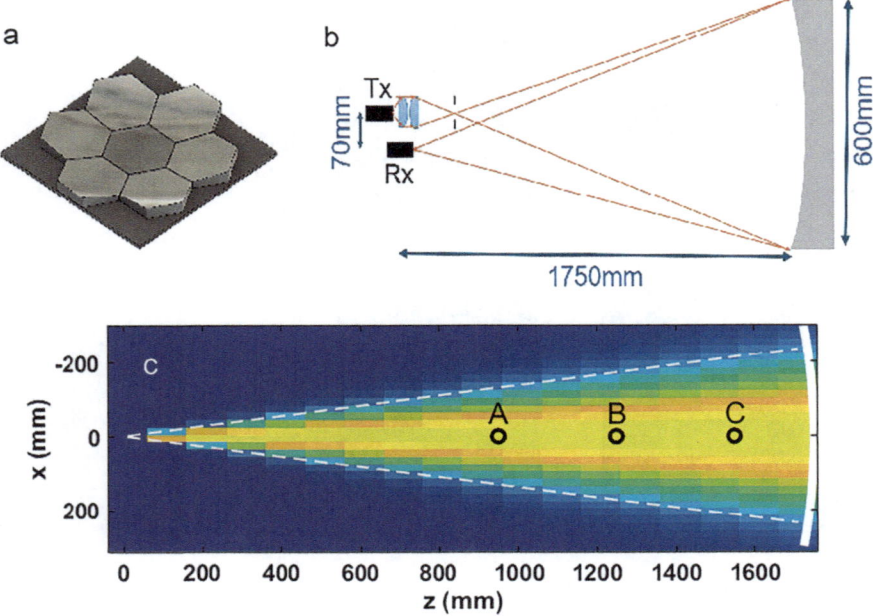

Fig. 32.5 (**a**) Render of the 7-hexagonal-segment mirror of 600 mm diameter and 1750 mm radius of curvature built for the experiment. (**b**) Schematic of the optical setup (not to scale). (**c**) Intensity beam profile from the focal position $z = 0$ to the mirror position at 1750 mm estimated from the measurement performed at 0.65 m form the focal position. Three black circles denote the position where the objects were placed during the measurements

objects in a defocused THz-broadband beam. Because of the ultra-broadband nature of the THz-TDS source, these data provide an indication of the relative effect of simple geometrical shadowing of the signal and diffraction as function of frequency. We employ a 7-hexagonal-segment spherical mirror with a radius of curvature of 1.75 m for these measurements. These segments form a mirror of approximately 60 cm diameter as shown in Fig. 32.5a. The transmitter and receiver were placed next to each other at a distance of 1.75 m from the large spherical mirror in such a way that the generated terahertz radiation was reflected by the mirror and refocused onto the receiver, with a total propagation distance of 2×1.75 m (Fig. 32.5b). Because the transmitter and receiver are laterally offset from each other by about 7 cm, the angle of approach to the large mirror is not precisely normal. This results in a small degree of spherical aberration in the focal plane at the receiver, but given the comparative length of the radius of curvature (175 cm), we estimate this effect to be negligible.

The terahertz beam waist was measured by using the razor-blade method at a distance of 0.65 m from the focal position resulting in a nearly Gaussian beam waist of 17.9 cm (standard deviation) which corresponds to a beam divergence semi-angle of $\sim7.8°$. From that measurement, we infer that the beam power distribution matches the color map shown in Fig. 32.5c. In the same figure, three circular marks are used to identify three positions where the objects under test were placed during the experiments. The beam waists at these three locations were 13.4 cm, 17.5 cm, and 20.0 cm (standard deviation).

Time-domain spectra were obtained as a reference, in the absence of objects, and for the objects placed in each position. The objects studied were a 66.2 mm diameter spherical ball, a 165 mm by 110 mm by 50 mm book, and a 272 mm long by 40.4 mm metallic cylindrical tube. The frequency-dependent transmittance was obtained from the experimental measurements as

$$T(\nu) = \frac{\tilde{E}_{\text{sam}}(\nu)}{\tilde{E}_{\text{ref}}(\nu)}, \tag{32.3}$$

where $\tilde{E}_{\text{sam}}(\nu)$ and $\tilde{E}_{\text{ref}}(\nu)$ are the Fourier transformed electric fields in the presence and absence of the sample in the beam path.

In order to model the transmittance of the objects in each position, we used the formalism presented in the previous section. We incorporate the effects of the mirror and the detector as spatial filters. The results from the measurements and the model are shown in Fig. 32.6. The continuous lines are the measurements, and

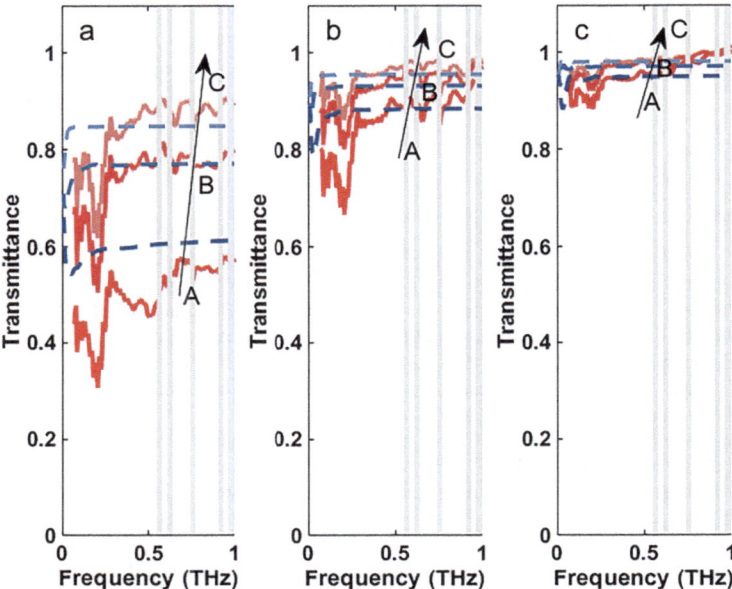

Fig. 32.6 Measured (continuous lines) and calculated (dashed lines) transmittance for a book (**a**), tube (**b**), and ball (**c**) as a function of the frequency for the positions denoted A, B, and C in Fig. 32.5. The frequencies corresponding to water vapor absorption lines are excluded and marked with gray bars

the dashed lines are the corresponding calculations from the Fresnel diffraction. The agreement between the predicted and measured frequency-dependent spectra is excellent, although not perfect. We attribute this to several factors, including imperfections in the placement of the objects relative to the mirror, which has an uncertainty in the order of a few centimeters, and the fact that the real beam is not perfectly Gaussian in transverse profile (as assumed for the calculated results). Despite these issues, the agreement is sufficient to validate the use of the Fresnel formalism for the estimation of the distribution of radiation after interacting with objects of this nature.

32.4 Summary

In this chapter, we have discussed the effects of shadowing and diffraction of millimeter and terahertz beams by objects that could be encountered in a typical indoor wireless link scenario. As graphically shown in Fig. 32.4, a human hand, or objects with comparable dimensions, does not cast deep shadows at low frequencies but can have a substantial impact on the received signal as the frequency increases, due to the formation of deep shadows. These phenomena can be modeled using Fresnel diffraction, which can be implemented even for quite complex scenarios with relatively low computational burden. We note that this tool is far less computationally expensive than a full finite element simulation of a real-scale scene and is therefore a valuable addition to the arsenal for computation of channel models.

References

1. C.A. Balanis, *Antenna Theory: Analysis and Design* (John Wiley & Sons, London, 2016)
2. R. Piesiewicz, T. Kleine-Ostmann, N. Krumbholz, D. Mittleman, M. Koch, J. Schoebel, T. Kürner, Short-range ultra-broadband terahertz communications: concepts and perspectives. IEEE Antennas Propagat. Mag. **49**(6), 24–39 (2007)
3. H.-J. Song, T. Nagatsuma, Present and future of terahertz communications. IEEE Trans. Terahertz Sci. Technol. **1**(1), 256–263 (2011)
4. T. Nagatsuma, G. Ducournau, C.C. Renaud, Advances in terahertz communications accelerated by photonics. Nat. Photon. **10**(6), 371–379 (2016)
5. W. Miller, Proceedings of the symposium on quasi-optics, vol. 14 Proc. IEEE **53**(7), 766–767 (1965)
6. M. Jacob, S. Priebe, R. Dickhoff, T. Kleine-Ostmann, T. Schrader, T. Kürner, Diffraction in mm and sub-mm wave indoor propagation channels. IEEE Trans. Microwave Theory Tech. **60**(3), 833–844 (2012)
7. K.-W. Kim, M.-D. Kim, J. Lee, J.-J. Park, Y. K. Yoon, Y. J. Chong, Millimeter-wave diffraction-loss model based on over-rooftop propagation measurements. ETRI J. **42**(6), 827–836 (2020)
8. J.M. Jornet, E.W. Knightly, D.M. Mittleman, Wireless communications sensing and security above 100 GHz. Nat. Commun. **14**(1), 841 (2023)

9. O. El Ayach, S. Rajagopal, S. Abu-Surra, Z. Pi, R.W. Heath, Spatially sparse precoding in millimeter wave MIMO systems. IEEE Trans. Wirel. Commun. **13**(3), 1499–1513 (2014)
10. F. Sohrabi, W. Yu, Hybrid analog and digital beamforming for mmWave OFDM large-scale antenna arrays. IEEE J. Sel. Areas Commun. **35**(7), 1432–1443 (2017)
11. D. He, B. Ai, K. Guan, L. Wang, Z. Zhong, T. Kürner, The design and applications of high-performance ray-tracing simulation platform for 5G and beyond wireless communications: a tutorial. IEEE Commun. Surv. Tutorials **21**(1), 10–27 (2018)
12. Y. He, Y. Chen, L. Zhang, S.-W. Wong, Z.N. Chen, An overview of terahertz antennas. China Commun. **17**(7), 124–165 (2020)
13. M.A. ElMossallamy, H. Zhang, L. Song, K.G. Seddik, Z. Han, G.Y. Li, Reconfigurable intelligent surfaces for wireless communications: Principles, challenges, and opportunities. IEEE Trans. Cogn. Commun. Netw. **6**(3), 990–1002 (2020)
14. J.-B. Gros, V. Popov, M.A. Odit, V. Lenets, G. Lerosey, A reconfigurable intelligent surface at mmWave based on a binary phase tunable metasurface. IEEE Open J. Commun. Soc. **2**, 1055–1064 (2021)
15. T.J. Cui, M.Q. Qi, X. Wan, J. Zhao, Q. Cheng, Coding metamaterials, digital metamaterials and programmable metamaterials. Light: Sci. Appl. **3**(10), e218–e218 (2014)
16. H. Guerboukha, B. Zhao, Z. Fang, E. Knightly, D.M. Mittleman, Curving THz wireless data links around obstacles. Commun. Eng. **3**(1), 58 (2024)
17. V. Petrov, M. Komarov, D. Moltchanov, J.M. Jornet, Y. Koucheryavy, Interference and SINR in millimeter wave and terahertz communication systems with blocking and directional antennas. IEEE Trans. Wirel. Commun. **16**(3), 1791–1808 (2017)

Chapter 33
Multipath Propagation in Rich Scattering

Carla Reinhardt, Diego Dupleich, Tobias Doeker, Giovanni Del Galdo, and Thomas Kürner

Abstract Rich scattering environments, characterized by numerous reflective and diffractive structures, play a crucial role in the propagation of THz signals. In industrial settings such as server rooms and production lines, metallic surfaces and small machine components contribute to complex multipath channels, introducing both opportunities and challenges for communication and sensing applications. The presence of specular and diffuse scattering generates diverse signal paths, enhancing spatial diversity while also causing interference and fading. Channel characterization in such environments requires sophisticated measurement techniques, including multidimensional wideband channel sounding and ray-tracing simulations. In this chapter, several measurements and models are presented for the characterization of propagation in industrial scenarios, showing that properly leveraging multipath propagation can enhance reliability and efficiency in industrial THz networks.

33.1 Rich Scattering Environments for THz Communication Systems

Rich scattering environments are characterized by the presence of multiple structures and objects with properties favorable to scattering. Therefore, a transmitted

C. Reinhardt (✉) · T. Doeker · T. Kürner
Technische Universität Braunschweig, Institut für Nachrichtentechnik, Braunschweig, Germany
e-mail: carla.reinhardt@tu-braunschweig.de; t.doeker@tu-braunschweig.de; t.kuerner@tu-braunschweig.de

D. Dupleich
Technische Universität Ilmenau, Institute of Information Technology, Ilmenau, Germany
e-mail: diego.dupleich@tu-ilmenau.de

G. Del Galdo
Technische Universität Ilmenau, Institute of Information Technology, Ilmenau, Germany

Fraunhofer Institute for Integrated Circuits (IIS), Ilmenau, Germany
e-mail: giovanni.delgaldo@tu-ilmenau.de

T. Kürner et al. (eds.), *Metrology for THz Communications*, Springer Series in Optical Sciences 256, https://doi.org/10.1007/978-3-032-01986-8_33

signal encounters numerous interaction points during its propagation toward the receiver. These interaction points describe a multipath channel leading to multiple copies of the transmitted signal at the receiver. The strength and number of these copies determine the degree of diversity of the channel which can be exploited.

Industrial environments, like server rooms, are prime examples of rich scattering environments due to the abundance of metallic structures, machinery, and other items. However, these spaces are often cluttered, leading to significant shadowing effects caused by penetration and diffraction losses at these frequencies.

Metallic objects with smooth surfaces scatter the impinging power in the direction of the specular component. A large amount of highly reflective objects in the environment increases the possibility of creating a multipath connection between a transmitter and receiver, either by single- or higher-order specular reflections. These objects typically create strong paths whose amplitude depends only on the free-space propagation path loss because of the low reflection losses (in case of metallic items). This contribution to the isotropic channel can be seen as a number of discrete strong paths from multiple directions with distinct delays and an amplitude decreasing with delay.

On the other hand, the size of the wavelength relative to the roughness of many surfaces generates diffuse scattering. The impinging waves are not propagated only in the specular direction, but also in a number of different directions that can be characterized by a scattering patterns [1], Chap. 31. Since the scattered energy is not concentrated in a single direction but distributed over a wide set of directions, leading to a lower energy of the received signal compared to the specular reflection case, therefore, the probability for the transmitter reaching the receiver via scattering is higher compared to reflections albeit with considerably lower signal strength. The contribution of this process to the isotropic channel can be observed as a dense set of multipath components arriving from a wide set of directions.

Small items such as rods, spindles, drill bits, etc., usually found in production lines, scatter significant power to be detected as multipath components, a consequence of the size of the wavelength relative to these objects. In addition, the increased carrier frequency produces higher Doppler shifts compared to lower bands, and small vibrations or movements of machine components can already be detected in the case of sensing or can also be harmful for communications by reducing the coherence time.

Good propagation conditions and a rich multipath channel can also be harmful because of interference generated by reaching undesired places and due to fading because of the constructive or destructive interactions of the multiple copies of the signal at the receiver. However, since highly directive radio interfaces need to be implemented at THz to compensate for the high path loss, many of these multipath components will be filtered out. In addition, the implementation of instantaneous large bandwidths reduces the total fading from the remaining multipath components.

The large resolution in delay and, possibly, angular domain obtained from the implemented bandwidths and radio interfaces, respectively, combined with the scattering properties of small objects relative to the wavelength makes THz systems suitable for sensing applications as well. For example, the position, velocity, and state of tiny components from machines can be inferred from the measured channel impulse responses.

In summary, the unique propagation characteristics at THz frequencies in rich scattering environments offer both opportunities and challenges. While the presence of multiple scatterers can complicate communication due to interference and fading, the high spatial resolution and sensitivity to small object movements present significant potential for integrated sensing and communication (ISAC) use cases. Properly designed THz systems can leverage these features to enhance industrial automation, improve precision monitoring, and enable new sensing capabilities.

33.2 Characterization of Propagation in Rich Scattering Environments

33.2.1 Channel Sounding

Channel sounding in rich scattering environments, and particularly considering also time-varying scenarios as industrial settings, comes with multiple challenges. Capturing the channel simultaneously in the different dimensions requires expensive hardware with MIMO architectures with fast enough switching schemes to capture the time variations and massive storage for the acquired data.

Architectures based on VNAs have the flexibility of achieving large SNR and measurement bandwidths by sweeping in frequency. However, this is at the expense of requiring a time-invariant channel. Therefore, one of the main limitations is that parameters as Doppler cannot be measured. On the other hand, architectures based on wideband signals achieve large instantaneous bandwidths that captures the time-varying characteristics of the channel. However, unless an antenna array is used during the measurements, the time-varying characteristic cannot be captured simultaneously with the angular information of the channel.

Therefore, with the current state of the art on channel sounding, most of the measurement setups target the simultaneous acquisition of at least two dimensions: delay/angle by using wideband excitation signals and high gain antennas to scan the environment [2–6] or delay/Doppler by using wideband signals and high sampling rates.

33.2.2 Simulation-Based Characterization

Performing extensive measurement campaigns in large environments is time consuming, expensive, and in some situations particularly challenging due to the size of the measurement equipment and locations. In addition, unless specific experiments are designed and conducted, the environment needs to be static.

Therefore, usually the result of measurement campaigns is still a limited set of channel impulse responses not sufficient for generalization, e.g., in the form of statistical characterization.

As measurements at THz in rich scattering environments have shown, the channel typically has a sparse characteristic and consists of a limited set of notable multipath components from specular reflections. Therefore, tools as ray tracing have gained a lot of notoriety in the latest years to generate realistic and precise channel impulse responses. In fact, ray tracing is particularly well suited to compute specular reflections. The latest advances on technologies to map and digitize environments and a rich database on electromagnetic properties of the construction materials allow the creation of very precise models in Chap. 10.

Ray tracing also allows to generate test situations that are expensive to measure, for example, the multidimensional time evolution of the channel under blockage by different vehicles or motion of different items.

However, ray-tracing models need first to be validated by measurements. This validation step can consist of a limited set of measurements, and once the model is validated, different situations other than the validation measurements can be readily simulated.

33.2.3   Cluster-Based Characterization

Another way to characterize the channel is by identifying clusters of MPCs. Those clusters are concentrated around specific scatterers in the environment. To identify the clusters, high angle-resolution double-directional rotational measurements need to be performed. By analyzing the power angular profile (PAP), scatters can be identified, and the main contributing MPCs can be pointed out. In Fig. 33.1, a PAP from measurements in a data center is presented [7]. The located MPCs are localized in a line around the LOS in the center of the APA. The MPCs come from the reflection at the server racks, which are in a corridor shape installation. These angles of interest can also be identified beforehand by ray-tracing simulations. This can

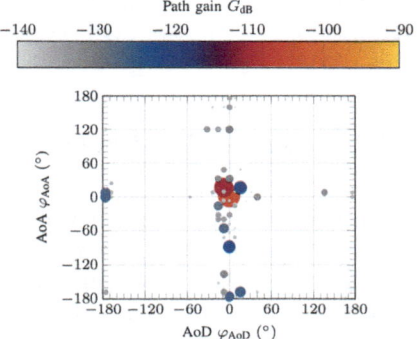

Fig. 33.1 PAP of double-directional channel measurements at Mid-High [7]

improve the measurement time, by concentrating the measurements around certain angles of interest. By combining the ray-tracing results and the measurements, which can be used to calibrate the ray-tracing simulations, a ray-tracing-based channel model can be improved.

33.3 Characterization of Propagation from Measurements in Industrial Environments

Production lines, industrial and machinery halls, are good examples of rich scattering environments. The presence of multiple metallic objects creates the necessary conditions for the presence of MPCs and allows communications even in NLOS situations. In this section, multiple measurement scenarios in industrial environments or machines are presented and discussed.

33.3.1 Measurements in Industrial Rooms

33.3.1.1 Small Machinery Room

The typical structure of the multipath channel from multiband wideband measurements of a NLOS link in a small machinery room [3] is shown in its multidimensional representation in Fig. 33.2.

Fig. 33.2 Multidimensional representation from measurements of the THz channel and comparison with 6 GHz (bottom-right)

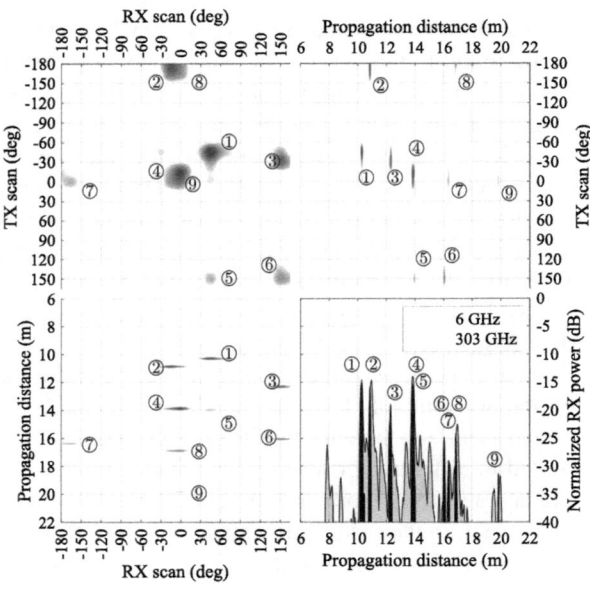

Strong single- and higher-order reflections can be observed in a wide range of departure and arrival angles. Many of these multipath component (MPC)s have similar angles of departure, e.g., ① and ③ or ⑤ and ⑥, but they are resolvable at the receiver because they have different angles of arrival. The opposite happens with ② and ④. Similarly, there are paths arriving with the same delay as ④ and ⑤ but with completely different angles of departure and arrival.

The presence of multiple scatterers also generates cases in which the same multipath reaches the receiver from opposite transmitting directions: ② and ④.

Compared to the 6 GHz channel, the THz channel appears to be more sparse with less diffuse components. The synthetic omnidirectional CIR from directive scans at 6 GHz and THz presented in Fig. 33.2 (bottom-right) shows a direct comparison between the channel structures at these two frequencies. A direct and a diffracted component on the edge of a CNC machine with metallic frames is only visible within the measurement DR at 6 GHz. In addition, in between the dominant reflections, many multipath components product of scattering are observed at lower frequencies. On the other hand, the dominant paths from strong reflections are present with similar gain in both bands.

33.3.1.2 Large Industrial Space

Another common use case for wireless communication in industrial spaces is the access point to user equipment (UE) communication. The UE can, e.g., be a robot that drives through the industrial space, either to transport packages or to execute a specific task. To investigate the influence of a highly reflecting and complex environment, a measurement campaign was conducted in an industrial like space [8]. Various different machine types, metal surfaces, and different heights levels can be found in this place. The radio channel measurements were performed with a UWB correlation-based channel sounder from Ilmsens. The channel sounder consists of a base unit, UWB modules, and frequency extensions. All of them are connected by cables. The base unit provides the power and a clock frequency at 9.22 GHz which is up-converted to 304.2 GHz in the frequency extensions. The UWB modules provide a pseudorandom M-Sequence of 12th order. The M-Sequence is chosen in such a way that the autocorrelation function is almost a Dirac impulse. By correlating the received signal with the M-Sequence, the channel impulse response can be derived. The data is calibrated with a back-to-back measurement (B2B). More information about the system can be found in [9, 10].

For the measurement campaign in the industrial like space, the transmitter was located at the middle level at a height of 5.25 m. Two receivers were placed on the ground level at a height of 1.25 m. Rx1 and Rx2 are separated by machines and located in different corridors, which leads to a more cluttered environment around Rx2 and a slightly more open around Rx1. A schematic view of the setup is shown in Fig. 33.3. In the case of alignment of Tx and Rx2, the value is − 100.6 dB, which is close to FSPL for the distance. For Rx1 the value stands at − 117.4 dB, which can be explained by looking at the placement of the machine. In this case, one part of a

Fig. 33.3 Schematic view of the measurement setup [8]

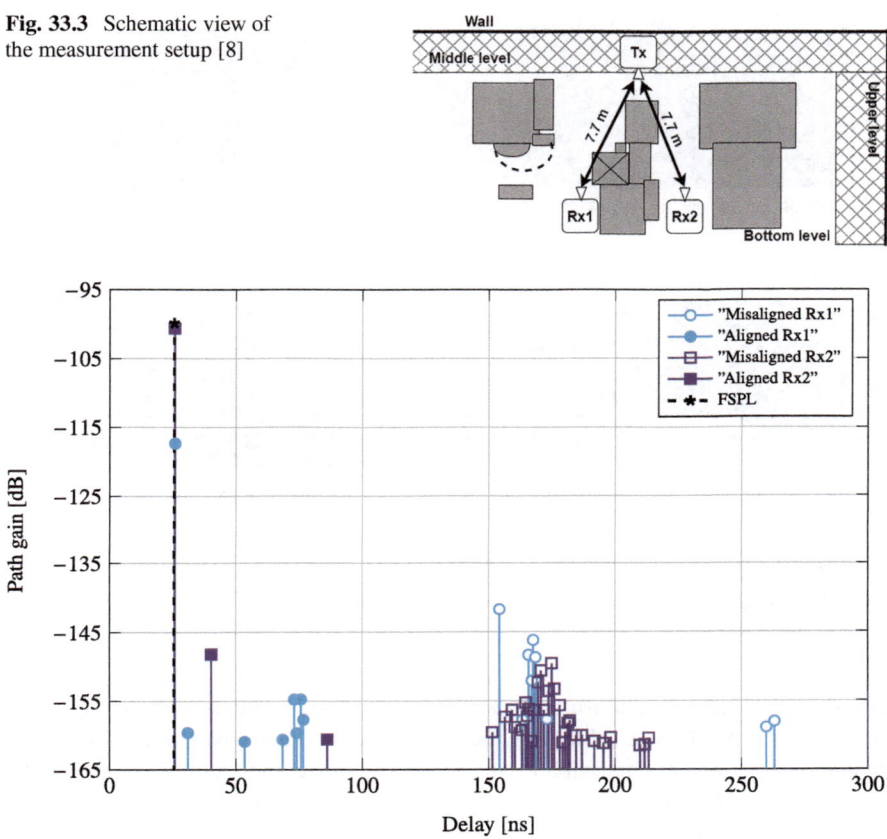

Fig. 33.4 Rx1 and Rx2 PDP comparison in extreme cases [8]

machine interrupts the LOS, so that we don't have a complete LOS condition, but an obstructed LOS (OLOS) condition which is causing a difference of 16.8 dB. One other interesting finding is that several multipath components appear in the case of misalignment. Clearly to identify is that for RX2 more of these components appear. This is due to a more cluttered space in the surroundings of Rx2 (Fig. 33.4).

33.3.2 Measurements in Industrial Machines

33.3.2.1 Measurements in a Milling Machine

Other good example of rich scattering environments are industrial machines. Usually built with metal frames and structures and consisting of multiple small highly reflective components, if they are completely closed, they can act as a good

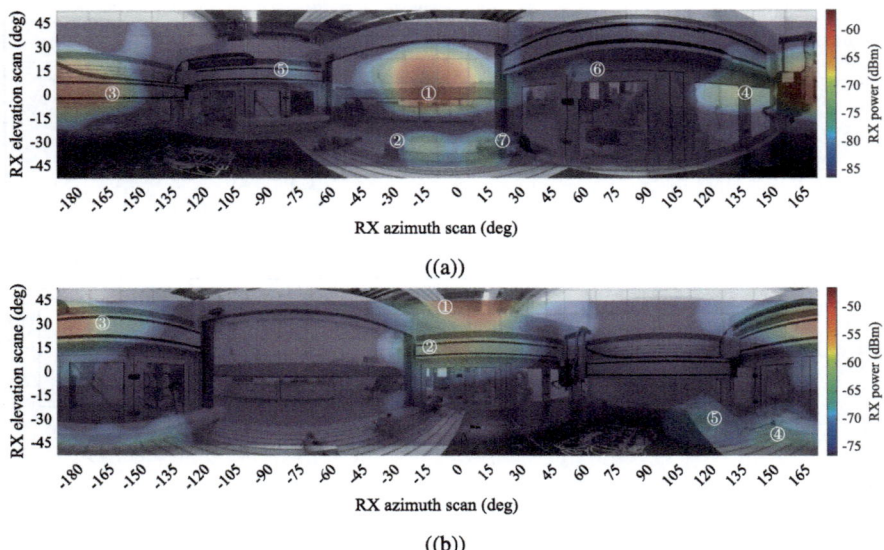

Fig. 33.5 Measured angular spectrum in (**a**) outside to inside of the machine and (**b**) inside to inside of the machine

reverberation environment for transceivers located inside. On the other hand, if they are open or partially accessible from the outside by apertures covered with protective materials with lower penetration losses, inner machine components and transceivers can still be reached with multipath components for communications or sensing from exterior transceivers.

Measurements from outside to inside of machines in a production line have shown that after an attenuation of approx. 2 to 4 dB by a protective transparent material such as Plexiglas^{TM}, the signal propagates generating multiple specular-like reflections from the different inner components [4]. The power angular profile in Fig. 33.5a shows the LOS component ①, the reflection on the ground plate of the machine ②, and multiple reflections on the inner frames ③, ④, ⑤, and ⑥ (some of them second- and third-order reflections). The channel can be easily predicted knowing the geometric properties of the machine, and inversely, the machine can be easily described by the geometrical properties of the multipath components.

The same has been observed when both transceivers are inside of the machine in Fig. 33.5b. In this example, the TX was located in the ceiling and the RX on the ground plate. Similarly, apart from the LOS component ①, multiple paths arriving from the metallic frames, ground plate, and other highly reflective structures are observed as single- and higher-order specular reflections.

33.3.2.2   Measurements Around a Robotic Arm Manipulator

Not only transceiver locations in closed machines like milling machines but also the location on machines like robotic arm manipulators is of interest. These robotic arms are more and more used in industrial settings either to perform repetitive and precise tasks or to assist workers in direct interaction. To investigate the impact of the positioning of robotic arm manipulators, measurements were performed in a workshop with commercially available programmable Franka Emika robotic arms [11].

It was an access point to UE (the robotic arm) scenario. The reference measurement was taken without any robotic arm near to the transmitter or receiver position. The MPCs of this measurement are in Fig. 33.6. Three different arm configurations were chosen. The gain of the LOS component doesn't change much over the different setups. The different positions don't impact the main component. If we compare them with the reference measurement, two main differences can be observed. First, the MPC at 41ns in the reference measurement changes by introduction of the arm. Two MPCs appear in close proximity. This can be explained by diffraction and interference effects due to the arm. Second, the components at 80 ns and 90 ns of the reference disappear when the arm is introduced. The arm blocks these components.

After studying mostly static setups and use cases to get a better understanding of the behavior of low-THz frequencies in such rich scattering environments, one other interesting part is the time-variant channel behavior. In this industrial

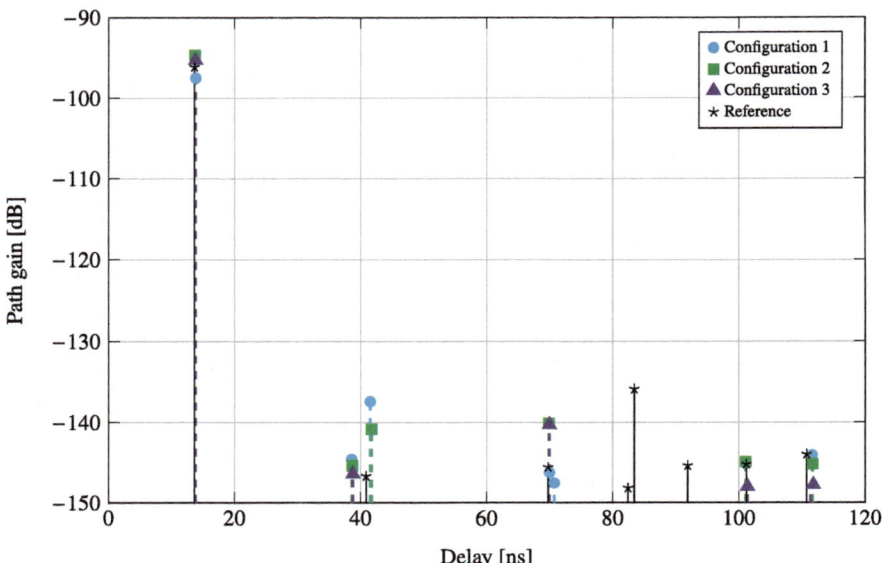

Fig. 33.6 MPCs for setups 1–3 and reference measurements [11]

environment, typical scenarios are inspired by construction lines, where multiple arms are working in one line and potentially influence the communication channel between each other. For more time-variant channel evaluation, see Chap. 34.

Even if it is a non-cluttered space, as the before-mentioned industrial spaces and machine surroundings, one other rich scattering environment is the inner of a ventilation shaft. The metallic surface and the long tunnel effect create a highly reflecting environment with multiple overlapping of propagation paths, which can lead to strong positive and negative interferences. A comparable experimental measurement environment is the reference structure in Chap. 18.

33.4   Modeling Rich Scattering Environments

In general, there are various approaches to model radio channels, with the most common distinctions being deterministic, stochastic, and hybrid channel models. Deterministic models, often based on ray-tracing techniques, offer high precision but require substantial computational power to run simulations. In addition, these models are based on detailed knowledge of the specific environment, which is not always available. Accurate simulations require a 3D model of the environment. The simulations are then performed for defined antenna configurations or scenarios. This method is particularly suitable for THz frequencies as they behave close to ray-optical channels in Chap. 10. The deterministic model produces outputs that include multipath components, characterized by parameters such as amplitude, delay, angle of departure (AoD), and angle of arrival (AoA). However, these models are typically tailored to specific use cases and environments.

In contrast, stochastic models provide a more general approach to channel modeling. These models capture the statistical properties of radio channels based on various factors, such as frequency, and whether the scenario is indoors or outdoors. While they may lack the accuracy of deterministic models, stochastic models can be applied across a broader spectrum of settings with similar environmental conditions. The channel is described in function of parameters like delay spread, the number and distribution of MPCs, and the angular spread.

A cross between deterministic and stochastic models is hybrid models which incorporate both modeling approaches. For communication links in a data center, such a model is described in [12]. The communication links are divided into three different categories: Top of Rack (ToR) LOS, ToR NLOS, and Medium Height (MH).

This model shows that rich scattering environments contribute to not negligible MPCs.

33.5   Summary

Rich scattering environments are challenging for THz communication and need particular investigation. The complex radio propagation properties due to multiple reflections, scattering, and diffraction can disturb the signal. Some effects can be counteracted through the communication system design, like the modulation scheme or the antenna properties, but the propagation in these environments needs to be characterized, to ensure a good design concept. The characterization is mainly done via simulations and measurements. Materials in industrial environments, such as metal or acrylic glass, cause reflections and transmission loss. OLOS also needs to be considered in complex environments. Objects can partly block the LOS link. Hybrid channel models based on measurements and simulations are a good approach for modeling rich scattering environments.

References

1. V. Degli-Esposti, F. Fuschini, E.M. Vitucci, G. Falciasecca, Measurement and modelling of scattering from buildings. IEEE Trans. Antennas Propagat. **55**(1), 143–153 (2007). https://doi.org/10.1109/TAP.2006.888422
2. Y. Lyu, P. Kyösti, W. Fan, Sub-THz VNA-based channel sounder structure and channel measurements at 100 and 300 GHz, in *2021 IEEE 32nd Annual International Symposium on Personal, Indoor and Mobile Radio Communications (PIMRC)* (2021), pp. 1–5. https://doi.org/10.1109/PIMRC50174.2021.9569702
3. D. Dupleich, A. Ebert, Y. Völker-Schöneberg, D. Sitdikov, M. Boban, L. Samara, G. Del Galdo, R. Thomä, Characterization of propagation in an industrial scenario from sub-6 GHz to 300 GHz, in *IEEE Globecom Workshops* (2023)
4. D. Dupleich, A. Ebert, Y. Völker-Schöneberg, L. Löser, M. Boban, R. Thomä, Spatial/temporal characterization of propagation and blockage from measurements at sub-THz in industrial machines, in *2023 17th European Conference on Antennas and Propagation (EuCAP)*, Florence, Italy (IEEE, Piscataway, 2023), pp. 1–5. https://doi.org/10.23919/EuCAP57121.2023.10133323
5. S. Ju, D. Shakya, H. Poddar, Y. Xing, O. Kanhere, T.S. Rappaport, 142 GHz sub-terahertz radio propagation measurements and channel characterization in factory buildings. IEEE Trans. Wirel. Commun. **23**(7), 7127–7143 (2024)
6. A. Schultze, M. Schmieder, S. Wittig, H. Klessig, M. Peter, W. Keusgen, Angle-resolved THz channel measurements at 300 GHz in an industrial environment, in *2022 IEEE 95th Vehicular Technology Conference: (VTC2022-Spring)* (2022), pp. 1–7. https://doi.org/10.1109/VTC2022-Spring54318.2022.9860598
7. J.M. Eckhardt, T. Doeker, T. Kürner, Channel measurements at 300 GHz for low terahertz links in a data center. IEEE Open J. Antennas Propagat. **5**(3), 759–777 (2024). https://doi.org/10.1109/OJAP.2024.3391798
8. C.E. Reinhardt, V.V. Elesina, J.M. Eckhardt, T. Doeker, L.C. Ribeiro, T. Kürner, Channel measurements in an industrial environment for access point-to-sensor communication at 300 GHz, in *2024 15th German Microwave Conference (GeMiC)* (2024), pp. 308–311. https://doi.org/10.23919/GeMiC59120.2024.10485341
9. J.M. Eckhardt et al., Uniform analysis of multipath components from various scenarios with time-domain channel sounding at 300 GHz. IEEE Open J. Antennas Propagat. **4**, 446–460 (2023). https://doi.org/10.1109/OJAP.2023.3263597

10. T. Doeker, J.M. Eckhardt, C.E. Reinhardt, T. Kürner, Time-domain channel sounder calibration at low terahertz band. IEEE Open J. Antennas Propagat. **5**(6), 1598–1611 (2024). https://doi.org/10.1109/OJAP.2024.3425915
11. V.V. Elesina, C.E. Reinhardt, T. Kürner, Channel measurements in workspace with robotic manipulators at 300 GHz and recent results, in *2024 18th European Conference on Antennas and Propagation (EuCAP)*, Glasgow, UK (IEEE, Piscataway, 2024), pp. 1–5
12. J.M. Eckhardt, T. Doeker, T. Kürner, Hybrid channel model for low terahertz links in a data center. IEEE Open J. Commun. Soc. **5**, 4731–4745 (2024). https://doi.org/10.1109/OJCOMS.2024.3433561

Chapter 34
Time-Variant Propagation Channel Including Metrics and Multi-link Connection

Carla Reinhardt, Mohanad Dawood Al-Dabbagh, Giovanni Del Galdo,
Thomas Kleine-Ostmann, and Thomas Kürner

Abstract Before introducing terahertz (THz) communications to use case with mobility involved, there is the necessity to investigate time-variant propagation channels. At sub-THz frequencies, one of the most significant challenges is the maintenance of the connection, even in the event of a loss of line-of-sight (LoS). To overcome this challenge, various methods can be employed in a time-variant propagation channel, including multi-link connections, relays, and the use of strong reflections. In this chapter, we will present measurement techniques for studying time-variant propagation channels and introduce a suitable metric to characterize them.

34.1 Time-Variant Propagation Channels in the Sub-THz Range

Time-variant channels are challenging for wireless communication. Change in environment or position of the receiver (RX) or transmitter (TX) can have a huge impact on a wireless channel. This can end up in time-varying received power, Doppler effects, or short stationarity intervals [1]. The so far discussed settings and investigations for the over-the-air (OTA) sub-THz application scenarios, like kiosk download, data management in data center or office scenarios were mainly

C. Reinhardt · T. Kürner (✉)
Institut für Nachrichtentechnik, Technische Universität Braunschweig, Braunschweig, Germany
e-mail: carla.reinhardt@tu-braunschweig.de; t.kuerner@tu-braunschweig.de

M. D. Al-Dabbagh · T. Kleine-Ostmann
Physikalisch-Technische Bundesanstalt (PTB), Braunschweig, Germany
e-mail: mohanad.al-dabbagh@ptb.de; thomas.kleine-ostmann@ptb.de

G. D. Galdo
Institute of Information Technology, Technische Universität Ilmenau, Ilmenau, Germany

Fraunhofer Institute for Integrated Circuits (IIS), Ilmenau, Germany
e-mail: giovanni.delgaldo@tu-ilmenau.de

© The Author(s) 2026
T. Kürner et al. (eds.), *Metrology for THz Communications*, Springer Series
in Optical Sciences 256, https://doi.org/10.1007/978-3-032-01986-8_34

done in static environments; see, e.g., [2, 3]. An overview of already conducted measurement campaigns can be found, for example, in [4, 5].

The high data rates and low latency goals for THz are predestined for use in industrial settings for either logistics or robot-assisted work [6, 7]. These environments are mostly non-static, as workers walk around, machines like forklifts are in mobility or even automated guided vehicles (AGV) are deployed. These person or machines can lead to a partly or completely blocked LoS connection. Taking in to consideration the short wavelength of a few millimeters or less at sub-THz frequencies, even small objects can lead to blockage and may influence the signal propagation channels.

Blocking effects by humans at 300 GHz have been investigated first by Jacob et al. [8]. In [9] first time-variant measurements at 300 GHz are reported, where the main focus was the effect of blockage on the mostly very focused sub-THz channel, due to the use of high gain antennas with a narrow half-power beamwidth (HPBW) of a few degrees. Keusgen and Eichler [10] reported on time-variant measurements at 160 GHz.

Time-variant measurements are challenging. Multiple aspects need to be considered including the difficulty of performing the measurements in a well-controlled environment allowing to investigate measurement repeatabilities accurately, which is the bases of establishing measurement uncertainties.

With respect to the measurement equipment, three methods are used for sub-THz channel sounding: time-domain spectroscopy (TDS), vector network analyser (VNA), and correlation-based channel sounder (CS). TDS and VNA measurement systems are mostly suitable for static scenarios. With these measurement equipment, pseudo time-variant setups are possible to investigate. In the Chap. 13, and in [11], we presented a compact time-varying measurement scenario using a VNA at sub-THz frequencies. The investigation involves characterizing the time intervals, the velocity, and the position of the moving object. Yet, a VNA measurement comes with challenge of measurement flexibility that a time-domain CS can offer.

A correlation-based CS (see also Chap. 8) is able to perform time-variant measurements. In Fig. 34.1, the architecture of the CS used for the investigations in this chapter is presented.

The CS can be operated in two different modes. An averaged one, where the send-out sequence is correlated with the received one directly in the hardware components. In this case, a predefined number of channel impulse response (CIR) is averaged and stored on the control laptop. This method is used for static scenarios, as the averaging mode is not suitable for time-variant channels.

The second mode is a real-time acquisition mode, where every CIR is directly stored on the SSD memory cards attached to the RX head. In this way, a real-time data saving mode can be used and changes in the radio channel originated from movements in the environment or of the TX/RX mobility can be captured. In this case, the correlation between the send-out M-Sequence and the received one is done in a later post-processing steps after the data have been saved.

The measured data are calibrated with the back-to-back (B2B) measurements, to eliminate effects of the measurement system. More information about the calibration

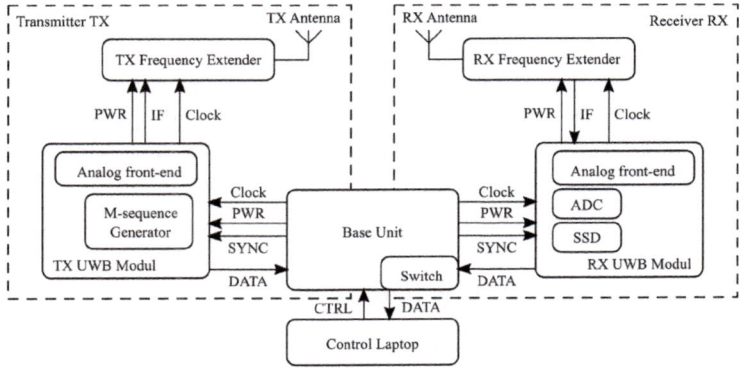

Fig. 34.1 The architecture of the correlational-based CS system operating at the 300 GHz [2]

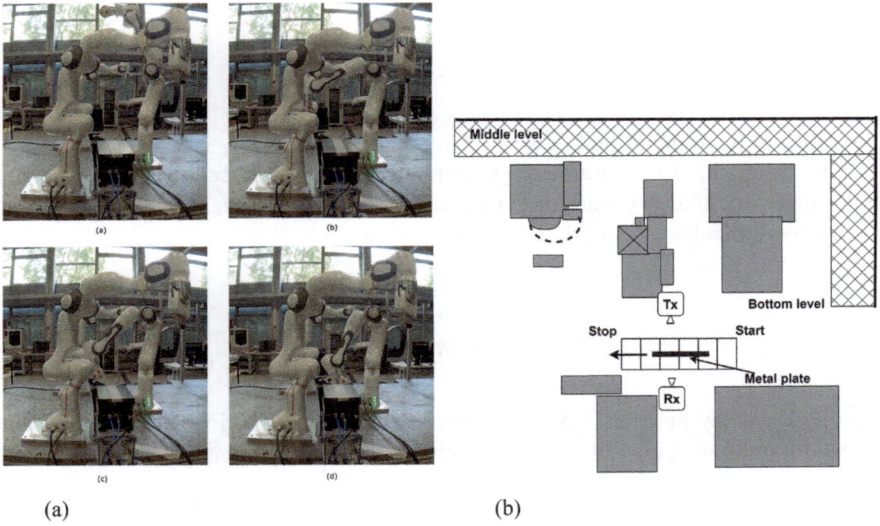

Fig. 34.2 Industrial-like environment measurement setup. (**a**) Blockage scenario robotic arm [13]. (**b**) Blockage scenario metal plate [13]

can be found in [12]. During post-processing, the data are structured as needed to calculate the needed parameters.

The remaining parts of this chapter are structured as follows: In Sect. 34.2, time-variant metrics are introduced and described. Exemplary evaluations are presented from measurements in two industrial environments; see Fig. 34.2. Section 34.3 introduces two concepts to measure multi-link connections. Conclusions are provided in Sect. 34.4.

34.2 Time-Variant Metrics

Multiple parameters can be used as a metric to describe time-variant radio channels. They are based on the time-variant impulse response of the channel, which are measured using the CS. In the following subsections, the most common metrics are presented.

34.2.1 Maximum Gain over Time

The maximum gain over time describes the changes in the strongest component of the CIR. In THz channels, the LoS component is mostly the one that has the highest impact on the communication channel and is mostly the strongest. The maximum gain is defined as

$$G_{\max}(t) = \max_{\tau} \{|h(t, \tau)|\}, \qquad (34.1)$$

with $h(t, \tau)$ the time-variant CIR.

By evaluating this parameter, time-related effects, caused, for example, by environmental changes, can be found and described. Often, it is used to detect moments of blockage or partly interruption of the LoS.

34.2.2 Time-Variant Power Delay Profile

The power delay profile (PDP) describes the power distribution of the signal with respect to the time delay of the multi-path signals. The time-variant PDP is defined as

$$P(t, \tau) = \sum_{i=1}^{N} |\alpha_i|^2 \cdot \delta(t - t_0) \cdot \delta(\tau - \tau_i), \qquad (34.2)$$

where A_i is the ith-MPC, δ denotes the Dirac impulse, and t_0 is the absolute time at which the PDP is considered.

Figure 34.3 shows the time-varying PDPs for a moving robotic arm and a moving metal plate, respectively. It can be seen that in case of the moving metal plate all signals are completely blocked with a difference of 40 dB between the blocked and non-blocked signals, whereas for the robotic arm, this difference is only 15 dB.

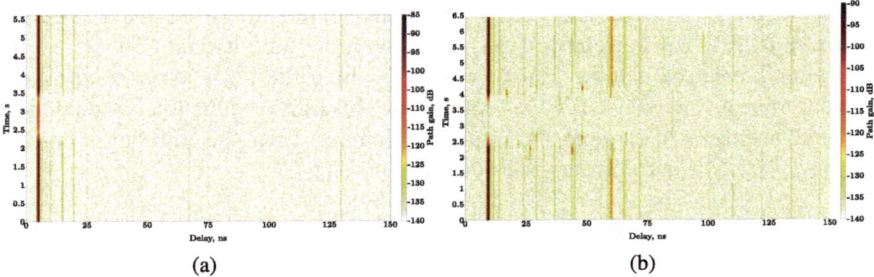

Fig. 34.3 Time-varying PDP measured in the two industrial environments shown in Fig. 34.2 [13]. (**a**) Moving robotic arm. (**b**) Moving metal plate

34.2.3 Doppler Shift

Another common time-variant parameter is the investigation of the Doppler shift. Due to the movement of either the surrounding or a TX or RX, a frequency shift f_d can occur:

$$f_d = \frac{v}{c} f_c, \tag{34.3}$$

where v is the velocity, c is the speed of light, and f_c denotes the carrier frequency.

Determining the Doppler shift requires a sampling rate, which is high enough to cover the velocity of movements. The CS used in the measurements presented allow the measurement of 17,900 impulse responses per second allowing the determination of the Doppler shift for typical velocities of walking humans.

34.2.4 Local Scatter Function Collinearity Metric

The wide-sense stationary and uncorrelated scattering (WSSUS) assumption is the base for most of the analysis in time-variant channel. As sub-THz applications are mainly discussed for indoor environments, the question of the validity of the uncorrelated scatering (US) criteria is of interest. Furthermore, the investigation of the regions where the wide-sense stationary (WSS) criteria can be assumed is of interest. In this subsection, two time-variant CS measurement campaigns and their metrics are presented based on [13].

Both measurements are blockage scenarios, where the LoS connection between RX and TX is partly or completely interrupted by a moving object, once a robotic arm (Fig. 34.2a) and once a metallic plate (Fig. 34.2b). Both scenarios were performed in an industrial-like environment.

For the analysis of the time-variant data, the overall data are split into smaller time segments, in which it can be assumed to be in a quasi-stationary region.

These segments are then analyzed with a discrete version on the local scattering function (LSF). The correlative CS has as output the time-variant CIR $h(t, \tau)$. By applying a discrete Fourier transform (DFT) along the delay domain, the time-variant transfer function $H(t, f)$ is computed. This transfer function is then divided in small segments in the time and frequency domain. Then, the correlation function $R_L(t, f, \Delta t, \Delta f)$ is calculated according to (34.4) [14]:

$$R_L(t, f; \Delta t, \Delta f) \triangleq \mathbb{E}\{H(t, f + \Delta f)H^*(t - \Delta t, f)\}, \tag{34.4}$$

where $\mathbb{E}$ denotes the expectation value, t and f denote the time and frequency, respectively, and Δt and Δf denote a time and frequency step, respectively.

The LSF is a time-frequency dependent scattering function of $H(t, f)$ [14]. The recorded signals are in a discrete format. Therefore, the discrete local scattering function (DLSF) can be described as

$$DLSF[n_t, n_f; \tau, \nu] = \sum_{n_{\Delta t}=0}^{N_{\Delta t}-1} \sum_{n_{\Delta f}=0}^{N_{\Delta f}-1} RL[n_t, n_f; n_{\Delta t}, n_{\Delta f}]$$

$$\exp\left(-j2\pi \left(\frac{\nu n_{\Delta t}}{N_{\Delta t}} - \frac{\tau n_{\Delta f}}{N_{\Delta f}}\right)\right), \tag{34.5}$$

where ν is the Doppler frequency, n_t and n_f are the indices for the time and frequency points, respectively, while $n_{\Delta t}$ and $n_{\Delta f}$ are the indices for the discrete time-lag and frequency-lag points. The $N_{\Delta t}$ and $N_{\Delta f}$ denote the number of discrete time-lag samples and discrete frequency-lag samples, respectively.

The LSF collinearity metrics in time and frequency domain are analyzed based on the method described in [15]:

$$\rho_{\text{LSF}}(n_{t_1}, n_{t_2}) = \frac{\sum_{0}^{N_\tau-1} \sum_{-N_\nu/2}^{N_\nu/2-1} \sum_{-N_f/2}^{N_f/2-1} C(n_{t_1}, n_f, \tau, \nu) \cdot C(n_{t_2}, n_f, \tau, \nu)}{||C(n_{t_1}, n_f, \tau, \nu)||_2 \cdot ||C(n_{t_2}, n_f, \tau, \nu)||_2}, \tag{34.6}$$

$$\rho_{\text{LSF}}(n_{f_1}, n_{f_2}) = \frac{\sum_{0}^{N_\tau-1} \sum_{-N_\nu/2}^{N_\nu/2-1} \sum_{-N_t/2}^{N_t/2-1} C(n_t, n_{f_1}, \tau, \nu) \cdot C(n_t, n_{f_2}, \tau, \nu)}{||C(n_t, n_{f_1}, \tau, \nu)||_2 \cdot ||C(n_t, n_{f_2}, \tau, \nu)||_2}, \tag{34.7}$$

where C denotes the LSF and $|| \cdot ||_2$ denotes the ℓ_2-norm. τ is the delay and ν is the Doppler frequency. Accordingly, N_ν, N_f, N_t, and N_τ is the number of Doppler frequency, frequency, time, and delay points, respectively.

Examples of measurements for the robotic arm scenario are presented in Fig. 34.4 for a slow and a fast movement of the robotic arm. As expected, the regions with

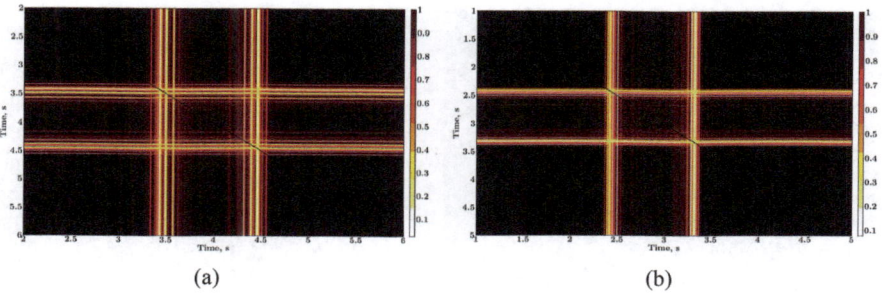

Fig. 34.4 Collinearity in the time domain for the movement of a robotic arm manipulator. (**a**) Slow movement. (**b**) Fast movement

high collinearity shrink both in time and frequency domain with increasing speed of the robotic movement.

Further evaluations of multipath propagation with a focus on static environments can be found in Chap. 33.

34.3 Multi-Link Connections

One strategy to better handle dynamic environments is the use of multi-link connections. They can be used to overcome the non-line-of-sight (NLoS) conditions due to blockage or environmental properties. Multi-link connections can be realized by multiple strategies:

- Using multiple TX
- The use of relays, either amplify and forward (AF) or decode and forward (DF) relays
- By using a strong reflection

In this section, an experimental setup that could be used to measure these different approaches is presented. In Sect. 34.3.1, a measurement scenario for the abovementioned strategies including exemplary measurements is presented. The following Sect. 34.3.2 described a measurement setup for measuring relay connections.

34.3.1 Multi-Hop Measurement

The chosen setup to investigate multi-hop communication was a NLoS corridor communication setup; see Fig. 34.5. The L-shape of the corridor makes sure to be in a NLoS communication case. To ensure a link between the two ends, an additional

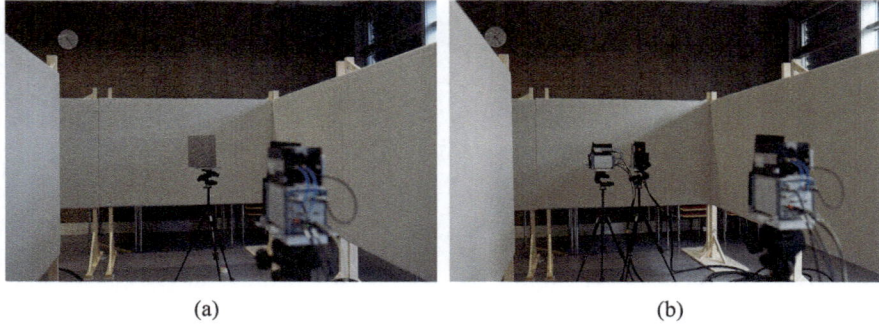

(a) (b)

Fig. 34.5 Multi-hop measurement scenario along the L-shape corridor. (**a**) Reflection scenario. (**b**) Multi-hop scenario

TX/RX combination or a metal plate as precise reflection was placed in the corner. In total, three different variants of the measurement setup were tested:

1. Only one TX and one RX at the ends of the corridor
2. A targeted reflection in the end of the corridor to forward the signal (Fig. 34.5a)
3. A multi-hop setup in which a second TX and RX were placed in the bend accordingly (Fig. 34.5b)

These three combinations were measured with and without interruption of the direct path. In the following, we will show exemplary results for variants (2) and (3).

The measurements were performed as point-to-point measurements at 300 GHz with two different types of antennas: high gain horn antennas with a HPBW of 8.5° and gain of 26 dBi and standard gain horn antennas with a HPBW of 30° and a gain of 15 dBi. The corridor was built out of plasterboard sheets, a material that is used in everyday construction for indoor walls. The L-shape was chosen to guarantee a NLoS connection between TX and RX on the ends of this corridor.

One of the limitations with these measurements is the HPBW of the used antennas. With the standard gain horn, the signal measured in variants (1) and (2) was too weak to emulate blocking due to the movements. Even with the high gain horn antenna, the signals were weak, as can be seen in Fig. 34.6, which shows the shadowing effect caused by a blocking effect in one of the corridors. In case of variant (2), only the strongest signal can be observed in the time-varying PDP, whereas the PDP for one of the two connections reveals several multipath signals.

34.3.2 Relay Measurement System

The measurement setup of variant (3) in Sect. 34.3.1 allows only the separate measurement of the two involved links.

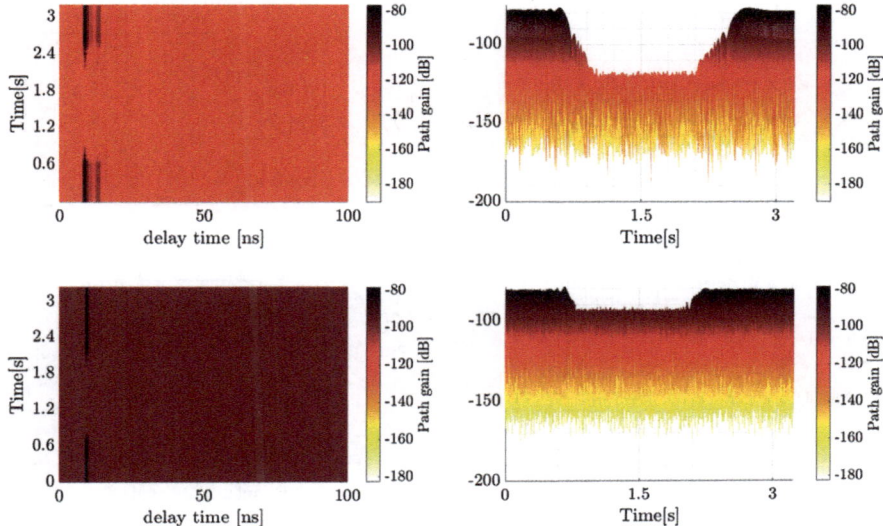

Fig. 34.6 Measurements for one of the links in variant 3) (see Fig. 34.5a) at the top and variant 2) (see Fig. 34.5b) at the bottom; time-varying PDPs are displayed on the left and time-varying amplitude of the first multipath signal are displayed on the right [16]

To allow an end-to-end measurement of both hops involved emulating a relay, a measurement system was built at Physikalisch-Technische Bundesanstalt (PTB). This allows the investigation of the feasibility of performing relay measurements using different CS systems. The measurements were conducted under controlled temperature and humidity conditions, with short separation distances ranging from 10 cm to 40 cm between the transmitting and receiving antennas.

This setup serves as a proof of concept for relay measurement possibilities between two different CS systems. The setup consisted of a correlational-based CS operating at a center frequency of 304 GHz with an 8 GHz bandwidth. This system generated a pseudorandom binary sequence (PRBS) waveform, transmitting and receiving signals through dedicated frequency nodes.

On the other side, a 300 GHz communication system was integrated using a (Keysight M8195A 65 GSa/s) arbitrary waveform generator (AWG), controlled via its designated software. The waveform was generated in MATLAB with suitable oversampling and uploaded in .csv format to the AWG.

Various waveforms were tested, including single-tone, modulated signals and a 12-bit PRBS waveform matching the built-in correlational-based CS sequence presented in Fig. 34.1.

The AWG output was up-converted to the sub-THz range using a frequency extension module from Virginia Diodes, Inc. A subharmonic mixer was employed to up-convert the AWG signal (DC—10 GHz) to 304 GHz. The local oscillator was generated by an external signal synthesizer at 16.89 GHz, amplified, and multiplied

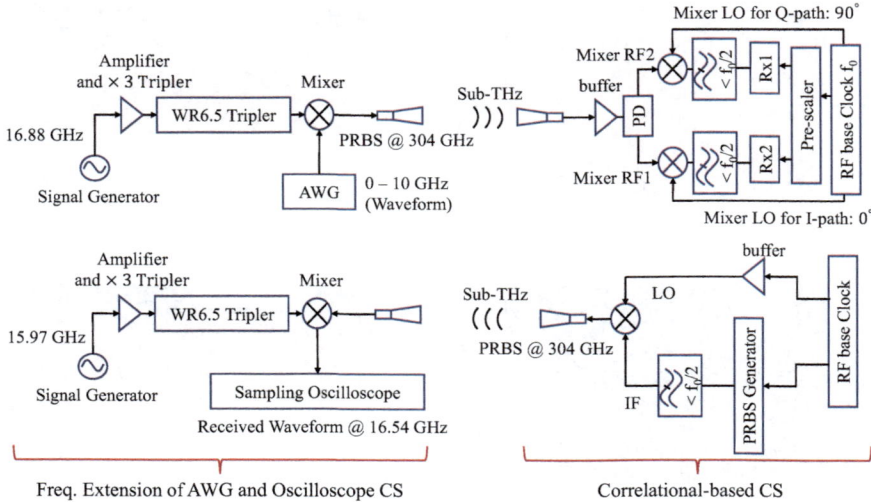

Fig. 34.7 Architecture of the relay measurement setup consisting of a correlation-based CS and an AWG and oscilloscope frequency extended CS

by a factor of 18 through sequential tripling stages, ensuring frequency alignment with the correlational-based CS. This frequency extension system was presented earlier for data transmission in [17].

For the reverse transmission, the correlational-based CS transmitted its sequence, which was down-converted using a receiving Virginia Diodes, Inc. module. The received sub-THz PRBS signal was down-mixed to an intermediate frequency (IF) and recorded using a (Tektronix DPO70000SX) sampling oscilloscope. The IF was set based on the signal synthesizer tuning at 15.97 GHz, resulting in an RF-LO difference of approximately 16.5 GHz, aligning with the oscilloscope's mid-band range. The relay measurement architecture is depicted in Fig. 34.7.

Although the measurements were performed over short distances, this setup provides a basis for understanding relay measurement principles. It facilitates the evaluation of path loss, hardware imperfections, and noise effects on transmitted and received signals. A photograph of the relay measurement setup is shown in Fig. 34.8.

During the measurements, several challenges arose, particularly in signal processing. One issue was the variation in sampling frequencies across different measurement instruments, requiring further investigation into their impact on received waveforms.

Additionally, minimizing noise in the retransmitted signal remains crucial, particularly in scenarios where extensive post-processing is not applicable. Investigations focusing on refining relay measurement techniques to improve performance and reliability have been ongoing at the time of the writing of this book.

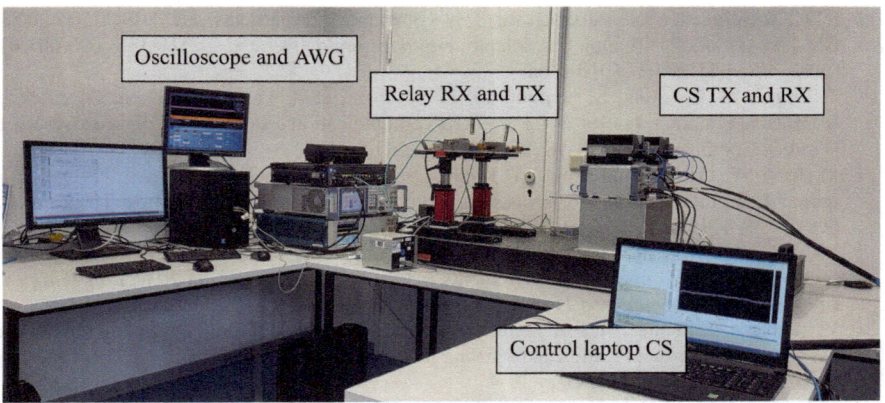

Fig. 34.8 Relay measurement setup, showing different CS equipment operating at 300 GHz range, used in this investigation

34.4 Conclusion

In this chapter, we presented concepts and metrics to measure time-varying channel at 300 GHz band, which have been applied to industry-alike environments. The measurements have revealed that blocking LoS is critical at these frequencies. Possible countermeasures to mitigate this effect are the setup of multi-link connections. A concept to perform measurements using this concept has been presented. First measurement results revealed a couple of challenges with this measurement concept, which need to be overcome.

The relay measurement investigation showed that reducing noise in the retransmitted signal represents a critical priority, especially in situations where there is limited scope for post-processing.

References

1. G.P. Sharma et al., Toward deterministic communications in 6G networks: state of the art, open challenges and the way forward. IEEE Access **11**, 106 898–106 923 (2023)
2. J.M. Eckhardt et al., Uniform analysis of multipath components from various scenarios with time-domain channel sounding at 300 GHz. IEEE Open J. Antennas Propag. **4**, 446–460 (2023)
3. D. He et al., Stochastic channel modeling for kiosk applications in the terahertz band. IEEE Trans. Terahertz Sci. Technol. **7**(5), 502–513 (2017)
4. A. Ghosh, M. Kim, THz channel sounding and modeling techniques: an overview. IEEE Access **11**, 17 823–17 856 (2023)
5. T. Kürner, T. Mittleman, D. M. Nagatsuma (eds.), *THz CommUnciations - Paving the Way Towards 1 Tbps.* Springer Series in Optical Sciences, Vol. 234 (2022)

6. ETSI, Identification of use cases for THz communication systems. ETSI ISG THz, ETSI GR THz 001 (2024) [Online]. Available: http://etsi.org/deliver/etsi_gr/THz/001_099/001/01.01.01_60/gr_THz001v010101p.pdf

7. T. Zugno, C. Ciochina, S. Sambhwani, P. Svedman, L.M. Pessoa, B. Chen, P.H. Lehne, M. Boban, T. Kürner, Use cases for terahertz communications: an industrial perspective. IEEE Wireless Commun. **32**(1), 90–98 (2025)

8. M. Jacob, S. Priebe, R. Dickhoff, T. Kleine-Ostmann, T. Schrader, T. Kürner, Diffraction in mm and sub-mm wave indoor propagation channels. IEEE Trans. Microwave Theory Tech. **60**(3), 833–844 (2012)

9. J.M. Eckhardt, C.E. Reinhardt, T. Doeker, E.A. Jorswieck, T. Kürner, Capacity analysis for time-variant MIMO channel measurements at low THz frequencies, in *2023 17th European Conference on Antennas and Propagation (EuCAP)*, Florence (2023), pp. 1–5

10. W. Keusgen, T. Eichler, Measurements of the time-variant indoor radio channel in the D-band at 160 GHz for communication and sensing, in *2024 4th URSI Atlantic Radio Science Meeting (AT-RASC)*, Meloneras (2024), pp. 1–4

11. M.D. Al-Dabbagh, D.A. Humphreys, T. Kleine-Ostmann, Investigating time-varying signal propagation at sub-THz frequencies using a VNA, in *2025 19th European Conference on Antennas and Propagation (EuCAP)* (2025), pp. 1–5

12. T. Doeker, J.M. Eckhardt, C.E. Reinhardt, T. Kürner, Time-domain channel sounder calibration at low terahertz band. IEEE Open J. Antennas Propag. **5**(6), 1598–1611 (2024)

13. V.V. Elesina, C.E. Reinhardt, L. Thielecke, T. Doeker, T. Kürner, Investigating the WSSUS assumption in 300 GHz time-variant channels in industrial environments. IEEE Open J. Veh. Tech. **5**, 1374–1385 (2024)

14. G. Matz, On non-WSSUS wireless fading channels. IEEE Trans. Wireless Commun. **4**, 2465–2478 (2005)

15. L. Bernadó, T. Zemen, F. Tufvesson, A.F. Molisch, C.F. Mecklenbräuker, The (in-) validity of the WSSUS assumption in vehicular radio channels, in *2012 IEEE 23rd International Symposium on Personal, Indoor and Mobile Radio Communications - (PIMRC)* (2012), pp. 1757–1762

16. C. Wächter, Zeitvariante Funkkanalmessung bei 300 GHz in einem Multi-hop Szenario (in German), Master's thesis, Technische Universität Braunschweig (2024)

17. C. Jastrow, K. Munter, R. Piesiewicz, T. Kürner, M. Koch, T. Kleine-Ostmann, 300 GHz transmission system. Electron. Lett. **44**(3), 213–214 (2008)

Chapter 35
Physical Layer Security

Tobias Doeker, Christoph Herold, and Daniel M. Mittleman

Abstract Due to the use of highly directional antennas in terahertz (THz) communication systems, it is often assumed that eavesdropping and attacks are not possible. However, various evaluations based on both measurements and simulations have shown the possibility of eavesdropping even with the use of highly directive antennas. In this chapter, eavesdropping aspects are discussed for three different scenarios: reflections within an indoor environment, reflections at scattering objects, and reflections caused by a moving person. For the analysis, the metrics of secrecy capacity and blockage are introduced.

35.1 Fundamentals of Eavesdropping

Eavesdropping and jamming are the most common attacks related to physical layer security. Jamming refers to an attack where a party outside the communication attempts to disrupt the communication between two interacting parties. Eavesdropping, on the other hand, refers to an attack where a third party attempts to intercept information being exchanged between two interacting parties. In this chapter, only eavesdropping will be discussed. In the context of eavesdropping, the interacting parties are typically referred to as "Alice" and "Bob," while the eavesdropper is called "Eve." In the following sections, two common metrics will be introduced first, before discussing various aspects and methods of eavesdropping.

T. Doeker (✉) · C. Herold
Institut für Nachrichtentechnik, Technische Universität Braunschweig, Braunschweig, Germany
e-mail: t.doeker@tu-braunschweig.de; c.herold@tu-braunschweig.de

D. M. Mittleman
School of Engineering, Brown University, Providence, RI, USA
e-mail: daniel_mittleman@brown.edu

© The Author(s) 2026
T. Kürner et al. (eds.), *Metrology for THz Communications*, Springer Series
in Optical Sciences 256, https://doi.org/10.1007/978-3-032-01986-8_35

35.1.1 Secrecy Capacity

The secrecy capacity (SC) illustrates the potential of an eavesdropping attack based on the signal-to-noise ratio (SNR). The normalized SC is given by Ma et al. [1]

$$\overline{c}_s = \frac{\log\left(1 + SNR_{\text{Bob}}\right) - \log\left(1 + SNR_{\text{Eve}}\right)}{\log\left(1 + SNR_{\text{Bob}}\right)} \tag{35.1}$$

where SNR_{Bob} and SNR_{Eve} denote the SNR for Bob and Eve, respectively. The normalized SC ranges between 0 and 1. An SC of 0 means that the SNR of Eve is equal to or higher than the SNR of Bob, implying that Eve can obtain the same information as Bob—the attack is successful. Conversely, an SC of 1 means that the SNR of Bob is significantly higher than the SNR of Eve, ensuring that Eve will receive no information—the attack fails. It should be noted that defining a precise threshold for a successful attack is challenging, as several other factors must be considered. However, this metric is effective for estimating the potential of an attack [1].

35.1.2 Blockage

It is obvious that the SC is 0 if Eve fully blocks the path between Alice and Bob. In this case, the SNR of Eve will definitely be higher than the SNR of Bob, as Bob does not receive any signal. However, in this scenario, Alice and Bob will notice the attack. Therefore, a second metric is used to illustrate the level of blockage at Bob, again referring to the SNR [1]

$$b = 1 - \frac{SNR_{\text{Bob}}^{\text{attack}}}{SNR_{\text{Bob}}^{\text{no object}}} \tag{35.2}$$

whereby the SNR during the attack $SNR_{\text{Bob}}^{\text{attack}}$ is compared to the SNR if no eavesdropper is present $SNR_{\text{Bob}}^{\text{no object}}$. The blockage b also has values between 0 and 1, where 1 means that the path between Alice and Bob is fully blocked and 0 means that the presence of an eavesdropper is not noticeable.

35.2 Eavesdropping Opportunities in Indoor Environment

With respect to THz communications and the corresponding usage of highly directive antennas, the assumption of a sparse wireless communication channel is

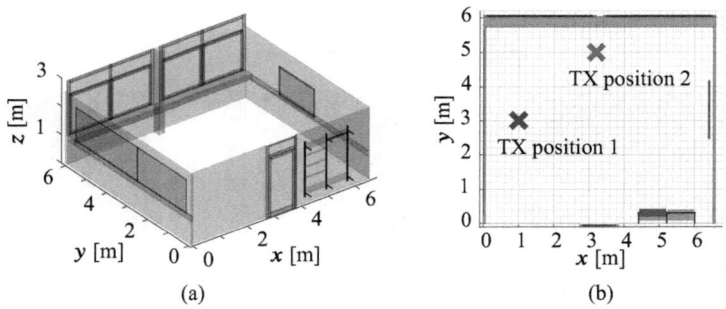

Fig. 35.1 Meeting room scenario [4]. (**a**) 3D view. (**b**) Top view

still prevalent. However, different investigations have shown that this assumption is not true [2, 3]. It is obvious that eavesdropping could therefore be possible due to the presence of multipath components (MPCs). For example, the opportunity for eavesdropping due to the environment is demonstrated in a meeting room scenario [4].

The scenario has dimensions of approximately $6.6\,\text{m} \times 6.25\,\text{m}$ and includes windows, a door with a metal frame, plasterboards, and a TV screen (see Fig. 35.1). To demonstrate the general possibility of eavesdropping, a ray-tracing simulation is first conducted to generate a map of the received power within the scenario. The simulations are conducted using the in-house developed Simulator for Mobile Networks of the Institute for Communications Technology at Technische Universität Braunschweig (TUBS) [5]. The transmitter (TX) is placed at two different positions within the scenario, and the received power at each position within the room is calculated. For the TX, a high-gain antenna with horizontal and vertical polarization is used, while the received power at each position within the room is calculated assuming an omnidirectional receiver (RX) antenna. The simulation results are shown in Fig. 35.2.

With horizontal polarization, it can be seen that many places within the room have quite high received power due to the sidelobes of the antenna. This is suppressed using vertical polarization, but even with vertical polarization, relatively high power can be received due to reflections within the room, as shown for position 2. In this case, the TX is facing the door, and strong reflections occur due to the metal frame of the door. Consequently, the eavesdropper could receive the signal from Alice even behind Alice.

To verify the simulation, the most promising positions are measured using the TUBS correlation-based time-domain channel sounder [2, 6]. For the TX, position 2 with horizontal polarization is chosen. For the RX, several positions along the y-axis, passing through the sidelobes of the TX antenna, as well as several positions along the x-axis behind the TX, are chosen (see Fig. 35.3). To emulate an omnidirectional RX, the sum of the highest path gains of each angular sampled measurement is calculated [4], following the state-of-the-art processing

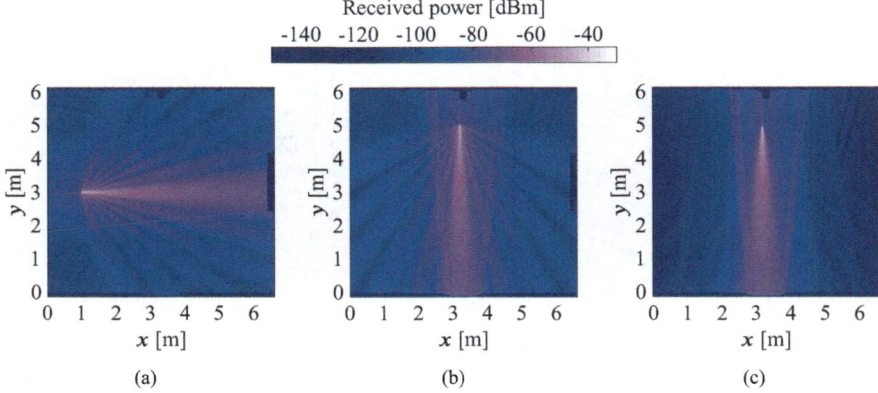

Fig. 35.2 Simulation results [4]. (**a**) Pos. 1, horizontal polarization. (**b**) Pos. 2, horizontal polarization. (**c**) Pos. 2, vertical polarization

Fig. 35.3 Measurement
points [4]

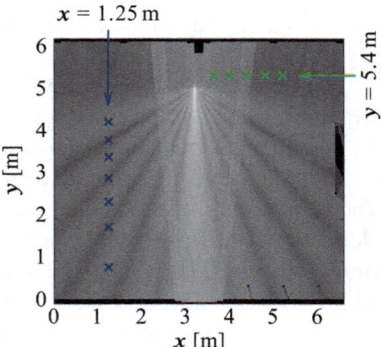

method for angular sampled measurements [7]. Figure 35.4 shows the additional path gain with respect to the free space path loss (FSPL) for both the simulation and the measurement. Along with this, Fig. 35.4 also shows the SC based on the measurement results.

It can be seen that the simulated and measured results fit very well. For the measurements along the y-axis, the SC goes down to 0.2. Even outside the sidelobes, the SC is lower than 0.4. According to the nearly linearly decreasing received power behind the TX, the SC nearly linearly increases from 0.2 up to almost 0.6. However, in both cases, it can be seen that Eve has a good potential to intercept the communication between Alice and Bob just because of the environment.

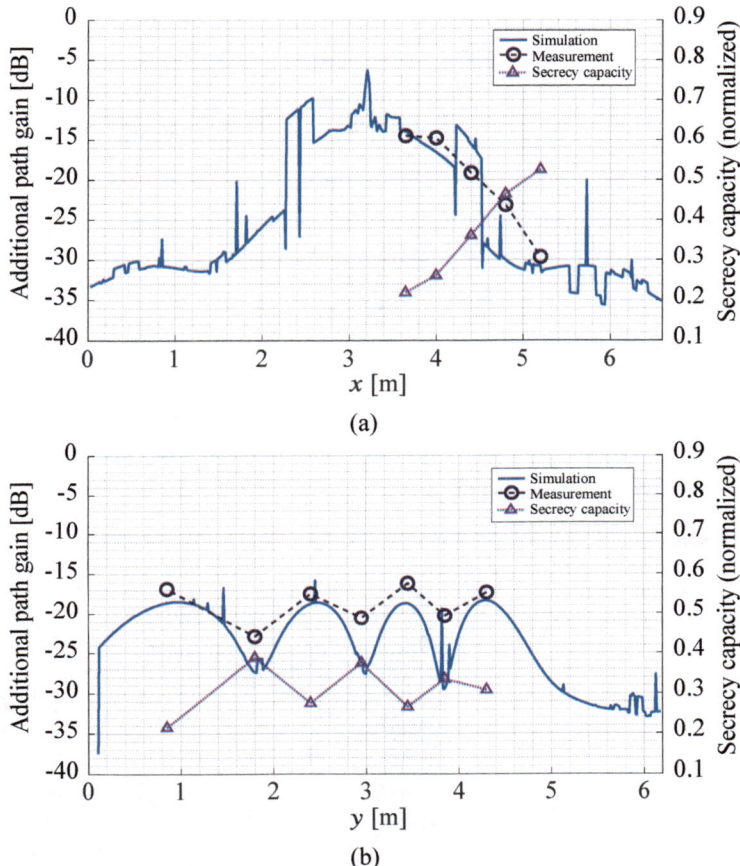

Fig. 35.4 Comparison between simulation vs. measurement and secrecy capacity [4]. (a) x-plane for $y = 5.4$ m. (b) y-plane for $x = 1.25$ m

35.3 Eavesdropping Opportunites with Scattering at Objects

To further develop the idea of the office scenario, the potential for eavesdropping will be emphasized by examining various objects [8], which are referred to in some literature as the "Coffee Cup Attack." Once more, the TUBS channel sounder is utilized to measure the received power at Bob and Eve. Alice and Bob are positioned on tables, representing, for example, two laptops communicating with each other. Eve is situated on a rail system next to the tables, occupying various positions along the line-of-sight (LOS)-axis from Alice to Bob while maintaining a constant distance from this axis. At nine different positions, which are grouped into three clusters of three with identical x-values but differing y-values, different scattering objects are placed. A schematic diagram of the setup, displaying the

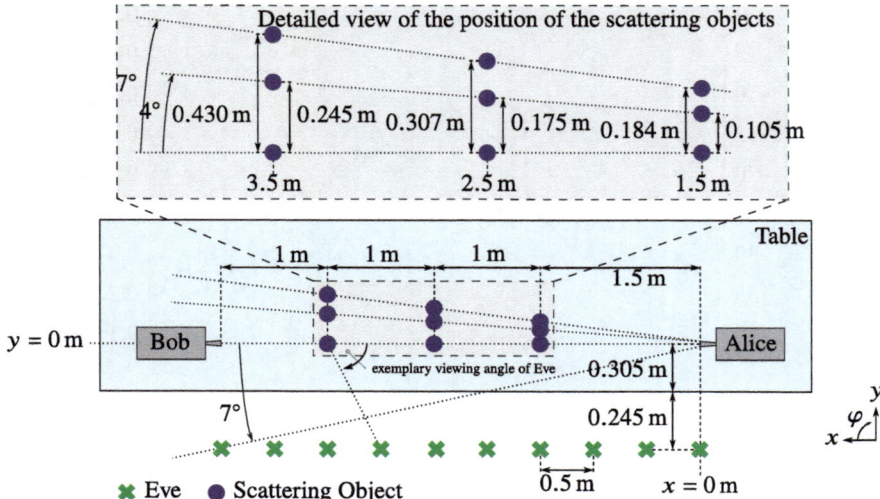

Fig. 35.5 Schematic top-down view of the measurement setup in the conference room [8]

Fig. 35.6 Partial illustration
of the measurement setup [8]

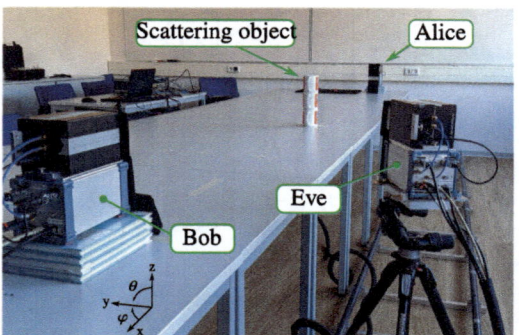

various positions, along with a representative photograph of the arrangement, is presented in Figs. 35.5 and 35.6, respectively. In this investigation, a laptop, a coffee cup, and a thermos are used as scattering objects.

It is evident that the blockage approaches nearly 1 when the scattering object is situated in the LOS path between Alice and Bob [8]. This observation holds true regardless of the positioning along the LOS-axis or the type of object used. When the object is placed outside the LOS path, the blockage reaches a maximum value of 0.14 and is 0 in most instances [8]. Here, "outside of the LOS path" signifies that the object is positioned at an angle equal to the half power beamwidth (HPBW) of the antenna concerning the LOS direction. Regarding the potential for eavesdropping, it is clear that Alice and Bob will undoubtedly notice the presence of the scattering object if it obstructs the LOS path. Conversely, they will not have the opportunity to

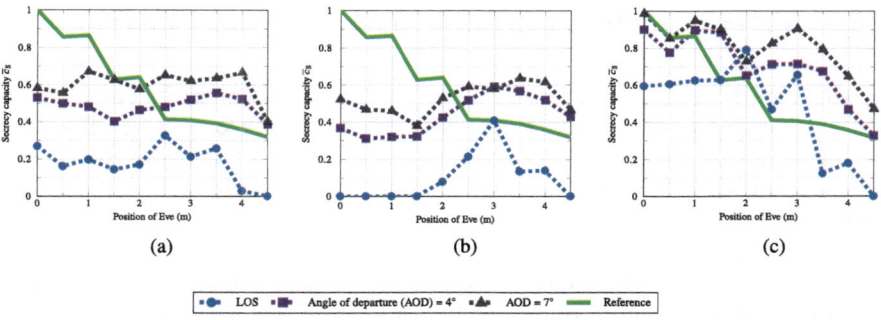

Fig. 35.7 SC for the different scattering objects at different positions based on the channel sounder measurements [8]. (**a**) Coffee cup placed at $x_{scat} = 1, 5$ m. (**b**) Thermos placed at $x_{scat} = 1, 5$ m. (**c**) Laptop 90° placed at $x_{scat} = 1, 5$ m

detect the scattering object if it is positioned outside the LOS path, which presents a favorable scenario for the eavesdropper.

In terms of the SC, it is observed that, in most cases, the SC is lower when Eve is oriented directly toward Alice compared to when Eve is directed toward the scattering object [8]. This implies that Eve will receive more power through the LOS component from Alice to Eve, even if Alice employs highly directive antennas. Notably, when the laptop serves as the scattering object, the scenario of "direct eavesdropping," where Eve is aligned with Alice, results in a lower SC than eavesdropping via reflection off the laptop [8]. Consequently, a laptop does not offer any advantage as a scattering object in terms of eavesdropping. Figure 35.7 illustrates the SC for all measured positions of Eve along the LOS-axis and for all three positions of the scattering object adjacent to Alice ($x = 1.5$ m). The plots for the remaining positions can be found in [8].

As described above, the SC value for the laptop is above the reference line in most cases, where the reference line indicates the SC in the absence of a scattering object and with Eve oriented directly toward Alice. Nevertheless, there are instances where the SC is 0, such as for the thermos, which occurs when the scattering object is positioned in the LOS path between Alice and Bob. It is apparent that in these situations, the received power at Eve exceeds that at Bob. However, for positions of Eve located less than 2 m away, the SC values are lower when the thermos and coffee cup serve as scattering objects, compared to the reference line—even when the scattering object is placed outside the LOS path. Thus, the potential for eavesdropping in the office scenario is enhanced by the presence of scattering objects, while simultaneously, Alice and Bob remain unaware of the object's existence.

35.4   Eavesdropping Opportunities due to Human Bodies

Extending the concept of scattering objects, this section examines the opportunities for eavesdropping involving moving objects within an office scenario. The most common moving object in such an environment is a human. Therefore, this section discusses the eavesdropping possibilities when a human body acts as a scattering object.

For this investigation, measurements were conducted using the TUBS channel sounder, in two different frequency ranges: 60 GHz and 300 GHz. Alice and Bob were positioned within an office scenario at a height of approximately 1 m for 60 GHz and about 1.26 m for 300 GHz. Perpendicular to the LOS path, Eve was positioned facing the middle of the LOS path between Alice and Bob, equidistant from this center point. A person was placed at the center point for the measurements. Figure 35.8 provides a schematic top-down view of the setup. It should be noted that the TXs and RXs for different frequency ranges are illustrated next to each other in Fig. 35.8, whereas in the actual setup, they are stacked vertically.

For 60 GHz, horn antennas with a HPBW of 35° and a maximum antenna gain of 13.8 dBi, referred to as *standard gain antenna (SGA)*, were used. For 300 GHz, two different types of antennas were employed: horn antennas with a HPBW of 30° and a maximum antenna gain of 14.9 dBi, also referred to as *SGA*, and horn antennas with a HPBW of 8.5° and a maximum antenna gain of 26.3 dBi, referred to as *high gain antenna (HGA)*.

Figure 35.9b shows the secrecy capacity, while Fig. 35.9a illustrates the blockage in the given scenario for various orientations of the human body relative to the coordinate system shown in Fig. 35.8. It is apparent that the human body fully blocks the LOS path between Alice and Bob, resulting in a blockage value of almost 1. However, for 60 GHz and human body orientations of 120° and 150°, the blockage value decreases to approximately 0.93. In these instances, the human body is oriented such that Bob still receives some signal from Alice due to diffraction

Fig. 35.8 Schematic top-down view of the setup

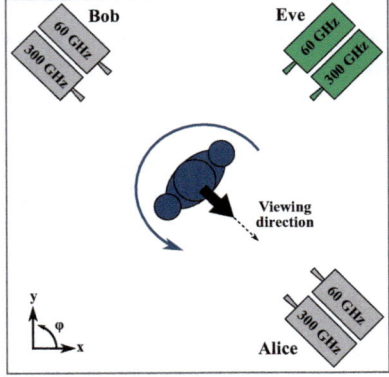

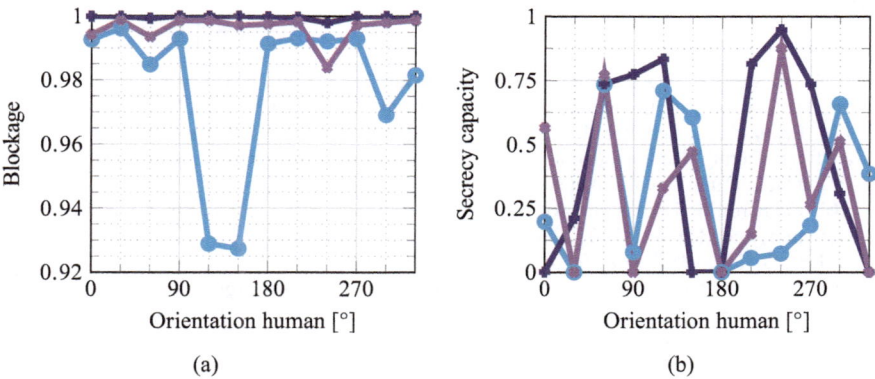

Fig. 35.9 Blockage and SC for different orientations of the human body (light blue: 60 GHz SGA; purple: 300 GHz HGA; pink: 300 GHz SGA). (**a**) Blockage. (**b**) Secrecy capacity

around the human body. This phenomenon is also reflected in the transmission loss data presented in [9].

Regarding the SC, no clear dependency on the orientation of the human body is observed. The SC values span almost the entire range from 0 to 1. For SC values of 0, it should be noted that the blockage is nearly 1, meaning the SC could be low because Bob did not receive any signal. Regardless of the frequency and antennas used, the SC drops to 0 when the human body is oriented at 180°. In this orientation, the reflection loss is minimal [9], allowing Eve to receive a relatively strong signal from Alice.

35.5 Summary

In the context of THz communications, it is often claimed that eavesdropping is nearly impossible due to the highly directional antennas used. However, the investigations presented here have shown otherwise. The potential for eavesdropping was explored within the framework of a meeting room scenario, examining possibilities through reflections off room components, common objects found in meeting rooms, and even the human body. Each of these areas provided potential eavesdropping opportunities. It is important to note, however, that in cases where, for example, the human body acts as a reflective object, Alice and Bob may indeed become aware of an attack due to significant blockage values. Nevertheless, the topic of physical layer security and eavesdropping remains highly relevant for THz communication, even with highly directional antennas.

References

1. J. Ma, R. Shrestha, J. Adelberg, C. Yeh, Z. Hossain, E. Knightly, J.M. Jornet, D.M. Mittleman, Security and eavesdropping in terahertz wireless links. Nature **563**(7729), 89–93 (2018)
2. J.M. Eckhardt, A. Schultze, R. Askar, T. Doeker, M. Peter, W. Keusgen, T. Kürner, Uniform analysis of multipath components from various scenarios with time-domain channel sounding at 300 GHz. IEEE Open. J. Antennas Propag. **4**, 446–460 (2023)
3. J.M. Eckhardt, T. Doeker, Lessons learned from a decade of THz channel sounding. IEEE Commun. Mag. **62**(2), 24–30 (2024)
4. C. Herold, T. Doeker, J.M. Eckhardt, T. Kürner, Investigation of eavesdropping opportunities In a meeting room scenario for THz communications, in *Proceedings of the 16th European Conference on Antennas and Propagation (EuCAP 2022)*, Madrid (2022), pp. 1–5
5. M. Schweins, L. Thielecke, N. Grupe, T. Kürner, Optimization and evaluation of a 3-D ray tracing channel predictor individually for each propagation effect. IEEE Open J. Antennas Propag. **5**(2), 495–506 (2024)
6. S. Rey, J.M. Eckhardt, B. Peng, K. Guan, T. Kürner, Channel sounding techniques for applications in THz communications: a first correlation based channel sounder for ultra-wideband dynamic channel measurements at 300 GHz, in *Proceedings of the 9th International Congress Ultra Modern Telecommunication and Control Systems Workshops (ICUMT)*, Munich (2017), pp. 449–453
7. W. Fan, F. Zhang, Z. Wang, O.K. Jensen, G.F. Pedersen, On angular sampling intervals for reconstructing wideband channel spatial profiles in directional scanning measurements. IEEE Trans. Veh. Technol. **69**(11), 13 910–13 915 (2020)
8. T. Doeker, C. Herold, J.M. Eckhardt, T. Kürner, Eavesdropping measurements for applications in office environments at low THz frequencies. IEEE Trans. Microw. Theory Tech. **71**(6), 2748–2757 (2023)
9. T. Doeker, D.M. Mittleman, T. Kürner, Scattering measurements with a moving human at 60 and 300 GHz, in *Proceedings of the 48th International Conference Infrared, Millimeter, and Terahertz Waves (IRMMW-THz)* (2023), pp. 1–2

Part V
Metrology for Simulation and Operation of THz Communications

Chapter 36
Simulation and Modelling of Electronic and Photonic Components

Dominik Wrana, Maxim Weizel, Simon Haussmann, Meysam Bahmanian, Ingmar Kallfass, and J. Christoph Scheytt

Abstract This chapter explores the crucial role of simulation and modelling of electronic and photonic components for terahertz (THz) systems. THz-related challenges already begin with setting up the signal generation and sampling parameters and continue with the realistic modelling of the electronic and photonic building blocks. Hereby, photonic components require not only the modelling of the optical signal propagation but also the modelling of the electronic interface in the THz regime. Furthermore, when advancing to the simulation of systems like fully integrated electronic transmit and receive frontends or photonically assisted analogue-to-digital converters (ADCs), it is up to the designer to find a suitable level of abstraction. Size, complexity, and available computational power versus accuracy must be taken into consideration and prioritized against each other.

36.1 Simulation Techniques for Linear and Nonlinear Components

An accurate prediction in simulation is essential for component and system design. The design procedure is either organized as top-down or bottom-up approach. In a top-down approach, realistic system performance is formulated and designed. The system simulation defines the performance of the individual components like filters, modulators, and amplifiers. However, in THz systems, often the bottom-up approach is followed, where components are developed first with respect to a certain goal. Once the performance of the individual component is known, the system is

D. Wrana · S. Haussmann (✉) · I. Kallfass
Universität Stuttgart, Institut für Robuste Leistungshalbleitersysteme, Stuttgart, Germany
e-mail: dominik.wrana@ilh.uni-stuttgart.de; simon.haussmann@ilh.uni-stuttgart.de; ingmar.kallfass@ilh.uni-stuttgart.de

M. Weizel · M. Bahmanian · J. C. Scheytt
Universität Paderborn, Schaltungstechnik (SCT)/Heinz Nixdorf Institut, Paderborn, Germany
e-mail: maxim.weizel@uni-paderborn.de; meysam.bahmanian@uni-paderborn.de; cscheytt@hni.upb.de

built. On hierarchical level, the top-down approach is more constructive because system performance is evaluated right from the start. However, bottom-up is applied often, as in THz-domain as cutting-edge technology, advances in single components already make substantial improvement to system performance.

36.1.1 Overview

Many electrical simulation techniques are available. Which simulator to choose is depending mainly on the field of activity. In simulation and modelling of optical and THz devices, the following simulators are commonly used:

- **S-parameter simulation** is a small-signal simulation and is valid for low signal amplitudes. The circuit is linearized in the present bias conditions. The linearized netlist is then solved with a closed mathematical solution. The simulation is non-recursive/not iterative and thus quite fast.
- **Transient simulation** is a large-signal time-domain simulation. When the input signal amplitudes would effectively change the bias conditions and prohibit the generation of a linearized netlist, transient simulation can be applied. The node voltages and currents are solved in the linearized initial condition, and iteratively, a solution for the consecutive timesteps is calculated. Although, for THz circuits, this is mostly not practicable, the simulation timestep should be chosen as a fraction of the smallest occurring signal period. In result, the computing time is too large.
- **Harmonic balance simulation** is a large-signal simulation. The user selects discrete frequencies, which are considered in the simulator. The linear elements of the netlist are solved in frequency domain, whereas the nonlinear elements are solved in time domain. An initial solution of the node voltages and mesh currents is assumed, and the solver minimizes the resulting error by iteratively adjusting the amplitudes and phases of the currents and voltages. After the error value falls below a given limit, the simulation is finished. If the chosen discrete frequencies do not reflect the complexity of the circuit, the simulator may not converge.
- **Envelope simulation** is a combination of transient simulation and harmonic balance simulation. Only the transients of the envelope of preselected discrete frequencies are considered. This simulation has become popular due to the possibility of representing modulation principles while maintaining moderate computational effort.
- **Electromagnetic simulations** (EM simulations) are used to simulate passive elements that are not available as lumped element models. As the structures are modelled as 3D-CAD geometries, meshed and solved with finite element method (FEM), high computational effort is present. Further, correct material characteristics and fabrication effects like edge roughness need to be accounted to correctly reflect the reality.

36.1.2 Complex Envelope Model

Assuming a signal at a carrier frequency of ω_c which is arbitrarily modulated in amplitude and phase, we can write this signal as having an in-phase $s_I(t)$ and a quadrature-phase $s_Q(t)$ components as

$$s(t) = s_I(t)\cos(\omega_c t) + s_Q(t)\sin(\omega_c t). \tag{36.1}$$

The signal $s(t)$ can be rewritten as

$$s(t) = \mathrm{Re}\left\{\left(s_I(t) - \mathbf{j}s_Q(t)\right)e^{\mathbf{j}\omega_c t}\right\}, \tag{36.2}$$

where $\mathrm{Re}\{\cdot\}$ denotes the real part of a complex number and $\mathbf{j}$ is the unit imaginary number (36.2) is useful in modelling and simulation of the signal, since it factorizes the signal into two components, a carrier at high frequencies and a complex envelope signal whose bandwidth is usually much lower than the carrier frequency. The complex envelope signal is defined as

$$\underline{s}_c(t) = s_I(t) - \mathbf{j}s_Q(t) = A(t)\cdot e^{\mathbf{j}\varphi(t)} \tag{36.3}$$

$$A(t) = \left|\underline{s}_c(t)\right| \tag{36.4}$$

$$\varphi(t) = \mathrm{phase}\left(\underline{s}_c(t)\right) \tag{36.5}$$

Co-simulation of optical and electronic components in a SPICE simulation can now be accomplished by converting the optical envelope behavior into a VerilogA model; see Sect. 36.4.

36.2 Electronic RF Frontends

This chapter discusses challenges and approaches for the modelling and simulation of monolithically integrated electronic THz transmit and receive frontends and their building blocks as well as system-level performance prediction. After an introduction to THz-related challenges, which gives an overview over the overall design procedure as well, we present simulation and characterization techniques for passives, active devices such as amplifiers and finally frequency-translating components like mixers or multipliers. In Sect. 36.3, we give an approach for the simulation of transmit or receive frontends with non-conformal layer stack.

36.2.1   THz-Related Challenges

In design of broadband systems, especially THz systems different hierarchical levels from device to system are pursued as indicated in Fig. 36.1. With increasing frequency, the effective wavelength becomes smaller, so even slight irregularities or small geometries are affecting the electrical performance. To reflect and accurately predict the circuit performance, full 3D electromagnetic (EM) simulation of the entire monolithic microwave integrated circuit (MMIC) is a crucial part of the design process. Exemplary coupling structures like combiners, but also passives, interconnects, matching networks, or transistor shells, are solved in EM 3D FEM simulators. The results of the 3D simulations are later incorporated as linear lumped elements in circuit design. The circuit design includes actives as well, which are included with, i.e., transistor models from a process design kit (PDK). Simulative characterization of linear as well as nonlinear behavior (see next subchapters) is conducted.

Exact prediction of the performance has the requirement for layout-based full 3D EM simulation for the MMICs. However, this is challenging due to the high complexity of MMICs, i.e., with large multiplication factors in the local oscillator (LO) chain. Two reasonable trade-offs are mentioned here: link-level (envelope) simulations of single components with broadband modulated data. Consequently, distortions induced by single components can be modelled. Secondly, based on the simulations of the components, abstracted/simplified models can be extracted to conduct system-level simulations. These do not reflect the complexity of the system but are able to predict the interaction and combination of the systems components and predict the susceptibility to individual impairments. Envelope system-level simulation can further be combined in co-simulation with transistor-level circuits when a more accurate analysis is needed.

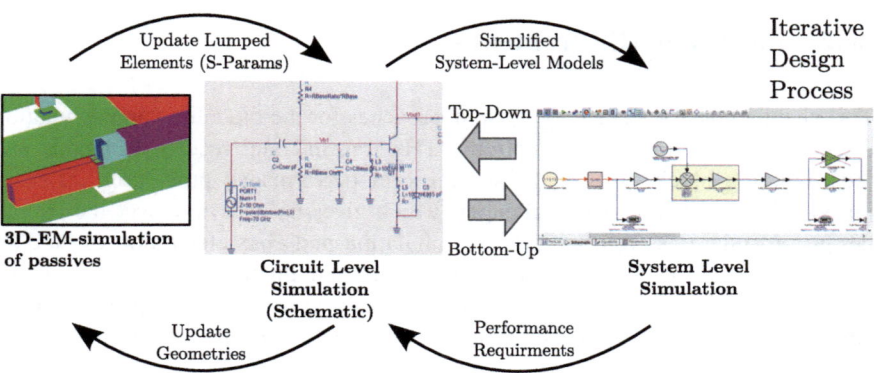

Fig. 36.1   Iterative design procedure for THz systems

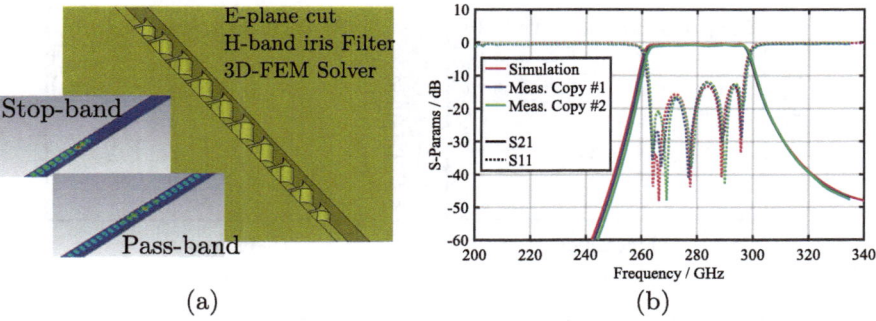

Fig. 36.2 Small-signal characterization obtained from EM 3D FEM simulation of a H-band waveguide filter. Comparison to measurements shows excellent agreement

36.2.2   *Passives*

Passive structures have linear characteristics. Small-signal S-parameters are used for simulations; other characteristics like gain/phase imbalance, group-delay ripple, etc. can be derived from that. As shown in Fig. 36.2, an excellent agreement between simulation and measurement can be achieved, when modelled correctly. However, for semiconductor processes, fabrication characteristics like the non-conformal layer stacks or under-etching have to be well known to reflect the geometries correctly in simulations.

For linear components, the superposition principle applies, so no intermodulation or harmonic generation is present. When the device is exited with superposition of input signals with frequency components, $f_1 - f_n$ only responses with the same frequency components can be observed at the devices outputs.

36.2.3   *Amplifiers*

Amplifiers are characterized in small-signal domain with S-parameters that capture gain variations, group-delay distortions, etc. Unlike passive elements, amplifiers are active, and nonlinear behavior can occur, either due to operation as class-B or -C amplifier or due to inherent nonlinearities such as trapping effects in the gate or similar physical effects. When exited with input signals with single or multiple frequency components, harmonics and/or intermodulations are generated. To incorporate these effects, additional metrices are used for characterization. Simple characterization method with single-tone excitation is the 1-dB compression point and saturated output power (single-tone excitation). The input signal can be described by

$$a_{1,\mathrm{cw}} = \hat{a} \cdot \sin(\omega \cdot t) . \tag{36.6}$$

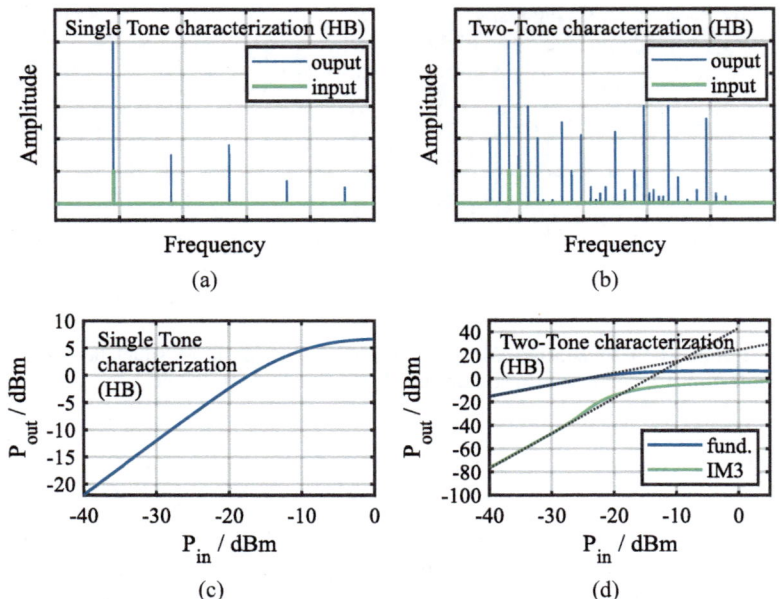

Fig. 36.3 Different amplifier characterization techniques. Single-tone characterization on the left results in 1-dB compression points and saturated output power. Two-tone characterization results in third-order intersect point by extrapolation of fundamental and IM3-curve

Based on this excitation with varying signal power, amplitude modulation (AM) of the output tone (AM-AM) and phase modulation (PM) of the output tone (AM-PM) can be obtained. However, only the fundamental tone is considered, as the harmonic multiples of the fundamental frequency, as indicated by Fig. 36.3a.

To include intermodulation distortion, multitone excitation, i.e., two-tone test can enhance the understanding of the design and distortion mechanisms, as visualized by Fig. 36.3b, c. The input port is exited with the superposition of two tones:

$$a_{1,\text{tt}} = \frac{\hat{a}}{2}\Big(\sin(\omega_1 \cdot t) + \sin(\omega_2 \cdot t)\Big). \tag{36.7}$$

The third-order intermodulation products, which are close to the input tones, are evaluated. By definition, third-order intermodulations are rising three times faster over the input power as the first-order intermodulation products. By linear extrapolation of the gain slopes, the intersection between both intermodulation products can be extracted and results in the third-order intersect point.

Approximated system-level models are often linear S-parameter models in conjunction with frequency-independent compression, polynomial model, Rapp-model, etc. These models can be derived from metrices like saturated output power, 1-dB compression point, third-order intersect point, etc.

36.2.4   Frequency Multipliers and Mixers

Frequency translating components like frequency multipliers or mixers are highly nonlinear devices, as the nonlinearity is exploited to produce intermodulation products.

Figure 36.4 shows exemplary different output products in a mixer. It is a simplified representation as at all ports intermodulation products can occur. Depending on the targeted performance, different frequency components are examined in order to reflect the component behavior in system simulations.

Conversion gain is the ratio of the incident wave at the input port with the input frequency and the outgoing wave at the output port with the desired output frequency. In up-converting mixers, intermediate frequency (IF) is the input and radio frequency (RF) is the output port. In down-converting mixers, it is vice versa.

Unwanted harmonics and intermodulation products are inherently produced as well. Especially in THz systems with large bandwidth and multiplication stages in the LO-path, spurious tones can lie in the band of operation. A versatile way of quantifying them is by relating the power of the spurious tone to the desired output signal. A representation in dBc is useful. **Leakage of LO and LO harmonics** are a special case of unwanted spurious signals, as they are often the most dominant leaked signals.

For the port reflections, **large-signal reflection coefficients** can be considered. In the large-signal reflection, the input power of the incoming wave (single tone) is put in relation with the power of the outgoing port wave, considering all frequency components. This is in contrary to the stand-alone S-parameter, which are just considering the a- and b-wave for a single frequency point, not considering possible intermodulation products.

Fig. 36.4 Exemplary representation of mixing and intermodulation products in a mixer simulation

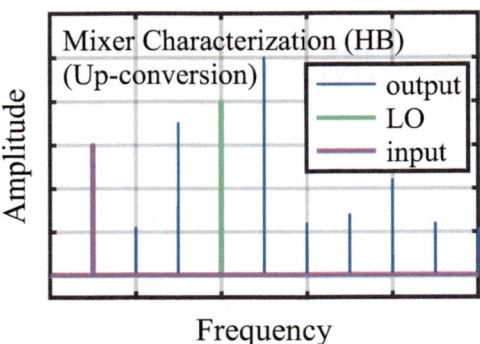

36.3   Conformal Stack EM Simulation for Nonplanar BEOL

With EM simulation being a crucial simulation technique during the design of THz circuits, an accurate replication of the technology layer stack and the resulting circuit geometries in the x, y, and z direction is ultimately necessary to receive precise results. Silicon-based semiconductor technologies such as CMOS or SiGe usually implement planar back-end-of-line (BEOL) processes with a multitude of different metal layers [1, 2]. In III–IV technologies, there are some such as [3], with a lower number of metal layers that do not incorporate planarization steps during wafer processing. The corresponding cross section of the BEOL including three variants of thin-film microstrip transmission lines (TFMSLs) is depicted in Fig. 36.5a. The omission of planarization steps leads to a topological relief of the wafer and the individual MMIC as indicated in Fig. 36.6, depending on whether a certain material is deposited at all or locally etched during fabrication of the BEOL. In [4], a scripted preprocessing methodology is introduced to accurately derive the relief of the BEOL based on the planar layer stack typically used in electronic design automation (EDA) tools as represented in Fig. 36.5b. The processed layout is the input to a FEM full 3D EM simulation tool.

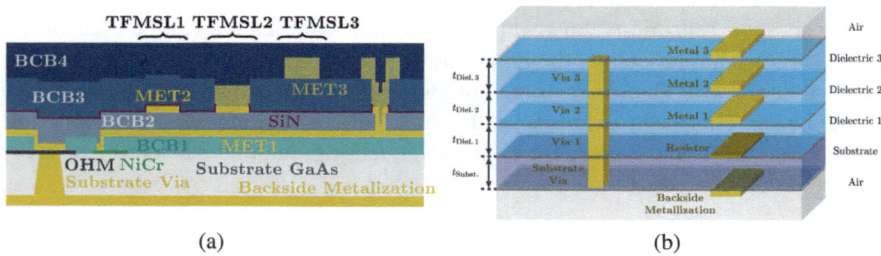

(a)   (b)

Fig. 36.5 (a) Exemplary BEOL cross section of a 35-nm InGaAs semiconductor technology used for implementation of THz frontend MMICs, adapted from [4], and (b) typical planar layer stack representation in EDA software

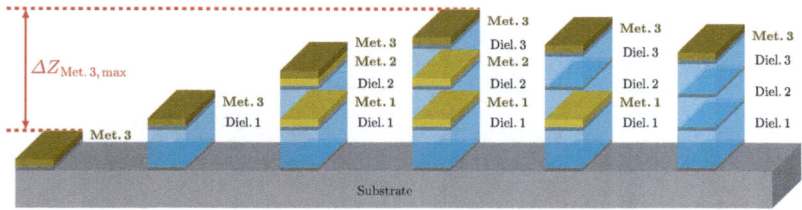

Fig. 36.6 Relief and height variations due to different layer combinations in a nonplanar BEOL

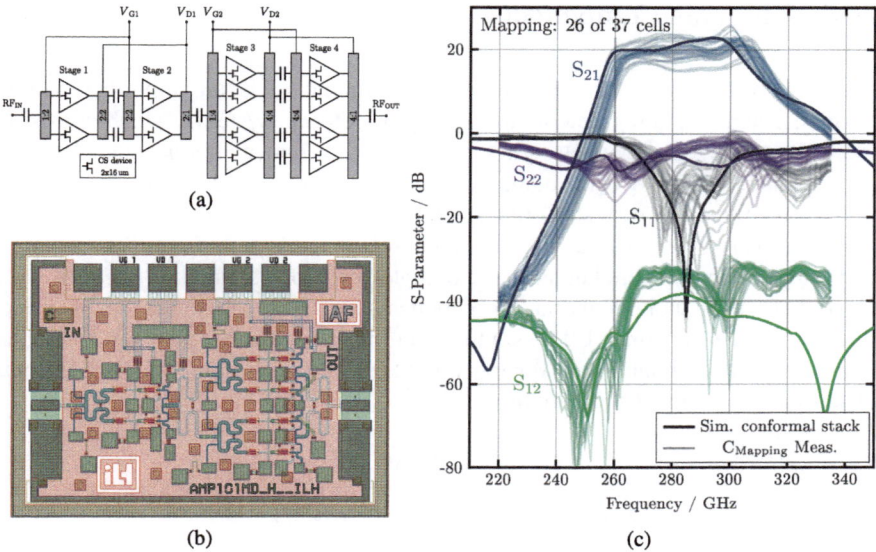

Fig. 36.7 Comparison of on-wafer measurement and 3D EM simulation results of a four-stage 300 GHz power amplifier MMIC

Exemplarily, Fig. 36.7 shows the block diagram and layout of a four-stage 300 GHz amplifier MMIC as well as its on-wafer characterization. The measurement results of in total 26 functional cells are displayed alongside the results obtained from 3D EM simulation of the entire MMIC using the conformal stack. The simulation takes into account all passive structures, including RF and direct current (DC) pads. Only the active transistor areas are cut from the layout, and ports for gate, drain, and source are inserted to later connect the obtained EM model with the transistor models on the schematic level. Overall, the results demonstrate an excellent agreement for all four S-parameters, which proves the accurate modelling of the conformal stack. Minor deviations are very well explained by process variations where certain parameters spread across the wafer. Furthermore, there are some specific effects which can not be represented by this simulation approach, like probe-to-probe coupling in the on-wafer measurement scenario. This among other things explains the steadily increasing deviation of the simulated and measured isolation S_{12} from 260 GHz on. However, structures outside of the MMIC are not easily implemented in the same tool which with its script-based geometry generation is bound to the EDA technology layer stack and does not allow for arbitrary geometries outside of it.

36.4   Photonic Components

Creating models for electronic-photonic co-simulation is an important part of designing and validating new innovative optoelectronic systems, which consist of more than a few components. The extremely high optical frequencies, 1550 nm corresponds to 193.41 THz, compared to the electronic signals in the 0 Hz to 1 THz range, require very short timesteps for the simulator used, if the full optical waveform is to be simulated. To circumvent this time-consuming process, a simple and effective way to simulate is to use complex envelope simulation (Sect. 36.1.2). Nevertheless, the individual optical components need to be simulated in a 3D finite-difference time-domain (FDTD) simulation tool first to extract their models. After which, e.g., a combination of VerilogA and SPICE models can be used to execute electronic-photonic co-simulations.

36.4.1   Setting Up the VerilogA Environment

To co-simulate photonic components with electronic components, new VerilogA natures and disciplines have to be defined in a file with, e.g., the name: "optodef.vams." The first is the E-field in V/m, which corresponds to a "potential," and the Poynting nature in dBm, which corresponds to a "flow." The optical signal is then combined into a vector with three elements, representing the real part of the electrical field, the imaginary part of the E-field, and the wavelength. With this setup, only transverse electric (TE) wave propagation is modelled, but there are components which can, e.g., create a TE-transverse magnetic (TM) mode conversion. To model this behavior, two more dimensions have to be added: real part of the TM wave and imaginary part of the TM wave, resulting in a five-component vector. Additionally, reflections can be introduced by also modelling backward propagating light, increasing the dimensions to nine. For simplicity, in the following, we will focus on components without or with negligible mode-conversion behavior and just describe the implementation procedure for the TE mode propagation discarding backward propagating light. The implementation is inspired by the works in [5–7], where especially in the appendix of [6] many code snippets of important optical elements can be found. The authors of [8] have included their code in the publication and uploaded it on GitHub.

36.4.2   Light Source

The light source converts an electronic input voltage V_{in} into an optical E-field with a proportional power. A reasonable choice is to limit V_{in} between 0 V and 1 V, where 0 V corresponds to a minimum laser power of P_{min} and 1 V corresponds to a maximum laser power P_{max}. The amplitude of the E-field (36.9) equals the square root of the light power in (36.8):

$$P_{out} = (P_{max} - P_{min}) \cdot V_{in} + P_{min} \tag{36.8}$$

$$E_{out} = \sqrt{P_{out}} \tag{36.9}$$

The variables, which can be set by the user, are P_{max}, P_{min}, the phase of the outcoming light φ_{rad} in radians (optionally in degree) and the wavelength λ. Real and imaginary part of the E-field can be calculated by (36.10) and (36.11). Together with the wavelength λ, this is the three-component vector, which will be passed to the next block:

$$\mathrm{Re}\{\underline{E}_{out}\} = E_{real} = E_{out} \cdot \cos(\varphi_{rad}) \tag{36.10}$$

$$\mathrm{Im}\{\underline{E}_{out}\} = E_{imag} = E_{out} \cdot \sin(\varphi_{rad}) \tag{36.11}$$

Additionally, a temperature-dependent shift in wavelength can be modelled (36.12), where $temperature is a VerilogA directive and returns the current simulation temperature T_{sim} in Kelvin. T_{nom} stands for the nominal temperature, the laser is specified at, λ_{nom} is the nominal wavelength, and k is a temperature coefficient:

$$\lambda = \lambda_{nom} + k \cdot (T_{sim} - T_{nom}) \tag{36.12}$$

36.4.3 Mode-Locked Laser

A mode-locked laser (MLL) can be modelled by applying the corresponding power envelope function $P(t)$ to the light source from Sect. 36.4.2. Usually, this would be a sech^2 pulse shape, as seen in (36.13). P_0 is the peak power and τ_p the pulse width. τ_p can be replaced by the more commonly used full width at half maximum (FWHM) τ_{FWHM} pulse width definition with the relation in (36.14) [9, p.72]:

$$P(t) = P_0 \cdot \mathrm{sech}^2 \left(\frac{t}{\tau_p} \right) \tag{36.13}$$

$$\tau_{FWHM} = 2 \ln \left(1 + \sqrt{2} \right) \tau_p \approx 1.7627 \tau_p \,. \tag{36.14}$$

Mostly, the user wants to specify the average pulse power P_{avg} and the repetition rate T_{rep}. Inserting these values into (36.15) gives the pulse energy, which in turn can be inserted into (36.16) to calculate the peak power:

$$E_p = P_{avg} \cdot T_{rep} = \int_{-\infty}^{\infty} P(t)\,dt = 2P_0\tau_p \tag{36.15}$$

$$P_0 = \frac{1}{2} \cdot \frac{E_p}{\tau_p}. \tag{36.16}$$

In the same way, any arbitrary envelope function can be defined in its own VerilogA block and applied as an input to the light source.

36.4.4 Phase Shifter

A phase shifter shifts the phase of the incoming light by a certain amount $\Delta\varphi_{shift}$, which is commonly modelled as a function of the applied voltage V_{in}. Additionally, an insertion loss IL is introduced, as can be seen in (36.17) and (36.18). $\Delta\varphi_{shift}(V_{in})$ can be an arbitrarily complicated function, so how should this function look like?

$$E_{real,out} = IL \cdot \sqrt{E_{real,in}^2 + E_{imag,in}^2} \cdot \cos\left(\varphi_{rad,in} + \Delta\varphi_{rad,shift}\right) \tag{36.17}$$

$$E_{imag,out} = IL \cdot \sqrt{E_{real,in}^2 + E_{imag,in}^2} \cdot \sin\left(\varphi_{rad,in} + \Delta\varphi_{rad,shift}\right) \tag{36.18}$$

In an integrated silicon photonic platform, the free-carrier plasma dispersion effect is used to tune the phase. The geometry of a phase shifter can be described as a p-n or p-i-n diode inside a rib waveguide. Therefore, in [10], we have proposed a novel concept, where in the optical domain the phase shift is modelled as a function of charges, which resemble actually the physical behavior in the waveguide. In the electrical domain, the p-n junction is modelled with its parasitics. As an example, a forward-biased phase shifter is chosen, and a VerilogA/SPICE block diagram can be seen in Fig. 36.8. The VerilogA code for the charge-to-phase converter $(q/\Delta\phi)$, the diffusion capacitance C_d, and waveguides can be found in [10].

36.4.4.1 Electrical Modelling

The equivalent electrical circuit can be seen in Fig. 36.8 in black. It models the p-i-n diode with an ideal main diode D_{main} and contact resistance R_{cont} in series. This would be quite enough to model DC and low-frequency operation. To add high-frequency behavior, a contact inductance L_{cont} and an RC network, which takes

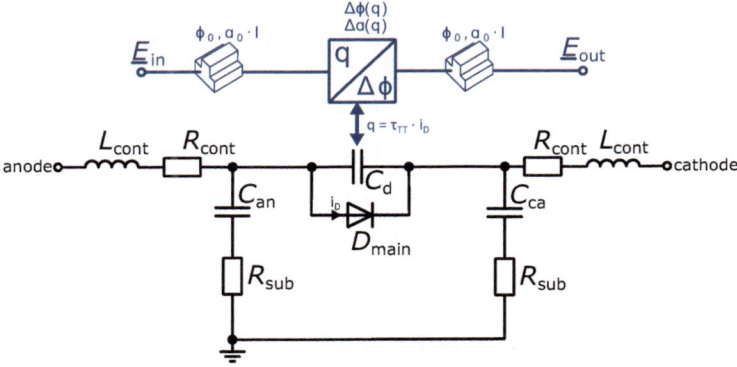

Fig. 36.8 SPICE- /VerilogA-based model of a forward-biased phase shifter. (Adapted from [10])

into account the parasitic capacitances C_{an} and C_{ca} from anode and cathode to the silicon substrate plus the substrate resistance R_{sub}, are added. Furthermore, a junction C_j and diffusion C_d capacitance are required to fully model a diode. Since, for a forward-biased diode, the junction capacitance does not occur, as long as no space charge region is built, it is omitted in the model and not shown in Fig. 36.8. The connecting element between the optical and electrical domain is the diffusion charge q, which is dependent on the diode current i_D and the forward transit time τ_{TT} [11]. τ_{TT} is voltage dependent and can be modelled with a tanh function, and i_D follows the Shockley equation, such that finally the charge is found by (36.19), where v_D is the applied voltage over the diode:

$$q = \tau_{TT} \cdot i_D = \tau_{TT}(v_D) \cdot i_D(v_D) \tag{36.19}$$

The diffusion capacitance can in turn now be calculated with (36.20) and the VerilogA code would incorporate (36.21). The full derivation can be found in [10]:

$$C_{diff} = \frac{dq}{dv_D} \tag{36.20}$$

$$i_{Cdiff} = C_{diff}(v_D) \cdot \frac{dv_D}{dt} \tag{36.21}$$

36.4.4.2   Photonic Modelling

In Fig. 36.8, the optical part of the model is depicted in red. The incident light undergoes a constant phase shift Φ_0 and length-dependent attenuation $\alpha_0 l$. A charge to phase conversion block introduces a linear charge-dependent change in the effective refractive index (36.22) of the waveguide, which leads to a linear charge-dependent change in phase (36.23). Furthermore, the block introduces a linear charge-dependent loss (36.24). Again, the details can be found in [10]:

$$\Delta n_{\text{eff}}(q) \propto q \tag{36.22}$$

$$\Delta \Phi_{\text{eff}}(q) \propto q \tag{36.23}$$

$$\Delta \alpha_{\text{eff}}(q) \propto q \tag{36.24}$$

36.4.5 Mach-Zehnder Modulator

An MZM can be built, e.g., out of two phase shifters driven by a differential signal (V_{s+}, V_{s-}), as shown in Fig. 36.9. R_G represents the generator impedance. Here, the forward-biased phase shifters from Sect. 36.4.4 are used. Additionally, an equalizer circuit, which consists of an RC-filter (R_{eq} and C_{eq}), is introduced. The voltage drop across the RC network reduces the voltage over the phase-shifter diode for low frequencies. With increasing frequency, the RC network impedance decreases, and the voltage drop reduces. Thereby, a high-frequency peaking is introduced [12]. This increases the bandwidth at the cost of reduced efficiency, which is tolerable due to the inherently large efficiency of forward-biased phase shifters.

36.4.6 Photodiode

Finally, special care has to be taken when modelling photodiodes. They are the connecting element between the optical and electrical world converting light into

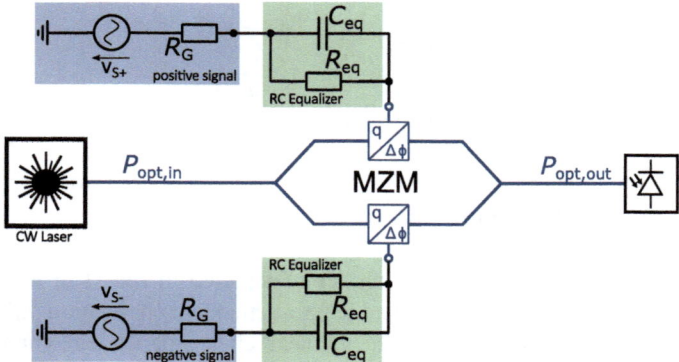

Fig. 36.9 MZM based on forward-biased phase shifters and RC-equalizer network

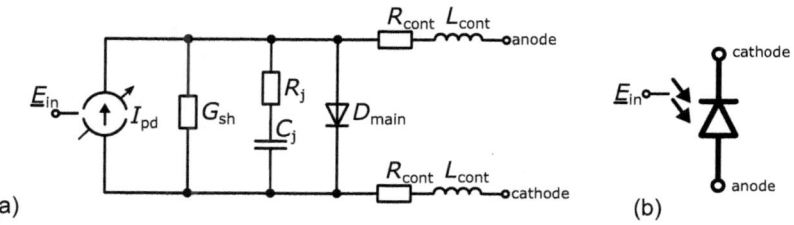

Fig. 36.10 (a) High-frequency model of a reverse-biased integrated p-i-n photodiode. (b) Top-Symbol as seen in the simulator

a current. In the simplest form, the current corresponds to the absolute value squared of the E-field times the responsivity of the photodiode R_{pd}; see (36.25). Obviously, this does not consider any nonlinear effects and frequency behavior of a real photodiode and can serve only as an approximation for low- or single-frequency applications:

$$I_{pd} = R_{pd} \cdot |\underline{E}_{in}|^2 = R_{pd} \cdot \left(E^2_{real,in} + E^2_{imag,in} \right) \qquad (36.25)$$

A more elaborate model can be seen in Fig. 36.10. It is partially adapted from [13]. R_{cont} and L_{cont} represent the contact resistance and inductance, R_j and C_j the junction resistance and capacitance, G_{sh} the shunt resistance, and D_{main} the p-n or p-i-n diode behavior, e.g., the Shockley diode equation. I_{pd} is a VerilogA block performing the operation shown in (36.25), where optionally the responsivity R_{pd} can be modelled wavelength dependent. In contrast to Sect. 36.4.4.1, since a high-speed photodiode is usually operated in reverse bias condition, the diffusion capacitance C_{diff} is omitted, and only the junction capacitance C_j needs to be modelled. An even more elaborate model can be found in [14].

36.5 Examples

36.5.1 Photonic ADC Simulation

As we have seen in this chapter, there are many ways to conduct electronic-photonic co-simulations. Looking at it from a system-level perspective, including hundreds of elements, the envelope simulation is considered the most efficient. Though we have sketched how to set up a VerilogA simulation in Sect. 36.4 before, it can be prone to errors and might not be the fastest prototyping solution. One has to rely on publications, write your own source code, or use open-source databases. But it seems like there is no code base with a large community yet. On the other hand, commercial tools are more likely to undergo strict quality control. The downside

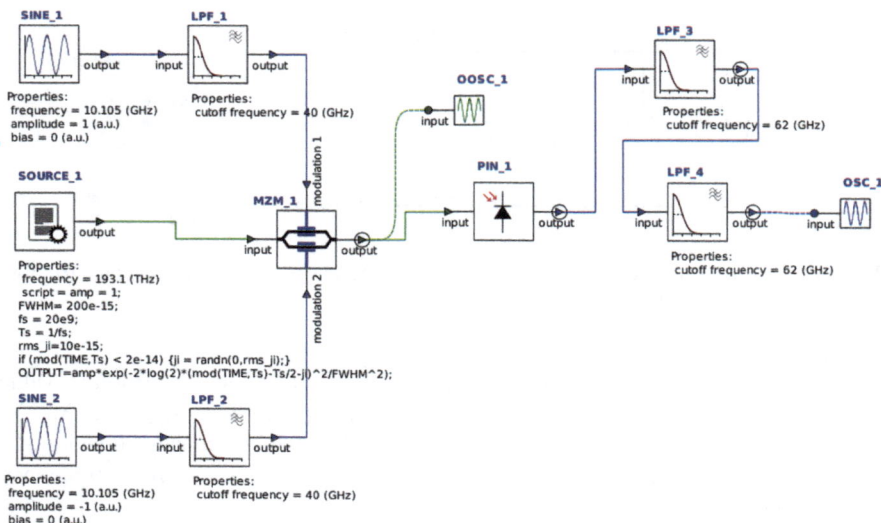

Fig. 36.11 Lumerical Interconnect model of an MZM sampler. (Adapted from [15])

compared to open-source solutions is a loss in flexibility and looking into the actual implementation.

Figure 36.11 shows a simple model of an MZM-based sampler frontend in Lumerical Interconnect [15]. The block *SOURCE_1* implements an MLL with, in contrast to (36.13), a Gaussian envelope, a FWHM of 200 fs, and a repetition rate of 20 GHz. Additionally, a root mean square (RMS) jitter of 10 fs is assumed. A differential single-tone input signal (*SINE_1*,*SINE_2*) is applied to the MZM (biased at quadrature), where *LPF_1* and *LPf_2* are modelling a 40 GHz single pole roll-off. This was corresponding to the fastest all-silicon MZMs available in a commercial large-scale silicon photonic technology by the time of the publication of [15]. A 50 Ω p-i-n photodiode is detecting the modulated input pulse train adding shot noise and thermal noise. Thereafter, a receiver circuit (*LPF_3*,*LPF_4*) is modelled in a simplified manner as a second-order low-pass filter with total 3 dB bandwidth of 40 GHz.

The time-domain output signal can be seen in Fig. 36.12a, and since the MLL pulse is extremely short in comparison to the receiver (RX) bandwidth, what can be seen is essentially the RXs weighted impulse response. This signal is then analyzed at the timesteps; the original MLL pulse train would have its maximum. Analytical derivations show that the maximum SNR (36.26) is limited by the shot noise (36.27), thermal noise (36.28), modulation index m, and the transimpedance amplifier (TIA) transfer function. Here, the modulation index m is defined as the percentage of the input signal amplitude relative to V_π of the MZM. R is the responsivity, ω_c the cutoff frequency of 62 GHz, E_p the pulse energy, P_{avg} the average optical power, q the

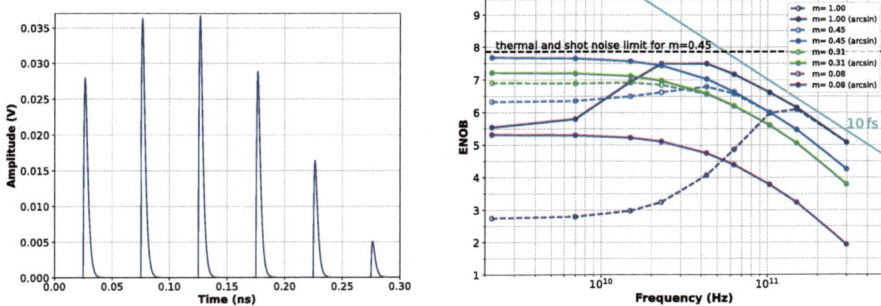

Fig. 36.12 (**a**) Time-domain waveform (**b**) ENOB. (Adapted from [15])

elementary charge, k_b the Boltzmann constant, Δf_{TIA} the receiver noise equivalent bandwidth, and Z the TIA transimpedance. Detailed derivation can be found in [15]:

$$\text{SNR} = \frac{\left(R \cdot \omega_c/e \cdot m \cdot E_p\right)^2 / 2}{q R E_p \cdot (\omega_c/e)^2 + 4k_b T \Delta f_{\text{TIA}} / \text{Re}(Z)} \tag{36.26}$$

$$\sigma_s^2 = 2q R P_{\text{avg}} \cdot \frac{1}{2T_s} T_s^2 (\omega_c/e)^2 = q R E_p (\omega_c/e)^2 \tag{36.27}$$

$$\sigma_{th}^2 = 4k_b T \Delta f_{\text{TIA}} / \text{Re}(Z) \tag{36.28}$$

Figure 36.12b shows the simulation results for different modulation depths, where the dashed lines represent the effective number of bits (ENOB) at the output without post-processing and solid lines the ENOB with arcs in post-processing. Post-processing is used to invert the nonlinear sine-shaped input to output characteristic of an MZM-biased in quadrature (36.26) gives an upper bound and matches really well with the simulation results obtained.

36.5.2 Transceiver Modelling

In the following, a system simulation of an electronic THz system, based on an envelope simulation, is presented. Figure 36.13 is showing the building blocks of a static point-to-point link. The TX consists of the random data generation and a modulator block, which converts the signal to an envelope signal. The modelled TX frontend components follow: IF-amplifier/attenuator, mixer stage consisting of a LO-chain with multipliers, a single-sideband mixer, and a power amplifier. In this example, the power amplifier is modelled with a memoryless polynomial AM-AM and AM-PM model. The path consists of a frequency-flat attenuator, an additive white Gaussian noise (AWGN) source, and a channel model, which is modelled

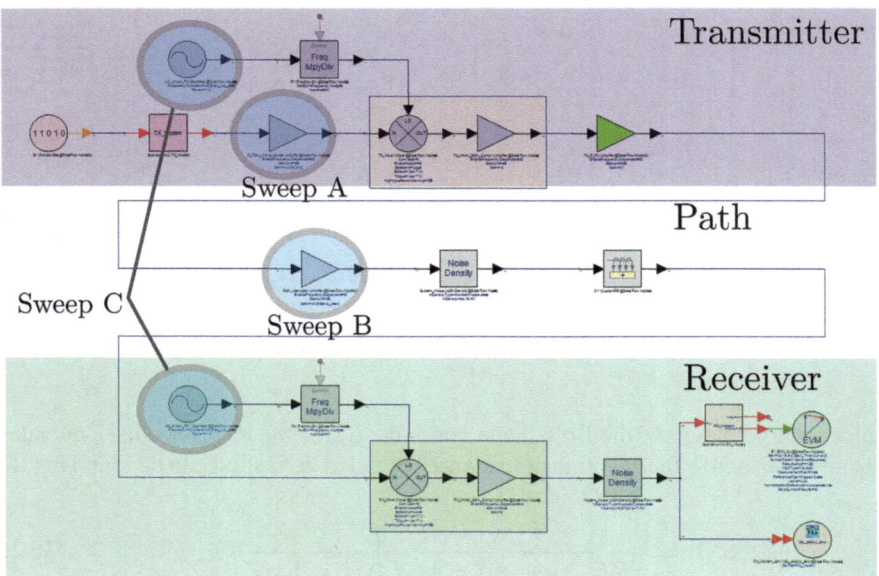

Fig. 36.13 Block diagram of a THz system in SystemVue. The model is divided in three parts: transmitter, path, and receiver. The blue circles indicate exemplary investigation of single impairments

by using a finite impulse response (FIR) filter implementation. The channel model itself was extracted from measurements. The receiver consists of the single-sideband down-conversion mixer, also with frequency multiplier in the LO-chain, and finally, the digital demodulator block. Based on this model, a sensitivity analysis to account for dominant impairments can be conducted. Selected investigations can be observed in Fig. 36.14. The first sweep (Fig. 36.14a) shows the influence of the TX compression on the signal quality. Other parameters are held constant. Major impairments are AWGN and TX linearity. The second sweep (Fig. 36.14b) sweeps the path loss of the wireless channel. Here, again the influence of AWGN as well as the RX linearity are present. Also visible is a region of constant SINR, which is defined by the remaining system impairments like PN, channel flatness, and spurious tones. The third sweep (Fig. 36.14c) is showing exemplarily the influence of the PN on the signal quality. The PN data, which is based on measurements of a dielectric resonator oscillator (DRO), is successively increased by 1 dB over all offset frequencies to retrieve the system sensitivity on the PN.

The system simulation already includes abstracted non-idealities. The system-level models can also include leakage of LO tones or harmonics, which contributes also to the system impairments. To reduce the level of abstraction, single building blocks can be replaced by a co-simulation block, where more detailed, transistor-

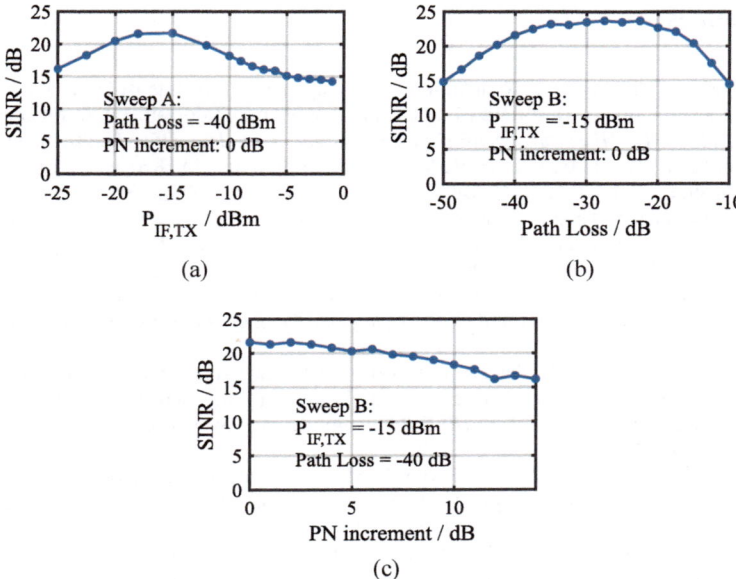

Fig. 36.14 The SINR is plotted with respect to selected parameter sweeps of the system simulation. (**a**) shows a sweep of the TX power resulting in PA compression. (**b**) shows a sweep of the path loss and (**c**) shows a variation of the PN

level schematics can be included in the simulation. However, a significant increase of the computing time is most likely.

36.6   Conclusion

In this chapter, we present simulation approaches for predicting the performance of THz components and systems. We begin by summarizing relevant simulation and characterization methods, followed by insights into specialized techniques that are particularly relevant to THz electronics, such as simulating nonplanar BEOL structures in EM simulations. Additionally, we explore the modelling of photonic components, including examples such as MZMs, phase-shifters, and nonlinear photodiode models, demonstrating their integration into co-simulation environments like Lumerical Interconnect and VerilogA-based frameworks.

By combining detailed component-level modelling with system-level abstraction, these methods provide a foundation for innovation in THz system design. Continued advancements in simulation fidelity and integration will drive future progress in this rapidly evolving field.

References

1. S. Andric, L. Ohlsson, L.-E. Wenrersson, Low-temperature front-side BEOL technology with circuit level multiline Thru-reflect-line kit for III-V MOSFETs on silicon, in *Proceedings of the 92nd ARFTG Microwave Measurement Conference (ARFTG)* (Orlando, 2019), pp. 1–4
2. A. Göritz, S.T. Wipf, M. Wietstruck, M. Kaynak, M. Fraschke, A. Krüger, M. Lisker, BEOL modifications of a 130 nm SiGe BiCMOS technology for monolithic integration of thin-film wafer-level encapsulated D-Band RF-MEMS switches, in *Proceedings of the Symposium on Design, Test, Integration & Packaging of MEMS and MOEMS (DTIP)* (2021), pp. 1–5
3. A. Leuther, A. Tessmann, M. Dammann, H. Massler, M. Schlechtweg, O. Ambacher, 35 nm mHEMT technology for THz and ultra low noise applications, in *Proceedings of the International Conference on Indium Phosphide and Related Materials (IPRM)* (Kobe, 2013), pp. 1–2
4. D. Wrana, C. Groetsch, B. Schoch, L. Gebert, T. Ufschlag, A. Leuther, R. Lozar, I. Kallfass, Methodology to accurately replicate a non-planar thin-film microstrip BEOL in 3D EM simulation, in *Proceedings of the IEEE Radio and Wireless Symposium (RWS)* (Los Angeles, 2024), pp. 71–74
5. P. Martin, F. Gays, E. Grellier, A. Myko, S. Menezo, Modeling of silicon photonics devices with Verilog-A, in *Proceedings of the 29th International Conference on Microelectronics Proceedings (MIEL)* (Belgarde, 2014), pp. 209–212
6. C. Sorace-Agaskar, J. Leu, M.R. Watts, V. Stojanovic, Electro-optical co-simulation for integrated CMOS photonic circuits with VerilogA. Opt. Express **23**(21), 27180–27203 (2015)
7. M.J. Shawon, V. Saxena, Rapid simulation of photonic integrated circuits using verilog-A compact models, in *Proceedings of the IEEE 62nd International Midwest Symposium on Circuits and Systems (MWSCAS)* (Dallas, 2019), pp. 424–427
8. K. Kawahara, T. Baba, Electro-optic co-simulation in high-speed silicon photonics transceiver design using standard electronic circuit simulator (2024). arXiv: 2410.02282
9. G. Agrawal, *Nonlinear Fiber Optics*, ser. Electronics & Electrical (Academic Press, New York, 2001)
10. T. Schwabe, C. Kress, S. Kruse, M. Weizel, H. Rhee, J.C. Scheytt, Forward-biased silicon phase shifter modelling for electronic-photonic co-simulation and validation in a 250 nm EPIC BiCMOS technology. J. Lightwave Technol. **43**(1), 255–270 (2025)
11. M. Reisch, *Halbleiter-Bauelemente*, ser. Springer-Lehrbuch (Springer, Berlin/Heidelberg, 2007)
12. T. Baba, S. Akiyama, M. Imai, T. Usuki, 25-Gb/s broadband silicon modulator with 0.31-Vcm VpiL based on forward-biased PIN diodes embedded with passive equalizer. Opt. Express **23**(26), 32950–32960 (2015)
13. J.-M. Lee, S.-H. Cho, and W.-Y. Choi, An equivalent circuit model for a Ge waveguide photodetector on Si. IEEE Photon. Technol. Lett. **28**(21), 2435–2438 (2016)
14. C. Mukherjee, D. Guendouz, M. Deng, H. Bertin, A. Bobin, N. Vaissiere, C. Caillaud, A.M. Arabhavi, R. Chaudhary, O. Ostinelli, C. Bolognesi, P. Mounaix, C. Maneux, SPICE modeling in Verilog-A for photo-response in UTC-photodiodes targeting beyond-5G circuit design. IEEE Trans. Comput.-Aided Design Integrated Circuits Syst. **42**(9), 3045–3052 (2023)
15. M. Weizel, F.X. Kaertner, J. Witzens, J.C. Scheytt, Photonic analog-to-digital-converters – comparison of a MZM-sampler with an optoelectronic switched-emitter-follower sampler, in *Proceedings of the Photonic Networks; 21th ITG-Symposium* (2020), pp. 1–6

Chapter 37
Link- and System-Level Simulation Toward Digital Twinning

Christoph Herold and Thomas Kürner

Abstract For the fundamental research on metrology and the development of communication technology, link- and system-level simulations provide opportunities to test hypotheses, explore new designs, and assess modifications of systems under development or under test. Realistic simulations can provide insights in channel characteristics while limiting the need for extensive and time-consuming measurement campaigns and provide channel models for further analysis. Link-level simulations allow for testing the performance of a system under different circumstances and scenarios that might be difficult to recreate, e.g., due to safety concerns or rare weather conditions. System-level simulations can provide an insight into network effects and interactions between components while protecting the physical hardware. In essence, findings from link- and system-level simulations pave the way toward successful digital twinning which describes the creation of a virtual replica of a physical system or a process. Through data integration and advanced algorithms, digital twins mimic real-world behavior providing insights into performance, maintenance, and potential failures. By offering possibilities for analysis, prediction, and optimization without impacting their real-world counterparts, digital twins enable safe and cost-effective research and experimentation.

37.1 Introduction

Simulation tools are an important asset in fundamental research for many reasons: Hardware is often sensitive, expensive, and hence rare. Having a well-calibrated simulation tool can therefore expand the possibilities to analyze scenarios independently from hardware availability and protect sensitive components. For research applications, it is common that components are developed by different partners. In the ThoR project, 12 different partners have contributed for a transmission link

C. Herold · T. Kürner (✉)
Institut für Nachrichtentechnik, Technische Universität Braunschweig, Braunschweig, Germany
e-mail: c.herold@tu-braunschweig.de; t.kuerner@tu-braunschweig.de

T. Kürner et al. (eds.), *Metrology for THz Communications*, Springer Series in Optical Sciences 256, https://doi.org/10.1007/978-3-032-01986-8_37

481

demonstration at 300 GHz, for example [1]. In the design process, simulation tools can give insights for parameter value requirements and interfacing issues before integration testing takes place. Identifying possible problems early in the process can make research more time- and resource efficient. The sparsity of hardware also makes it difficult to analyze interactions between different entities of a system. In the field of wireless communications, it can impair the analysis of interference behavior, for example.

It is important to identify the scope and requirements for the simulation itself. For communication systems, link-level and system-level simulations are the two most common types of simulations. A link-level simulation considers a point-to-point communication link. It can give insights into the performance of a communication system by analyzing transmitted bits or frames, coding and modulation schemes, as well as the influence of channel characteristics. Performance metrics such as bit error rates (BERs) or effective data rates can be estimated using link-level simulations. System-level simulations take a more global approach and consider the system as a whole. In case of communication systems, this can mean looking at a whole wireless communication network rather than a single link. Interference phenomena, overall network performance, and coverage of wireless networks are common examples for this simulation type.

Simulation frameworks try to combine the different level of simulations. This offers numerous advantages: The usage of a common base set of input data and models facilitates consistent simulation results. Problems caused by interfacing issues and incompatible models can be minimized. Digital twinning takes this concept even a step further. It is a trend that arose during recent years. The idea of a digital twin is the creation of a virtual copy that recreates a real-world system as closely as possible and integrate real-time data. Digital twins, or cyber-physical systems, can be useful to experiment, test, and analyze solutions and concepts that might not be possible to perform in the real-world twin.

Within the Meteracom project, the simulation framework SiMoNe has been extended and used for the research on metrology for terahertz (THz) communications. Fundamental concepts for simulations and requirements for simulation tools as well as Simulator for Mobile Networks (SiMoNe)'s implementation of them will be discussed in this chapter.

37.2   SiMoNe: A Simulation Framework for Mobile Communications

The Simulator for Mobile Networks has been developed at the Institute for Communications Technology at Technische Universität Braunschweig since 2013. Originally started as a simulation tool for radio network planning, it has become a simulation framework combining simulation modules for various frequency ranges, from Long Term Evolution (LTE) frequencies to the THz frequency range. Today,

SiMoNe contains a ray tracing, a link-, and system-level module; all operate on a joint database of input data and models in order to ensure consistency across all use cases.

A modular block concept is used throughout the whole simulation framework. These blocks handle tasks such as database access, computations, visualizations, and preparation of results. They are composed to form so-called simulation flows that can be executed, edited, and saved. This concept enables developers to encapsulate functionalities and exchange single tasks for simulation and is thus common practice for simulation tools. It provides the flexibility to change and replace functions in order to react to further developments, technology changes, and prototypes for new concepts.

SiMoNe has provided simulations and predictions to different projects in THz research. In Terapod, SiMoNe's simulations focused on a data center use case, providing interference studies and performance metrics for flexible, high data rate wireless links between server racks [2]. Weather conditions and THz backhaul applications were of interest for the ThoR project [3]. Within the Meteracom project, SiMoNe has been extended to include advanced link-level simulations for THz communication links [4], updated material parameters derived from measurements [5], and implications of high gain antennas on the physical layer security [6].

In the following, the fundamentals of the link- and system-level simulation modules will be explained in more detail. Examples of their usage within the THz communications research will be given. As the ray-tracing module provides input data for both simulation modules, fundamentals from Chap. 10 on ray tracing for metrology will be refreshed and extended for this specific use case.

37.2.1 Ray Tracing

Ray-tracing simulations are commonly used for radio propagation predictions. Simplifications based on the Maxwell equations enable the consideration of radio waves by ray-optical methods. Ray-tracing applications for metrology for THz communications are highlighted in Chap. 10. In this section, however, it is discussed in more detail how ray-tracing simulations using the example of SiMoNe can serve as an input for other simulation modules.

Scenarios under investigation must be digitized and modelled within the framework, and this data serves as a joint database for all simulation modules. Input data includes 3D maps of the scenario, antenna positions and properties, material parameter, weather conditions, and height maps. Simulation parameters are selected before the execution of a ray-tracing simulation. Interaction effects and interaction order and the choice of input data and export settings define the simulation and influence run-time and resolution. Simulations are either executed for specific points, e.g., for device-to-device communication as shown in Fig. 37.1, or for maps, e.g., for coverage and capacity analysis of base stations.

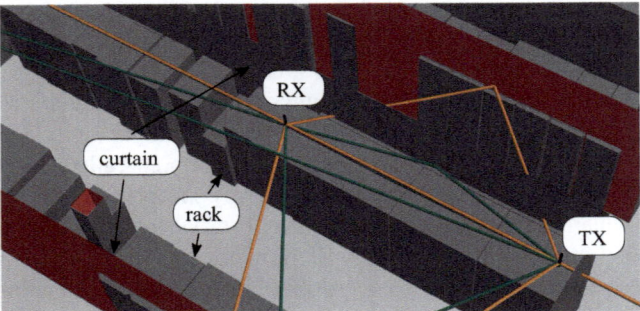

Fig. 37.1 Ray-tracing analysis of a data center communication scenario: Reflected rays (green) and transmitted rays (orange) between TX and RX. (From [4], used under a Creative Commons CC–BY license)

The simulation results are saved in two different formats so that they can be used as flexible as possible. Isotropic path loss data saves information such as interaction points, angles, and path loss but without the influence of antenna diagrams and directional communication. This can be useful for scenarios with varying conditions, e.g., due to reconfigurable intelligent surfaces (RISs) or self-organizing network functions. Masked path loss on the other hand already includes antenna pattern and their influence. For microscopic simulations, this enables lower run-times as coverage for fixed base stations do not have to be computed again for different time steps. From both data representations, tapped delay line models and values such as receive signal strength can be derived.

37.2.2 Link-Level Simulation

Link-level simulators focus on point-to-point communication links and evaluate performance metrics such as BER and data rate for a given channel. This kind of simulation can help to evaluate components of communication systems and their interoperability and performance. Modulation schemes and channel codecs can be tested under various conditions and for different signal types and scenarios, for example. A brief overview of the link-level simulation module is given in the following. An even more comprehensive description of the link-level simulator's blocks and features can be found in [4].

In order to perform point-to-point link simulations effectively and meet the requirements for interoperability, run-time performance, and accuracy, some architectural considerations are taken: A modular composition is advantageous so that components can be swapped out easily, providing a test bed for newly developed features. This approach is deeply anchored in SiMoNe's software design and led to a modular block structure for the link-level simulation module as shown in

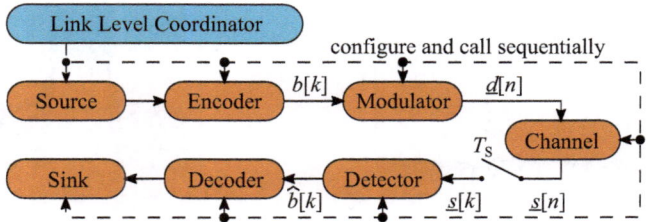

Fig. 37.2 Block diagram of the link-level simulation process. (From [4], used under a Creative Commons CC–BY license)

Fig. 37.2. Having a coordinator allows for iterative computations and regular checks of abortion criteria in order to reach statistical significance but not waste run-time. For BER simulations, a certain number of bits has to be simulated for statistical significance (as a rule of thumb, a BER of $1 \cdot 10^{-6}$ requires $4.61 \cdot 10^6$ bits for an error-free transmission and a certainty of 0.99).

The IEEE Std 802.15.3d-2017 standard provided a base for the selection of modulation schemes (BPSK, QPSK, 8-PSK, 8-APSK, 16-QAM, 64-QAM), coding schemes (Reed Solomon (240,224), 11/15 LDPC, 14/15 LDPC), and further simulation parameters [7]. Apart from statistical channel models, such as additive white Gaussian noise (AWGN) channels, deterministic channel models can be imported based of measurement results or ray-tracing simulations. A collection of amplitude and delay of the signal is used to create a tapped delay line representation of the channel with the impulse response:

$$h_C(t) = \sum_i A_i \delta(t - \tau_i), \qquad (37.1)$$

where A_i and τ_i describe amplitude and delay of tap i. The modulation channel is modelled using a convolution of the band-limited transmit pulse with amplitude A and the impulse response of the propagation channel leading the following representation with sampling frequency f_s:

$$h_{TX,C}[n] = (g * h_C[n]) = \sum_i A_i A T_S h_{TX}(n\Delta t - \tau_i). \qquad (37.2)$$

A specific block for multi-carrier communication allows for the concurrent simulation of multiple channels at once, enabling the analysis of channelization concepts and their inter-carrier interference as well as performance of multi-carrier systems. After detection and decoding, the transmitted bit sequence is compared to the received bit signal for the processing of the numerical metrics, such as error vector magnitude (EVM), BER, and effective data rate.

Apart from numerical metrics, visualizations such as constellation diagrams, spectra, and eye diagrams can provide valuable insights into phenomena and

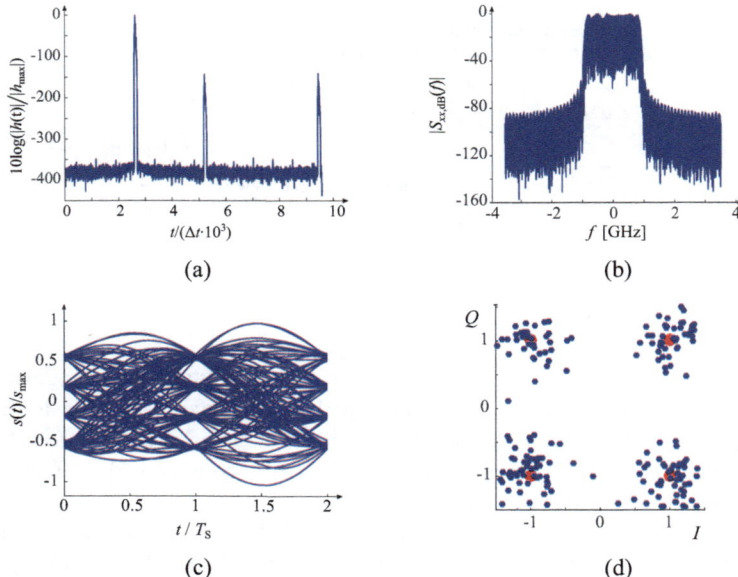

Fig. 37.3 Example visualizations of SiMoNe's link-level simulator. (From [4], used under a Creative Commons CC–BY license). (**a**) Channel impulse response. (**b**) Power spectral density. (**c**) Eye diagram. (**d**) Constellation diagram

problems of a communication link. In the simulation, the digital representation of the bits and signals is available at every step and can therefore be plotted without significant effort, even at points of the communication that would not be available in the real-world system. A small variety of available visualizations can be found in Fig. 37.3.

37.2.2.1 Applications of Link-Level Simulations

The detailed information about channel parameters and frequency- and time-domain behavior as well as the possibility to visualize the current channel state make the link-level tool an important asset for research in metrology. Within the Meteracom project, SiMoNe's link-level simulation module has been used for the following tasks:

Machine learning algorithms require a significant amount of training data representing different conditions and scenarios in order to avoid over-fitting. Measurements of such a variety of scenarios can be prohibitively time- and resource-consuming. Simulation tools can be helpful by completing sparse datasets from measurements or by creating entire sets based on simulation results. The feasibility of this approach has been shown in Chap. 38.

High data rate transfer is one of the key applications of THz communications technology. The high data rates challenge hardware encoders and decoders. Simulation tools can provide an assessment of network codecs performance for THz communications even if no suitable hardware is currently available. SiMoNe's link-level simulator provided a test bed and has been used to demonstrate network codecs' capabilities. More details on this work can be found in Chap. 39.

For fundamental research in new frequency spectrum, such as the THz range, the availability of hardware can be a limiting factor for the investigation of concepts such as multi-carrier transmissions or interference studies. A simulation tool, such as SiMoNe's link-level simulator, can provide insights in scalability.

37.2.3 System-Level Simulation

System-level simulations for communication systems usually consider a complete communication network. Especially during the early stages of technology development, this is the only possibility to investigate network effects and interference patterns as hardware is generally sparse and multiple TXs and RXs are difficult to obtain for testing purposes. Target metrics for these simulations usually receive signal strength and signal-to-noise-and-interference ratios, as well as metrics that can be derived from them such as throughput and coverage.

Several different kinds of input data serve for system-level simulations: Measurement data, path loss predictions from a ray tracer, or hybrids thereof. SiMoNe, for example, derives throughput data from ray-tracing results based on a so-called throughput mapping. With the predicted available receive signal strength, a certain throughput can be expected. These mappings are usually stepwise defined functions and commonly used for system-level simulations. Commonly, they are either defined by a standard or derived from measurements or link-level simulations. Receive signal strength maps and the locations of TXs and RX can be used to consider interference or handover behavior.

By providing insight into the complex behaviors of entire networks, system-level simulation can provide guidance for the development by offering rapid prototyping and a sandbox setting for parameter selection. Self-optimization algorithms and newer concepts such as RIS-assisted or relay-assisted communication can be evaluated for various scenarios, seeing what applications and implementations are promising.

37.2.3.1 Applications of System-Level Simulations

System-level simulations focus on the interaction of various different components and their interactions. For THz backhaul applications, system-level simulations have helped to develop and verify network topologies and network performances [8–10]. The simulations have further shown that even under adverse weather conditions, wireless backhaul connections are feasible and can help to reduce the need for fiber connections while maintaining the required data rates for backhaul applications [11].

Interference on various levels is also of interest for the metrology of THz communication systems. System-level simulations could show that the channelization concept of the IEEE Std 802.15.3d-2017 standard is feasible with state-of-the-art hardware [12]. In the data center use case, it could be identified that link coordination is required to avoid interference for certain communication pairs, e.g., along a row of server racks [2].

Possibilities for eavesdropping based on propagation effects have been measured and simulated within the Meteracom project. It could be shown that even everyday objects such as door frames can have a detrimental effect on secure communication at THz-frequencies. More details about the studies can be found in Chap. 35.

37.2.4 Synergy Effects of Simulation Frameworks

Being integrated into a single simulation framework, link- and system-level simulation in SiMoNe operates on the same database. This creates synergy effects and opens up opportunities for different simulation types. For THz backhaul applications, wireless network planning, system-level simulation for the impact of weather conditions, and link-level simulations for the impact on BER and data rate have been combined [13]. In the system-level simulation, different weather conditions can be selected, and their performance will be displayed in the map overview. By selecting a specific link, a more detailed link-level simulation can be triggered, and the weather impact on constellation and eye diagram can be seen as shown in Fig. 37.4.

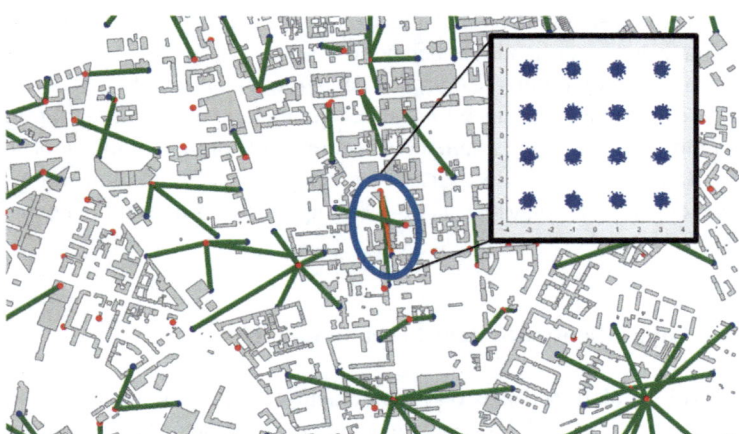

Fig. 37.4 System-level visualization with a detailed link-level simulation of a link of interest (orange line). Red dots are fiber-connected base stations and blue dots wirelessly connected ones. Green lines show wireless connections

37.3   Verification and Validation of Simulation Tools

Simulation results can only be helpful if their results are trustworthy and reliable. Verification and validation of simulation processes are therefore important parts of the development of respective tools. While the term verification refers to the simulation being suitable for the requirements, the term validation refers to the correctness of the simulation results. For both processes, standard practices from software engineering can provide guidance. The processes have been executed during the development of the link-level simulation module of the SiMoNe framework and will be explained in more detail based on this example.

The establishment of requirements is a necessary first step for the development of a piece of software. For the link-level simulator, the IEEE Std 802.15.3-2023 standard [14] served as a guidance. In order to be compliant with the standard, certain modulation (e.g., BPSK, QPSK, 16QAM) and coding schemes (e.g., Reed-Solomon (240,224), 11/15 LDPC, 14/15 LDPC) have to be implemented and rules for spectrum masks and channelization are to be followed, for example. Automated testing by unit and high-level test has been used to verify that the requirements are met and implemented methods perform as expected meeting spectrum masks, etc.

A first reference for the simulation can be found in problems that have analytical solutions. In case of the link-level simulator, the BER of an AWGN channel served this purpose. Formulas for various modulation and coding schemes have been derived, e.g., in [15] and implemented in MATLAB's BER tool [16]. These calculated values can be compared to the simulation results as seen in Fig. 37.5 for uncoded and coded AWGN channels. Small deviations can occur due to the statistical processes involved; however, in general, a good agreement should be reached.

Measurements are a crucial part of the development of models for propagation effects and material interactions. Specifically designed measurement campaigns isolate phenomena in order to model behaviors and effects. In the Meteracom project, a measurement campaign has been conducted to analyze complex building

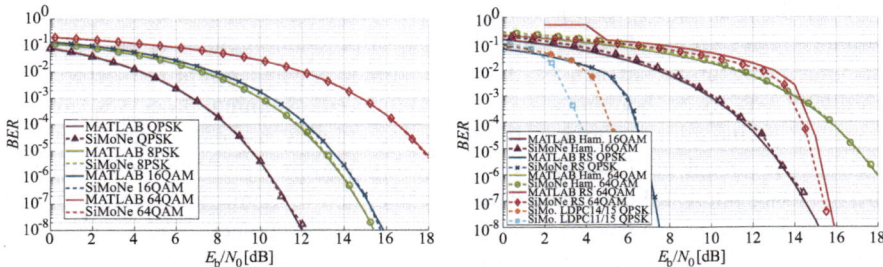

Fig. 37.5 Comparison between link-level simulation and theoretical reference for uncoded and coded transmission via AWGN channel. (From [4], used under a Creative Commons CC–BY license)

materials and provide insight into their behavior when interaction with radio signals is presented in Chap. 31. The influence of temperature and humidity has been investigated by another measurement setup using a climate-controlled chamber. Models derived from these and similar measurements can be implemented in simulation tools to imitate characteristics of the specific effect. More details can be found in Chap. 30.

More advanced measurements using reference scenarios or structures as presented in [17] can provide comparable results for measurements and simulations. As these references usually include a variety of propagation effects, their evaluation is usually the last step in the validation process. The presented structure has a variable second reflection path and can therefore create artificial interference pattern between both paths. After verifying the single components using measurements and their interactions in complex scenarios, it can be assumed that the simulation tool provides reliable results. More information can be found in Chap. 18.

37.4 Digital Twinning: A Look Ahead

Simulations help to understand certain properties of real-world systems. Digital twinning goes one step further. Its ultimate goal is to have a virtual representation of a process or physical object mirroring its properties as a whole while integrating real-time data. It is commonly considered a key technology for modern industry [18]. Today's digital twins strive to integrate complex system models and provide a common format for data exchange. Simulation models with real-time data can offer a sandbox environment to plan and optimize processes in the virtual space. One reason for its rising importance is the growing hardware capabilities. Multicore central processing units (CPUs) and graphics processing units (GPUs) enable complex simulations and computations in reasonable time. Abstraction of processes is still important to save run-time; however, the expanding possibilities, fueled by machine learning, artificial intelligence, and computer vision, provide excellent opportunities to analyze real-world use cases.

Digital twinning as a tool is applied to processes and systems in various domains. They are most prevalent in the field of production engineering. Complex manufacturing processes can be designed, simulated, and optimized before any hardware has to be bought, set up, or configured. Running processes are not impacted by the digital planning. In the service industry, use cases such as remote monitoring, visual guidance during troubleshooting, and predictive maintenance are of strong interest. For technical development and fundamental research, digital twins can provide guidance for various development choices. The exchange of information between subcomponents can be analyzed, informed parameter choices can be taken, and scalability test can be conducted. Especially for sparse, expensive, or critical components, a digital twin can be a way to investigate problems, improve performance, and visualize processes without endangering the real-world object.

Visualization capabilities can create views that are not available in real life, e.g., a cut through of a running machine or data streams at various stages of a process.

For the data center communication use case, SiMoNe has shown that a digital twin can give valuable insights into the performance of wireless links in such a scenario. The detailed 3D model and propagation predictions allowed research to analyze interference and timing between links for different setups and configurations. Recommendations about positions for possible communications without interference could be given [2].

References

1. T. Kürner et al., THz Communications and the demonstration in the ThoR–backhaul link. IEEE Trans. Terahertz Sci. Technol. **14**(5), 554–567 (2024)
2. J.M. Eckhardt, C. Herold, B. Friebel, N. Dreyer, T. Kürner, Realistic interference simulations in a data center offering wireless communication at low terahertz frequencies, in *Proceedings of the International Symposium on Antennas and Propagation (ISAP)*, (Taipei, 2021), pp. 1–2
3. B.K. Jung, T. Kürner, Performance analysis of 300 GHz backhaul links using historic weather data. Adv. Radio Sci. **19**, 153–163 (2021)
4. J.M. Eckhardt, C. Herold, B.K. Jung, N. Dreyer, T. Kürner, Modular link level simulator for the physical layer of beyond 5G wireless communication systems. Radio Sci. **57**(2), 1–15 (2022)
5. F. Taleb, G.G. Hernandez-Cardoso, E. Castro-Camus, M. Koch, Transmission, reflection, and scattering characterization of building materials for indoor THz communications. IEEE Trans. Terahertz Sci. Technol. **13**(5), 421–430 (2023)
6. C. Herold, T. Doeker, J.M. Eckhardt, T. Kürner, Investigation of eavesdropping opportunities in a meeting room scenario for THz communications, in *Proceedings of the 16th European Conference on Antennas and Propagation (EuCAP)* (Madrid, 2022), pp. 1–5
7. V. Petrov, T. Kürner, I. Hosako, IEEE 802.15.3d: first standardization efforts for sub-terahertz band communications toward 6G. IEEE Commun. Mag. **58**(11), 28–33 (2020)
8. B.K. Jung, T. Kürner, Automatic planning algorithm of 300 GHz backhaul links using ring topology, in *Proceedings of the 15th European Conference on Antennas and Propagation (EuCAP)* (2021), pp. 1–5
9. T. Kürner, B. K. Jung, Automatic planning of NLOS backhaul links at 300 GHz arranged in star topology, in *Proceedings of the XXXIVth General Assembly and Scientific Symposium of the International Union of Radio Science (URSI GASS)* (Rome, 2021), pp. 1–3
10. Y. Liu, B.K. Jung, T. Kürner, Automatic planning algorithm of 300 GHz backhaul links using mesh topology, in *Proceedings of the 18th European Conference on Antennas and Propagation (EuCAP)* (Glasgow, 2024), pp. 1–5
11. J.M. Eckhardt, C. Herold, B.K. Jung, T. Kürner, Performance analysis of a wireless backhaul network at terahertz frequencies, in *Proceedings of the International Symposium on Antennas and Propagation (ISAP)* (Sydney, 2022), pp. 179–180
12. J.M. Eckhardt, C. Herold, T. Kuerner, Intercarrier interference at terahertz frequencies for IEEE Std 802.15.3d multiband transmissions, in *Proceedings of the 26th International ITG Workshop on Smart Antennas (WSA) and 13th Conference on Systems, Communications, and Coding (SCC)* (Braunschweig, 2023), pp. 1–6
13. B.K. Jung, C. Herold, J.M. Eckhardt, T. Kürner, Link-level and system-level simulation of 300 GHz wireless backhaul links, in *Proceedings of the International Symposium on Antennas and Propagation (ISAP)* (Taipei, 2021), pp. 619–620
14. IEEE Standard for Wireless Multimedia Networks. IEEE Std 802.15.3-2023 (Revision of IEEE Std 802.15.3-2016) (2024), pp. 1–684

15. J.G. Proakis, *Digital Communications*, 5th ed., ser. McGraw-Hilll Higher Education (McGraw-Hill, 2008)
16. The MathWorks Inc., Matlab version: 24.1.0 (r2024a). Natick, Massachusetts, US, 2024. https://www.mathworks.com
17. T. Doeker, C.E. Reinhardt, C. Herold, U. Hellrung, D. Mittleman, T. Kürner, Variable reference structure for channel sounding in the low terahertz range, in *Proceedings of the 19th European Conference on Antennas and Propagation (EuCAP)* (Stockholm, 2025), pp. 1–4
18. S. Haag, R. Anderl, Digital twin—proof of concept. Manuf. Lett. **15**, 64–66 (2018)

Chapter 38
Trustworthy Modulation Recognition

Anouar Nechi, Christoph Herold, and Mladen Berekovic

Abstract AI is recognized as a powerful tool for addressing the complexities of 6G technology. However, trust in artificial intelligence (AI)'s decision-making processes must be established to ensure its successful and responsible integration. A practical trustworthiness evaluation approach is proposed for AI-driven 6G applications, explicitly focusing on automatic modulation recognition (AMR). This approach centers on three core attributes: data robustness, ensuring the reliability of the models despite variations in input data; parameter sensitivity, where minor adjustments to model settings should not lead to drastic performance changes; and security against adversarial examples, demonstrating the resilience of the models to intentionally crafted malicious inputs. This focus on trustworthiness evaluation contributes to developing trustworthy AI solutions, facilitating AI's secure and dependable deployment in future wireless communication systems.

38.1 Introduction

The emerging 6G network technology aims to outperform the current wireless capabilities by leveraging frequencies exceeding 100 GHz [1]. Developing effective communication systems at these frequencies is far more complex than those at lower frequencies. 6G technology relies significantly on two main pillars: terahertz (THz) communications and machine learning (ML). While 6G has taken another step toward THz communications according to the IEEE 802.15.3d standard [2], ML is endorsed as an innovative approach for optimizing 6G performance [1]. In other words, the ultrawide THz band is regarded as a promising candidate for 6G

A. Nechi (✉) · M. Berekovic
Universität zu Lübeck, Institut für Technische Informatik, Lübeck, Germany
e-mail: anouar.nechi@uni-luebeck.de; mladen.berekovic@uni-luebeck.de

C. Herold
Technische Universität Braunschweig, Institut für Nachrichtentechnik, Braunschweig, Germany
e-mail: c.herold@tu-braunschweig.de

© The Author(s) 2026
T. Kürner et al. (eds.), *Metrology for THz Communications*, Springer Series in Optical Sciences 256, https://doi.org/10.1007/978-3-032-01986-8_38

technology. Meanwhile, ML has proven effective in addressing technical challenges in existing communication systems [3].

Additionally, ML ambiguously addresses various challenges in wireless communication systems. For instance, deep neural networks (DNNs) have been applied in a black-box manner for AMR [4]. Thus, it is essential to understand the risks of deploying such an AI algorithm. As emphasized by an Independent High-Level Expert Group on AI [5], ensuring AI trustworthiness throughout its lifecycle, from development to deployment, is paramount to fully harness its benefits. The concept of trustworthy AI aims to address the risks associated with AI deployment [6]. Several studies have proposed definitions of system trustworthiness [7, 8] and trustworthy AI [5, 6]. However, these definitions remain general and present principles more than practical approaches [9].

This research investigates the deployment of DNNs for AMR in 6G systems. Two customized DNNs are presented, accompanied by a novel trustworthiness model designed to rigorously evaluate their reliability within the unique challenges of the 6G environment. This analysis underscores the critical need to prioritize trustworthiness to ensure the safe and effective integration of AI in future 6G infrastructure

38.2 Modulation Recognition for THz Communication

Effective training of DNNs for 6G AMR systems depends on having comprehensive and representative datasets. To this end, we introduce a simulator designed to generate synthetic datasets that capture the unique characteristics of THz communication signals. Utilizing this tool, we create a dataset specifically for training and evaluating two customized DNNs aimed at addressing the AMR challenge. We then analyze the performance of these DNNs, demonstrating their ability to accurately classify modulation schemes within the 6G environment.

38.2.1 Simulator for Mobile Networks (SiMoNe)

Fundamental research often uses newly developed, expensive, or sensitive hardware. Large-scale measurements or experiments are often limited by this. Simulation tools can help alleviate this situation, but it is essential that they are properly developed and calibrated to accurately represent reality. Within the Meteracom project, the simulation framework SiMoNe has been extended to allow simulating various aspects of THz communications. The simulator contains a ray-tracing module for the computation of channel impulse responses based on realistic 3D models of scenarios [10] and a link-level simulation module that provides statistics and signals of point-to-point communication links [11].

For the modulation recognition task at THz frequencies, hardware-specific impairments and propagation characteristics are important to consider. Material parameters and water vapor density, for example, influence propagation [12], are considered during the ray-tracing simulation. Hardware impairments, such as I/Q-Imbalance or phase noise, can be included as frequency responses [11]. More information about the general concepts, the architecture, and the capabilities of the simulation framework SiMoNe can be found in Chaps. 10 and 37.

38.2.2 Synthetic Dataset

For the training of the AMR function, a synthetic dataset has been generated using SiMoNe. Synthetic data allows to include a variety of hardware and propagation effects that are difficult to measure and recreate reliably.

Communication links in an additive white Gaussian noise (AWGN) channel have been simulated without hardware impairments. The modulation schemes OOK, BPSK, QPSK, 8-PSK, 8-APSK, 16-QAM, and 64-QAM, all included in the IEEE 802.15.3d standard [2], have been considered over a signal-to-noise ratio (SNR) range from -20 dB to 30 dB in 2 dB-increments. This results in a total of 182 parameter combinations. The transmitted data of each parameter combination has been divided up into 4096 samples of 1024 consecutive symbols each.

As the tool chain for the generation of training data based on simulations is set up, the investigation of many more scenarios can follow. Based on detailed 3D models in the ray-tracing module, communication links in other common scenarios, such as meeting rooms, hallways, or industrial areas, can be simulated and analyzed.

38.2.3 Deep Learning-Based Modulation Recognition

In modern communication systems, transmitters (TXs) utilize various modulation schemes to manage data rates and optimize bandwidth usage efficiently. While TXs can dynamically adapt the modulation type, receivers (RXs) may not always be aware of this selection. This poses a challenge for RXs, as they need to recognize the modulation type accurately. One common solution involves embedding modulation information within each signal frame so that the RX can identify the modulation type and adjust its demodulation strategy accordingly. However, as wireless networks become more diverse and the number of users increases, this approach becomes increasingly intricate, potentially decreasing spectrum efficiency due to the additional information embedded in each frame [13].

AMR has emerged as a promising alternative. It enables the detection of the modulation scheme without adding extra overhead to the network protocol, allowing for accurate demodulation and successful data recovery. Traditional AMR techniques often require significant computational resources or expert knowledge for feature extraction [14].

deep learning (DL) provides a compelling solution to overcome these challenges, as it has the potential to offer high classification accuracy for AMR tasks without the need for prior preprocessing or feature extraction. For instance, convolutional neural networks (CNNs) have been effectively used to extract features directly from raw I/Q data and perform accurate classification [15–17]. Recurrent neural networks (RNNs) have demonstrated their capability in modulation recognition by leveraging sequential correlations within I/Q components and amplitude and phase signal components. This approach enhances modulation recognition accuracy [4]. Additionally, RNNs have been employed for signal parameter estimation and distortion correction [3]. These efforts have shown that recurrent neural network (RNN) models excel in signal distortion estimation and surpass many other DL methods in classification accuracy.

CNNs have proven highly effective in computer vision tasks due to their ability to learn directly from raw data, eliminating manual feature extraction or preprocessing. To emphasize this strength for AMR, a CNN architecture comprising three convolutional layers was developed. Each layer is sequentially followed by batch normalization, a rectified linear unit (ReLU) activation function, and a MaxPooling layer. Raw I/Q samples of each radio signal are input into the CNN model. The extracted features are then passed to the dense layers of the network for classification. This process utilizes the scaled exponential linear unit (SeLU) activation function and alpha dropout regularization. The proposed CNN classifier consists of 555,287 parameters and achieves an average accuracy of 68.8% across all SNR levels. While it demonstrates superior performance for lower-order modulation schemes (84.6% for BPSK, 93.1% for OOK), reduced accuracy is exhibited in predicting higher-order modulation schemes (56.2% for 16QAM, 55.9% for 64QAM) as detailed in Fig. 38.1.

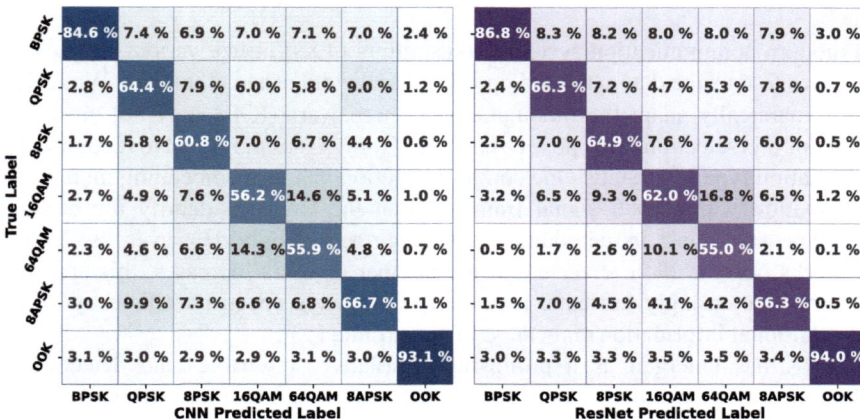

Fig. 38.1 CNN and ResNet-6 confusion matrices for modulation recognition

Deep residual networks (ResNets), an enhanced variant of CNNs, incorporate skip connections to facilitate processing features at multiple scales and depths within the network. This architectural innovation enables the utilization of wider layers, efficient training with fewer epochs, and the attainment of superior results compared to traditional CNNs. Inspired by [18], a ResNet architecture was constructed for radio signal classification. It comprises six residual units, each incorporating two skip connections, followed by a fully connected region analogous to the proposed CNN but with a reduced parameter count of 159,015. The ResNet classifier achieves an overall accuracy of 70.8% across all SNR levels demonstrating a 2% improvement over the CNN while utilizing fewer parameters. This result underscores the effectiveness of ResNets over conventional CNN classifiers. As depicted in Fig. 38.1, the confusion matrix for the ResNet reveals a 16.8% confusion rate between 16QAM and 64QAM, with a slight improvement in accuracy observed for the remaining modulation schemes.

38.3 Trustworthiness Model and Attributes

AI trustworthiness is crucial for reliable THz communications. This section explores this concept, focusing on DNN-based modulation recognition. A novel trustworthiness model is introduced to evaluate the trustworthiness of DNNs in 6G communication systems.

38.3.1 Artificial Intelligence Trustworthiness

The definition of trustworthiness and dependability has been the subject of extensive study, with a focus on identifying their key characteristics. In system design, attributes like availability, reliability, safety, integrity, and maintainability are widely regarded as essential aspects of dependability. However, this definition may not fully encompass all security attributes, particularly confidentiality. In contrast, another study presents trustworthiness as a counterpart to dependability, encompassing reliability, safety, maintainability, availability, integrity, and confidentiality, thus incorporating security within the dependability framework.

When it comes to AI-based system design, these definitions of trustworthiness may not sufficiently address the unique requirements of modern AI systems. Given AI's heavy reliance on data, specific attributes are needed to ensure its trustworthiness. Consequently, new trustworthiness attributes have emerged, primarily focusing on security, robustness, safety, transparency, and fairness. Nevertheless, these attributes still lack specificity for particular AI applications.

A comprehensive examination and elucidation of the interaction between the DNN and its host environment are essential for accurately pinpointing the trustworthiness attributes of DNNs in the context of AMR within THz communications-based 6G technology.

38.3.2 Trustworthiness Model

To identify the key trustworthiness attributes of DNNs when applied to AMR in THz communications-based 6G technology, it is imperative to begin by formulating DNN as a function of its various inputs and parameters. Subsequently, we link this formulation and the relevant trustworthiness attributes.

A layer ℓ within a DNN can be represented as an operation $f_\ell[p_\ell](x_{\ell-1})$, where p_ℓ denotes the parameters of layer ℓ. This set includes the weights W_ℓ^j and bias b_ℓ. Additionally, $x_{\ell-1}$ represents the output of the previous layer. The DNN classifier, denoted as f_{DNN}, is defined by the combined operations of these layers:

$$f_{\text{DNN}}(x_{\text{in}}; p) = f_l[p_l] \circ \cdots \circ f_2[p_2] \circ f_1[p_1](x_{\text{in}})$$

where x_{in} is an input signal, p is a set of DNN parameters $p = \{p_1, \ldots, p_L\}$, and L is the number of DNN layers. The DNN parameters and the model hyperparameters are established during the training phase based on the dataset. In the prediction phase, the DNN classifier $f_{\text{DNN}}(x_{\text{in}}; p)$ can be perceived as a function of two inputs: the trained parameters p and the signal x_{in} like an input variable.

Consequently, the proposed trustworthiness framework for a DNN only takes into account the input signals x_{in} and the DNN parameters $p = \{p_1, \ldots, p_L\}$. Other components of the DNN, such as activation functions, are assumed to be reliable and trustworthy. This framework clarifies how DNNs interact with the THz communication environment and the user. Figure 38.2 shows the three critical attributes of trustworthiness that need to be considered in DNN-based AMR. These attributes are explained as follows:

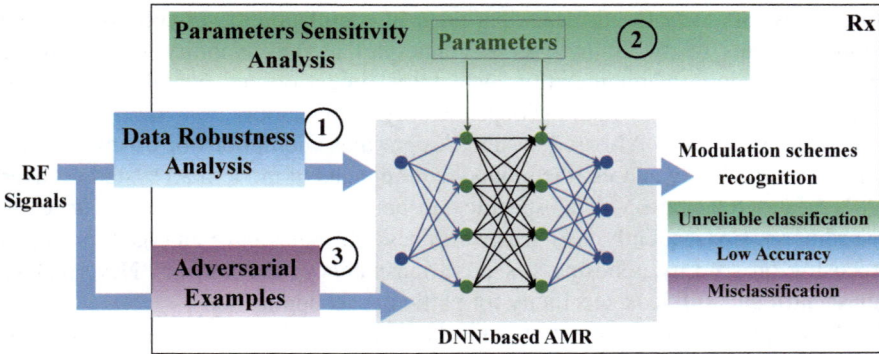

Fig. 38.2 The proposed trustworthiness model for AMR in 6G technology

1. **Data robustness analysis** helps to understand when DNN classifiers show reduced accuracy due to environmental variations. It assesses these variations and their influence on the performance of DNN classifiers [9]. Specifically, the analysis of DNN robustness examines how a noisy environment affects the input signals x_{in} and its repercussions on DNN classification accuracy. In this context, various levels of SNR are applied to the input signal x_{in}, and the resulting decline in DNN accuracy is monitored and evaluated.
2. **Parameter sensitivity analysis** provides a thorough insight into the reliability of DNNs, particularly regarding the factors that contribute to unreliable classification. The reliability can be assessed by examining the sensitivity of the DNN parameters for specific signals. *Reliability* signifies that the DNN classifier should achieve accuracy levels consistent with its design without any failures (unreliable classification occurs when accuracy is below 50%). The proposed sensitivity analysis utilizes a model based on random bit-flipping [19].
3. **Adversarial examples** demonstrate how deliberate input signal changes can impact DNN classifiers' performance. Attackers can exploit this vulnerability to compromise system security by generating inputs designed to mislead the DNN during the inference stage, ultimately leading to incorrect classifications [20].

Our trustworthiness model omits DNN transparency from its attributes as AMR does not utilize private data, and it disregards DNN fairness since the employed dataset is balanced. Misclassification results in the selection of an incorrect scheme, rendering the received signals unable to be demodulated. This event and its consequences are already included in the reliability attribute. Consequently, DNN safety can be viewed as a subset of reliability in our specific application [29].

38.4 Trustworthiness Analysis of DL-Based Automatic Modulation Recognition

In this part, we evaluate the trustworthiness of the suggested DNN models by applying our trustworthiness model and its attributes.

38.4.1 Data Robustness

A key factor in ensuring the reliability of a trained DNN model is its ability to handle variations in the environment. The model should be robust enough to accommodate diverse data distributions encountered under different environmental conditions [9]. In this context, the effects of a noisy environment on DL-based modulation recognition are analyzed. This issue is of paramount importance, as it directly affects the data robustness of the DNN classifier.

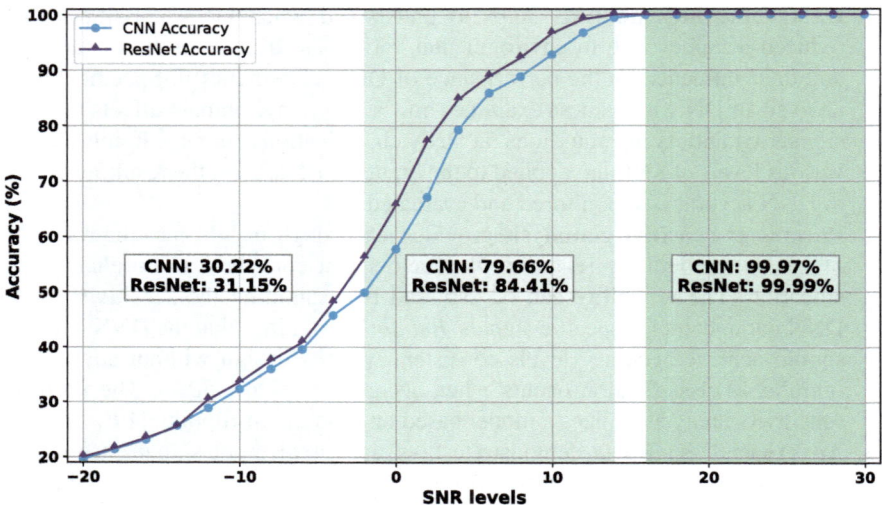

Fig. 38.3 Data robustness analysis of both CNN- and ResNet-based AMR on THz dataset

SNR represents an essential metric within any communication system. SNR assesses the quality of the signal in relation to the noise inherent in the communication channel, effectively reflecting the impact of environmental variations. To evaluate the data robustness of DL-based modulation recognition, the following steps are carried out: (1) The dataset is split into a training and testing set with consideration of the various SNR levels to maintain a balanced dataset, (2) DNN models are trained based on the resulting dataset, and (3) the accuracy of the proposed DNN models is evaluated considering the various SNR levels.

Additionally, we apply the aforementioned steps on the suggested DNN models. Figure 38.3 illustrates that data samples with low SNR ranging from -20 to -4 dB are hard to classify, achieving a maximum accuracy of 50%. At such noise levels, the constellation of the received signals appears random and fails to create distinct clusters necessary for differentiating between various modulation schemes. Notably, as the SNR rises from -2 to 10 dB, the model's accuracy improves. The model accuracy attains 99% as SNR approaches 10 dB. The highest model accuracy is achieved starting from 10 dB. Moreover, the ResNet model demonstrates better accuracy compared to the CNN model within the SNR range of -2 to 10 dB. The accuracies for both models show a correlation outside this range. Consequently, the ResNet-based AMR proves to be more robust than the CNN-based AMR regarding noise variations.

38.4.2   Sensitivity Analysis

AI sensitivity analysis identifies vulnerable bits that, when altered, significantly undermine classification accuracy. This analysis employs a bit-flipping model of AI parameters. The objective of sensitivity analysis is to provide a comprehensive understanding of AI behavior and to offer insights that elucidate the decision-making processes of AI systems.

To perform the sensitivity analysis of the CNN and ResNet classifiers, a random single-bit flip is applied to the parameters of the DNN [19]. Both the parameter locations and the bit positions are uniformly distributed. Initially, we inject single-bit flips 1000 times at various bit positions and parameter locations across each layer of the CNN model. For the ResNet model, we introduce single-bit flips randomly within the residual block, convolution, and dense layers. Nevertheless, the injected faults in the convolution and dense layers follow a similar approach to that of the CNN. In contrast, faults are injected randomly at different bit positions, parameter locations, and layers for the entire residual block. The aforementioned fault injection tests are carried out during the inference process.

Injecting single-bit faults into 32-bit floating-point (FP) parameters reveals that the exponent bits (positions 23 to 30) are more susceptible to errors than the mantissa bits (positions 0 to 22). This observation aligns with established findings in existing research. The injection of single-bit faults into 32-bit FP parameters demonstrates that the exponent bits (specifically, those ranging from the 23rd to the 30th position) exhibit greater sensitivity compared to the mantissa bits (from the 0th to the 22nd position). In order to enhance our understanding of exponent sensitivity, we categorize the susceptible exponent bits into two distinct groups: The first group comprises the bits that, when compromised, lead to misclassification, thereby resulting in an unreliable classifier with an accuracy of less than 50%. The second group consists of vulnerable bits causing accuracy degradation.

Figure 38.4 demonstrates how single-bit faults affect the CNN classifier, specifically in the convolution layers C1, C2, C3 and the dense layers D1, D2, D3. Unreliable classifications are detected at 25 in C1 and 30 in C1, C2, C3, D2, D3. However, flipping bit 30 in D1 only leads to a decrease in accuracy. It is important to mention that faults in the other layers exhibit minimal accuracy loss. Additionally, the figure presents the influence of bit-flipping across all layers of the ResNet classifier. The 30th bit (i.e., bit 31) is particularly sensitive, as it leads to unreliable classifications throughout all layers. The other vulnerable bits result in mere accuracy drops. Consequently, flipping the critical 30th bit leads to misclassification in both classifiers, while the other vulnerable bits cause only a decrease in accuracy.

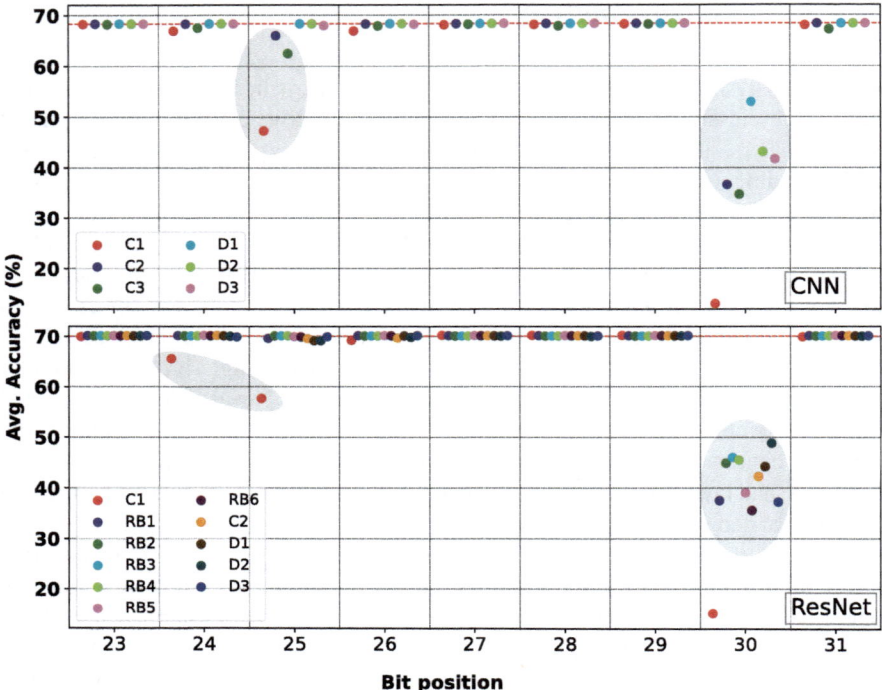

Fig. 38.4 Weight sensitivity analysis of CNN and ResNet classifiers: Assessing the impact of bit flips on classification accuracy. The gray spots indicate the vulnerability of layers to bit flips

38.4.3 Security Robustness

In [20], a variety of neural network models are shown to be vulnerable to adversarial examples, wherein an attacker generates inputs that result in misclassification. These inputs are slightly different from the original inputs that are correctly classified, yet they have the potential to cause misclassification. The phenomenon of adversarial examples primarily arises from a certain *"linear behavior in high-dimensional spaces"* [21]. This observation has given rise to numerous efficient methods for generating adversarial example attacks, such as the fast gradient method (FGM) [21] and projected gradient descent (PGD) [22].

To evaluate the effects of adversarial examples, we conduct a series of eight distinct attacks to generate adversarial examples against the studied CNN and ResNet classifiers utilizing the Adversarial Robustness Toolbox (v1.2.0) [28]. Each attack is configured according to a predetermined set of parameters. The results of the attacks, which include the FGM [21], PGD [22], NewtonFool [23], DeepFool [24], HopSkipJump [25], zeroth order optimization (ZOO) [26], and the Carlini & Wagner method (C&W) [27] L_2 and L_∞, are summarized in Table 38.1. The adversarial example resistance (AER) quantifies the models' capabilities to

Table 38.1 Results of
adversarial example attacks

Attack	AER_{CNN}	AER_{ResNet}
FGM [21]	18.83%	22.12%
PGD [22]	6.11%	5.60%
NewtonFool [23]	26.84%	28.50%
DeepFool [24]	26.07%	15.43%
HopSkipJump [25]	14.92%	8.04%
Zoo [26]	26.57%	19.25%
C&W [27] L_2	53.90%	59.25%
C&W [27] L_∞	21.57%	14.57%

maintain accurate classifications despite the generation of adversarial examples. For instance, both classifiers demonstrate comparably high AER when faced with the C&W attack under L_2, whereas the adversarial examples produced by PGD significantly impair both classifiers. In general, the models exhibit various AER values in relation to the specific attacks employed. As a result, the determination of the more resilient model against adversarial examples will depend on the chosen attacks.

38.4.4   *Trustworthiness Evaluation*

Based on the previous analysis, we can assess the trustworthiness of the investigated models. First, within the SNR range of 0 to 30 dB, both classifiers are robust against environmental changes, with ResNet demonstrating superior robustness compared to CNN. Second, the sensitivity analysis of the parameters indicates that altering the vulnerable 30th bit leads to unreliable classifications. Finally, both classifiers exhibit different levels of resistance to specific adversarial example attacks without a clear verdict for either DNN. Future research could pinpoint appropriate attacks to deliver the necessary metrics.

38.5   Conclusion

A methodology for constructing a practical trustworthiness model for DNNs was introduced, specifically applied to the demands of 6G technology. Such a model is crucial given the heightened reliability and security requirements of 6G compared to previous generations. The application of AMR within a THz communications-based 6G environment was the focus, with the development of two DNN classifiers for this purpose. A trustworthiness model was then employed to analyze these classifiers, considering key attributes relevant to this environment: robustness, DNN parameter reliability, and resistance to adversarial examples. Experimental results demonstrate that the proposed trustworthiness model offers a suitable approach for evaluating the trustworthiness of DNNs used for AMR in THz communications-based 6G technology.

References

1. V. Petrov, T. Kurner, I. Hosako, IEEE 802.15.3d: first standardization efforts for sub-terahertz band communications toward 6G. IEEE Commun. Mag. **58**(11), 28–33 (2020)
2. IEEE 802 LAN/MAN Standards Committee et al., IEEE standard for high data rate wireless multi-media networks–amendment 2: 100 Gb/s wireless switched point-to-point physical layer, in *IEEE Std 802.15.3d-2017 (Amendment to IEEE Std 802.15.3-2016 as amended by IEEE Std 802.15.3e-2017)* (2017), pp. 1–55
3. S. Hanna, C. Dick, D. Cabric, Combining deep learning and linear processing for modulation classification and symbol decoding, in *Proceedings of the IEEE Global Communications Conference (GLOBECOM)* (Taipei, 2020), pp. 1–7
4. Y. Guo, H. Jiang, J. Wu, J. Zhou, Open set modulation recognition based on dual-channel LSTM model. arXiv: 2002.12037 (2020), pp. 1–4
5. AI HLEG, High-Level Expert Group on Artificial Intelligence, Ethics Guidelines for Trustworthy AI. AI HLEG, Tech. Rep., 2019
6. S. Thiebes, S. Lins, A. Sunyaev, Trustworthy artificial intelligence. Electron. Markets **31**(2), 447–464 (2021)
7. A. Avizienis, J.-C. Laprie, B. Randell, C. Landwehr, Basic concepts and taxonomy of dependable and secure computing. IEEE Trans. Depend. Secure Comput. **1**(1), 11–33 (2004)
8. B. Bauer, M. Ayache, S. Mulhem, M. Nitzan, J. Athavale, R. Buchty, M. Berekovic, On the dependability lifecycle of electrical/electronic product development: the dual-cone V-model. Computer **55**(9), 99–106 (2022)
9. B. Li, P. Qi, B. Liu, S. Di, J. Liu, J. Pei, J. Yi, B. Zhou, Trustworthy AI: from principles to practices. ACM Comput. Surv. **55**(9), 1–46 (2023)
10. N. Dreyer, T. Kürner, An analytical raytracer for efficient D2D path loss predictions, in *Proceedings of the 13th European Conference on Antennas and Propagation (EuCAP)* (Krakow, 2019), pp. 1–5
11. J.M. Eckhardt, C. Herold, B.K. Jung, N. Dreyer, T. Kürner, Modular link level simulator for the physical layer of beyond 5G wireless communication systems. Radio Sci. **57**(2), 1–15 (2022)
12. F. Taleb et al., Propagation of THz radiation in air over a broad range of atmospheric temperature and humidity conditions. Sci. Rep. **13**(20782), 1–13 (2023)
13. S. Chen, Y. Zhang, Z. He, J. Nie, W. Zhang, A novel attention cooperative framework for automatic modulation recognition. IEEE Access **8**, 15673–15686 (2020)
14. W. Xiao, Z. Luo, Q. Hu, A review of research on signal modulation recognition based on deep learning. Electronics **11**(17), 1–29 (2022)
15. J. Shi, S. Hong, C. Cai, Y. Wang, H. Huang, G. Gui, Deep learning-based automatic modulation recognition method in the presence of phase offset. IEEE Access **8**, 42841–42847 (2020)
16. H. Gu, Y. Wang, S. Hong, G. Gui, Blind channel identification aided generalized automatic modulation recognition based on deep learning. IEEE Access **7**, 110722–110729 (2019)
17. Y. Wang, M. Liu, J. Yang, G. Gui, Data-driven deep learning for automatic modulation recognition in cognitive radios. IEEE Trans. Veh. Technol. **68**(4), 4074–4077 (2019)
18. T.J. O'Shea, T. Roy, T.C. Clancy, Over-the-air deep learning based radio signal classification. IEEE J. Sel. Topics Signal Process **12**(1), 168–179 (2018)
19. G. Li, S.K.S. Hari, M. Sullivan, T. Tsai, K. Pattabiraman, J. Emer, S.W. Keckler, Understanding error propagation in deep learning neural network (DNN) accelerators and applications, in *Proceedings of the International Conference for High Performance Computing, Networking, Storage and Analysis (SC17)* (Denver, 2017), pp. 1–12

20. C. Szegedy, W. Zaremba, I. Sutskever, J. Bruna, D. Erhan, I.J. Goodfellow, R. Fergus, Intriguing properties of neural networks. arXiv: 1312.6199 (2014), pp. 1–10
21. I.J. Goodfellow, J. Shlens, C. Szegedy, Explaining and harnessing adversarial examples, in *Proceedings of the 3rd International Conference on Learning Representations (ICLR)* (San Diego, 2015)
22. A. Kurakin, I.J. Goodfellow, S. Bengio, Adversarial examples in the physical world, in *Artificial Intelligence Safety and Security* (Chapman and Hall/CRC, 2018), pp. 99–112
23. U. Jang, X. Wu, S. Jha, Objective metrics and gradient descent algorithms for adversarial examples in machine learning, in *Proceedings of the 33rd Annual Computer Security Applications Conference* (Orlando, 2017), pp. 262–277
24. S. Moosavi-Dezfooli, A. Fawzi, P. Frossard, DeepFool: a simple and accurate method to fool deep neural networks, in *Proceedings of the Conference on Computer Vision and Pattern Recognition (CVPR)* (Las Vegas, 2016), pp. 2574–2582
25. J. Chen, M.I. Jordan, M.J. Wainwright, HopSkipJumpAttack: a query-efficient decision-based attack, in *Proceedings of the IEEE Symposium on Security and Privacy (SP)* (San Francisco, 2020), pp. 1277–1294
26. P.-Y. Chen, H. Zhang, Y. Sharma, J. Yi, C.-J. Hsieh, ZOO: zeroth order optimization based black-box attacks to deep neural networks without training substitute models, in *Proceedings of the 10th ACM Workshop on Artificial Intelligence and Security* (Dallas, 2017), p. 15–26
27. N. Carlini, D. Wagner, Towards evaluating the robustness of neural networks, in *Proceedings of the IEEE Symposium on Security and Privacy (SP)* (Los Alamitos, 2017), pp. 39–57
28. M.-I. Nicolae, M. Sinn, M.N. Tran, B. Buesser, A. Rawat, M. Wistuba, V. Zantedeschi, N. Baracaldo, B. Chen, H. Ludwig, I. Molloy, B. Edwards, Adversarial robustness toolbox v1.2.0. Computing Research Repository (CoRR); arxiv: 1807.01069 (2018)
29. A. Nechi, A. Mahmoudi, C. Herold, D. Widmer, T. Kürner, M. Berekovic, S. Mulhem, Practical Trustworthiness Model for DNN in Dedicated 6G Application, in *2023 19th International Conference on Wireless and Mobile Computing, Networking and Communications (WiMob)*, pp. 312–317 (2023). https://doi.org/10.1109/WiMob58348.2023.10187759

Chapter 39
THz Networks with Intelligent Control and Digital Twining

Zied Ennaceur, Cao Vien Phung, Mounir Bensalem, André Drummond, and Admela Jukan

Abstract 6G campus networks are expected to include various new features compared to traditional 4G and 5G private network deployment, such as exploiting terahertz (THz) frequency bands and optimizing real-time networks. Furthermore, to enable future-proof network control, the network digital twin (NDT) reference architecture can be exploited to create an autonomic network system that provides intent-based interfaces and offers a complete adaptive network control tailored for campus network applications. This chapter presents the THz network control challenges and introduces a digital twin (DT)-based architecture to support THz systems. Some application scenarios are presented and discussed, ranging from direct (LOS) communication to multi-hop reconfigurable intelligent surface (RIS)-assisted mesh networks.

39.1 Introduction

In recent years, the DT concept has grown increasingly popular across various application domains and industries, signaling a significant shift toward digital integration. Specifically, in the realm of networking, the Internet Research Task Force (IRTF) has developed foundational concepts and a reference architecture for NDT, as detailed in their publication [1]. This architecture has become a critical reference for developing modern networks, particularly those that tackle research

Z. Ennaceur (✉) · C. V. Phung · M. Bensalem · A. Jukan
Technische Universität Braunschweig, Institut für Datentechnik und Kommunikationsnetze, Braunschweig, Germany
e-mail: zied.ennaceur@tu-bs.de; c.phung@tu-bs.de; mounir.bensalem@tu-bs.de; a.jukan@tu-bs.de

A. Drummond
Technische Universität Braunschweig, Institut für Datentechnik und Kommunikationsnetze, Braunschweig, Germany

University of Brasilia, Federal District, Brasilia, Brazil
e-mail: andre.drummond@tu-bs.de; andred@unb.br

© The Author(s) 2026
T. Kürner et al. (eds.), *Metrology for THz Communications*, Springer Series in Optical Sciences 256, https://doi.org/10.1007/978-3-032-01986-8_39

challenges related to creating artificial intelligence (AI)-driven DTs. The Internet draft from the IRTF highlights that NDT can be a crucial enabler for implementing intent-based networking (IBN) concepts. As DTs provide dynamic virtual replicas of physical networks, they offer an ideal platform for integrating IBN, which relies on defining network goals and objectives rather than specific configurations. By combining NDT with IBN, networks can achieve higher levels of automation, adaptability, and intelligence. There is a broad consensus among researchers and industry experts that NDT and IBN will play a pivotal role in developing future 6G mobile networks. These technologies promise to revolutionize how networks are managed and operated by providing real-time insights, predictive capabilities, and enhanced decision-making processes. However, seamlessly integrating 6G technology into the NDT framework presents a significant challenge for high-tech research and industry.

One of the most prominent use cases for 6G is campus networks for industry verticals applications, such as smart factories. Several key factors are critical for ensuring optimal performance in the context of a smart factory campus network. Among these, efficient indoor radio coverage, high bandwidth, and reliability are essential requirements. Additionally, mitigating impairments at the physical layer is crucial for maintaining robust communication [2]. Given these demands, a communication system operating in the THz frequency band, ranging from 0.1 to 10 THz, is expected to be highly effective [3]. This frequency band offers the potential to meet the stringent needs of such environments by providing the necessary data rates and coverage. Managing a THz communication system presents significant challenges, particularly in achieving robust control over the network. Additionally, incorporating THz systems into an NDT framework poses further complexities. The complexity primarily derives from the necessity to synchronize various transmission, reception, and reflection devices throughout the network. NDT enables the seamless integration and control of various network components by creating a virtual replica of the physical network. This allows for real-time simulation and optimization of network performance. With the proper use of intents, physical layer configurations can be automated, eliminating the need for standard manual adjustments in current THz communication systems. IBN allows operators to specify desired outcomes and performance criteria without detailing the specific configuration steps, enabling the network to adjust its settings to meet these goals autonomously.

This chapter presents the challenges of integrating the NDT Internet draft with a THz physical layer network. Key NDT components are incorporated into THz campus networks to address this, introducing an intent-based adaptive control module and a control plane module [4]. The NDT framework, functioning as an abstraction layer for the physical network, enables the intent-based control to interact exclusively with the DT while allowing it to explore and leverage various network scenarios to refine its decision-making process. This approach facilitates adaptive system configuration in response to channel state variations, such as those caused by mobility or transmission conditions, to enhance throughput and fault tolerance in a THz system [5].

39.2   Background

This section lays the conceptual foundations for realizing a NDT architecture for THz systems.

39.2.1   THz Systems

THz band communication (ranging from 0.1 to 10 THz) is considered a key technology for 6G and beyond to address the exponential growth of data traffic and the increasing demand for faster and more extensive wireless networks [6]. Although millimetre-wave (mm-wave) communications (30–300 GHz) below 100 GHz have been officially adopted in 5G, offering significantly more bandwidth than traditional microwave communication systems, they still fall short of meeting the future requirements of wireless communications growing data traffic [7, 8]. Therefore, to achieve the extremely high data rates needed for ultrafast wireless communication, THz spectral bands are crucial [9].

Despite its potential, THz communication faces several challenges due to the short wavelengths of THz signals, leading to high propagation losses and hardware limitations. THz signals experience significantly higher energy loss than signals in more commonly used frequency bands. This attenuation is caused by both free space propagation and atmospheric conditions, primarily molecular absorption, which substantially reduces the maximum range for THz signals [10]. As a result, the effective distance over which THz waves can travel is considerably shorter, making it necessary to find innovative solutions to maintain signal integrity over longer distances. Significant advancements have been made in addressing these challenges, leading to promising results in developing THz communication technologies. Researchers are exploring various techniques to mitigate propagation losses, such as employing beamforming, advanced modulation schemes, and RIS to effectively reflect and focus THz signals. These innovations aim to enhance signal strength and reduce attenuation, extending the practical range of THz communication.

The progress made in overcoming the inherent challenges of THz communication opens up new possibilities for achieving terabit-per-second data rates. Such high data rates can revolutionize various applications, including ultra-high-definition video streaming, real-time augmented and virtual reality experiences, and massive machine-type communications in Internet of Things (IoT) networks. Additionally, the extremely short wavelengths of THz waves enable precise and high-resolution imaging and sensing applications, which could be valuable in fields like medical imaging, security screening, and industrial quality control.

39.2.2 Network Digital Twin

While concept of NDT for communication networks is still evolving, it is generally understood as a DT specifically applied to communication networks. NDT is a vital enabler for the effective control and management of modern communication networks [11], creating dynamic virtual replicas that allow for risk-free analysis, diagnosis, and emulation of the physical network. Leveraging these capabilities, NDT can autonomously generate control decisions for the physical network [12]. When applied to cellular networks, the NDT concept involves constructing a comprehensive virtual model of the entire network infrastructure. This virtual model mirrors real-world cellular networks' physical, operational, and performance characteristics, enabling operators to optimize, monitor, and predict network behaviors and outcomes.

NDT significantly enhances the efficiency of network management and optimization. By providing insights into current network utilization, resource usage, and reconfiguration costs [13], NDT enables the implementation of optimal configurations to meet both present and future traffic demands. To achieve this, five key elements are essential: data, models, mapping, interfaces, and logic [1]. Unlike traditional offline network simulations, NDTs facilitate real-time decision-making based on network conditions and service requirements. This dynamic capability revolutionizes network management by providing a continuously updated virtual replica of the physical network that accurately reflects current operational states, traffic patterns, and user behaviors. As a result, NDTs have become increasingly valuable in evaluating network performance, a topic that has gained considerable attention in recent years. By offering a safe and cost-efficient environment for performance evaluation, NDTs provide accurate assessments of "what-if" scenarios [14]. Conventional simulations, often used in "what-if" performance evaluation, lack the flexibility and accuracy required by modern networks, making them unsuitable for NDT [14]. In contrast, data-driven methods are considered highly promising for achieving the necessary flexibility, fidelity, and efficiency, enabling more adaptable and responsive network evaluations.

Recognizing the potential of NDT, standardization organizations have initiated efforts to integrate NDT concepts into the industry. For instance, the IRTF has drafted a document outlining the NDT concept and architecture. These efforts aim to establish a standardized framework for implementing and utilizing NDT across various communication networks, ensuring consistency, interoperability, and scalability. Complementing these initiatives, the ITU-T released the ITU-T Y.3090 standard [15], which provides detailed requirements, architecture, and critical considerations for constructing a DT of the physical network. Additionally, the ITU-T standard includes specific use cases for DT networks, further enhancing NDT's practical application and relevance in the industry.

39.2.3 Intent-Based Networking

Intelligent networks need a clear definition of their core functionalities to operate effectively. One approach that has been outlined is the use of intents, which serves as a method for expressing these requirements and enabling a higher level of automation. Unlike traditional systems that follow rigid, predefined instructions, intents empower the network to decide on achieving desired outcomes. The concept of intent is gaining momentum and recognition within the industry, with established standards organizations like IRTF defining and endorsing its use. IRTF states, "Intent is defined as a set of operational goals (that a network is supposed to meet) and outcomes (that a network is supposed to deliver) defined in a declarative manner without specifying how to achieve or implement them" [16]. This definition emphasizes that intent involves expressing desired achievements, outcomes, or conditions to avoid without dictating the specific actions or strategies to be employed. Essentially, intent conveys what should be accomplished or preferred, leaving the method of achieving these goals open to interpretation [17].

Intent-driven management has been introduced to address the complexities of managing 5G services, reflecting the growing interest in more intelligent and autonomous management methods, with IBN emerging as one of the leading trends [18]. Intents serve as a way to articulate the specific requirements and expectations that an autonomous system must fulfil. They facilitate communication between management stakeholders, such as consumers and producers, by clearly conveying the requirements or constraints that must be considered by the management subsystem or system [16].

Developing fully autonomous networks is challenging due to the need to integrate diverse AI techniques [16] into a unified framework that can understand and act on intents. This requires creating advanced algorithms capable of interpreting and prioritizing intents in real time, enabling the network to adjust its behavior accordingly. The dynamic nature of network environments further complicates this task, as intents must be flexible enough to adapt to changing conditions and evolving user demands. Additionally, ensuring the security and reliability of autonomous networks is crucial, as they need to be resilient to potential threats and failures.

39.2.4 Data and Measurements

An NDT is an advanced platform for network emulation, enabling the prediction of network behavior, analysis of impacts, and planning for various scenarios. Complementing this, simulation plays a crucial role in DT technology, where the virtual model interacts bidirectionally with the physical entity in real time, ensuring that network emulation not only mirrors but also responds to real-world dynamics [19]. Unlike traditional simulations, NDT simulations offer a seamless connection between virtual and physical networks, allowing changes to be tested in a controlled environment. This interactive mapping distinguishes NDTs from conventional simulations, as it integrates model-based (e.g., emulation) analysis into real network operations, even supporting automated adjustments.

However, real-time interaction is not always necessary for all NDT use cases. For instance, when evaluating configuration changes or testing innovative techniques, the DT can function as an isolated simulation platform without relying on real-time telemetry data, as demonstrated in the THz use case. Nevertheless, regardless of the need for real-time interaction, robust data collection remains a critical enabling technology for building a NDT. It is the foundation for establishing a data repository essential to its operation.

Data is fundamental to building a NDT system. Large volumes of network data gathered from the physical network can be stored within the virtual twin network, serving as a unified data repository. This repository acts as the single source of truth, offering timely and accurate data to support various models [15]. To maximize the effectiveness of this data, a target-driven approach should be employed to gather data from diverse sources. The type, frequency, and method of data collection must align with the requirements of the NDT application. Various existing tools and techniques, such as SNMP, NETCONF, and telemetry, can gather different types of network data [1].

39.3   Network Digital Twin Architecture for THz Networks

We adopted the NDT reference architecture as presented in [1]. This reference architecture consists of three principal layers: real network, NDT, and application layers. The real network, which serves as the physical counterpart of an NDT, can include various components such as a mobile access network, transport network, mobile core, or backbone. This physical network layer forms the bottom layer in the NDT architecture. All network elements within the physical network exchange extensive data and control information with the NDT entity through southbound interfaces [15]. The middle layer, the NDT layer, serves as the core of the NDT system. The top layer is the network application layer. Here, network applications provide input to the NDT layer via northbound interfaces and deploy services through modelled instances. The envisioned application is a 6G THz Campus Network that supports smart factories.

Our proposed architecture, illustrated in Fig. 39.1, aligns with the NDT reference architecture. Still, it has been expanded beyond the conventional NDT framework to include two additional layers: intent management and the THz control plane layers. Our previous work introduced the latter as a foundational element for THz measurements, as detailed in [5]. We have integrated this layer into the NDT architecture by incorporating NDT capabilities and management mechanisms driven by newly defined intents. Consequently, the proposed architecture now comprises four distinct layers: the THz physical network layer, the THz control plane layer, the digital twin layer, and the intent management layer. While it is essential to

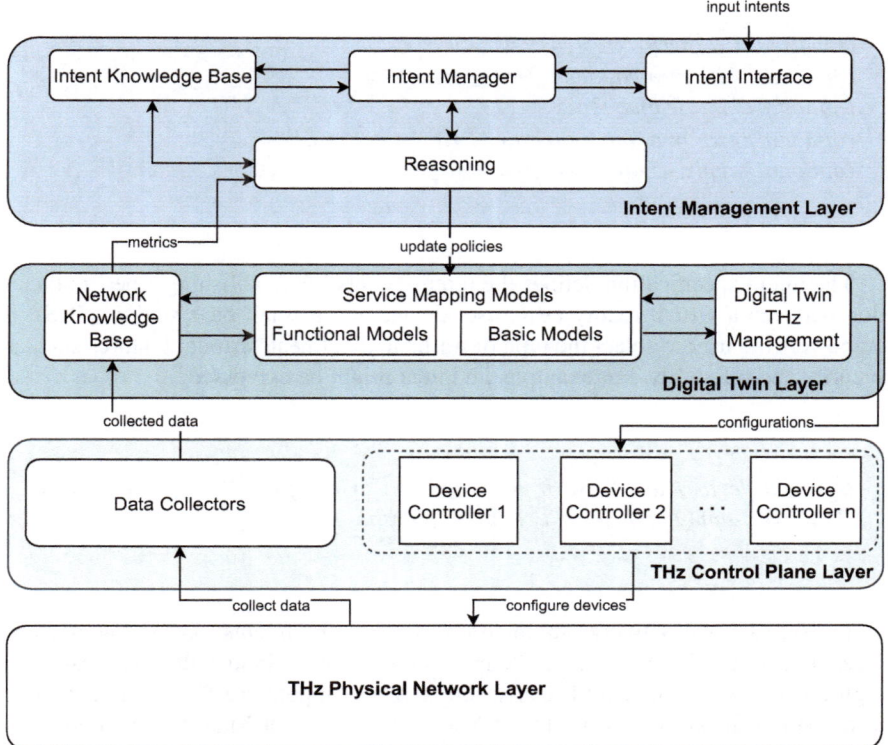

Fig. 39.1 The NDT architecture, proposed in [4]

acknowledge that this architecture is not the first to utilize DTs in THz systems, its integration into the NDT Internet draft architecture represents a novel approach. This innovation contributes to the existing body of knowledge by (a) significantly enhancing the capabilities of the adaptive control plane to mitigate physical layer impairments and (b) developing a novel intent-based management system tailored specifically for THz networks.

39.3.1   Intent Management Layer

The intent management layer is designed to automate the control and configuration of systems [20], providing a seamless interface between user commands and system operations. Intents are always customized to the particular application being considered. In the THz network case study, intent types are predefined objects that can be further extended to meet specific user requirements. The following outlines the design of a general intent input:

> *(required)* **Connect** *<src>* **to** *<dst>*
> *(optional)* **with min capacity** *<min_capacity>*
> *(optional)* **for** *<traffic_type>*
> *(optional)* **with max bit error rate (BER)** *<max_BER>*
> *(optional)* **apply** *<requirement><feature>*

The intent specification defines the requirements for establishing a new connection between a virtual reality (VR) user device and a small base station (SBS). It empowers the user to select the type of traffic and delineate critical features such as security and reliability. For example, an intent might be expressed:

> **Connect** *device1* **to** *BS1* **with min capacity** *3 Gbps* **for** *VR_traffic* **with max BER** 10^{-5} **apply** *high security* **apply** *high reliability*

In Fig. 39.2, the network administrator receives the intents via the Intent Interface. The Intent Interface is the entry point for user inputs through restricted high-level language or user-friendly forms. These inputs are then communicated to a component known as the Intent Manager. The Intent Manager has two main

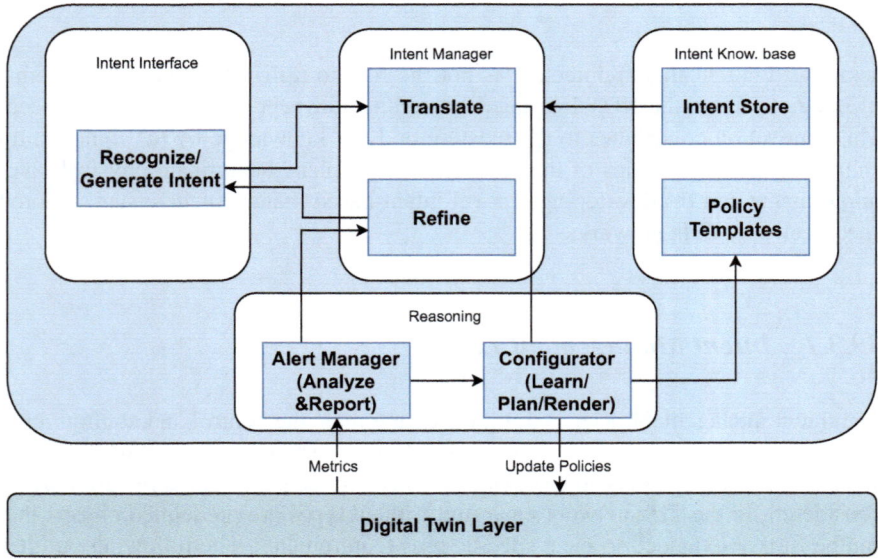

Fig. 39.2 Intent management layer schematic, proposed in [4]

functionalities: translate and refine, which are critical in translating and parsing these user commands into a structured and actionable format. This process utilizes advanced natural language processing techniques to ensure accurate interpretation of user intentions. By leveraging an enriched Intent Knowledge Base containing a comprehensive set of intent vocabularies, the Intent Manager can effectively process a wide range of user requests. The Intent Knowledge Base processes intents and stores policy templates. These templates are essential for the Reasoning component, which uses them to update decision-making policies dynamically. Subsequently, the manager activates the Reasoning component to evaluate the necessary actions. The Reasoning component evaluates various metrics from the Network Knowledge Base to make informed adjustments. It represents the system's intelligence, enabling automatic optimization of network services without needing explicit intervention from a network administrator.

For a clearer understanding of the information flow within this layer, we refer to the Intent Life Cycle from the [21] standard, which illustrates its primary functions, organized into two horizontal functional planes: fulfilment and assurance. Additionally, these functions are distributed across three (vertical) spaces. The User Space includes functions that interface between the network and the human user, such as the Intent Interface and Alert Manager in the proposed use case. The Translation or Intent-based System Space comprises functions that bridge the gap between intent users and the network operations infrastructure, specifically the Intent Manager and the Intent Knowledge Base. The Network Operations Space encompasses orchestration, configuration, and monitoring functions responsible for implementing the rendered intent and observing its effects on the network. In the proposed architecture, this corresponds to the Configurator component.

Going back to Fig. 39.2, upon receiving a report or alert from the Reasoning component, the Intent Manager informs the platform user and suggests a list of intents for further refinement. The Reasoning component has two primary functions: configuring policies (learning from past executions, planning the necessary actions, and making decisions) and managing alerts, which analyze the network telemetry and convey relevant information to the user. Considering the intent example expressed earlier, some sample output policies could be defined as follows:

Policy 1: Allocate a dedicated bandwidth of at least 3 Gbps on the link between Device1 and BS1 specifically for VR traffic.

Policy 2: Implement Quality of Service (QoS) policies that guarantee latency, jitter, and packet loss parameters are within the acceptable range for VR applications.

Policy 3: Enable forward error correction (FEC) mechanisms and adjust the modulation and coding schemes to maintain the BER below the specified threshold.

When a new intent is introduced, the Configurator validates the translated intent by verifying the existence of the relevant connection and its associated constraints. If the connection is found and requires modification, such as an updated capacity constraint, the Configurator within the Reasoning component will request an update to the appropriate modules in the DT layer below. Additionally, when new data–such as the measured BER—is obtained, the Alert Manager reviews the data to detect any violations of the predefined intents. This allows the Reasoning component to continuously assess whether the desired BER for the transmission is being achieved.

39.3.2 Digital Twin Layer

In Fig. 39.3, we propose three primary components within the DT layer: the Network Knowledge Base, service mapping models (SMMs), and digital twin THz management. Each component plays a vital role in creating a comprehensive

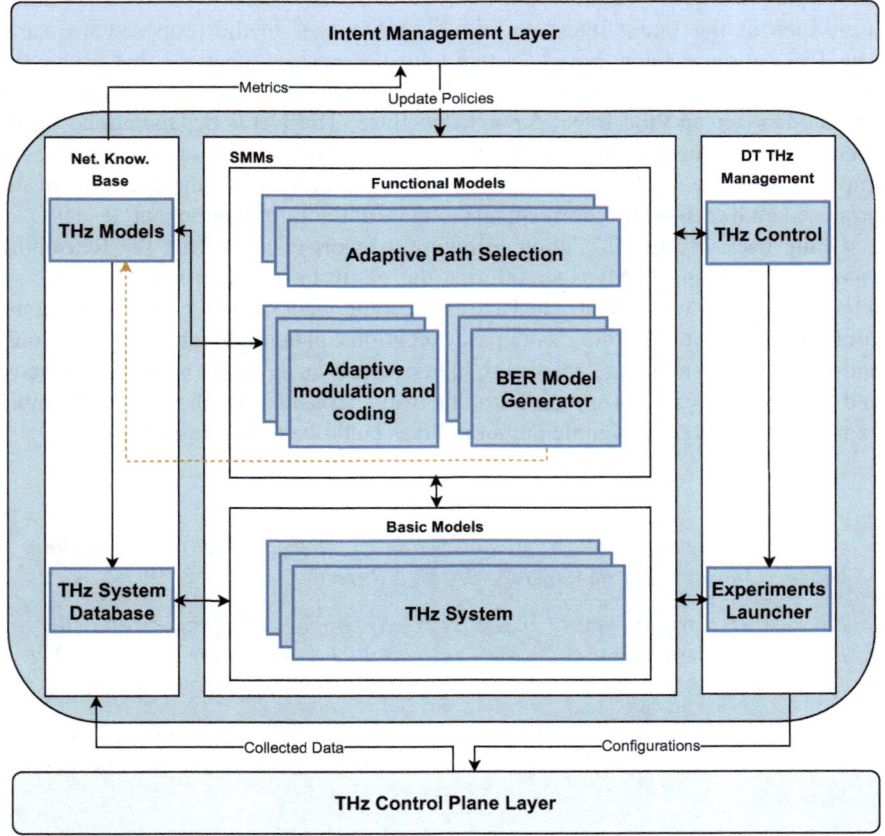

Fig. 39.3 Digital twin layer schematic, based on [4]

digital network representation. The Network Knowledge Base is a repository for telemetry data collected from various Data Collectors in the control plane layer. This data serves multiple purposes, providing valuable insights for different SMMs and supporting the Reasoning component in the intent management layer. The Network Knowledge Base enables better decision-making and optimizes network performance by storing and organizing telemetry data.

SMMs are divided into two categories: functional models and basic models. Functional models are network services that provide data model instances for various network applications, enhancing network flexibility and programmability, and allowing for dynamic adaptation to changing requirements and conditions. In this case, three functional models contribute uniquely to the network's adaptability and performance. The first model involves adaptive modulation and coding, where we utilize an algorithm proposed by [5] to calculate predicted BER values using pre-calculated THz models from the Network Knowledge Base. This helps generate the most optimal adaptive coding and modulation schemes with the highest throughput and code rate, thereby minimizing the BER. The generated parameters are then returned to the transmitter and receiver controllers, allowing them to configure the devices in the physical layer accordingly. The second model is the BER Model Generator, which requires input parameters such as the center THz frequency, modulation schemes, and the transmission distance between the transmitter and receiver. It also includes essential embedded parameters like transmission power, antenna gain, and simulated thermal noise. The generator outputs BER values across various transmission distances, modulation schemes, bandwidths, and THz frequencies. Finally, the third model involves adaptive path selection in a THz mesh network, where the SBS transfers data to VR users through intermediate RISs. As VR users move, the THz system adapts the transmission paths to optimize scheduling time and avoid device interference, resulting in improved network throughput.

Basic models refer to the virtual representation of network element models, providing a foundational structure that supports the overall DT architecture. These models ensure that each network element is accurately represented within the DT, enabling precise simulations and analyses. In this architecture, this component serves as a concrete and real-time representation of the physical network, defining and constructing the network topology models, link topology, and other relevant information of the THz physical network.

Digital twin THz management oversees the life cycle of services operating within the DT environment. This component is essential for managing advanced THz communication services, such as position prediction, adaptive modulation, adaptive coding, beamforming, and device discovery. These services are integral to achieving optimal performance and efficiency in THz networks. For instance, the adaptive control proposed in [5] would be implemented within this component, allowing for real-time adjustments and enhancements. Two key interfaces are defined to facilitate seamless integration and interaction between the DT and other network layers: Collected Data encompasses both real-time telemetry data and other data gathered over time and Configurations. These interfaces enable effective communication and information exchange between the DT and the control plane layers. The intent management layer also shares updated policies and metrics to ensure coherent operation and alignment with overall network objectives.

The DT layer provides a robust framework for modelling, managing, and optimizing network operations. Leveraging real-time data and advanced modelling techniques enables proactive decision-making and enhances the network's ability to adapt to evolving demands.

39.3.3   THz Control Plane Layer

Figure 39.4 shows the control plane layer, which is crucial in managing all the signaling required to monitor and control devices within the physical network layer. This layer includes the Device Controller modules, specifically designed to interface with various network hardware, ensuring seamless communication and coordination across the network. Additionally, it consists of the Data Collectors module, which is responsible for connecting to and gathering telemetry data from all the deployed sensing devices.

In traditional network architectures, the control plane layer often carries the burden of complex control loops and decision-making processes. However, the NDT architecture introduces a significant improvement by offloading these complex tasks to the DT layer and the intent layer. Unlike architectures that solely rely on the control plane layer, such as those described in [5], the NDT architecture enables us to simplify the control plane by removing intricate control loops like the adaptive control plane architecture. The NDT architecture shifts the service and decision-making module to the DT layer, allowing for a more streamlined and

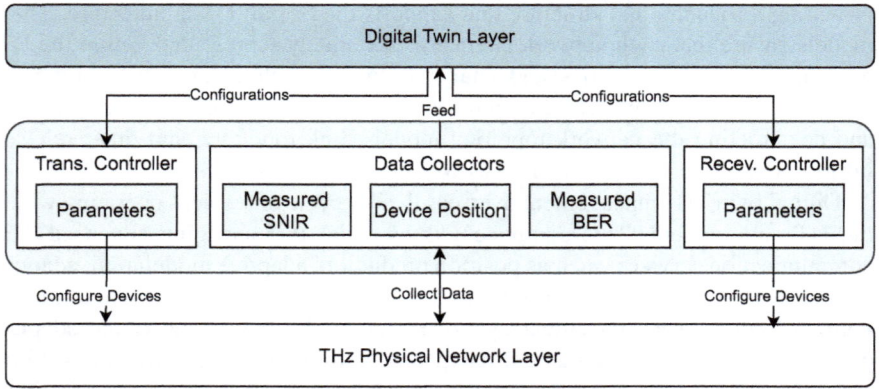

Fig. 39.4 THz control plane layer schematic, based on [4]

efficient control plane. This separation of responsibilities reduces the control plane's complexity and enhances the overall network performance.

39.3.4   THz Physical Network Layer

The THz physical network layer network shown in Fig. 39.5 comprises various mobile communication devices, including robots and factory workers who utilize high-resolution augmented or virtual reality (augmented reality (AR)/VR) devices as receivers. A SBS serves as the transmitter in this setup. Establishing a THz link requires careful consideration of several critical physical layer characteristics, such as the extremely wide bandwidth (ranging from tens to hundreds of gigahertz), significant propagation loss caused by molecular absorption, and the vulnerability to LOS blockages [3].

In this scenario, we incorporate RISs, advanced devices capable of reflecting and enhancing the THz signal [22]. These devices are crucial in maintaining robust communication links, especially in challenging environments. We assume that users within the network are mobile, meaning they can move around freely within the

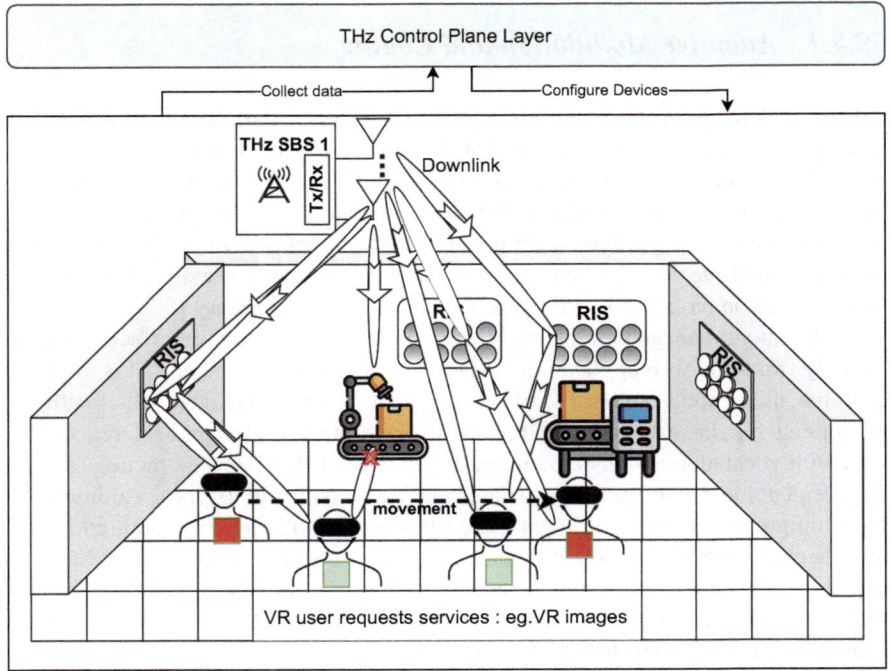

Fig. 39.5   THz physical network layer schematic, based on [4]

coverage area. Each user is primarily serviced directly by the SBS. However, if the direct link is weak or obstructed, communication is maintained through a multi-hop path using an RIS device. This ensures that users continue to receive reliable connectivity even in the presence of physical obstacles. The network environment is characterized by various obstacles, such as machines, walls, and human operators, which can block the direct communication line between the SBS and the users. Overall, the THz physical layer network is designed to accommodate the complexities of modern communication needs, leveraging advanced technologies like RIS to overcome the inherent challenges of high-frequency transmission.

39.4 Network Digital Twin Architecture Use Cases

Figure 39.6 illustrates the schematic workflow of the proposed architecture. We have adopted two scenarios: one where modulation and coding adjustments are necessary due to changes in channel conditions and another where a THz mesh network is considered. In the latter scenario, the SBS transfers data to VR users through intermediate RISs.

39.4.1 Adaptive Modulation and Coding

Advanced VR applications are anticipated for future smart factories, relying on high-speed wireless connections within the THz frequency band. This technology will allow workers to monitor factory operations in real time through immersive 360° video streams. Unfortunately, THz communication is vulnerable to disruptions and environmental influences, leading to reduced data transfer rates. To mitigate this issue, a system capable of dynamically adjusting THz transceiver settings for coding and modulation based on varying channel conditions is necessary [5, 23].

In the intent management layer shown in Fig. 39.6, the Intent Interface is a user-friendly front-end web application designed to interpret user intents. This interface captures user inputs and forwards them to the Intent Manager, which utilizes predefined regular expressions (Regex) to parse the text. It cross-references this data with vocabularies registered in the Intent Knowledge Base, extracting specific requirements and constraints associated with different intent types. Additionally, the Configurator oversees all intent requests, ensuring that newly submitted intents are checked against previously established intents stored in the Intent Store to manage redundancy and maintain coherence. When the Configurator determines that a connection creation or update is needed, it generates a hypertext transfer protocol (HTTP) request to the THz Control.

In the physical network layer, we consider an example of one static THz SBS that transfers its data to one dynamic VR user. The data source is coded with FEC and then is modulated before sending it over THz channel to the VR user. The original

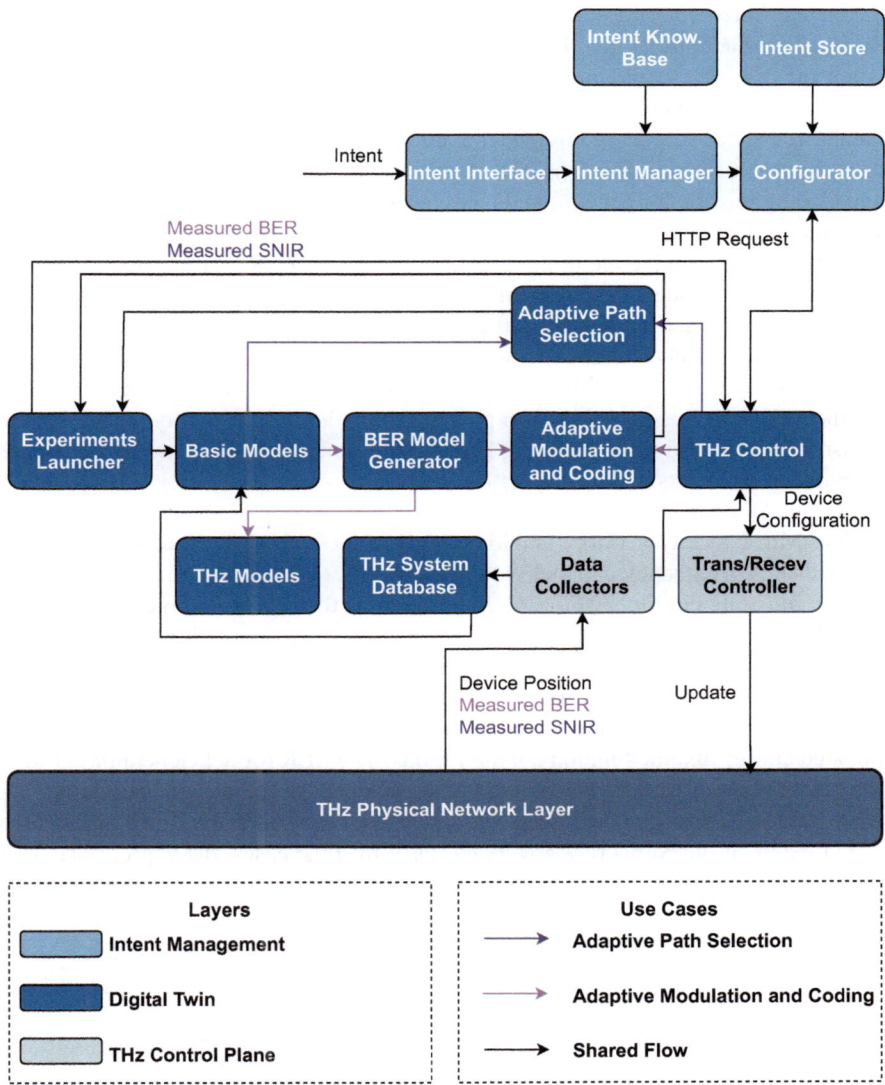

Fig. 39.6 Schematic workflow of the proposed architecture

data is recovered after the demodulation and coding processing on the VR user side. To configure adaptively the THz system with coding and modulation techniques when the VR user is moving, Data Collectors periodically gather data regarding Device Position and measured BER, storing this information in the THz system database. Device discovery components notify the control plane whenever the VR user changes their position. The THz control will require the Adaptive Modulation and Coding module for generating the best adaptive coding and modulation schemes with the highest throughput and code rate to reduce the BER based on the current device positions and computing a set of predicted BER values by using pre-calculated THz models. Suppose any new configuration of coding and modulation schemes is generated. In that case, the new configuration parameters are sent to the transmitter and receiver controller, which can adaptively configure the devices at the physical network layer. Simultaneously, any new device configurations generated by the DT THz Management prompt updates to the device configuration by the transmitter and receiver controllers. This ensures that the system remains up to date with user movements and device settings.

In the DT layer, upon receiving a connection request with specified parameters, such as Min Capacity = 3 Gbps and Max_BER = 10^{-5}, the THz Control may opt for a 294.84 GHz frequency channel with a bandwidth of 2.16 GHz, drawing upon the dataset illustrated in Fig. 39.7a [24]. Subsequently, the system initiates the adaptive modulation and coding process, which leverages the requested bandwidth and the dataset provided by the BER Model Generator, as depicted in Fig. 39.7b [24], to estimate the distance between the VR user and the SBS. As the VR user moves, causing variations in the THz channel conditions, the module dynamically generates updated coding and modulation configurations. The BER Model Generator, through the deployment of parallel instances of basic models, computes various sets of THz BER values. These generated datasets are then systematically stored within the THz models of the Network Knowledge Base. In some cases, the appropriate BER

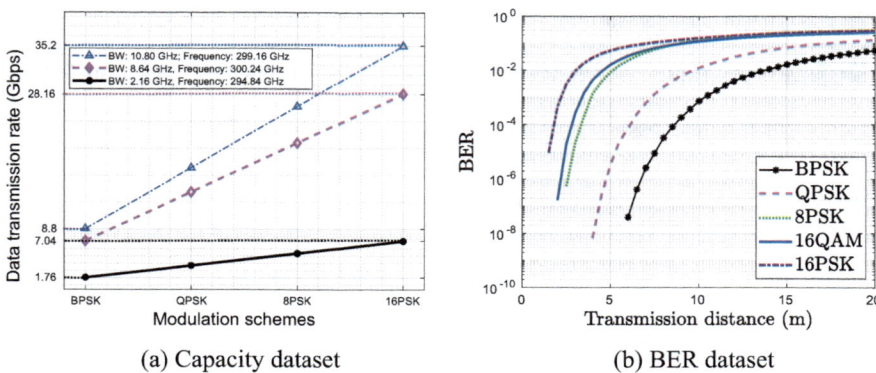

(a) Capacity dataset   (b) BER dataset

Fig. 39.7 Sample THz models for several modulation schemes, from [4]. (**a**) Capacity dataset. (**b**) BER dataset

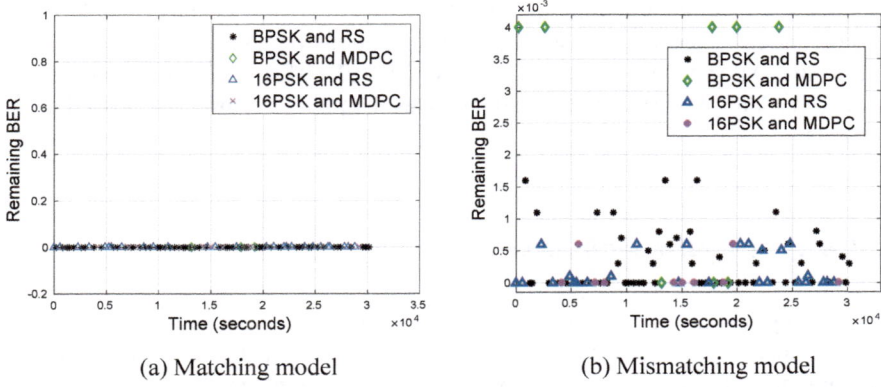

(a) Matching model   (b) Mismatching model

Fig. 39.8 Remaining BER for selected modulation and coding schemes, from [4]. (**a**) Matching model. (**b**) Mismatching model

dataset may not be available in the THz models when a new intent arrives, resulting in a mismatch with the requested parameters. In such instances, NDT will operate using the available dataset until the BER Model Generator produces the appropriate BER model. Once the new dataset is generated, the THz Control initiates the adaptive modulation and coding process to generate updated recommendations for modulation and coding schemes. This process is crucial, as demonstrated in Fig. 39.8a, where the correct application of BER models leads to optimal performance. Conversely, when mismatched models are used, the remaining BER after decoding is not optimized and, as illustrated in Fig. 39.8b, may sometimes exceed zero.

When a suitable THz model that matches the actual bandwidth in use is available, the adaptive modulation and coding control can function effectively, achieving a BER of zero. However, if only mismatched models are available, the system should select a model with a higher bandwidth [4] to optimize transmission, potentially achieving a BER close to zero, which may suffice to meet the required intent. If this adjustment is not possible, the observed BER values will be relatively high, compelling the THz system to operate under suboptimal conditions until the DT generates a new model. It is important to emphasize that without the use of the DT, there is no guarantee of matching models.

The basic models employ a network simulator to virtually represent all network elements, providing the flexibility to create multiple instances and evaluate "what-if" scenarios. This capability is essential for predicting network behavior. The basic models are periodically updated with the latest state information from the THz System Database to maintain near real-time synchronization with the physical layer. This ensures that the virtual representations remain accurate and reflect current network conditions.

The proposed coding and modulation configurations undergo evaluation by the Experiments Launcher using an instance of the DT models, which incorporates all the application programming interfaces (APIs) from the basic models. The Experiments Launcher assesses these configurations by analyzing metrics derived from the DT instance, such as the measured BER. This data is then fed back to the THz Control, which checks and validates the policy requirements based on the results.

39.4.2 Adaptive Path Selection

In the physical layer, we consider a THz mesh network, as shown in Fig. 39.5. The SBS transfers its data to the VR users via the intermediate RISs. For simplicity, assume that if one VR user is moving, the THz systems need to change adaptive transmission paths with the VR user to optimize the transmission scheduling time to avoid interference among devices, resulting in improved network throughput. To this end, and similar to the use case described in Sect. 39.4.1 and illustrated in Fig. 39.6, Data Collectors periodically gathered data on device positions and measured signal-to-interference-and-noise ratio (SINR) among the SBS, the VR user, and the possible interference sources, storing this information in the THz system database. Whenever the VR user changes their position, device discovery components notify the control plane, the THz control will require the Adaptive Path Selection module for generating the best transmission path with less interference, based on the current device positions and computing SINR values between the SBS, the VR user, and the possible interference sources. Before generating the new transmission path configurations, the THz systems must generate an adaptive interference graph, defined in [25].

The interference graph summarizes all possible interference paths in the network based on the measured SINR values and the SINR threshold values defined for each VR user. The adaptive path selection configurations with less interference are generated with that interference graph, which the authors in [26] compute by the integer linear programming (ILP) method. Suppose any new configuration of adaptive path selection is generated. In that case, the new path selection parameters are sent to the transmitter and receiver controller, which can adaptively configure the devices at the physical network layer.

An example of the performance evaluation with interference graphs is presented in Fig. 39.9, whereas the conflict complexity is the number of conflicts among devices in the interference graphs. The higher the conflict complexity is, the less the network throughput is. Different strategies can be considered to build an interference graph. A simple, naive solution that only considers the physical overlap between communication pairs can be used. But it disregards the physical impairments (SINR measurements) in the THz systems. For example, the circles (ZIM) in Fig. 39.9 represent the naive solution with the highest conflict complexity. In contrast, the remaining strategies (DCS/ICS) will consider the SINR calculations of the VR users and compare them with the threshold values for building the interference graph. Hence, there exists a large number of redundant conflicts in the naive solution.

Fig. 39.9 Conflict mapping complexity for several strategies (less is better)

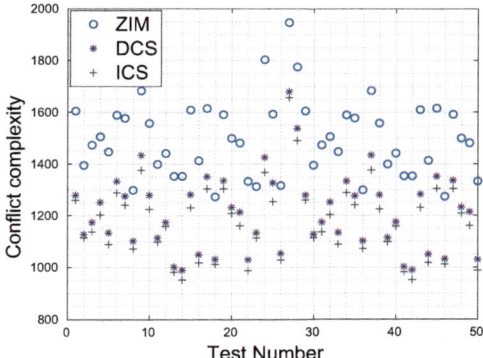

This analysis fits with devices operated on the THz frequency bands, on which the interference is sometimes insignificant due to the high path loss over longer transmission distances, whereas the naive solution still considers that interference as conflict, which is unnecessary. Creating redundant conflicts can result in fewer path selections and increase device transmission scheduling time. As a result, the network throughput gets worse. Based on the data analyzed in Fig. 39.9, the control plane can select a suitably adaptive interference graph solution based on the positions of the VR users and the SINR values measured. It is crucial to note that without the use of a DT, there is no way to evaluate different strategies to ensure an adaptive path selection procedure.

39.5   Summary and Conclusion

This chapter introduces an innovative architecture to facilitate intelligent control of 6G campus networks by integrating the NDT Internet draft into a THz campus network environment. The architecture enhances the existing NDT reference framework by incorporating intent-based management that is aligned with the latest standards, aiming to provide a more adaptive and responsive network control mechanism.

Two use cases have been presented: one focuses on adaptive coding and modulation techniques for THz transmissions, and the other examines the path selection on THz mesh networks. In the latter case, the SBS transmits data to VR users through intermediate RISs. A specific intent is proposed, along with implementing an adaptive control loop. This control loop is engineered to continuously evolve and optimize performance in response to the dynamic conditions typically encountered in THz communication. This approach demonstrates the feasibility of intent-based management in future 6G networks and highlights its potential to significantly enhance the efficiency and reliability of network operations in complex environments.

References

1. C. Zhou, H. Yang, X. Duan, D. Lopez, A. Pastor, Q. Wu, M. Boucadair, C. Jacquenet, *Digital Twin Network: Concepts and Reference Architecture*, Internet Engineering Task Force, Internet-Draft (2022), work in Progress
2. M. Wen, Q. Li, K.J. Kim, D. López-Pérez, O.A. Dobre, H.V. Poor, P. Popovski, T.A. Tsiftsis, Private 5G networks: concepts, architectures, and research landscape. IEEE J. Sel. Top. Sign. Proces. **16**(1), 7–25 (2022)
3. C. Han, Y. Wang, Y. Li, Y. Chen, N.A. Abbasi, T. Kürner, A.F. Molisch, Terahertz wireless channels: a holistic survey on measurement, modeling, and analysis. IEEE Commun. Surv. Tutorials **24**(3), 1670–1707 (2022)
4. Z. Ennaceur, M. Bensalem, C.V. Phung, A.C. Drummond, A. Jukan, Enabling 6G campus networks intelligent control with digital twin: a case study, in *Proceedings IEEE Network Operations and Management Symposium (NOMS)*, Seoul (2024), pp. 1–6
5. C.V. Phung, A. Jukan, Increasing fault tolerance and throughput with adaptive control plane in smart factories, in *Proceedings of IEEE Global Communications Conference (GLOBECOM)*, Rio de Janeiro (2022), pp. 1831–1837
6. A.J. Seeds, H. Shams, M.J. Fice, C.C. Renaud, TeraHertz photonics for wireless communications. J. Lightwave Technol. **33**(3), 579–587 (2015)
7. Z. Chen, X. Ma, B. Zhang, Y. Zhang, Z. Niu, N. Kuang, W. Chen, L. Li, S. Li, A survey on terahertz communications. China Commun. **16**(2), 1–35 (2019)
8. I.F. Akyildiz, C. Han, Z. Hu, S. Nie, J.M. Jornet, Terahertz band communication: an old problem revisited and research directions for the next decade. IEEE Trans. Commun. **70**(6), 4250–4285 (2022)
9. X.-W. Yao, C.-C. Wang, W.-L. Wang, C. Han, Stochastic geometry analysis of interference and coverage in Terahertz networks. Nano Commun. Netw. **13**, 9–19 (2017)
10. M. Pengnoo, M.T. Barros, L. Wuttisittikulkij, B. Butler, A. Davy, S. Balasubramaniam, Digital twin for metasurface reflector management in 6G terahertz communications. IEEE Access **8**, 114 580–114 596 (2020)
11. P. Almasan et al., Network digital twin: context, enabling technologies, and opportunities. IEEE Commun. Mag. **60**(11), 22–27 (2022)
12. I. Vil`a, O. Sallent, J. Pérez-Romero, On the design of a network digital twin for the radio access network in 5G and beyond. Sensors **23**(3), 1–17 (2023)
13. C. Zhou, J. Gao, M. Li, X. Sherman Shen, W. Zhuang, Digital twin-empowered network planning for multi-tier computing. J. Commun. Inf. Netw. **7**(3), 221–238 (2022)
14. L. Hui, M. Wang, L. Zhang, L. Lu, Y. Cui, Digital twin for networking: a data-driven performance modeling perspective. IEEE Netw. **37**(3), 202–209 (2023)
15. ITU-T, Digital twin network – Requirements and architecture, The International Telecommunication Union Telecommunication Standardization Sector (ITU-T), Recommendation ITU-T Y.3090 (2022)
16. P.H. Gomes, M. Buhrgard, J. Harmatos, S.K. Mohalik, D. Roeland, J. Niemöller, Intent-driven closed loops for autonomous networks. J. ICT Stand. **9**(2), 257–290 (2021)
17. Tm forum introductory guide – intent in autonomous networks v1.3.0 (ig1253), https://www.tmforum.org/resources/how-to-guide/ig1253-intent-in-autonomous-networks-v1-3-0/. Accessed 05 Aug 2024
18. M. Bensalem, J. Dizdarević, A. Jukan, Benchmarking various ML solutions in complex intent-based network management systems, in *Proceedings of 45th Jubilee International Convention on Information, Communication and Electronic Technology (MIPRO)*, Opatija (2022), pp. 476–481
19. M. Liu, S. Fang, H. Dong, C. Xu, Review of digital twin about concepts, technologies, and industrial applications. J. Manuf. Syst. **58**, 346–361 (2021)

20. M. Bensalem, J. Dizdarević, F. Carpio, A. Jukan, The role of intent-based networking in ICT supply chains, in *Proceedings of IEEE 22nd International Conference on High Performance Switching and Routing (HPSR)*, Paris (2021), pp. 1–6
21. A. Clemm, L. Ciavaglia, L.Z. Granville, J. Tantsura, Intent-based networking – concepts and definitions. Internet Eng. Task Force RFC – Informational (2022). https://datatracker.ietf.org/doc/rfc9315/
22. S. Dash, C. Psomas, I. Krikidis, I.F. Akyildiz, A. Pitsillides, Active control of thz waves in wireless environments using graphene-based RIS. IEEE Trans. Antennas Propag. **70**(10), 8785–8797 (2022)
23. C.V. Phung, Z. Ennaceur, A. Drummond, A. Jukan, On the adaptive THz system for mobile VR users in smart factories, in *Proceedings of 47th MIPRO ICT and Electronics Convention (MIPRO)*, Opatija (2024), pp. 779–784
24. C.V. Phung, C. Herold, D. Humphreys, T. Kürner, A. Jukan, Performance analysis of MDPC and RS codes in two-channel THz communication systems, in *Proceedings of 45th Jubilee International Convention on Information, Communication and Electronic Technology (MIPRO)*, Opatija (2022), pp. 482–487
25. C.V. Phung, A. Drummond, A. Jukan, Enhancing path selections with interference graphs in multihop relay wireless networks, in *IEEE Globecom: 5th Workshop on Emerging Topics in 6G Communications*, Cape Town (2024)
26. C.V. Phung, A. Drummond, A. Jukan, Maximizing throughput with routing interference avoidance in RIS-assisted relay mesh networks, in *Proceedings of 47th MIPRO ICT and Electronics Convention (MIPRO)*, Opatija (2024), pp. 736–741

Chapter 40
Device Discovery

Tobias Doeker, Anouar Nechi, Mladen Berekovic, and Thomas Kürner

Abstract Due to the use of highly directional antennas in future terahertz (THz) communication systems, precise alignment between the transmitter (TX) and the receiver (RX) is mandatory. Thus, the process by which the TX and the RX find each other is a key challenge for these systems, which is called device discovery. In this chapter, the challenges of device discovery are explained in more detail, followed by a possible solution to overcome the challenges: a device discovery approach supported by well-known compressed sensing techniques. To make this approach suitable for future systems, a hardware-based acceleration of the approach is discussed in addition.

40.1 Alignment Challenges in THz Communications

THz communications in the low THz frequency range (0.1 to 1 THz) offer significant potential for high data rate transmission, owing to the substantial bandwidths available in this spectrum. However, these benefits come with challenges, particularly due to high penetration and path losses. To address these issues, the use of high-gain antennas is anticipated [1]. Nevertheless, this solution presents new challenges, as high-gain antennas are typically highly directional, with a half-power beamwidth (HPBW) of less than $10°$. For successful communication, it is crucial that the TX and RX are accurately aligned to maximize received power and thereby enhance the signal-to-noise ratio (SNR). This alignment is especially challenging with narrow beams, as even slight misalignments can lead to significant reductions in received power [2]. The task of locating and aligning the TX and RX, known as

T. Doeker (✉) · T. Kürner
Technische Universität Braunschweig, Institut für Nachrichtentechnik, Braunschweig, Germany
e-mail: t.doeker@tu-braunschweig.de; t.kuerner@tu-braunschweig.de

A. Nechi · M. Berekovic
Universität zu Lübeck, Institut für Technische Informatik, Lübeck, Germany
e-mail: anouar.nechi@uni-luebeck.de; mladen.berekovic@uni-luebeck.de

© The Author(s) 2026
T. Kürner et al. (eds.), *Metrology for THz Communications*, Springer Series in Optical Sciences 256, https://doi.org/10.1007/978-3-032-01986-8_40

device discovery, is vital for wireless communications, particularly in the context of THz communications, where high precision is essential.

Device discovery has always been a cornerstone of wireless communications, with several methods developed and implemented for existing systems. However, as previously mentioned, THz communications present new challenges that make some existing methods unsuitable. In current systems, the initial rough alignment is often accomplished through omnidirectional power reception [3]. However, given the limitations on the maximum output power of existing transceivers in the low THz range, omnidirectional approaches are not feasible. Currently, iterative search algorithms—which scan the entire angular range in various directions—are considered the most promising approach for THz communications [4]. Nonetheless, these methods can be time-consuming. This chapter introduces an approach that builds on the fundamental concept of iterative search but improves its efficiency by reducing the number of required measurement steps. Additionally, further improvements in speed are achieved through hardware acceleration of the proposed method presented afterward.

40.2 Compressed Sensing-Assisted Device Discovery

The following approach is based on an iterative search of the TX and/or RX in all directions, as described, for example, in [4]. In this method, the TX and/or RX measures the path gain for each direction—typically in angular steps equal to the HPBW of the antenna—and selects the direction with the highest path gain, which corresponds to the lowest path loss.

40.2.1 Basic Concept

The base of the following approach is the so-called power angular profile (PAP), which contains the path gain for each measured angle of depature (AOD)-angle of arrival (AOA) combination. In this case, the corresponding path gain is calculated as the sum over the measured channel impulse response (CIR), so that the PAP is given by

$$PAP_{\varphi_{AOD},\varphi_{AOA}} = \sum_n |h(n\Delta t, \varphi_{AOD}, \varphi_{AOA})|^2 \tag{40.1}$$

where $h(n\Delta t, \varphi_{AOD}, \varphi_{AOA})$ denotes the CIR for the corresponding AOD and AOA. Assuming the same step size for both AOD and AOA, the total number of measurements is given by k^2, where the number of steps k is given by

$$k = \left\lceil \frac{360°}{\Delta\varphi} \right\rceil \tag{40.2}$$

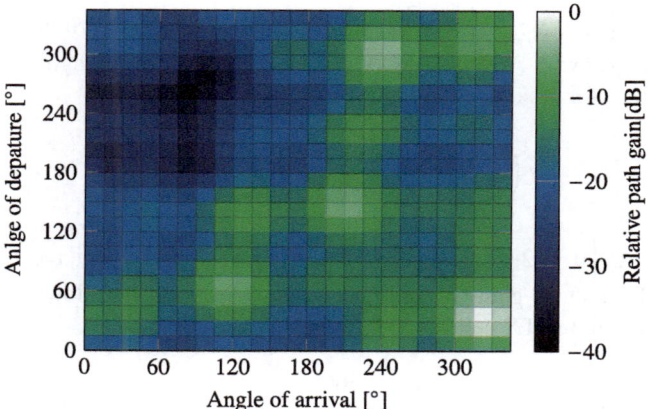

Fig. 40.1 Exemplary PAP based on ray-tracing simulations in an meeting room scenario

where $\Delta\varphi$ denotes the step size. Figure 40.1 illustrates an exemplary PAP based on ray-tracing simulations.

Referring to (40.2), it can be seen that the total number of measurements increases quadratically, leading to a large number of measurements needed for high precision, i.e., high angular resolution and small step size $\Delta\varphi$. Therefore, the goal of this approach is to reduce the required measurements without compromising precision.

For each measurement step, the measured CIR, which forms the basis for the cumulative path gain in the corresponding measurement step, is given by the multipath components (MPCs) weighted by the antenna gain for the corresponding AOD and AOA. However, in a static scenario, the MPCs are fixed relative to the TX and the RX, so the measured path gain in the PAP is related to the paths within the scenario by the antenna radiation pattern. In [5], it is shown that this relation can be reformulated as a linear system:

$$\mathbf{y} = \mathbf{A}\mathbf{x} \tag{40.3}$$

where $\mathbf{y} \in \mathbb{R}^m$ denotes the vectorized form of the PAP, $\mathbf{A}$ is the $m \times n$ measurement matrix related to the antenna radiation pattern, and $\mathbf{x} \in \mathbb{R}^n$ denotes the vectorized form of the paths.

Assuming that the PAP is measured with an angular resolution of β, the output vector is given by [5]

$$\mathbf{y}\left(1 + j_2 + \left\lceil \frac{360°}{\beta} \right\rceil i_2\right) = PAP_{i_2\beta, j_2\beta} \tag{40.4}$$

with $i_2, j_2 \in \left\{ 0, 1, 2, \ldots, \left\lceil \frac{360°}{\beta} \right\rceil - 1 \right\}$. The paths are also distributed over a two-dimensional grid with AOD and AOA $MPC_{\varphi_{AOD}, \varphi_{AOA}}$ with an angular resolution α in a discretized form, so that the input vector is given by [5]

$$\mathbf{x} \left(1 + j_1 + \left\lceil \frac{360°}{\alpha} \right\rceil i_1 \right) = MPC_{i_1\alpha, j_1\alpha} \tag{40.5}$$

with $i_1, j_1 \in \left\{ 0, 1, 2, \ldots, \left\lceil \frac{360°}{\alpha} \right\rceil - 1 \right\}$. It can be shown that the relation with respect to the antenna gain G—in linear values—is given as follows, assuming the same antennas at the TX and RX [5]:

$$\mathbf{A} \left(1 + j_2 + \left\lceil \frac{360°}{\beta} \right\rceil i_2, 1 + j_1 + \left\lceil \frac{360°}{\alpha} \right\rceil \right) = G(i_2\beta - i_1\alpha) \cdot G(j_2\beta - j_1\alpha) . \tag{40.6}$$

Due to the measurement of the PAP, the output vector $\mathbf{y}$ is known, as well as the used antennas, and therefore the antenna gain G. The vector containing the MPCs $\mathbf{x}$ is unknown. However, instead of measuring the PAP with high angular resolution, reconstructing the unknown vector containing the MPCs provides more detailed information about the direction of the MPCs in a faster way. Therefore, the linear system given in (40.3) must be solved in terms of the variable $\mathbf{x}$. For this to be efficient, the angular resolution of the measured PAP—and thus the number of values in $\mathbf{y}$, m—should be much lower than the angular resolution of the reconstructed MPCs, i.e., the number of values in $\mathbf{x}$, n. In this case, the direction of the MPCs can be reconstructed with high angular resolution using low angular resolution measurements. However, this results in an underdetermined linear system, making conventional techniques unsuitable for solving the problem. Nonetheless, assuming that not many dominant paths exist in scenarios involving THz communications, the vector $\mathbf{x}$ can be considered sparse. In this case, compressed sensing techniques can be employed to resolve the problem.

Here, the ℓ_1-norm regularized least squares method [6] is used with

$$\min_{\mathbf{x}} ||\mathbf{y} - \mathbf{A}\mathbf{x}||_2^2 + \lambda ||\mathbf{x}||_1 \text{ such that } \mathbf{x} \succeq \mathbf{0} \tag{40.7}$$

where λ is a regularization parameter, $||\cdot||_2^2$ describes the square of the ℓ_2-Norm, $||\cdot||_1$ is the ℓ_1-Norm, and $\succeq$ stands for the element-wise $\geq$ inequality.

The PAP shown in Fig. 40.1 is generated via ray-tracing simulation in an exemplary meeting room scenario using a horn antenna at 300 GHz, with an antenna radiation pattern shown in Fig. 40.2. It should be noted that, the PAP in this case pertains to the azimuth plane. Thus, Fig. 40.2 only illustrates the antenna radiation pattern for the azimuth plane.

For reference, Table 40.1 summarizes the AOD and AOA of the five strongest MPCs that occur within the given scenario, along with their corresponding absolute

Fig. 40.2 Antenna radiation pattern of a horn antenna at 300 GHz, azimuth plane, horizontal polarization; antenna gain in [dBi]

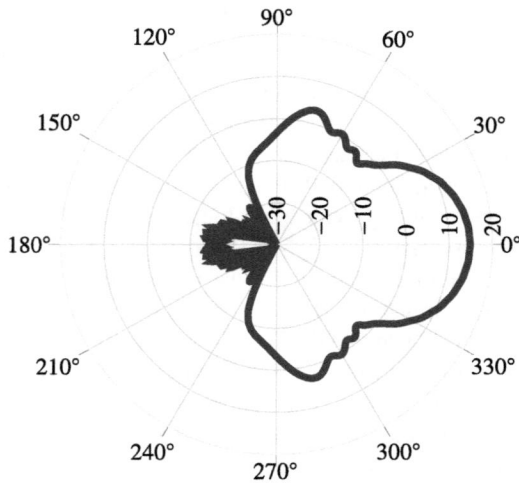

Table 40.1 Exact orientation and path gain of the five strongest MPCs in the exemplary PAP

AOD [°]	35	300	148	64	313
AOA [°]	325	240	212	116	313
Path gain [dB]	−107	−110	−112	−113	−120
Relative path gain [dB]	0	−3.6	−5.2	−5.6	−13

and relative path gain. As the scenario is a non-line-of-sight (NLOS) scenario, the path gains are consistently lower than the corresponding free space path loss (FSPL) due to the additional reflection loss that must be considered. The dominant path, which is preferably the orientation for the alignment of the TX and the RX, occurs at an AOD of $\varphi_{AOD} = 35°$ and an AOA of $\varphi_{AOA} = 325°$.

Referring to the reformulation as a compressed sensing problem, the **A** matrix is calculated based on the antenna radiation pattern shown in Fig. 40.2 using (40.6), and the **y** vector is calculated based on the PAP given in Fig. 40.1 according to (40.4). The proposed procedure then leads to two dominant paths:

1. The strongest path at $\varphi_{AOD} = 35°$ and $\varphi_{AOA} = 325°$.
2. The second strongest path at $\varphi_{AOD} = 300°$ and $\varphi_{AOA} = 240°$.

This exactly matches the two strongest paths in the simulation (see Table 40.1). However, the other MPCs are not predicted clearly. An enhancement of the procedure is therefore provided in the following subsection.

40.2.2 Enhancement for Device Discovery

Compressed sensing unfolds its full potential when the signal to be reconstructed is sparse. However, in the case of the PAP, the reconstructed PAP will not contain only the MPCs, as illustrated by the vectorized representation in Fig. 40.3a. Instead, the reconstructed PAP will include a smooth density gradient, with the maxima corresponding to the exact locations of the MPCs, as demonstrated by the vectorized representation in Fig. 40.3b.

Therefore, the PAP to be reconstructed is not sparse. In such cases, a discrete cosine transform (DCT) is a commonly used technique to address this issue [7]. It can be shown that the PAP to be reconstructed is sparse if the corresponding vector is transformed from the standard basis to the DCT basis. Figure 40.4 shows exemplary the vectorized representation given in Fig. 40.3b in the DCT basis.

Referring to the equations given for the reformulation as a compressed sensing problem in the prior section, this implies the following: if we assume that the vector

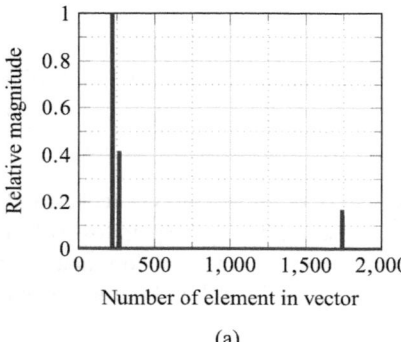
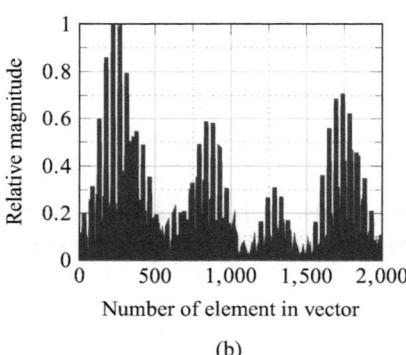

(a)   (b)

Fig. 40.3 Vectorized representation of the reconstructed PAP; corresponds to the **x** vector of the compressed sensing problem. (**a**) Ideal example: Only the MPCs are included. (**b**) Realistic example: The MPCs are equal to the maxima along with smooth density gradients

Fig. 40.4 Vectorized representation of the reconstructed PAP in the DCT basis; corresponds to the **x** vector of the compressed sensing problem

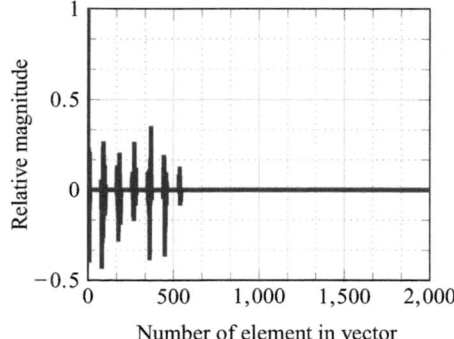

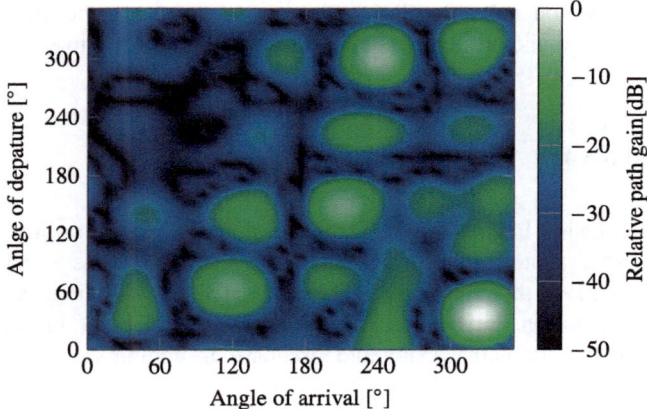

Fig. 40.5 Reconstructed PAP with high angular resolution of the exemplary meeting room scenario

Table 40.2 Orientation of the five strongest predicted MPCs

AOD [°]	35	300	148	63	313
AOA [°]	324	240	211	117	314
Relative path gain [dB]	0	− 3.2	− 4.7	− 5.3	− 6.7

$\mathbf{x}$ is sparse in the DCT basis, then the vector $\mathbf{x}$ is given by

$$\mathbf{x} = \mathbf{D} \cdot \mathbf{x}_{\text{sparse}} \qquad (40.8)$$

where $\mathbf{D}$ denotes the DCT matrix and $\mathbf{x}_{\text{sparse}}$ represents the sparse $\mathbf{x}$ vector in the DCT basis. (40.3) can then be reformulated as

$$\mathbf{y} = \mathbf{A}\mathbf{x} = \mathbf{A} \cdot \mathbf{D} \cdot \mathbf{x}_{\text{sparse}} \ . \qquad (40.9)$$

For the solver of the compressed sensing problem, $\mathbf{y}$ and $\mathbf{A} \cdot \mathbf{D}$ are then used to calculate $\mathbf{x}_{\text{sparse}}$. Finally, the vector $\mathbf{x}$ is calculated following (40.8). For the exemplary PAP provided in Fig. 40.1 with an angular resolution of 15°, Fig. 40.5 shows the reconstructed PAP with an angular resolution of 1° using the enhanced procedure. Furthermore, Table 40.2 summarizes the five strongest maxima of the predicted PAP.

It can be observed that the reconstructed PAP predicts the MPCs very well, along with a smooth density gradient that aligns with the given PAP of low angular resolution (Fig. 40.1). Moreover, the predicted maxima (see Table 40.2) match the expected maxima (see Table 40.1) with a deviation of only up to 1°. The comparison of the relative path gain based on the simulation (see Table 40.1) with the predicted relative path gain (see Table 40.2) shows that the enhanced procedure also predicts the first four path gains almost correct. The deviation is less than 0.5 dB. Thus, the extended method not only allows precise conclusions to be drawn about the orientation but also about the relative relationship of the MPCs to each other.

40.3   Hardware Acceleration

The THz frequency band, with its immense potential for ultra-broadband communication, necessitates high-throughput processing capabilities to deliver optimal services. To achieve this, advanced techniques such as pipelining, parallelization, and data quantization are being explored as key enablers of high-performance computing in this domain. A notable use case is compressed sensing-assisted device discovery, where leveraging graphics processing units (GPUs) and field programmable gate arrays (FPGAs) for acceleration has yielded promising results, albeit with certain challenges yet to be overcome.

40.3.1   Algorithmic Analysis

As given above, compressed sensing utilizes the ℓ_1-norm regularized least squares method to recover sparse signals. Formally, this is represented by the optimization problem:

$$\min_{\mathbf{x}} ||\mathbf{y} - \mathbf{Ax}||_2^2 + \lambda ||\mathbf{x}||_1 \ \text{ such that } \ \mathbf{x} \succeq \mathbf{0} \qquad (40.10)$$

where $\mathbf{y}$ represents the vectorized form of the measured PAP and $\mathbf{A}$ denotes the measurement matrix constructed based on the antennas' directivity. λ serves as the regularization parameter, balancing data fidelity and sparsity, and its value is typically tuned for optimal performance. Finally, $\mathbf{x}$ is the sparse vectorized form of the MPCs directions, which is the unknown quantity that compressed sensing aims to recover.

A primal interior-point method is presented, tailored for solving large-scale ℓ_1-regularized least squares problems. The algorithm traverses the central path, defined by the minimizer of a convex function augmented with a logarithmic barrier term, as the parameter t progressively increases.

A truncated Newton method is employed instead of directly solving the computationally demanding Newton system. This approach approximates the search direction using the preconditioned conjugate gradient (PCG) method. The preconditioner is strategically designed to enhance computational efficiency by approximating the first term of the Hessian with its diagonal entries while retaining the second term.

The algorithm dynamically adapts the PCG relative tolerance parameter based on the current duality gap, ensuring computational efficiency throughout the iterative process. Termination occurs when the duality gap, serving as a measure of

---

Algorithm 1 ℓ_1 regularized least square

Require: $A \in \mathbb{R}^{m \times n}$; $y \in \mathbb{R}^m$; $\lambda > 0$; $\epsilon_{rel} > 0$
Initialise : $t = 1/\lambda$; $x = 0$; $u = (1..1) \in \mathbb{R}^n$

Repeat
1. Determine an approximate search direction (Δ x, Δ u) by solving the Newton system.
2. Employ a backtracking line search to determine the appropriate step size "s."
3. Update (x,u) = (x,u) + s(Δ x, Δ u).
4. Construct a dual feasible point υ.
5. Evaluate the duality gap η.
6. Quit if $\eta / G(\upsilon) \leqslant \epsilon_{rel}$.
7. Update the update factor t.

---

suboptimality, falls below a predefined threshold relative to the dual objective value. This method offers a practical and effective solution for large-scale ℓ_1-regularized least squares problems, particularly in compressed sensing applications. It balances computational efficiency and accuracy by judiciously employing approximations and adaptive parameters, making it suitable for handling the large-scale, underdetermined systems often encountered in compressed sensing scenarios.

40.3.2 Computational Profiling

Having gained a thorough understanding of the ℓ_1-regularized least squares algorithm's internal mechanisms, a comprehensive profiling of its original implementation was undertaken. This analysis aimed to quantify the algorithm's runtime across a spectrum of problem shapes and regularization parameter values, shedding light on its computational performance characteristics. As shown in Fig. 40.6, the most computationally intensive scenario encountered during profiling resulted in a runtime of 176 seconds for the central processing unit (CPU)-based execution of the algorithm, underscoring the potential for performance bottlenecks, mainly when dealing with large-scale problems.

A preliminary examination of the profiling results highlighted the pivotal role played by both the regularization parameter and the problem shape in dictating the overall runtime. Notably, in certain cases, an inadequately chosen regularization parameter λ was found to be a major contributing factor to the prolonged runtime of the Newton system approximation. This suboptimal parameter choice can induce stagnation within the iterative PCG solver, necessitating additional computational effort to approximate the search direction [8, 9].

Further investigation revealed that the algorithm's runtime is mainly attributed to three major components: the Newton system approximation, the duality gap calculation, and the backtrack line search. A more granular breakdown of the runtime distribution unveiled the Newton system approximation as the most computationally demanding phase, commanding an average of nearly 89% of the total algorithm

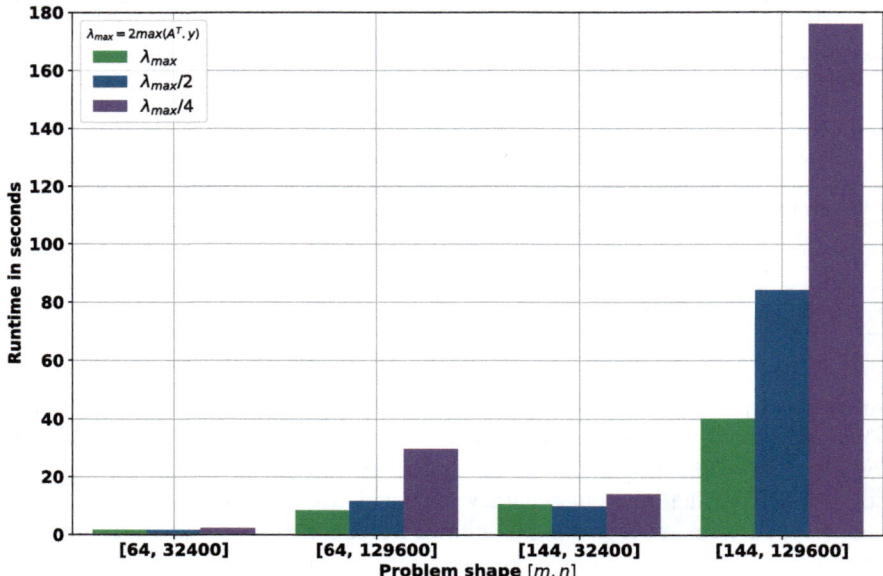

Fig. 40.6 Runtime profiling with different regulariation parameter λ and problem shapes on Intel i9 − 10th Gen

runtime. The remaining computational burden is shared between the duality gap calculation, accounting for 5.82%, and the backtrack line search, consuming 4.81%.

40.3.3 Hardware Acceleration

40.3.3.1 GPU Acceleration

Based on the profiling observations, we proceeded with GPU acceleration of the compressed sensing algorithm, which scored 9 seconds in the worst-case runtime as observed in Fig. 40.7. We re-implemented the original code in Python using the CuPy package [10], an open-source array library for GPU-accelerated computing with Python. The computational requirements of the ℓ_1-regularized least squares algorithm, particularly for tasks like normalization and matrix multiplications, were efficiently handled through the use of this library. This was achieved by leveraging the parallel processing power of GPUs.

The linear operator technique, also known as the matrix-free method, was employed to mitigate the significant memory requirements of the execution runtime, primarily for approximating the Newton system. This technique avoids explicit storage of matrix coefficients by accessing the matrix through the evaluation of matrix-vector products. This approach minimized the impact of memory access

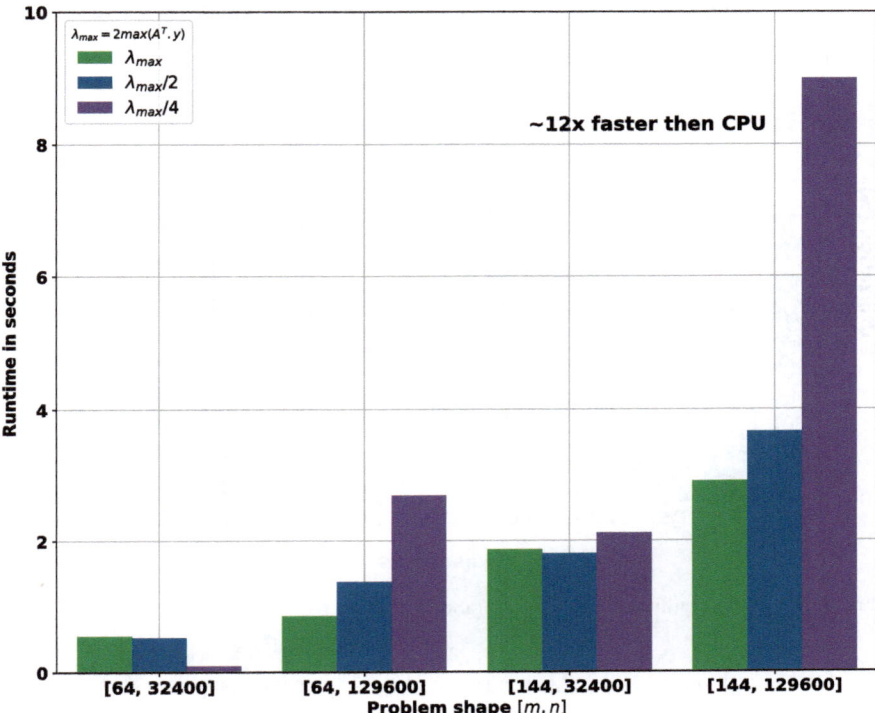

Fig. 40.7 GPU runtime with different regularization parameter λ and problem shapes on Nvidia A100

and enhanced overall computational efficiency. Through increased parallelism and optimized memory access, a 12 times improvement in runtime was achieved compared to the CPU-based implementation.

40.3.3.2 FPGA Acceleration

An FPGA-based acceleration of compressed sensing has been investigated. The hardware design consists of three independent intellectual property (IP) modules, each developed using high-level synthesis to represent a core algorithmic component of the compressed sensing framework: *Duality Gap*, *Newton Step*, and *Backtrack Line Search*. These IP modules are interconnected with an ARM processing system via an AXI High-Performance Bus, facilitating efficient data transfer and control.

Due to the significant data demands associated with real-world compressed sensing applications, a demonstrative approach has been adopted to evaluate the accelerator's performance using a moderate problem size. The FPGA design accepts

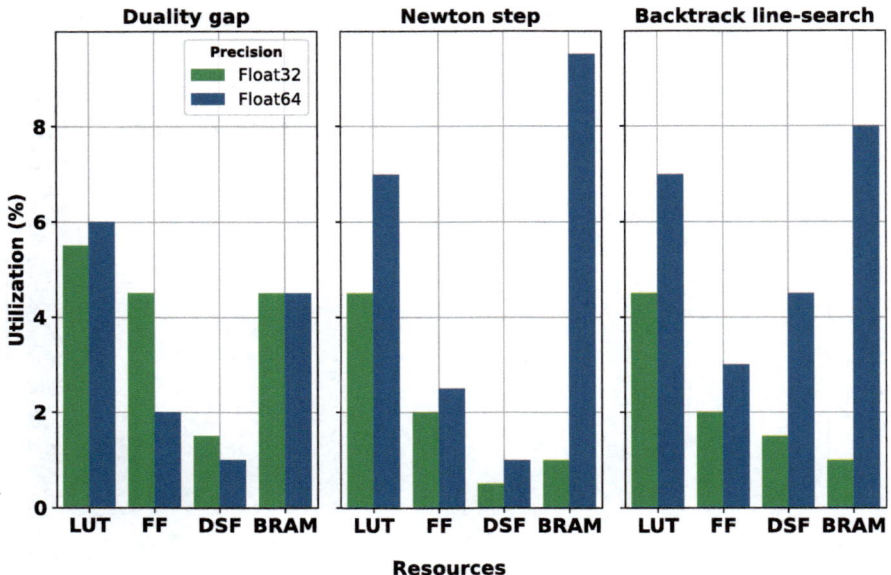

Fig. 40.8 Resource utilization on Zynq UltraScale+ FPGA

a matrix **A** with dimensions (64, 1024) and a vector **y** of size 64 as inputs. The matrix **A** is stored in external DDR memory to manage its size. At the same time, the vector **y** and the result vector **x** are allocated within the programmable logic, specifically in Block RAM (BRAM) cells. This strategic memory organization optimizes data access and reduces latency.

An assessment of resource utilization, illustrated in Fig. 40.8, indicates that BRAM constitutes the most heavily utilized resource, reflecting the algorithm's inherent memory intensity. Furthermore, this study analyzes the impact of data type precision on resource consumption by comparing implementations using Float32 and Float64 data types. As anticipated, the Float64 implementation consistently necessitates a greater allocation of FPGA resources across all IP modules, thereby underscoring the trade-off between numerical precision and resource utilization.

As depicted in Fig. 40.9, the accelerator's performance evaluation reveals a substantial speedup compared to CPU-based implementations. The *Duality Gap* IP achieves approximately a 5 times speedup, the *Newton Step* IP demonstrates a 3 times speedup, and the *Backtrack Line Search* IP realizes an impressive 12 times speedup. These findings highlight the effectiveness of hardware acceleration in significantly reducing execution times for computationally intensive tasks associated with compressed sensing.

Additionally, the study identifies challenges linked to scaling the problem size. Enlarging the problem size increases data transfer latency from the external DDR memory, adversely affecting overall runtime. Moreover, increasing the problem size

Fig. 40.9 Hardware
acceleration ratios of the
different hardware IPs on
FPGA

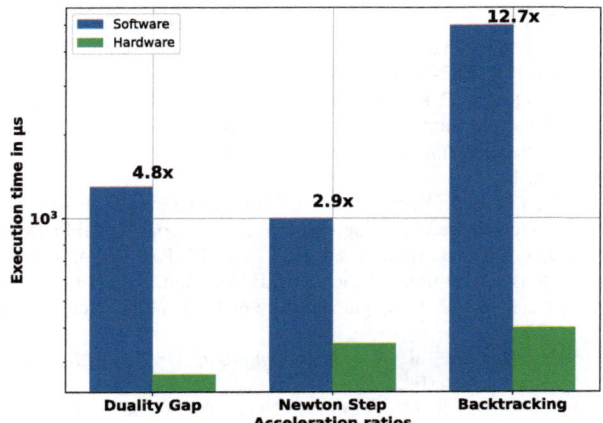

may surpass the available FPGA resources, thereby limiting the feasible dimensions
of the problem that can be addressed.

40.4   Conclusion

Device discovery is crucial for future THz communication systems, as the significant path loss needs to be compensated for by the use of highly directive antennas.
In this context, an iterative search-based approach for device discovery shows
considerable promise. This chapter presents an enhancement to such an iterative
search-based device discovery method by predicting the precise orientation of the
MPC through compressed sensing techniques.

Furthermore, this chapter introduces an acceleration approach for the proposed
algorithm through hardware enhancements. After analyzing the algorithm, we
implement it using GPU and FPGA technologies, resulting in speedups of up to
12 times.

References

1. T.S. Rappaport, Y. Xing, O. Kanhere, S. Ju, A. Madanayake, S. Mandal, A. Alkhateeb, G.C.
   Trichopoulos, Wireless communications and applications above 100 GHz: opportunities and
   challenges for 6G and beyond. IEEE Access **7**, 78729–78757 (2019)
2. S. Priebe, M. Jacob, T. Kürner, Affection of THz indoor communication links by antenna
   misalignment, in *Proceedings of the 6th European Conference on Antennas and Propagation
   (EUCAP)* (Prague, 2012), pp. 483–487
3. T. Nitsche, A.B. Flores, E.W. Knightly, J. Widmer, Steering with eyes closed: mm-wave beam
   steering without in-band measurement, in *Proceedings of the IEEE Conference on Computing
   and Communication (INFOCOM)* (Kowloon, 2015), pp. 2416–2424

4. B. Peng, K. Guan, S. Rey, T. Kürner, Power-angular spectra correlation based two step angle of arrival estimation for future indoor terahertz communications. IEEE Trans. Antennas Propag. **67**(11), 7097–7105 (2019)
5. T. Doeker, P. Reddy Samala, P.S. Negi, A. Rajwade, T. Kürner, Angle of arrival and angle of departure estimation using compressed sensing for terahertz communications, in *Proceedings of the 15th European Conference on Antennas and Propagation (EuCAP 2021)*, virtual (2021), pp. 1–5
6. S.-J. Kim, K. Koh, M. Lustig, S. Boyd, D. Gorinevsky, An interior-point method for large-scale ℓ_1-regularized least squares. IEEE J. Sel. Top. Signal Process. **1**(4), 606–617 (2007)
7. J.W. Choi, B. Shim, Y. Ding, B. Rao, D.I. Kim, Compressed sensing for wireless communications: useful tips and tricks. IEEE Commun. Surv. Tut. **19**(3), 1527–1550 (2017)
8. J.L. Nazareth, Conjugate gradient method. Wiley Interdiscip. Rev. Comput. Stat. **1**(3), 348–353 (2009)
9. N. Andrei et al., *Nonlinear Conjugate Gradient Methods for Unconstrained Optimization* (Springer, Berlin, 2020)
10. R. Nishino, S.H.C. Loomis, CuPy: a NumPy-compatible library for NVIDIA GPU calculations, in *Proceedings of the Conference on Neural Information Processing Systems (NIPS)* (Long Beach, 2017), pp. 1–7

Chapter 41
Beam Tracking and Switching

Tobias Doeker, Anouar Nechi, Mladen Berekovic, and Thomas Kürner

Abstract In addition to the initial precise alignment between the transmitter (TX) and receiver (RX), called device discovery, a constant alignment between the TX and RX is mandatory for highly directional antennas. Therefore, beam tracking and/or switching is required in time-varying scenarios. In this chapter, a beam-tracking approach based on antenna pattern information is explained. This approach is verified using channel sounder measurements and improved using artificial intelligence (AI)/reinforcement learning (RL) techniques.

41.1 Adaptive Precision: Understanding Beam Tracking in THz Communication

For terahertz (THz) communications, highly directive antennas are anticipated to compensate for the significant path loss at these frequencies [1]. Precise alignment between the TX and the RX is mandatory, making initial device discovery, where the TX and RX locate each other, crucial [Chap. 40]. However, when the TX and/or RX are not stationary and are in motion, maintaining this alignment becomes challenging. Due to the nature of highly directive antennas, even minor changes in the angle of departure (AOD) and/or angle of arrival (AOA) can lead to a substantial drop in received power [2]. Thus, ensuring precise alignment between the TX and RX is essential, even when the TX and/or RX are in motion. In addition to device

T. Doeker (✉)
Technische Universität Braunschweig, Institut für Nachrichtentechnik, Braunschweig, Germany
e-mail: t.doeker@tu-braunschweig.de

A. Nechi · T. Kürner
Universität zu Lübeck, Institut für Technische Informatik, Lübeck, Germany
e-mail: nechi@iti.uni-luebeck.de; t.kuerner@tu-braunschweig.de

M. Berekovic
Technische Universität Braunschweig, Institut für Nachrichtentechnik, Braunschweig, Germany
e-mail: berekovic@iti.uni-luebeck.de

© The Author(s) 2026
T. Kürner et al. (eds.), *Metrology for THz Communications*, Springer Series in Optical Sciences 256, https://doi.org/10.1007/978-3-032-01986-8_41

discovery, beam tracking and beam switching, in the event of shadowing effects, are also vital for THz communication systems.

Beam tracking and switching are not issues exclusive to THz communications, and various approaches have been developed for different frequency ranges [3–5]. Nonetheless, existing algorithms are not always suitable for THz communications due to constraints such as the impracticality of omnidirectional antennas [4] or the use of antenna arrays [5]. In this chapter, we present an alternative option for beam tracking, specifically designed for THz communications, which leverages knowledge of the antenna pattern [6, 7].

In the second part of this chapter, we also look at the possibilities for improving beam tracking methods using reinforcement learning.

41.2 Antenna Pattern-Based Beam Tracking

41.2.1 Basic Concept

The concept of the antenna pattern-based tracking algorithm was first introduced in [6]. Assuming perfectly aligned TX and RX at the beginning of a movement process in a line-of-sight (LOS) setup, the received power can be calculated based on Friis' equation. After movement, the distance, and therefore the free space path loss (FSPL), as well as the AOD and AOA, will have changed. The ratio between the received power before $P_{RX,0}$ and after the movement $P_{RX,1}$ is given by

$$\frac{P_{RX,1}}{P_{RX,0}} = \frac{P_{TX} \cdot G_{TX}(\varphi_{AOD,1}) \cdot G_{FSPL}(d_1) \cdot G_{RX}(\varphi_{AOA,1})}{P_{TX} \cdot G_{TX}(0°) \cdot G_{FSPL}(d_0) \cdot G_{RX}(0°)} \quad (41.1)$$

where P_{TX} denotes the transmit power, G_{TX} and G_{RX} denote the antenna gain of the TX and the RX antenna, and G_{FSPL} denotes the FSPL for the distance between TX and RX before d_0 and after the movement d_1. However, due to the time of flight (TOF) of the signal, the distance and the corresponding FSPL can be calculated, so that the antenna gain after the movement is the only unknown value in (41.1), leading to

$$\frac{P_{RX,1}}{P_{RX,0}} \cdot \frac{G_{FSPL}(d_0)}{G_{FSPL}(d_1)} \cdot G_{TX}(0°) \cdot G_{RX}(0°) = G_{TX}(\varphi_{AOD,1}) \cdot G_{RX}(\varphi_{AOA,1}) . \quad (41.2)$$

If the initial orientation of the TX and RX does not change during the movement, the AOD remains equal to the AOA after the movement. Furthermore, for each value of this angle, the superposition of the antenna gain of the TX and RX can be determined—this is introduced as the combined antenna gain. Therefore, the change in the AOD and AOA can be found by searching for an angle in the combined antenna gain that leads to the same gain as indicated by the right side of (41.2). An exemplary combined antenna gain is shown in Fig. 41.1, using a horn antenna with

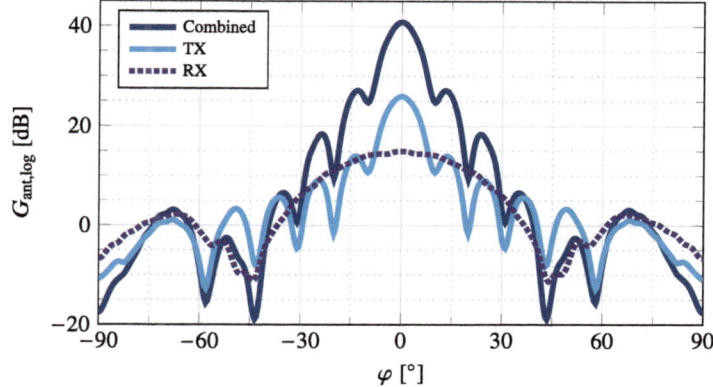

Fig. 41.1 Example for the combined antenna gain

8° half-power beamwidth (HPBW) as well as a horn antenna with 35° HPBW, both with an initial orientation of 0°; in other words, the main lobe of both antennas is at $\varphi = 0°$.

However, there is the issue of nonunique solutions, meaning there are several angles for which the superposition of the antenna gains results in the same combined antenna gain. Thus, different combined antenna gains must be used. Let's assume that the TX is equipped with one antenna, while the RX is equipped with two different antennas. This results in two different combined antenna gains and two different received power values. The received power with the first antenna combination might lead to potential angles of $\varphi_1 = \{-10, -5, 5, 10\}°$. For the same setup, the received power with the second antenna combination might lead to potential angles of $\varphi_2 = \{-20, -5, 5, 20\}°$. As the setup remains unchanged, the correct angle for both combinations must be the same, so the intersection of potential angles is the correct value, namely, -5 and $5°$ in this example. The addition of a second antenna at the RX thus reduces the number of potential angles from four to two. Using multiple different antennas at the TX and/or RX increases the number of combined antenna gains and decreases the pool of potential angles, ideally leading to a unique solution, as demonstrated by ray-tracing simulations in [6].

The antenna pattern-based algorithm can be summarized as follows: By using several antennas (with different radiation patterns or initial orientations) at the TX and RX, different received power values can be measured. Each antenna combination yields its own combined antenna gain. After a movement, comparing the measured values for each antenna combination with the corresponding combined antenna gain results in several potential angles. The intersection of the potential angles reveals the change in the AOD and AOA.

In principle, the algorithm can also be applied in a non-line-of-sight (NLOS) case. However, additional considerations and assumptions must be factored in. Firstly, the received power in this scenario is affected by reflection losses, introducing an additional loss term into (41.1) and (41.2), which must be consequently reformulated to

$$\frac{P_{RX,1}}{P_{RX,0}} = \frac{G_{TX}(\varphi_{AOD,1}) \cdot G_{FSPL}(d_1) \cdot G_{RX}(\varphi_{AOA,1}) \cdot \dfrac{1}{L_{refl,1}}}{G_{TX}(0°) \cdot G_{FSPL}(d_0) \cdot G_{RX}(0°) \cdot \dfrac{1}{L_{refl,0}}} \qquad (41.3)$$

and

$$\frac{P_{RX,1}}{P_{RX,0}} \cdot \frac{G_{FSPL}(d_0)}{G_{FSPL}(d_1)} \cdot G_{TX}(0°) \cdot G_{RX}(0°) \cdot \frac{L_{refl,1}}{L_{refl,0}} = G_{TX}(\varphi_{AOD,1}) \cdot G_{RX}(\varphi_{AOA,1})$$

$$(41.4)$$

where $L_{refl,0}$ and $L_{refl,1}$ denote the reflection loss before and after the movement. As previously discussed, in (41.2) the left side of the equation consists solely of known values, equating to the sought-after combined antenna gain. For the algorithm to function in the NLOS case, reflecting losses must also be known (left side of (41.4) has to be known); hence, initial measurements with perfectly aligned TX and RX can be used to extract the reflection loss before the movement $L_{refl,0}$ from received power measurements. Thus, only the post-movement reflection loss remains unknown on the left side of (41.4). The assumption is made that the reflection loss does not change significantly between tracking intervals, so $L_{refl,1} = L_{refl,0}$. After a tracking step, the TX and RX are realigned to reevaluate the reflection loss. However, should significant changes occur in the reflection loss, for example, due to abrupt changes in the reflection coefficient, the algorithm will fail, necessitating an initial device discovery.

41.2.2 Measurement-Based Verification

To verify the applicability of the antenna pattern-based tracking algorithm, channel sounder measurements at 300 GHz were conducted using a correlation-based time domain channel sounder available at Technische Universität Braunschweig (TUBS). The measurements and the performance of the algorithm were evaluated in both LOS and NLOS scenarios. The measurement campaign and the results presented here are published in [7]. Figure 41.2 shows exemplary photos of the setup.

In the LOS case, the TX was fixed in position, while the RX was moved on a rail system. Different initial alignments between the TX and RX were realized

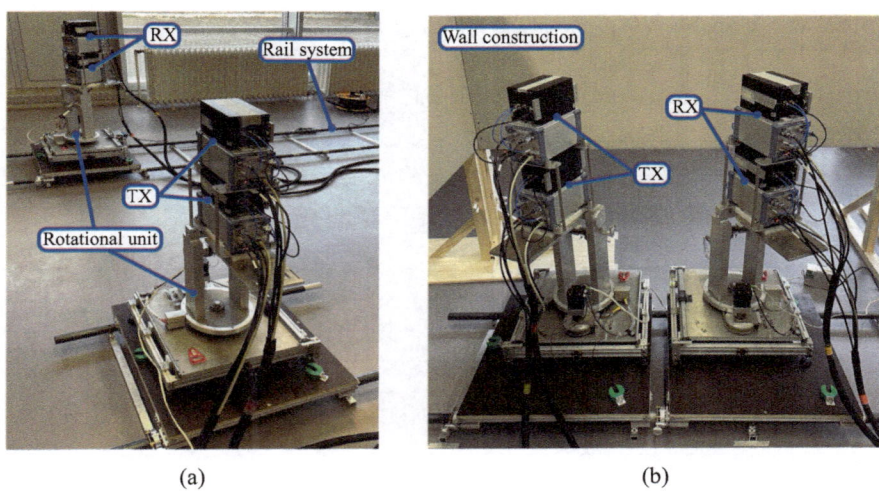

Fig. 41.2 Exemplary photos of the measurement setup [7]. (a) LOS case. (b) NLOS case

using rotational units, emulating various antennas and enabling different antenna combinations for different combined antenna gains. The RX was moved along the rail system to realize differences in the AOD and AOA from 0° up to 70°, in 5° increments.

For the NLOS case, the TX and RX were placed on the same rail system. The dominant multipath component (MPC) in the channel impulse response was created by a reflection off a wall. It should be noted that the material characteristics in this measurement were assumed consistent over the entire distance, as the same material was used for the wall. Although the LOS component through side lobes was not explicitly avoided during measurements, only the reflected component was considered in the post-processing. Again, the TX and RX were placed on rotational units to enable different initial antenna orientations. In the NLOS case, the RX was moved on the rail system to achieve changes in the AOD and AOA from 0° up to 50°, in 5° steps.

In both scenarios, the RX was moved according to the specified steps, and for each position, the received power for all possible antenna combinations (achieved through varying antenna orientations) was measured. It should be noted that the algorithm had to be adapted for the measurements, compared to its operation with simulation results. Since the algorithm is based on the antenna radiation pattern, which is provided only for discrete values, the calculated values based on the measurement (corresponding to the left side of (41.2) and (41.4)) will not exactly match a discrete value of the combined antenna gain. Therefore, the value of the combined antenna gain with the lowest deviation between the measured value and the discrete pattern value is selected. To increase the accuracy of the algorithm,

Table 41.1 Comparison of predicted and target angles (exemplary LOS case) [7]

Target angle	First set		Second set	
	Prediction	Deviation	Prediction	Deviation
0°	0.6°	+ 0.6°	0.4°	+ 0.4°
5°	5.3°	+ 0.3°	5.0°	−
10°	9.1°	− 0.9°	9.1°	− 0.9°
15°	14.5°	− 0.5°	15.0°	−
20°	19.3°	− 0.7°	19.1°	− 0.9°
25°	29.9°	+ 4.9°	23.1°	− 1.9°
30°	25.3°	− 4.7°	29.9°	− 0.1°
35°	33.3°	− 1.7°	33.6°	− 1.4°
40°	43.5°	+ 3.5°	39.0°	− 1.0°
45°	43.5°	− 1.5°	44.4°	− 0.6°
50°	48.2°	− 1.8°	49.1°	− 0.9°
55°	55.8°	+ 0.8°	53.1°	− 1.9°
60°	59.8°	− 0.2°	61.4°	+ 1.4°
65°	67.6°	+ 2.6°	66.1°	+ 1.1°
70°	70.5°	+ 0.5°	71.9°	+ 1.9°

several values with the lowest deviation are chosen. Furthermore, the intersection of the potential angle candidates from each antenna combination may not have a common value in this case. Thus, the most likely value is chosen based on calculating the standard deviation of all possible angle combinations from different antenna configurations. Further details are provided in [7].

It can be shown that multiple antenna combinations lead to a correct prediction of the angles, achieving a target accuracy of ±1° in the range of 0° to 20° and ±2° in the range of 0° to 70° in the LOS scenario. Table 41.1 presents exemplary results calculated with the proposed algorithm based on the measured power of four different antenna combinations.

In addition, the measurements demonstrate that the algorithm can also accurately predict angular changes in the NLOS case, achieving an accuracy of ±1° for ranges from 0° to 20°, and even within 0° to 45° with an accuracy of ±4°. Exemplary results given in Table 41.2 are also based on the received power from four different antenna combinations.

The measurement campaign demonstrates that the antenna pattern-based algorithm is effective in real-life applications using horn antennas at 300 GHz. It is shown that with four different antenna combinations, for instance, two antennas at the TX and two antennas at the RX, the algorithm can accurately predict changes in the AOD and AOA within an accuracy that is less than half of the HPBW of the antennas used.

Table 41.2 Comparison of predicted and target angles (exemplary NLOS case) [7]

Target angle	± 1° accuracy		± 4° accuracy	
	Prediction	Deviation	Prediction	Deviation
0°	0.8°	+ 0.8°	0.3°	+ 0.3°
5°	5.8°	+ 0.8°	3.3°	− 1.7°
10°	10.7°	+ 0.7°	10.3°	+ 0.3°
15°	14.5°	− 0.5°	18.2°	+ 3.2°
20°	19.2°	− 0.8°	17.7°	− 2.3°
25°	38.3°	+ 13.3°	28.5°	+ 3.5°
30°	38.8°	+ 8.8°	30.0°	−
35°	37.2°	+ 2.2°	36.3°	+ 1.3°
40°	40.0°	−	40.9°	+ 0.9°
45°	39.5°	− 5.5°	48.5°	+ 3.5°
50°	56.5°	+ 6.5°	80.0°	+ 30.0°

41.3 Beam Tracking Enhancement Through Reinforcement Learning

41.3.1 General Concept

Adapting wireless communication systems through learning and interaction resonates strongly with device discovery and tracking challenges in the THz band. With its immense bandwidth and potential for high-speed data transfer, THz communication introduces unique complexities due to its sensitivity to blockages and the highly directional nature of THz signals [8]. These characteristics make discovering and tracking devices in THz networks particularly challenging. The traditional reliance on predefined beam patterns or scanning mechanisms may prove inadequate in dynamic THz environments where devices move and channel conditions fluctuate rapidly. Employing RL to enable the system to learn and adapt its strategies in real time offers a promising solution [9]. By treating the system as an agent interacting with its environment, taking actions (such as adjusting beam directions or transmission power), and receiving rewards (based on successful device discovery or tracking accuracy), RL can enable the system to explore and exploit the THz channel intelligently. The agent can learn to make decisions that maximize its cumulative reward, improving performance in discovering and tracking devices even in challenging THz environments. The ability of RL to adapt to the specific environment and hardware limitations without relying solely on explicit channel knowledge is particularly valuable in the THz context, where channel modelling and estimation can be complex and computationally expensive.

41.4 Reinforcement Learning-Based Device Tracking and Discovery

The adaptive capabilities of RL present a promising avenue for enhancing device discovery and tracking within the THz communication landscape. THz environments' dynamic and often unpredictable nature, coupled with the inherent challenges of channel estimation and the highly directional characteristics of THz signals, underscores the need for intelligent and adaptable mechanisms in this domain. The RL-based methodology proposed in [10] can be effectively tailored to address these challenges. In this context, the THz system assumes the role of an agent, learning to optimize its actions, such as beam steering adjustments or transmission power control, based on the feedback it receives from the environment, such as successful device discovery or the accuracy of device tracking.

41.4.1 The Wolpertinger Architecture: The Foundation for Adaptive Learning

The Wolpertinger architecture, an actor-critic model designed to handle large discrete action spaces, provides a robust foundation for implementing RL in THz device discovery and tracking. The actor network within this architecture proposes actions, such as adjustments to the beam steering direction. In contrast, the critic network evaluates the value of these actions based on feedback from the environment. The k-nearest neighbor (KNN) classifier, an integral component of the Wolpertinger architecture, effectively manages the vast discrete action space associated with beam steering in THz systems. It maps the continuous "proto-action" suggested by the actor network to the nearest feasible beam direction, ensuring that the system's actions remain practical and implementable. The critic network then assesses the efficacy of this chosen action in achieving the desired goal, such as successfully discovering or tracking a device. The critic network's feedback is then utilized to update both the actor and critic networks, facilitating their continuous learning and refinement of decision-making capabilities.

41.4.2 The Reinforcement Learning Mechanism in Action: Device Discovery and Tracking

The RL mechanism, illustrated in Fig. 41.3, can be directly applied to device discovery and tracking tasks in THz communication. The system's current configuration, including beam direction and transmission power, defines its state. The action taken by the system involves adjusting this configuration, and the reward is contingent upon the success of device discovery or the precision of device tracking. Within

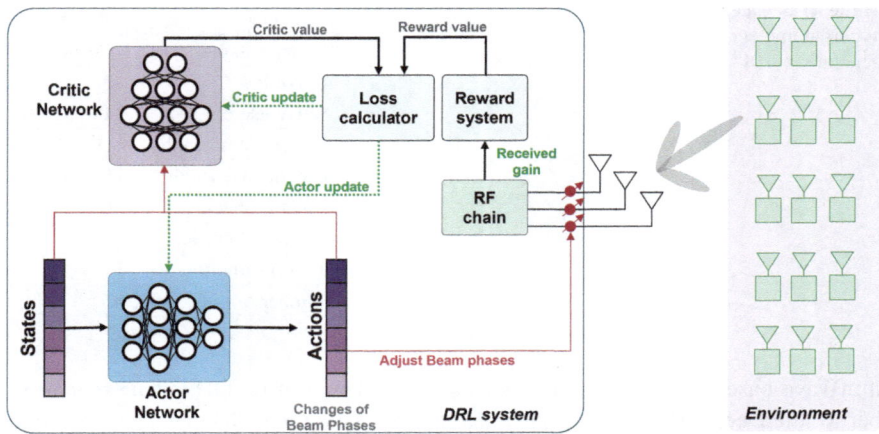

Fig. 41.3 RL framework for beam pattern design using deep RL, illustrating the agent architecture and its interaction with the environment

the RL framework, the THz system functions as an agent, interacting with its environment by taking actions and receiving corresponding feedback. This iterative process enables the agent to learn and optimize its actions to maximize cumulative reward. Exploration of the action space is promoted by introducing noise into the proto-action generated by the actor network. Subsequently, the KNN classifier ensures the validity and implementability of the final action. The observed reward and the critic's evaluation serve as the foundation for updating the actor and critic networks. This adaptive mechanism allows the system to adjust its beam steering and transmission power strategies dynamically, effectively discovering and tracking devices within a dynamic THz environment.

Furthermore, the concept of clustering and assignment can be extended to device discovery and tracking as presented in the context of beam codebook learning. Devices can be clustered based on their channel characteristics or location information, and different RL agents can be assigned to track or discover devices within their assigned clusters. This approach can lead to more efficient and targeted device discovery and tracking in complex THz environments, as each agent can focus on a specific group of devices and learn specialized strategies for that group.

41.5 Experimental Results

41.5.1 Communication Scenario and Dataset

The scenario used for performance evaluation is an outdoor LOS scenario. In this scenario, all users have a direct equal gain combining (EGC) connection with the

Table 41.3 DeepMIMO
hyperparameters for channel
generation [10]

Parameter	Value
Scenario name	O1 60
Active BS	3
Active users	1101–1400
Number of antennas	(1, 32, 1)
System bandwidth	0.5 GHz
Antenna spacing	0.5
OFDM subcarriers	1
OFDM sampling	1
Number of multipaths	5

mmWave base station, which operates at a frequency of 60 GHz. This scenario is part of the DeepMIMO dataset [11], and the corresponding channel set is generated using DeepMIMO scripts based on the configuration depicted by Table 41.3. The datasets incorporating hardware impairments are also based on this EGC scenario, considering factors like antenna spacing and phase mismatches.

41.5.2 Model Architecture

The learning architecture is based on the deep deterministic policy gradient (DDPG) framework [12]. It comprises two neural networks: the actor and the critic. The actor network receives the state, represented by the phases of the phase shifters, as input, with a dimensionality of (M). This input is processed through two hidden layers, each composed of 16(M) neurons and utilizing rectified linear unit (ReLU) activation functions. The output of the actor network is the predicted action, also with a dimensionality of (M), followed by hyperbolic tangent (tanh) activation functions scaled by π. In contrast, the critic network receives the concatenated state and action as input, resulting in a dimensionality of 2(M). It similarly possesses two hidden layers, each with 32(M) neurons and ReLU activation functions. The critic network outputs a real scalar value representing the predicted Q-value for the given state-action pair. The described RL algorithm is trained based on the hyperparameters defined in Table 41.4.

41.5.3 Results

In [10], the aforementioned RL is experimented. The target users are indicated by blue dots in Fig. 41.4. While classical beam steering codebooks generally perform well in EGC scenarios [13], the proposed method achieves a higher beamforming gain than the best beam in the codebook after only 2100 iterations. Notably, with fewer than 40k iterations, the solution achieves over 90% of the EGC upper bound.

Table 41.4 Hyperparameters for RL model training

Parameter	Value	
Models	Actor	Critic
Replay buffer	8192	
Mini-batch size	1024	
Optimizer	Adam	
Learning rate	10^{-3}	
Weight decay	10^{-2}	
Noise start/end variance	1/0.05	
Noise decay period	10^5	

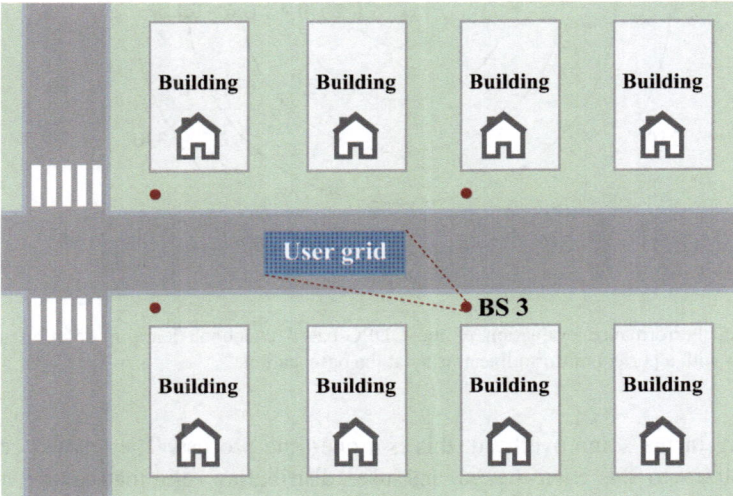

Fig. 41.4 Outdoor EGC scenario at 60 GHz "O1_60"

It is important to highlight that the EGC upper bound is typically only achievable with perfect channel knowledge and unquantized phase shifters.

At the beginning of the learning process, the beam pattern is characterized by strong side lobes, which limit the main lobe's gain. The main lobe's gain is progressively enhanced through the learning process, effectively suppressing the side lobes. This yields a highly focused beam with the main lobe significantly stronger than the other side lobes. Furthermore, the learned beam pattern closely approximates the EGC beam pattern, accounting for its exceptional performance. The slight mismatch is primarily attributed to quantized phase shifters with only three-bit resolution.

The average beamforming gain increases with the number of beams in the codebook. Notably, the proposed solution achieves comparable performance to a 32-beam codebook with only six beams and surpasses it with eight beams. This highlights the advantage of the adaptive beam allocation based on user distribution, which minimizes overhead by efficiently targeting users. While the initial codebook

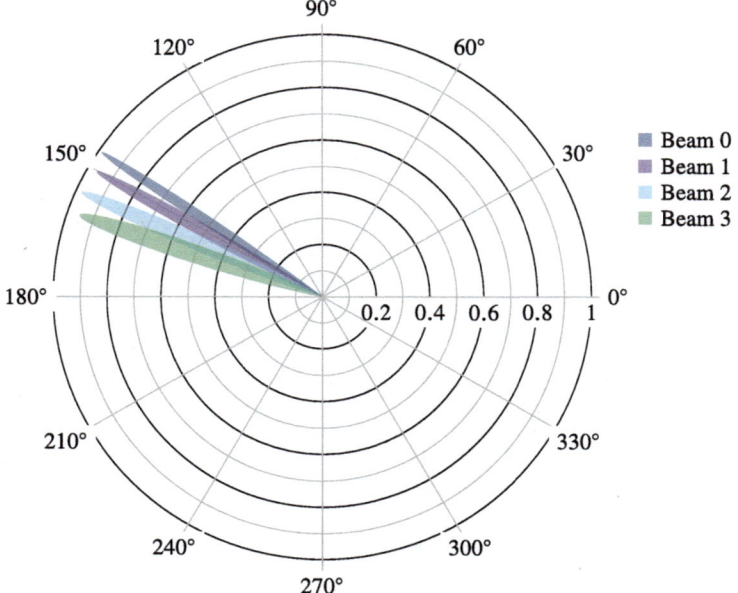

Fig. 41.5 Performance evaluation of the DDPG-based codebook learning solution in a LOS scenario with a perfect uniform linear array at the base station

learning incurs some overhead, this is a one-time process. The learned codebook generalizes to the environment and user distribution, eliminating the need for repeated learning cycles with new users. This translates to significant long-term savings in beam training overhead.

It is worth noting that a smaller codebook size, like the 4-beam codebook depicted in Fig. 41.5, reduces initial access latency, which is vital for delay-sensitive applications. This codebook effectively clusters users with similar channel conditions and generates beams that optimally cover the user grid, maximizing beamforming gain. The ability to achieve high performance with a compact codebook underscores the efficiency and practicality of the proposed solution.

41.6   Conclusion

Due to the necessity for highly directive antennas in THz communications—where high path loss is prevalent—beam tracking and switching are crucial for the effectiveness of these communication systems. In this chapter, we introduce a novel beam tracking algorithm suitable for THz communications, which relies on the knowledge of the antenna's radiation pattern. Our results demonstrate that this approach achieves an accuracy of $\pm 1°$ within an angular range of up to $20°$.

Furthermore, this chapter explores the potential of RL in the context of beam tracking and switching. Its applicability is demonstrated through an example involving an existing codebook-based tracking algorithm.

References

1. T.S. Rappaport, Y. Xing, O. Kanhere, S. Ju, A. Madanayake, S. Mandal, A. Alkhateeb, G.C. Trichopoulos, Wireless communications and applications above 100 GHz: opportunities and challenges for 6G and beyond. IEEE Access **7**, 78729–78757 (2019)
2. S. Priebe, M. Jacob, T. Kürner, Affection of THz indoor communication links by antenna misalignment, in *Proceedings of the 6th European Conference on Antennas and Propagation (EuCAP 2012)* (Prague, 2012), pp. 483–487
3. M. Qurratulain Khan, A. Gaber, P. Schulz, G. Fettweis, Machine learning for millimeter wave and terahertz beam management: a survey and open challenges. IEEE Access **11**, 11880–11902 (2023)
4. J. Tan, L. Dai, Wideband beam tracking in THz massive MIMO systems. IEEE J. Sel. Areas Commun. **39**(6), 1693–1710 (2021)
5. D. Zhang, A. Li, M. Shirvanimoghaddam, P. Cheng, Y. Li, B. Vucetic, Codebook-based training beam sequence design for millimeter-wave tracking systems. IEEE Trans. Wirel. Commun. **18**(11), 5333–5349 (2019)
6. L.H.W. Löser, T. Doeker, T. Kürner, Antenna pattern tracking algorithm for low terahertz communications, in *Proceedings of the 6th European Conference on Antennas and Propagation (EuCAP 2024)* (Glasgow, 2024), pp. 1–5
7. T. Doeker, L.H. W. Löser, T. Kürner, Measurements and verification of an antenna pattern-based tracking algorithm at 300 GHz. IEEE Trans. Terahertz Sci. Technol. **15**(3), 359–369 (2025)
8. D. Serghiou, M. Khalily, T.W. Brown, R. Tafazolli, Terahertz channel propagation phenomena, measurement techniques and modeling for 6G wireless communication applications: a survey, open challenges and future research directions. IEEE Commun. Surv. Tutorials **24**(4), 1957–1996 (2022)
9. Y.J. Tan, C. Zhu, T.C. Tan, A. Kumar, L.J. Wong, Y. Chong, R. Singh, Self-adaptive deep reinforcement learning for THz beamforming with silicon metasurfaces in 6G communications. Opt. Express **30**(15), 27763–27779 (2022)
10. Y. Zhang, M. Alrabeiah, A. Alkhateeb, Reinforcement learning of beam codebooks in millimeter wave and terahertz MIMO systems. IEEE Trans. Commun. **70**(2), 904–919 (2021)
11. A. Alkhateeb, DeepMIMO: a generic deep learning dataset for millimeter wave and massive MIMO applications (2019). arXiv: 1902.06435
12. T.P. Lillicrap et al., Continuous control with deep reinforcement learning (2015). arXiv: 1509.02971
13. M. Alrabeiah, Y. Zhang, A. Alkhateeb, Neural networks based beam codebooks: learning mmWave massive MIMO beams that adapt to deployment and hardware. IEEE Trans. Commun. **70**(6), 3818–3833 (2022)

GPSR Compliance

The European Union's (EU) General Product Safety Regulation (GPSR) is a set of rules that requires consumer products to be safe and our obligations to ensure this.

If you have any concerns about our products, you can contact us on ProductSafety@springernature.com

In case Publisher is established outside the EU, the EU authorized representative is:

Springer Nature Customer Service Center GmbH
Europaplatz 3
69115 Heidelberg, Germany

Batch number: 10121690

Printed by Printforce, the Netherlands